Agro-Food Chain By-Products and Plant Origin Food to Obtain High-Value-Added Foods

Agro-Food Chain By-Products and Plant Origin Food to Obtain High-Value-Added Foods

Guest Editors

Gianluca Nardone
Rosaria Viscecchia
Francesco Bimbo

Basel • Beijing • Wuhan • Barcelona • Belgrade • Novi Sad • Cluj • Manchester

Guest Editors

Gianluca Nardone
University of Foggia
Foggia
Italy

Rosaria Viscecchia
University of Foggia
Foggia
Italy

Francesco Bimbo
University of Foggia
Foggia
Italy

Editorial Office
MDPI AG
Grosspeteranlage 5
4052 Basel, Switzerland

This is a reprint of the Special Issue, published open access by the journal *Foods* (ISSN 2304-8158), freely accessible at: https://www.mdpi.com/journal/foods/special_issues/E8WD1ODKKG.

For citation purposes, cite each article independently as indicated on the article page online and as indicated below:

Lastname, A.A.; Lastname, B.B. Article Title. *Journal Name* **Year**, *Volume Number*, Page Range.

ISBN 978-3-7258-2661-2 (Hbk)
ISBN 978-3-7258-2662-9 (PDF)
https://doi.org/10.3390/books978-3-7258-2662-9

© 2024 by the authors. Articles in this book are Open Access and distributed under the Creative Commons Attribution (CC BY) license. The book as a whole is distributed by MDPI under the terms and conditions of the Creative Commons Attribution-NonCommercial-NoDerivs (CC BY-NC-ND) license (https://creativecommons.org/licenses/by-nc-nd/4.0/).

Contents

Gianluca Nardone, Rosaria Viscecchia and Francesco Bimbo
The Use of Agro-Food Chain By-Products and Foods of Plant Origin to Obtain High-Value-Added Foods
Reprinted from: *Foods* **2024**, *13*, 3154, https://doi.org/10.3390/foods13193154 1

Fredrick Nwude Eze, Rattana Muangrat, Sudarshan Singh, Wachira Jirarattanarangsri, Thanyaporn Siriwoharn and Yongyut Chalermchat
Upcycling of Defatted Sesame Seed Meal via Protein Amyloid-Based Nanostructures: Preparation, Characterization, and Functional and Antioxidant Attributes
Reprinted from: *Foods* **2024**, *13*, 2281, https://doi.org/10.3390/foods13142281 4

Jordan Richards, Amy Lammert, Jack Madden, Iksoon Kang and Samir Amin
Physical Treatments Modified the Functionality of Carrot Pomace
Reprinted from: *Foods* **2024**, *13*, 2084, https://doi.org/10.3390/foods13132084 31

Victoria Baggi Mendonça Lauria and Luciano Paulino Silva
Green Extraction of Natural Colorants from Food Residues: Colorimetric Characterization and Nanostructuring for Enhanced Stability
Reprinted from: *Foods* **2024**, *13*, 962, https://doi.org/10.3390/foods13060962 46

Spasoje D. Belošević, Danijel D. Milinčić, Uroš M. Gašić, Aleksandar Ž. Kostić, Ana S. Salević-Jelić, Jovana M. Marković, et al.
Broccoli, Amaranth, and Red Beet Microgreen Juices: The Influence of Cold-Pressing on the Phytochemical Composition and the Antioxidant and Sensory Properties
Reprinted from: *Foods* **2024**, *13*, 757, https://doi.org/10.3390/foods13050757 61

Nikoo Jabbari, Mohammad Goli and Sharifeh Shahi
Optimization of Bioactive Compound Extraction from Saffron Petals Using Ultrasound-Assisted Acidified Ethanol Solvent: Adding Value to Food Waste
Reprinted from: *Foods* **2024**, *13*, 542, https://doi.org/10.3390/foods13040542 88

Kaitlyn Burghardt, Tierney Craven, Nabil A. Sardar and Joshua M. Pearce
Towards Sustainable Protein Sources: The Thermal and Rheological Properties of Alternative Proteins
Reprinted from: *Foods* **2024**, *13*, 448, https://doi.org/10.3390/foods13030448 109

Sílvia Petronilho, Manuel A. Coimbra and Cláudia P. Passos
Quality Characteristics of Raspberry Fruits from Dormancy Plants and Their Feasibility as Food Ingredients
Reprinted from: *Foods* **2023**, *12*, 4443, https://doi.org/10.3390/foods12244443 132

Georgios A. Papadopoulos, Styliani Lioliopoulou, Nikolaos Nenadis, Ioannis Panitsidis, Ioanna Pyrka, Aggeliki G. Kalogeropoulou, et al.
Effects of Enriched-in-Oleuropein Olive Leaf Extract Dietary Supplementation on Egg Quality and Antioxidant Parameters in Laying Hens
Reprinted from: *Foods* **2023**, *12*, 4119, https://doi.org/10.3390/foods12224119 148

Mariana-Atena Poiana, Ersilia Alexa, Isidora Radulov, Diana-Nicoleta Raba, Ileana Cocan, Monica Negrea, et al.
Strategies to Formulate Value-Added Pastry Products from Composite Flours Based on Spelt Flour and Grape Pomace Powder
Reprinted from: *Foods* **2023**, *12*, 3239, https://doi.org/10.3390/foods12173239 170

Cristina Anamaria Semeniuc, Floricuța Ranga, Andersina Simina Podar, Simona Raluca Ionescu, Maria-Ioana Socaciu, Melinda Fogarasi, et al.
Determination of Coenzyme Q10 Content in Food By-Products and Waste by High-Performance Liquid Chromatography Coupled with Diode Array Detection
Reprinted from: *Foods* **2023**, *12*, 2296, https://doi.org/10.3390/foods12122296 . . . . **196**

Reham Abdullah Sanad Alsbu, Prasad Yarlagadda and Azharul Karim
An Empirical Model for Predicting the Fresh Food Quality Changes during Storage
Reprinted from: *Foods* **2023**, *12*, 2113, https://doi.org/10.3390/foods12112113 . . . . **208**

Ghanya Al-Naqeb, Cinzia Cafarella, Eugenio Aprea, Giovanna Ferrentino, Alessandra Gasparini, Chiara Buzzanca, et al.
Supercritical Fluid Extraction of Oils from Cactus *Opuntia ficus-indica* L. and *Opuntia dillenii* Seeds
Reprinted from: *Foods* **2023**, *12*, 618, https://doi.org/10.3390/foods12030618 . . . . **224**

Charles Stephen Brennan
Regenerative Food Innovation: The Role of Agro-Food Chain By-Products and Plant Origin Food to Obtain High-Value-Added Foods
Reprinted from: *Foods* **2024**, *13*, 427, https://doi.org/10.3390/foods13030427 . . . . **240**

Chee Kong Yap and Khalid Awadh Al-Mutairi
A Conceptual Model Relationship between Industry 4.0—Food-Agriculture Nexus and Agroecosystem: A Literature Review and Knowledge Gaps
Reprinted from: *Foods* **2024**, *13*, 150, https://doi.org/10.3390/foods13010150 . . . . **253**

Editorial

The Use of Agro-Food Chain By-Products and Foods of Plant Origin to Obtain High-Value-Added Foods

Gianluca Nardone, Rosaria Viscecchia and Francesco Bimbo *

Department of Agricultural Sciences, Food, Natural Resources and Engineering, University of Foggia, 71121 Foggia, Italy; rosaria.viscecchia@unifg.it (R.V.)
* Correspondence: francesco.bimbo@unifg.it

The increased consumer demand for sustainable, health-promoting foods has propelled research into plant-based products and the valorization of food by-products. This Special Issue compiles pioneering research that focuses on the development of novel foods of plant origin and the incorporation of agro-food chain by-products into functional food products. These works not only address environmental concerns but also explore innovative ways to utilize underused resources [1,2].

Several articles in this Special Issue delve into the potential use of food by-products as valuable ingredients. Fredrick Nwude Eze et al. [Contribution 1] explore the upcycling of defatted sesame seed meal into protein amyloid-based nanostructures, showcasing its functional potential as a novel food ingredient. Similarly, Jordan Richards et al. [Contribution 2] investigate carrot pomace, a by-product of juice production, and the enhancement of its functional properties via physical treatments, thus contributing to waste reduction and an improvement in the sustainability of food. Victoria Baggi Mendonça Lauria and Luciano Paulino Silva [Contribution 3] further highlight how food residues can be transformed into natural colorants using green extraction methods, enhancing their stability through nanotechnology.

Research on plant-based products is also represented in this Special Issue. Spasoje D. Belošević et al. [Contribution 4] examine the phytochemical composition and antioxidant properties of cold-pressed microgreen juices, positioning them as rich sources of bioactive compounds for functional food markets. Nikoo Jabbari et al. [Contribution 5] optimize the extraction of bioactive compounds from saffron petals using ultrasound-assisted methods, demonstrating the value of saffron by-products in the production of natural antioxidants and pigments.

Contributing to the advancement of alternative protein sources, Kaitlyn Burghardt et al. [Contribution 6] investigate the thermal and rheological properties of spirulina, soy, pea, and brown rice proteins; they thus offer insights regarding the processing of plant-based meat substitutes, which is critical for reducing environmental harm. Sílvia Petronilho et al. [Contribution 7] analyze raspberry fruits discarded during dormancy, presenting strategies for their incorporation into food formulations while preserving their color and flavor. Papadopoulos et al. [Contribution 8] highlight the nutritional benefits of olive leaf extract in laying hens, demonstrating improvements in the quality of eggs and antioxidant parameters, which contribute to animal feed innovations that promote sustainability and enhance the quality of food. Similarly, Semeniuc et al. [Contribution 10] explore the presence of coenzyme Q10 in food by-products and waste, detecting it using high-performance liquid chromatography. This study underscores the potential of food by-products to be used as sources of valuable bioactive compounds. Additionally, Mariana-Atena Poiana et al. [Contribution 9] explore the integration of grape pomace into pastry formulations, revealing how the incorporation of by-products into functional food products can enhance functionality without compromising sensory appeal.

Sanad Alsbu et al. [Contribution 11] present an empirical model to predict the changes in the quality of fresh food during storage, providing a practical tool that can be employed

Citation: Nardone, G.; Viscecchia, R.; Bimbo, F. The Use of Agro-Food Chain By-Products and Foods of Plant Origin to Obtain High-Value-Added Foods. *Foods* 2024, 13, 3154. https://doi.org/10.3390/foods13193154

Received: 30 September 2024
Accepted: 2 October 2024
Published: 3 October 2024

Copyright: © 2024 by the authors. Licensee MDPI, Basel, Switzerland. This article is an open access article distributed under the terms and conditions of the Creative Commons Attribution (CC BY) license (https://creativecommons.org/licenses/by/4.0/).

by the food industry to reduce spoilage and food loss during storage. In a similar vein, Al-Naqeb et al. [Contribution 12] study the extraction of oils from cactus seeds by using supercritical fluid extraction, a technique that offers an efficient and sustainable method for extracting valuable components from agricultural by-products.

This Special Issue concludes with two reviews by Brennan et al. [Contribution 13] and Kong Yap and Al-Mutairi [Contribution 14]; these reviews explore regenerative food innovation and the role of agro-food chain by-products and foods of plant origin in obtaining high-value-added foods. They also discuss a conceptual model relationship between Industry 4.0, the food–agriculture nexus and the agroecosystem.

This collection of papers provides significant insights into how plant-based ingredients and by-products can contribute to sustainable food systems, highlighting the intersection of health, innovation, and environmental responsibility in the modern food industry [1,2].

Funding: This research received no external funding.

Acknowledgments: As Guest Editors of this Special Issue entitled "Agro-Food Chain By-Products and Plant Origin Food to Obtain High-Value-Added Foods", we would like to express our gratitude to all the authors whose valuable work was published in this Special Issue and thus contributed to the completion of this edition.

Conflicts of Interest: The authors declare no conflicts of interest.

List of Contributions

1. Eze, F.N.; Muangrat, R.; Singh, S.; Jirarattanarangsri, W.; Siriwoharn, T.; Chalermchat, Y. Upcycling of Defatted Sesame Seed Meal via Protein Amyloid-Based Nanostructures: Preparation, Characterization, and Functional and Antioxidant Attributes. *Foods* **2024**, *13*, 2281. https://doi.org/10.3390/foods13142281.
2. Richards, J.; Lammert, A.; Madden, J.; Kang, I.; Amin, S. Physical Treatments Modified the Functionality of Carrot Pomace. *Foods* **2024**, *13*, 2084. https://doi.org/10.3390/foods13132084.
3. Baggi Mendonça Lauria, V.; Paulino Silva, L. Green Extraction of Natural Colorants from Food Residues: Colorimetric Characterization and Nanostructuring for Enhanced Stability. *Foods* **2024**, *13*, 962. https://doi.org/10.3390/foods13060962.
4. Belošević, S.D.; Milinčić, D.D.; Gašić, U.M.; Kostić, A.Ž.; Salević-Jelić, A.S.; Marković, J.M.; Đorđević, V.B.; Lević, S.M.; Pešić, M.B.; Nedović, V.A. Broccoli, Amaranth, and Red Beet Microgreen Juices: The Influence of Cold-Pressing on the Phytochemical Composition and the Antioxidant and Sensory Properties. *Foods* **2024**, *13*, 757. https://doi.org/10.3390/foods13050757.
5. Jabbari, N.; Goli, M.; Shahi, S. Optimization of Bioactive Compound Extraction from Saffron Petals Using Ultrasound-Assisted Acidified Ethanol Solvent: Adding Value to Food Waste. *Foods* **2024**, *13*, 542. https://doi.org/10.3390/foods13040542.
6. Burghardt, K.; Craven, T.; Sardar, N.A.; Pearce, J.M. Towards Sustainable Protein Sources: The Thermal and Rheological Properties of Alternative Proteins. *Foods* **2024**, *13*, 448. https://doi.org/10.3390/foods13030448.
7. Petronilho, S.; Coimbra, M.A.; Passos, C.P. Quality Characteristics of Raspberry Fruits from Dormancy Plants and Their Feasibility as Food Ingredients. *Foods* **2023**, *12*, 4443. https://doi.org/10.3390/foods12244443.
8. Papadopoulos, G.A.; Lioliopoulou, S.; Nenadis, N.; Panitsidis, I.; Pyrka, I.; Kalogeropoulou, A.G.; Symeon, G.K.; Skaltsounis, A.-L.; Stathopoulos, P.; Stylianaki, I.; et al. Effects of Enriched-in-Oleuropein Olive Leaf Extract Dietary Supplementation on Egg Quality and Antioxidant Parameters in Laying Hens. *Foods* **2023**, *12*, 4119. https://doi.org/10.3390/foods12224119.
9. Poiana, M.-A.; Alexa, E.; Radulov, I.; Raba, D.-N.; Cocan, I.; Negrea, M.; Misca, C.D.; Dragomir, C.; Dossa, S.; Suster, G. Strategies to Formulate Value-Added Pastry Products from Composite Flours Based on Spelt Flour and Grape Pomace Powder. *Foods* **2023**, *12*, 3239. https://doi.org/10.3390/foods12173239.
10. Semeniuc, C.A.; Ranga, F.; Podar, A.S.; Ionescu, S.R.; Socaciu, M.-I.; Fogarasi, M.; Fărcaș, A.C.; Vodnar, D.C.; Socaci, S.A. Determination of Coenzyme Q10 Content in Food By-Products and Waste by High-Performance Liquid Chromatography Coupled with Diode Array Detection. *Foods* **2023**, *12*, 2296. https://doi.org/10.3390/foods12122296.

11. Sanad Alsbu, R.A.; Yarlagadda, P.; Karim, A. An Empirical Model for Predicting the Fresh Food Quality Changes during Storage. *Foods* **2023**, *12*, 2113. https://doi.org/10.3390/foods12112113.
12. Al-Naqeb, G.; Cafarella, C.; Aprea, E.; Ferrentino, G.; Gasparini, A.; Buzzanca, C.; Micalizzi, G.; Dugo, P.; Mondello, L.; Rigano, F. Supercritical Fluid Extraction of Oils from Cactus *Opuntia ficus-indica* L. and Opuntia dillenii Seeds. *Foods* **2023**, *12*, 618. https://doi.org/10.3390/foods12030618.
13. Brennan, C.S. Regenerative Food Innovation: The Role of Agro-Food Chain By-Products and Plant Origin Food to Obtain High-Value-Added Foods. *Foods* **2024**, *13*, 427. https://doi.org/10.3390/foods13030427.
14. Yap, C.K.; Al-Mutairi, K.A. A Conceptual Model Relationship between Industry 4.0—Food-Agriculture Nexus and Agroecosystem: A Literature Review and Knowledge Gaps. *Foods* **2024**, *13*, 150. https://doi.org/10.3390/foods13010150.

References

1. Mateos-Aparicio, I. Plant-based by-products. In *Food Waste Recovery*; Academic Press: Cambridge, MA, USA, 2021; pp. 367–397.
2. Amoah, I.; Taarji, N.; Johnson, P.N.T.; Barrett, J.; Cairncross, C.; Rush, E. Plant-based food by-products: Prospects for valorisation in functional bread development. *Sustainability* **2020**, *12*, 7785. [CrossRef]

Disclaimer/Publisher's Note: The statements, opinions and data contained in all publications are solely those of the individual author(s) and contributor(s) and not of MDPI and/or the editor(s). MDPI and/or the editor(s) disclaim responsibility for any injury to people or property resulting from any ideas, methods, instructions or products referred to in the content.

Article

Upcycling of Defatted Sesame Seed Meal via Protein Amyloid-Based Nanostructures: Preparation, Characterization, and Functional and Antioxidant Attributes

Fredrick Nwude Eze [1,2], Rattana Muangrat [2,3,*], Sudarshan Singh [4], Wachira Jirarattanarangsri [2], Thanyaporn Siriwoharn [2] and Yongyut Chalermchat [2]

1. Office of Research Administration, Chiang Mai University, Chiang Mai 50200, Thailand; fredrickeze10@gmail.com
2. Faculty of Agro-Industry, Chiang Mai University, Chiang Mai 50100, Thailand; wachira.j@cmu.ac.th (W.J.); thanyaporn.s@cmu.ac.th (T.S.); yongyut.c@cmu.ac.th (Y.C.)
3. Department of Food Process Engineering, Faculty of Agro-Industry, Chiang Mai University, Chiang Mai 50100, Thailand
4. School of Medical & Allied Sciences, K.R. Mangalam University, Gurugram 122103, India; sudarshansingh83@hotmail.com
* Correspondence: rattanamuangrat@yahoo.com or rattana.m@cmu.ac.th

Abstract: Herein, the possibility of valorizing defatted sesame seed meal (DSSM) as a viable source for valuable plant proteins and amyloid-based nanostructure was investigated. Sesame seed protein isolate (SSPI) and the major storage protein globulin (SSG) were prepared by alkaline extraction–isoelectric point precipitation as well as fractionation in the case of SSG. The protein samples were characterized for their physicochemical attributes. SSPI and SSG were also evaluated for their ability to form amyloid structures under heating (90 °C) at low pH (2.0). Additionally, the functional attributes, antioxidant activity, and biocompatibility of the proteins and amyloid nanostructures were also examined. SSPI and SSG were both successfully prepared from DSSM. The data showed that the physicochemical attributes of both protein samples were quite similar, except for the fact that SSG was mostly composed of 11S globulin, as evinced by Tricine-SDS-PAGE analysis. TEM micrographs revealed that SSG was able to form curly-shaped fibrillar amyloid structures, whereas those derived from SSPI were mostly amorphous. Thioflavin-T assay and Tricine-SDS-PAGE analysis indicated that acidic heating promoted protein hydrolysis and self-aggregation of the hydrolyzed peptides into a β-sheet rich amyloid structure. Importantly, the amyloid preparations displayed commendable solubility, superior water and oil holding capacities, and antioxidant activity against DPPH and ABTS. The protein amyloid nanostructures were found to be non-toxic against RAW264.7 cells, HaCaT cells, and red blood cells. These findings indicate that DSSM could be upcycled into valuable protein amyloid structures with good potentialities as novel food ingredients.

Keywords: sesame seed protein isolate; sesame seed globulin; plant-based proteins; food protein amyloids; antioxidant activity; agri-food residue valorization

1. Introduction

Amyloid-based nanostructures have continued to command intense research interest due to their intriguing structure and versatile roles in areas as diverse as materials, biomedicine, and food science. Evidence emerging from the previous decade does not only demonstrate that food protein amyloid nanostructures are safe for human nutrition [1] but also indicates their potential multifarious applications as transporters for the delivery of valuable bioactive compounds and nutrients [2] and sustainable food coating materials [3,4], as well as in the development of various novel food products such as cultivated meat [5]. Food protein amyloid nanostructures, by virtue of their unique physicochemical attributes, have been found to enhance the solubility, thermal and photostability, and biological properties of various functional ingredients and food products [6,7]. Interestingly, it has also been noticed that protein amyloid nanostructures exhibit superior techno-functional characteristics and antioxidative attributes when compared to their precursor proteins [8,9], underscoring their potential importance in the modulation of food quality and design of food with desirable attributes. Nonetheless, widespread application remains elusive because animal proteins used as source materials are expensive. Thus, there is a clear need for cheaper, more abundant, and more sustainable sources of food proteins capable of producing amyloid nanostructures. In this regard, plant-derived protein appears to be very promising.

Recent investigations have largely been focused on the preparation and characterization of amyloid nanostructure from more conventional food protein sources such as whey [9], soy [5], rice [8], and oats [10]. The successful formation of protein amyloid-based nanostructures was accomplished through the process of aggregation facilitated by thermal-induced denaturation and polypeptide hydrolysis, often at temperatures around 80–95 °C and low pH (2.0). Less-conventional plant protein sources, such as amaranth protein-rich waste [11] and defatted hempseed [12], have also been explored. The challenge often encountered in using plant-derived proteins for amyloid formation is that they are complex and of low purity. In addition, the presence of small bioactive molecules in the protein concentrate or isolate may hamper amyloid fibril formation. The source of the plant protein and the preparation strategy may improve the outcome.

Defatted sesame seed meal (DSSM) is an interesting but unexplored resource that could be potentially utilized as a sustainable, cheap, and abundant plant protein source for amyloid-based nanostructures. DSSM is an agri-food industry residue obtained from the extraction of sesame seed oil. DSSM could also be used as organic nitrogenous fertilizer [13]. Remarkably, DSSF is exceedingly rich in proteins (33–40%, depending on the source) [14], and as a result, it is often used in feed formulation for cattle, poultry, and fish [15]. Sesame seed protein is mainly composed of four storage proteins, namely prolamin (1.3%), glutelin (6.9%), albumin (8.6%), and globulin (67.3%). In other words, globulin (conventionally called α-globulin) and, to a lesser extent, albumin (conventionally called β-globulin) make up the majority of the proteins in sesame seed, together accounting for more than 90% of sesame seed storage proteins [16,17]. This implies that a large amount of sesame seed globulin (SSG) could be obtained from DSSM without much difficulty. Additionally, sesame seed protein isolate (SSPI) obtained from DSSM is regarded as high quality and capable of meeting the essential amino acid requirement for adults, according to FAO/WHO (2007) [18]. SSPI is unique in its rich content of sulfur-containing amino acids (3.8–5.5%), especially methionine (2.5–4%) [17], making it a reliable complementary source for balancing most plant proteins, which are limited in these amino acids, such as soy and ground rice. Sesame protein and peptides have also exhibited commendable biological properties (antioxidative, antihyperlipidemic, antihypertensive) [19–21] and functionality (surface properties and hydration) [16], underscoring their potential in the development of food products for health improvement. DSSM is very cheap (0.27–0.41 USD/kg) and abundant [22], thus, it can be an inexpensive, readily available, and sustainable source for not only valuable plant proteins but also plant protein amyloid nanostructures. At the moment, there is a paucity of reports on the preparation of food amyloid nanostructures from plant proteins. Particularly, there is no report on the preparation and characterization of protein amyloid nanostructure from proteins derived from DSSM.

Thus, this work is focused on investigating the viability of proteins derived from defatted black sesame seed meal (DSSM) as a source material for the production of plant protein amyloid-based nanostructures. To this end, sesame seed protein isolate (SSPI) was prepared from DSSM using the alkaline extraction–isoelectric point precipitation (AE-IP) method. Being the most abundant protein group, sesame seed globulin fraction (SSG) was also prepared from DSSM via alkaline extraction, fractionation, and the isoelectric point precipitation (AE-F-IP) method. The ability of both SSPI and SSG to form amyloid nanostructures under thermal-induced aggregation in acidic conditions (90 °C, pH 2.0) was investigated. Furthermore, the amyloid nanostructures from both protein samples were characterized, and their functional, antioxidant properties and biocompatibility were examined. The findings from this work are expected to open an additional frontier for the valorization of defatted black sesame seed meal in the food industry.

2. Materials and Methods

2.1. Materials

Black sesame seeds sourced from the Huai Siao Royal Project Development Center in Ban Pong Subdistrict, Hang Dong District, Chiang Mai Province, Thailand, were subjected to roasting at 80 °C prior to extraction using supercritical CO_2 (at 45 °C and 225 bars for an extraction duration of 3 h). The resultant defatted black sesame seed meal (DSSM) obtained after supercritical CO_2 extraction was subsequently utilized for the preparation of sesame seed protein isolate (SSPI) and sesame seed globulin (SSG). Acrylamide was obtained from ACROS ORGANICS (Geel, Belgium), bis-acrylamide was obtained from TOKU-E Company (Bellingam, WA, USA), glycine was obtained from Fisher Scientific Ltd. (Leicester, UK), sodium Lauryl sulfate was purchased from Sigma-Aldrich (St. Louis, MO, USA), Tris (molecular biology grade) was obtained from Vivantis Technologies Sdn Bhd (Selangor Darul Ehsan, Malaysia), Tricine was purchased from USB Corporation (Cleveland, OH, USA). Except stated otherwise, all chemical reagents used in this work were of analytical grade.

2.2. Preparation of Sesame Seed Protein Isolate (SSPI)

Sesame seed protein isolate was prepared using the so-called 'universal alkaline extraction and isoelectric point precipitation' (AE-IP) method (Figure 1). Put succinctly, the whole defatted black sesame seed meal (DSSM) was pulverized into fine powder. DSSM powder was dispersed into distilled water at a solid-to-solvent ratio of 1:10 w/v. The pH of the dispersion was adjusted to 10.0 by adding a few drops of 2 M NaOH. The extraction was carried out by stirring for 2 h at room temperature. The mixture was then centrifuged at $10,000\times g$ for 15 min at 4 °C. The supernatant was collected while the pellet was discarded The supernatant was adjusted to pH 4.5 by adding a few drops of 5 M HCl to enable the isoelectric point precipitation of sesame seed proteins. The solution was left for 12 h at 4 °C for complete precipitation. Thereafter, sesame seed protein was collected by centrifuging the solution at $15,000\times g$ for 15 min at 4 °C. The protein was collected as the pellet. The pellet was washed twice by re-dispersing it in distilled water, followed by centrifugation Finally, the pH of the protein solution was adjusted to 7.0 using 1 M NaOH solution, frozen at −20 °C for 24 h, and freeze-dried using Labconco FreeZone 12 freeze dryer (Labconco Corp., Kansas City, MO, USA) to obtain a sesame seed protein isolate (SSPI). The coarse freeze-dried SSPI flakes were crushed into powder and stored at 4 °C.

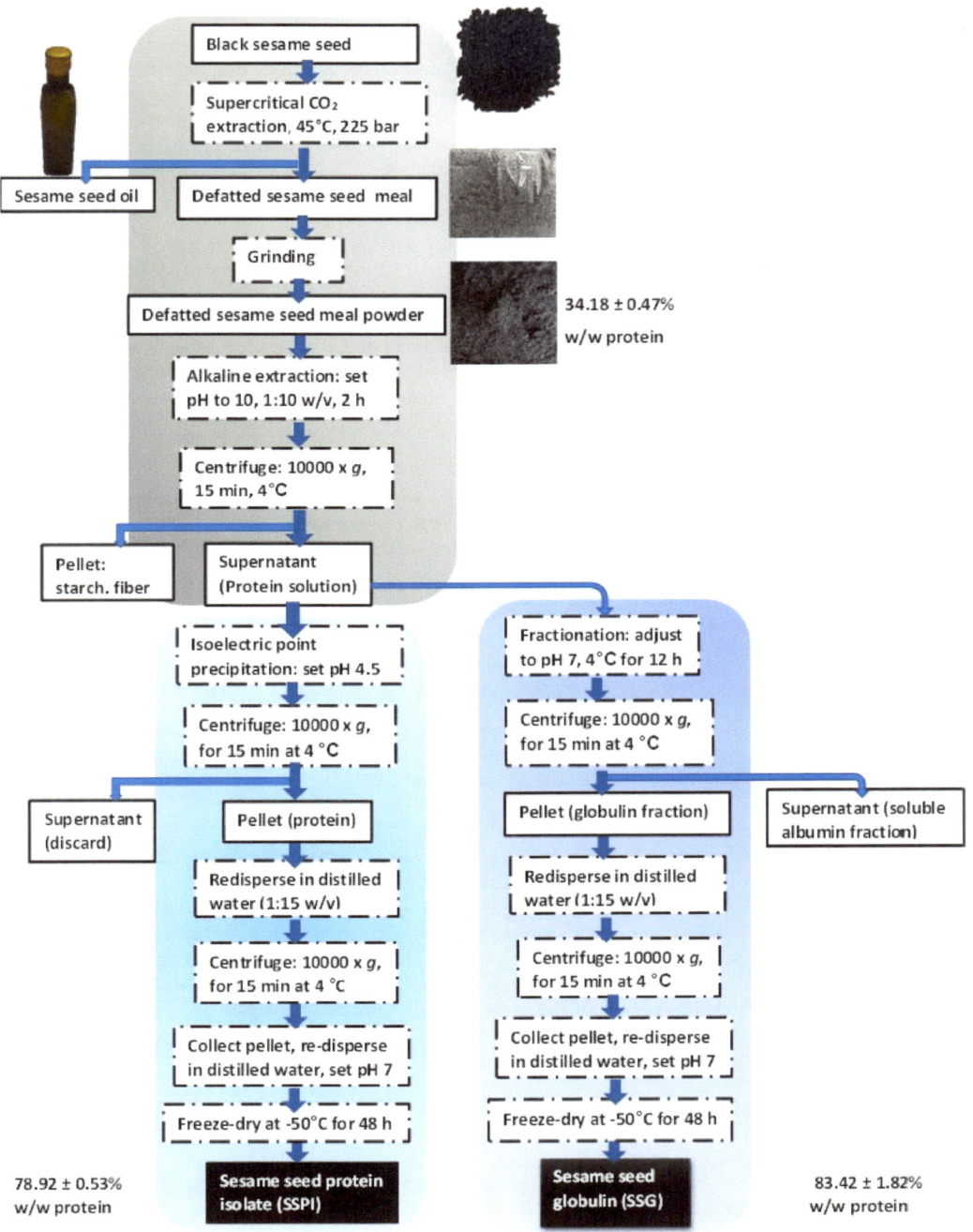

Figure 1. Schematic depiction of the preparation of sesame seed protein isolates via alkaline extraction and isoelectric point precipitation (AE-IP) and sesame seed globulin via alkaline extraction, fractionation, and isoelectric point precipitation (AE-F-IP).

2.3. Preparation of Sesame Seed Globulin (SSG)

Firstly, DSSM powder was dispersed into distilled water at a ratio of 1:10 w/v. After adjusting the pH to 10.0 using 2 M NaOH solution, the mixture was stirred at room temperature for 2 h. Then, the supernatant was collected after centrifuging at 10,000× g for 15 min at 4 °C. Subsequently, the pH of the supernatant was adjusted to 7.0 using a chilled HCl solution. The protein solution was kept at 4 °C for 12 h. After centrifugation of the solution, crude sesame seed globulin was collected as a pellet, whereas the supernatant, which is rich in water-soluble albumin, was discarded. The collected pellet was further re-dispersed in distilled water (containing 5% NaCl) at 1:15 w/v. The dispersion was stirred for 2 h, after which the supernatant was collected via centrifugation. The pH of the supernatant was adjusted to 4.5 by adding a few drops of 5 M HCl solution to trigger the precipitation of sesame seed globulin (SSG). After centrifugation, the SSG precipitate was collected, washed twice with distilled water, and adjusted to pH 7.0 using 1 M NaOH solution. Salt and other small molecule impurities were removed by dialyzing for 24 h against distilled water using a snakeskin dialysis membrane (MWCO: 14 kDa). The retentate was freeze-dried to obtain SSG powder, which was stored at 4 °C awaiting further use.

2.4. Determination of Protein Yield and Recovery Rate

The total protein content of the samples was determined using the Kjeldahl method for total nitrogen content. The protein content in the samples was then calculated using a nitrogen-to-sesame seed conversion factor of 6.25. The protein yield and protein recovery rate were determined thusly:

$$\text{Protein yield (\%)} = [M2/M1] \times 100 \quad (1)$$

$$\text{Protein recovery rate (\%)} = [(M2 \times P2)/(M1 \times P1)] \times 100 \quad (2)$$

where M2 and M1 are the weight in grams of sesame seed protein and defatted sesame seed meal, respectively, whereas P2 and P1 are the respective purity in percent of sesame seed protein and defatted sesame seed meal.

2.5. Preparation of Sesame Seed Protein Amyloid-Based Nanostructures

Amyloid fibrils from sesame seed globulin were prepared via acid hydrolysis as previously described. SSG solution (3% w/w) was prepared by dispersing the protein powder in distilled water, pH 2.0. After stirring for 2 h, the solution was kept at 4 °C for 12 h for complete hydration. The protein dispersion was then centrifuged at 10,000× g for 15 min at 4 °C, and the supernatant was collected. Protein aggregation was initiated by adjusting the pH of the solution to 2.0 using 5 M HCl solution followed by heating at 90 ± 5 °C while stirring at 300 rpm. Aliquots of the protein solution were collected at specific time intervals (0, 4, 8, 12, and 24 h), quenched by inserting in an ice-water bath for 20 min, and kept at 4 °C awaiting further analysis.

2.6. Color Analysis

The physical appearance of all the protein samples was examined using a Konica Minolta CR-400 chroma meter manufactured by Konica Minolta Optics, Inc. (Tokyo, Japan). Color parameters were recorded as L*, a*, and b* values, where the L* value represents lightness, the a* value (red/green), and the b* value (yellow/blue) [23]. The color difference, ΔE, was calculated as follows:

$$\Delta E = \sqrt{\left[(L_0^* - L_1^*)^2 + (a_0^* - a_1^*)^2 + (b_0^* - b_1^*)^2\right]} \quad (3)$$

where L_0^*, a_0^* and b_0^* represent the color parameters of DSSM and L_1^*, a_1^*, and b_1^* represent those of the different samples.

2.7. Tricine SDS-PAGE

Tricine SDS-PAGE was performed following the method outlined by [24,25] with minor modifications. Firstly, protein sample solutions were added into 1.5-mL Eppendorf tubes. GE Healthcare's low molecular weight marker, which contained standard proteins with molecular weight in the range of 14.4–97 kDa, was prepared alongside and used as a protein marker. SDS sample loading buffer (free of β-mercaptoethanol) was added to the protein solutions, followed by mixing. Each sample of protein solution was boiled briefly (5 min) and separated on Tricine SDS-PAGE system with a 5% polyacrylamide gel stacking layer and a 12% resolving gel layer under low voltage using Atto AE-6450 Dual Mini Slab Kit for electrophoresis, Atto Corporation (Tokyo, Japan) connected to EC-105 LINE VOLTAGE 230VAC 50–60 HZ 80 W power source, E-C Apparatus Corporation (Marietta, OH, USA). Subsequently, the gels were gently removed from the glass plates, rinsed with distilled water, and immediately stained with 0.2% Coomassie brilliant blue G-250 dye in methanol: distilled water: acetic acid (45:45:10; $v/v/v$) staining solution for 12 h. After, the gels were destained with the same solvent system used in the staining solution but without the dye to enable the visualization of the protein bands of the samples.

2.8. Scanning Electron Microscopy Analysis

Freeze-dried protein samples were dispersed in ethanol. The samples were then spotted on an adhesive copper tape and allowed to dry. SEM micrographs of the samples were obtained using a Hitachi SU3800 scanning electron microscope (Hitachi Ltd., Tokyo, Japan) operated at 3–5 kV.

2.9. Transmission Electron Microscopy Analysis

TEM images of the amyloid fibrils were acquired using the JOEL JEM-2100Plus transmission electron microscope, JOEL Ltd. (Tokyo, Japan). Samples (20 µL) were initially dispersed in ultrapure water (1 mL). The protein solution (5 µL) was then carefully dropped onto the surface of the TEM grid and allowed to rest for 2 min. Afterward, the tip of a filter paper was used to gently blot off excess fluid from the grid. Then, the grid was rinsed twice, each time with a drop of ultrapure water. Subsequently, the proteins on the grid were negatively stained by dropping 10 µL of 2% uranyl acetate onto the grid's surface. The TEM grids were allowed to air dry prior to visualization of the proteins with the microscope at 100 kV.

2.10. FT-IR Analysis

FTIR analysis was performed using a JASCO FT/IR-4700 spectrometer, JASCO Ltd. (W. Yorkshire, UK). Each sample powder was mixed with KBr powder and pressed into a thin translucent disc. The KBr disc was inserted into the sample holder, and spectra were captured with 120 scans at 4 cm^{-1} resolution over a range of 4000 to 400 cm^{-1}. Baseline, deconvolution, and peak fitting treatments were subsequently performed using OriginPro software version 10.1.0.170, OriginLab Corporation (Northampton, MA, USA).

2.11. Thioflavin-T Fluorescence Spectroscopy

Thioflavin-T (ThT) stock solution (2 mM) was prepared by dissolving the powder in ultrapure water. The stock solution was filtered using a 0.22 µm syringe filter to remove any residual precipitates. ThT working solution (50 µM) was prepared from the stock solution by diluting it with ultrapure water. Then, aliquots of the protein solution (15 µL) were introduced into a 96-well microplate. ThT working solution (200 µL) was added to each protein solution and mixed well. After 10 min of reaction, the ThT fluorescence intensity of the samples was recorded using a microplate reader at an excitation wavelength of

440 nm and an emission wavelength of 480 nm. Fluorescence of ThT working solution (200 µL) with ultrapure water, pH 2.0 (15 µL) was used for background correction.

2.12. Surface Hydrophobicity

The surface hydrophobicity of sesame protein samples was determined via the bromophenol blue binding assay [26]. Protein dispersion (2 mg/mL) was prepared in phosphate-buffered saline (PBS), pH 6.0, by vortex mixing. Bromophenol blue dye solution (1 mg/mL) was also prepared in PBS buffer. Then, 10 mL of protein sample (2 mg/mL) was mixed with 2 mL of bromophenol blue solution. The control solution was prepared by replacing the protein solution with only PBS buffer in the protein-dye mixture. Thereafter, all samples were centrifuged for 15 min at 6000× g, and supernatants were collected. The supernatants were diluted 10-fold, and absorption of the samples was recorded at 595 nm.

$$\text{Bromophenol blue bound (µg/mg)} = 200 \times [(A_c - A_s)/(A_c \times 2)] \quad (4)$$

where A_c and A_s represent the absorbance of the control and samples, respectively.

2.13. Functional Attributes

2.13.1. Solubility

Protein solubility was ascertained following the description given by [27] with minor modifications. Firstly, protein dispersions were prepared from the powder in a phosphate-buffered saline solution (PBS, 10 mM). The pH of the dispersions was adjusted using either 1 M NaOH or 1 M HCl to obtain dispersions with pH values from 2.0 to 10.0. Thereafter, the dispersions were stirred at 100 rpm for 2 h. After centrifuging at 8000× g for 20 min, the supernatants were collected. A modified Bradford assay [28] was used to determine the protein content of the supernatant (Conc. sample). The supernatant obtained from protein dispersion prepared using 0.1 M NaOH was regarded as the control (Conc. control). The protein solubility (%) was calculated thusly:

$$\text{Protein solubility (\%)} = [\text{Conc. sample}/\text{Conc. control}] \times 100 \quad (5)$$

2.13.2. Fluid Holding Capacity

The water holding capacity (WHC) and oil holding capacity (OHC) of the protein powder samples were determined following the method described by [29]. Each protein powder (1 g) was introduced to 50-mL centrifuge tube. Distilled water (10 mL) was added to the protein sample in the case of WHC, whereas soybean oil (10 mL) was added to the powder in the case of OHC. The protein mixture was vortexed and mixed for 2 min. After 30 min, the protein dispersions were centrifuged for 20 min at 3000× g. The supernatants were carefully decanted to allow the weight of the wet pellet or paste to be measured. The WHC or FAC was calculated thusly:

$$\text{WHC or OHC} = (W_2 - W_1)/W_0 \quad (6)$$

where W_2 is the weight of the wet pellet or paste plus tube (g), W_1 is the weight of the dry powder plus tube (g), and W_0 is the weight of the dry powder (g).

2.14. Antioxidant Activity

The antioxidant activity of the protein samples was evaluated using DPPH and ABTS assays [30]. Protein dispersions (2 mg/mL) were prepared in distilled water. DPPH solution (0.1 mM) was prepared using methanol. The DPPH assay was performed by adding 100 µL of the protein dispersion into a 96-well microplate, followed by 100 µL of the DPPH solution. Control solution consisted of DPPH solution with distilled water. The solution mixture was mixed briefly and incubated for 30 min at room temperature in the dark.

Absorbance of the samples was read at 517 nm. The antioxidant activity was presented as radical scavenging activity (RSA%).

$$\text{RSA (\%)} = [(A_c - A_s)/A_c] \times 100 \quad (7)$$

where A_c and A_s in this equation represent absorbance of the control and sample, respectively.

The antioxidant activity of the protein samples was also determined using ABTS assay [8]. Firstly, the $ABTS^{\cdot+}$ stock solution was prepared by mixing 5 mL of ABTS solution (7 mM) with 5 mL of potassium persulfate solution (2.45 mM). This stock solution was incubated in the dark for 12 h. The stock solution was diluted with distilled water to obtain an absorbance of 0.7 ± 0.02 as the working solution. Thereafter, protein samples (30 µL) were reacted with 250 µL of the ABTS working solution for 6 min. The absorbance of the samples was recorded at 734 nm using a microplate reader. Samples with distilled water in place of the protein solution were used as a control. Result obtained was presented as radical scavenging activity (%) calculated as described supra.

2.15. Biocompatibility

2.15.1. In Vitro Cytotoxicity

In vitro cytotoxicity of the prepared amyloid samples was evaluated on mouse macrophage (RAW264.7) and human keratinocyte (HaCaT) cells as previously described [31] with minor modification. RAW264.7 cells were obtained from ATCC (Manassas, VA, USA), whereas HaCaT cells were obtained from CLS Cell Lines Service GmbH (Eppelheim, Germany). The cells were seeded into 96-well plates at a density of 1×10^4 cells/cm^2 for 24 h. Thereafter, the cells were treated with SSPAN and SSGAN solution and incubated for 24 h in a CO_2 incubator at 37 °C. Then, the exhausted media from the wells was gently aspirated. This was replaced with 200 µL fresh medium (without fetal bovine serum) containing 0.5 mg/mL MTT reagent. The plates were then incubated at 37 °C for 3 h. Subsequently, media was removed, and DMSO was added to the wells to dissolve the formazan crystals that were formed. The absorbance of the wells was measured at 560 nm using a Thermo Scientific microplate reader (Themo Scientific, Waltham, MA, USA).

$$\text{Cell viability (\%)} = (AbsT/AbsC) \times 100 \quad (8)$$

where AbsT and AbsC depict the absorbance of the treated sample and the absorbance of the control sample, respectively.

2.15.2. Hemolytic Effect

The hemolytic effect of the protein amyloid-based nanostructures was evaluated on rat erythrocytes, according to a previous report [32]. Fresh rat blood (10 mL) was commercially obtained from the National Laboratory Animal Center, Mahidol University, Thailand. Erythrocytes were collected by centrifuging the blood at 1500 rpm for 10 min. The pellet containing erythrocytes was collected and washed thrice by resuspending in PBS, followed by centrifugation. Then, fresh erythrocyte suspension was prepared from the cell pellet (2% v/v) by adjusting with PBS. Protein dispersion was also prepared in PBS, pH 7.4. The protein dispersion (2 mL) was added to the erythrocyte suspension (2 mL) and gently mixed. The negative control consisted of PBS solution added into the erythrocyte suspension, while the positive control contained distilled water. The resultant mixture was incubated at room temperature for 30 min. The mixture was subjected to 10 min of centrifugation at a speed of 1500 rpm and at 4 °C. Supernatants of the samples were collected and diluted 4-fold with distilled water. Then, the absorption of the diluted solutions was recorded at 540 nm using a UV-vis spectrophotometer. Erythrocyte hemolysis induced by the samples was calculated thusly:

$$\text{Hemolysis (\%)} = [(AS - AN)/(AP - AN)] \times 100 \quad (9)$$

where AS, AN, and AP represent the absorbance of the sample, the negative control, and the positive control, respectively.

2.16. Statistical Analysis

All data were collected at least in triplicate. The data were analyzed using one-way ANOVA followed by the Tukey test for post hoc multiple comparisons. p value was set at less than 0.05 for statistical significance. All data analysis was carried out using GraphPad Prism software version 10, GraphPad Software, Inc. (San Diego, CA, USA).

3. Results and Discussion

3.1. Preparation and Characterization of Sesame Proteins from DSSM

3.1.1. Protein Extraction and Recovery

In view of the growing importance of sustainability and promoting a circular economy around the world to meet the challenges posed by resource constraints, it is absolutely imperative to re-examine hitherto underexplored agri-food industry residue as a potential feedstock for the generation of value-added bioproducts and food ingredients. Herein, defatted black sesame seed meal, a highly promising protein-rich byproduct of the oil extraction process from roasted black sesame seeds using supercritical CO_2 extraction, was examined. The purpose was to determine whether it could serve as a viable and sustainable source for the production of food proteins and amyloid nanostructures for potential use in nutritious and functional food products.

Sesame seed protein isolate (SSPI) and sesame seed globulin (SSG) fraction were successfully obtained from the defatted sesame seed meal. An overview of the workflow involved in the preparation of both SSPI and SSG is depicted in the schematic (Figure 1).

As shown in Table 1, the protein content of the defatted sesame meal powder was (34.18%). This value was within the range of crude protein content previously reported for sesame seed cake by other authors [33]. The recovery rate for both SSPI and SSG using the AE-IP method was modest, approximately 50% and 40%, respectively. Considering the fact that the preparation was based on an orthodox and simple extraction technique, the recovery rate was actually commendable. This notion is buttressed by the findings from analogous studies by Yang et al., wherein extraction of proteins by AE-IP technique from cold-pressed sesame cake with or without ultrasonic pre-treatment exhibited recovery of 22.24–25.95% and solid yield of 15.57–18.62% [33]. The protein content (%w/w) of SSPI and SSG was equally impressive for a conventional extraction process. This is particularly notable since the source material was extracted only once and without any pre-treatment step in contrast to other reports where multiple extractions were carried out on the crude material. Often, the protein content is considered as a rough indicator of purity. Apparently, SSG had better purity compared to SSPI (protein content of 83% vs. 79%), which is credited to the additional fractionation steps included during the preparation of SSG (Figure 1). Also, it could be said that both SSPI and SSG were of comparable purity to protein isolates obtained from Turkish local beans (protein content of 80–85%) by AE-IP method [34] but of superior purity to protein proteins extracted from pea flour by salt extraction, micellar precipitation or AE-IP techniques (protein content of 73–75%) [35] and protein isolate (protein content of 77.1%) obtained from sesame meal powder AE-IP approach. Although this may not strictly be an ideal comparison given the nuances sometimes involved, such as differences in the source of the raw material as well as the extraction procedure, the results in Table broadly confirmed that the protein preparation technique applied herein was efficacious.

Table 1. Protein extraction, yield, and recovery rate of defatted black sesame seed protein samples.

Sample	Protein Content (%)	Yield (%)	Recovery Rate (%)
DSSM	34.18 ± 0.47 [a]	-	-
SSPI	78.92 ± 0.53 [b]	21.53 ± 1.22 [c]	49.71 ± 0.98 [b]
SSG	83.42 ± 1.81 [c]	16.33 ± 0.91 [d]	39.86 ± 0.76 [a]

Different superscript letters within the same column represent mean values that are significantly different ($p < 0.05$).

3.1.2. Physical Appearance of Sesame Protein Samples

The physical appearance of protein products, such as isolates, hydrolysates, or concentrates, has a major influence on consumer appeal and acceptance. This is because of the use of appearance by many as a first approximation of product quality. The visual appearance and corresponding color parameters of DSSM and protein samples prepared and derived from it are presented in Figure 2 and Table 2, respectively. It can be seen that both SSPI and SSG were similar in color but darker when compared to DSSM.

Figure 2. Physical appearance of defatted sesame seed meal (DSSM), sesame seed protein isolate (SSPI), and sesame seed globulin (SSG) powder samples.

Table 2. Colorimetric parameters of defatted black sesame seed protein samples.

Sample	L*	a*	b*	ΔE
DSSM	44.92 ± 0.31 [a]	0.40 ± 0.05 [c,d]	5.02 ± 0.08 [a]	-
SSPI	33.39 ± 0.07 [b]	0.11 ± 0.04 [c,d,e]	4.40 ± 0.40 [b]	11.66 ± 0.37 [a]
SSG	33.24 ± 0.29 [b]	−0.38 ± 0.01 [c,e]	3.37 ± 0.05 [c]	11.83 ± 0.32 [a]

Different superscript letters within the same column represent mean values that are significantly different ($p < 0.05$).

The color parameters in terms of L*, a*, and b* values were consistent with the visual appearance. The L* value denotes lightness, i.e., the higher the L* value (closer to 100), the lighter the color, and vice versa. Compared to both SSPI and SSG, the crude sesame seed flour was significantly lighter in appearance, as evinced by the higher L* value. The L* values of both SSPI and SSG did not only signify that both were darker, but also that there was no variability in their dark color. The same trend was reflected in their a* (greenness to redness) and b* (blueness to yellowness) values. With respect to the overall color, there was a large difference in the color of the protein samples, which could be visually perceived (ΔE value greater than 6.0) [23] relative to the DSSM sample.

Pigmentation, such as the dark color of sesame seed, had been credited to metabolites in the seed coat [18], such as flavonoids and terpenoids, as well as amino acids and their derivatives. In fact, studies have reported a greater preponderance of these valuable bioactives in black sesame seeds compared to other color varieties such as brown or white [36]. According to Dossou et al., the dark color or pigmentation is mainly due to melanin from the seed coat or hull of the sesame seed. The melanin from black sesame seed was similar to melanin from other natural sources. Polyphenols were proposed as precursors to the pigment [37]. Meanwhile, the 'black sesame pigment' isolated from black sesame seed was reported to exhibit strong antioxidant, reducing, and anti-nitrosating properties [38]. On the basis of the ostensible increase in the dark color of the protein samples, a proposition could be made. That is, processing the DSSM into SSPI and SSG led to the release and greater availability of bioactive pigment components ending up in the obtained protein sample. This implies that better activity of the protein samples could be obtained, thereby enriching their value as potential functional food ingredients. Meanwhile, from an aesthetic perspective, SSPI and SSG could serve an auxiliary purpose as natural colorants to improve the visual appeal of the functional food products, such as bread, cake, cookies, beverages, etc., into which they are incorporated.

3.1.3. Microstructure of Sesame Seed Protein Samples

The morphology and structure of sesame protein microparticles were examined using SEM. Figure 3 depicts representative micrographs of SSPI and SSG. It can be seen that particles from both samples broadly feature the same morphology, although SSG seemed a bit denser and more compact than SSPI. The protein particles manifested a rugged rock-like appearance. This quintessential morphology of sesame protein particles can be credited to the use of the freeze-drying technique during the drying of the aqueous dispersion in order to obtain solid protein samples. Freeze-drying typically takes a long time, during which ice crystals are formed. The long duration allows greater molecular interaction of the protein molecules, enabling them to coalesce and form clumps. Sublimation of the ice crystals facilitates the formation of the pores in the microstructure [39]. The ridges observed on the surface of the particles can be attributed to the rate of water loss as a consequence of sublimation during lyophilization. The rapid rate of water loss on the surface of the sample during the freeze-drying process can cause the formation of wrinkles on the surface of particles, giving rise to the ridge-like structure and rough appearance. It is apparent that the microparticles of both SSPI and SSG are irregular in shape as a result of the freeze-drying. Similar rough surfaces and non-uniformly shaped particles were observed for freeze-dried protein isolates obtained from unprocessed alfalfa seeds [40].

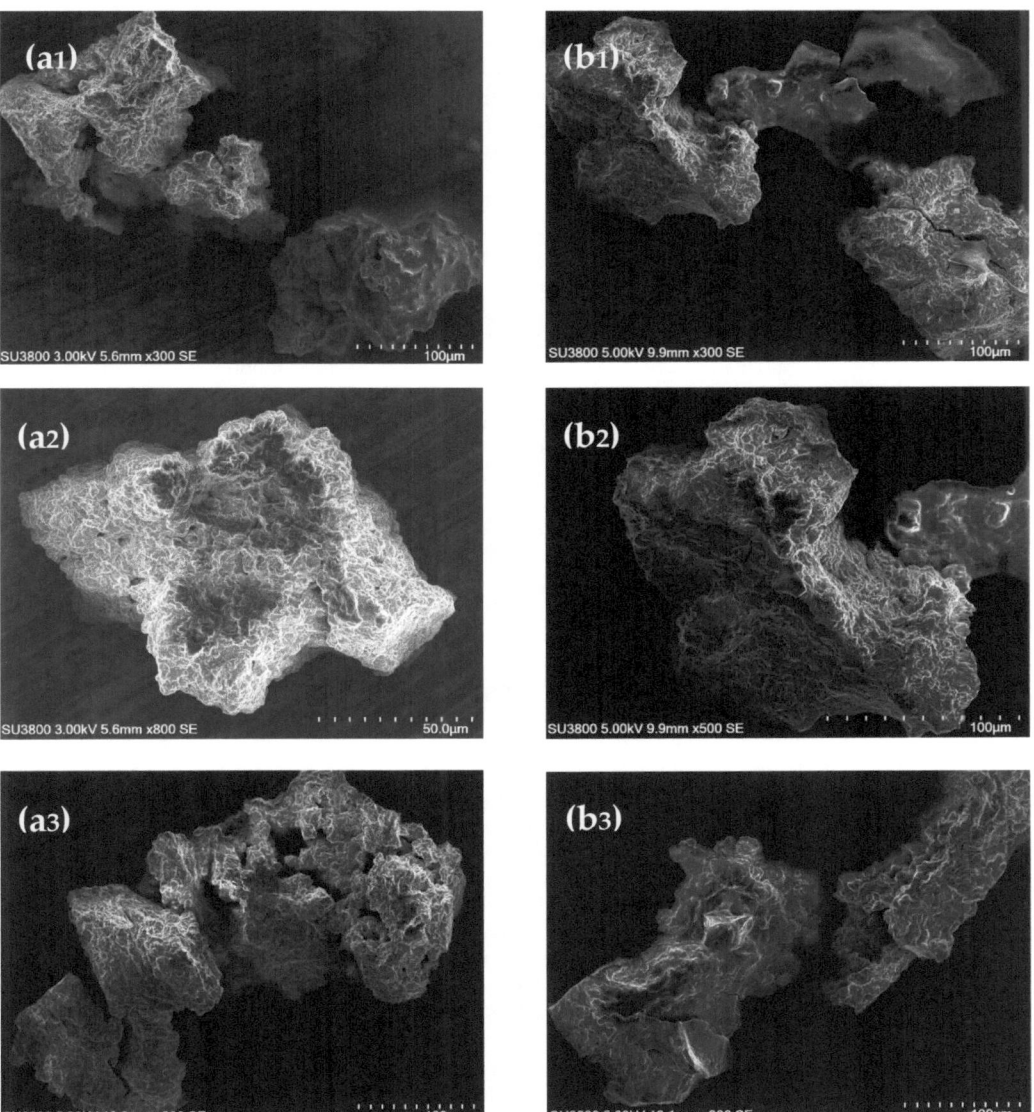

Figure 3. SEM images of sesame seed protein isolate, SSPI (**a1–a3**) and sesame seed globulin, SSG (**b1–b3**). Both SSPI and SSG were prepared from defatted black sesame seed meal.

3.1.4. Electrophoretic Mobility of Sesame Protein Samples

The protein profile of DSSF, SSPI, and SSG were ascertained via Tricine-SDS-PAGE. Illustrated in Figure 4 is the electrophoretic pattern of the different protein samples resolved under non-reducing conditions (no β-mercaptoethanol). It can be seen that the most intense and prominent protein band was present in the range of 45–66 kDa (Figure 4). As mentioned earlier, 11S globulin makes up the majority of protein in sesame seeds, with a reported relative abundance of 78.68% [41]. The native protein exists as an oligomeric structure consisting of six subunits held together by noncovalent bonds. Each of these subunits is further composed of an acidic (30–34 kDa) and a basic (20–25 kDa) linked together by a disulfide linkage. Thus, under non-reducing (i.e., absence of disulfide bond chemical reducing agents) conditions, 11S globulin is expected to be resolved with apparent MW around 50 kDa. According to Orruño and Morgan, under non-reduced SDS-PAGE, sesame 11S globulin fraction exhibited four protein bands around 36, 41, 46.6, and 51.8 kDa [42]. The protein band around 45–60 kDa was the most intense. A similar pattern of protein bands can be observed in the SSG prepared in this, indicating that 11S globulin was the dominant component. Other authors have also noted 11S globulin in this same position under non-reducing SDS-PAGE condition. Nouska and colleagues reported an identical SDS-PAGE profile for sesame protein isolates obtained via micellization and isoelectric precipitation. The authors noted the presence of a major 11S globulin band at 50–53 kDa position when the protein isolates were resolved under non-reducing conditions. In the presence of reducing agents, the widely reported ≈20 kDa and ≈30–33 kDa polypeptides become prominent [43]. The electrophoretic pattern of 11S globulin in SSG is also similar to that of oat globulin, which presents an 11S monomer with apparent molecular weight around 55 kDa position under non-reducing conditions because of its component α (32 kDa) and β (22 kDa) polypeptides linked by a disulfide bond [44]. Meanwhile, Chen et al., using Tricine-SDS-PAGE, found that the sesame 11S globulin acidic polypeptide unit and basic polypeptide unit were located at 33 kDa and 20 kDa positions, respectively [41]. Taken together, the result seen in the present electrophoretogram suggested that the sesame 11S globulin contains disulfide bonds linking the two main polypeptide subunits, resulting in intense protein bands in the region between 45 and 60 kDa, and the prepared protein has not been altered. In addition to the 11S globulins, another protein band could be seen around position 97 kDa, which had been ascribed to sesame globulin. Other authors have also spotted this globulin fragment in sesame seed proteins at similar positions [45,46]. It has also been reported that sesame seed contains a 7S globulin, which constitutes about 1–5% of the total protein content [42,47]. Previously, the 7S globulin monomers were seen as polypeptides devoid of disulfide linkages between 12.5 and 60 kDa. The most prominent of these had a band around 45 kD [42]. Because of the low relative abundance of sesame 7S globulin, it is presumed that the protein bands around 45 kDa were overshadowed by the 11S globulin in Figure 4. Importantly, it should be pointed out that while both SSPI and SSG protein profiles were largely similar, an interesting difference emerges on closer examination. Around the 20 kDa range, it can be seen that the SSPI profile manifested an intense protein band. This protein band was less prominent in the SSG protein profile. Given that the 11S globulin basic polypeptide unit has a molecular weight of about 20 kDa, at first it might be assumed that it is representative of this band. Alternatively, it is conceivable that this protein band belongs to albumin fraction. The latter notion is supported by the absence of a disulfide reducing agent which is necessary to produce the globulin fraction at this location, thus pointing to the band being albumin. Secondly, the diminished intensity of the band in the SSG further supports that it is albumin and its lack of prominence in SSG was due to their removal during the fractionation procedure. Albumins have been previously reported to appear in similar position in the electrophoretogram of sesame protein [45] as well as in other plant proteins such as mung bean, Bambara groundnut and pea [48]. In essence, the result of the protein profiles supports the notion that the facile fractionation approach adopted herein, indeed produced a protein fraction rich in sesame globulins.

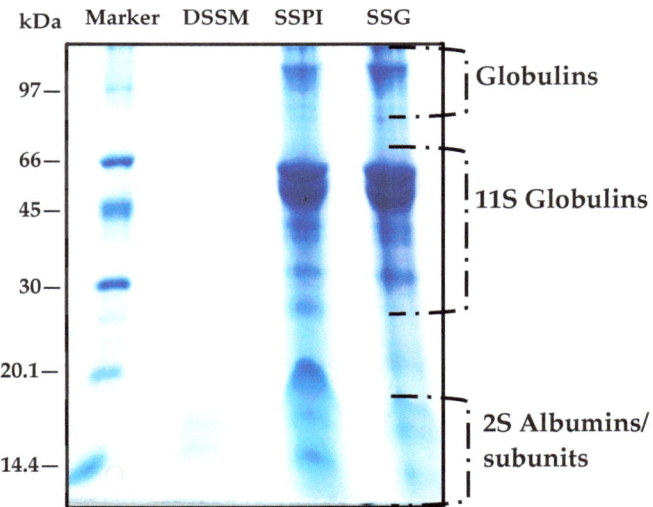

Figure 4. Tricine SDS-PAGE gel image of sesame proteins separated under non-reducing conditions and stained with Coomassie brilliant blue G. Marker: low molecular weight protein standard, DSSM: defatted black sesame seed meal, SSPI: sesame seed protein isolate, and SSG: sesame seed protein globulin.

3.2. Formation of Sesame Protein Amyloid Nanostructures

The feasibility and extent to which the prepared sesame proteins could generate amyloid nanostructures were investigated under acid-induced hydrolysis coupled with heat treatment. This condition facilitates the partial unfolding of the globular entity, release of the polypeptide chains, and cleavage into smaller peptides, which are more amyloidogenic. In this context, the tendency of the protein species to aggregate and undergo fibrillation is significantly enhanced. Insights into the morphology of the protein species following exposure to aggregation conditions after 24 h was obtained using transmission electron microscopy. The electron micrographs of SSPI and SSG after 24 h of incubation are presented in Figure 5. The TEM micrograph of SSPI revealed the presence of non-fibrillar amyloid aggregates in the form of clusters (Figure 5a,b). Some of these structures were spherical, but most of the aggregates were amorphous in shape and of varied sizes. There have been a number of reports on the protein fibrillation-enhancing effect of ultrasonication treatment [2,49]. This notion was briefly tested by applying ultrasonication treatment (power of 750 Watts, frequency of 20 kH at 50 amplitude, and pulse of 5 s ON and 5 s OFF) for 30 min at a temperature below 35 °C, CPX750 ultrasonic processor, Cole-Parmer Instruments (Vernon Hills, IL, USA) to SSPI after hydrating at pH 2.0 followed by heating. The result was somewhat interesting. Instead of the amyloid fibril aggregates that were expected, spherical amyloid nanoparticles were observed (Figure 5c). Meanwhile, the micrograph of the SSG sample subjected to aggregation conditions but without ultrasound treatment revealed aggregates with morphology that was mostly fibrillar, curly, short, and worm-like in shape (Figure 5d). The process of protein aggregation under acidic heating is an intricate phenomenon with a continuum of oligomeric intermediates having various morphologies. Kinetic and thermodynamic competition between these intermediate species would determine the predominant form of amyloid post-incubation. Additionally, many other factors could influence the amyloid formation process, including the composition of the protein isolate, the amino acid composition of the polypeptides and the hydrolyzed peptides, reaction conditions, and protein concentration, among others. Apparently, on the basis of TEM analysis, it seems SSG was more amenable to formation of amyloid with a well-defined structure relative to SSPI under the same condition. A possible explanation is that the additional fractionation steps during the preparation of SSG removed small soluble bioactive molecules, which could restrict amyloid fibril formation.

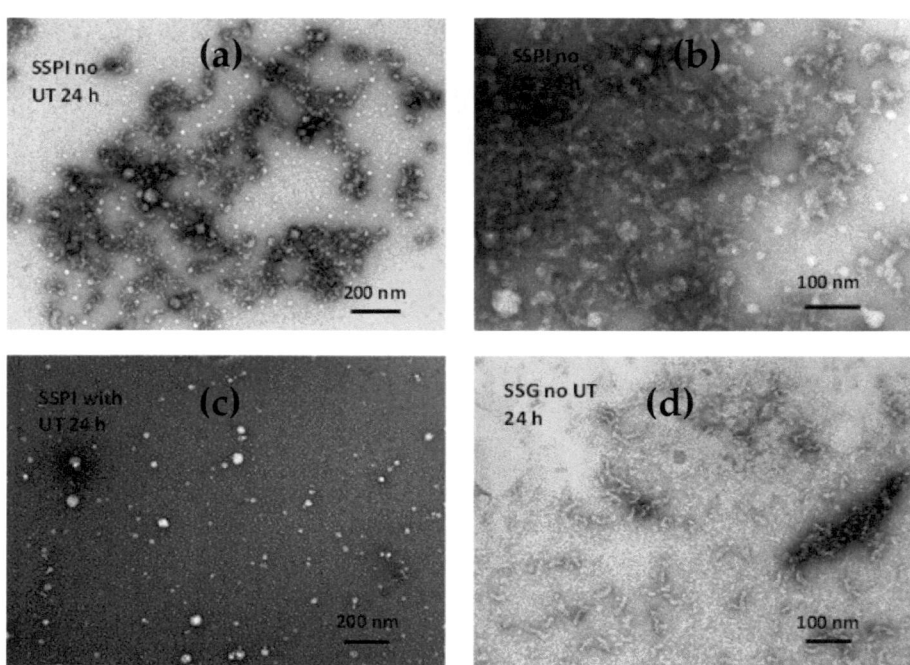

Figure 5. TEM images of sesame protein amyloid species after fibrillation for 24 h at pH 2 and 80 °C: (**a**) SSPI amyloid species prepared from 3% w/w protein solution without ultrasonication treatment; (**b**) Image of SSPI amyloid species at higher magnification; (**c**) SSPI amyloid species prepared from 3% w/w protein solution with ultrasonication treatment; (**d**) SSG amyloid species prepared from 3% w/w protein solution without ultrasonication treatment.

Hydrolysis of proteins into their subunits and peptides is an essential preceding step prior to heat-induced aggregation under acidic conditions (pH, 2.0) [50]. Tricine-SDS-PAGE was used to monitor the extent of SSG and SSPI hydrolysis during the course of the aggregation process (Figure 6a,b). Gel images of both SSG and SSPI aliquots during aggregation were resolved under non-reducing conditions. It is noticeable from both gel images that the 11S globulin band (≈55 kDa) intensity became narrow or greatly depleted at time 0 h in comparison to the same band for the native proteins that were not subjected to aggregation (as shown in Figure 4). This reduction in 11S globulin band intensity at time 0 h revealed that even before thermal treatment, a marked amount of the 11S globulin in SSG was hydrolyzed merely by exposing the protein solution to acidic conditions (pH, 2.0). A similar pattern was observed for SSPI, albeit the hydrolysis of 11S globulin appears to be greater. The 11S globulin from sesame seed is known to be structurally similar to the typical 11S globulin from seed proteins such as soy protein. The acid-induced hydrolysis of sesame 11S globulin observed here was in accord with reports from the research of Yu et al., who noticed that incubation of soy protein isolates at pH 2.0 caused a considerable hydrolysis of the 11S globulin subunit [51]. Application of thermal treatment (90 °C) to the protein solution under acidic milieu caused a gradual hydrolysis of the major medium molecular weight proteins (>35 kDa) with time. In fact, it can be seen that all the 11S globulin had been hydrolyzed by the 8th hour in both SSG and SSPI, and by the 12th hour, most of the proteins that remained were of low molecular weight ≤20 kDa.

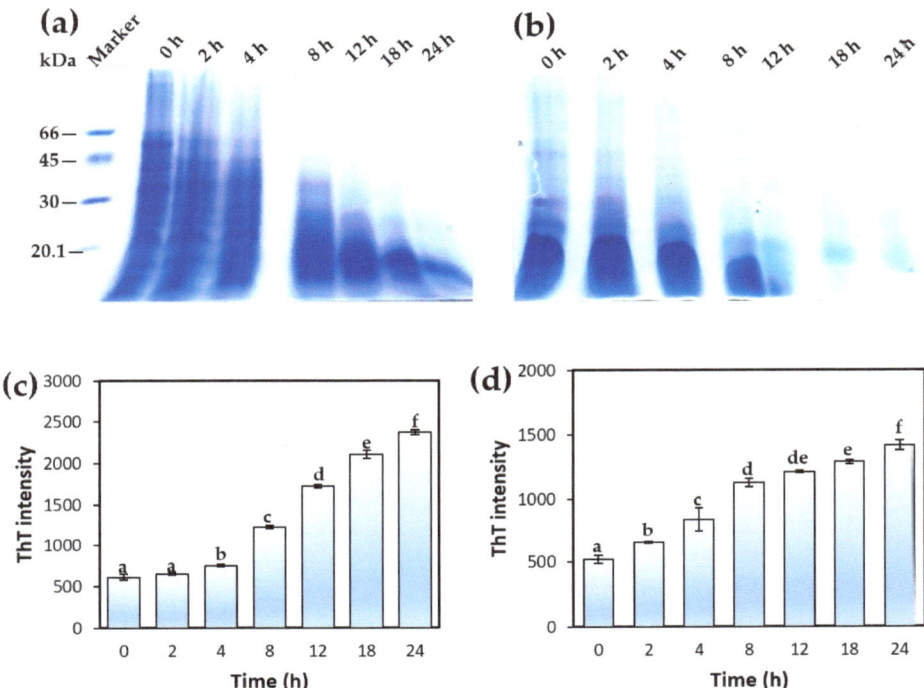

Figure 6. Tricine-SDS-PAGE profiles of (**a**) sesame seed globulin and (**b**) sesame seed protein isolate incubated at pH 2.0 and 90 °C depicting the gradual hydrolysis of the proteins with the increase in time (0 to 24 h). ThT fluorescence intensity of (**c**) SSG and (**d**) SSPI solutions, pH 2.0, subjected to aggregation by heating (90 °C) for 2, 4, 8, 12, 18, and 24 h. Different lowercase letters represent values that are significantly different ($p < 0.05$).

The evolution of amyloid nanostructures from sesame protein solution under aggregation induced by acidic heating was monitored using a ThT fluorescence assay. ThT is a small fluorescent dye with a remarkable ability for preferential binding to the cross-β-sheet rich motifs typically found in amyloid structures. This binding interaction with amyloid proteins results in increased fluorescence intensity [52]. Thus, the ThT assay has become a valuable probe of choice for examining the presence and extent of protein amyloid formation based on its variation in fluorescence intensity [8,52,53]. The ThT fluorescence intensity of SSG and SSPI solutions at pH 2.0 subjected to aggregation under different heating times (0–24 h) is presented in Figure 6c,d. As shown in Figure 6, there was a marginal but insignificant increase in the ThT fluorescence intensity of the SSG solution after heating for 2 h at pH 2.0. This is probably an indication of the relatively modest content of β-sheet structures in the protein solution at that time because protein hydrolysis was still quite limited, and the rate of self-assembly of the peptides to amyloid species was low. In other words, the SSG aggregating solution was at a lag phase. A marked increase in the ThT fluorescence intensity of the SSG solution was observed after heating for 4 h at pH 2.0, signifying a relative increase in the content of β-sheet structures and amyloid formation. This increase in ThT fluorescence intensity was progressively enhanced with an increase in heating time from 4 to 24 h. This can be credited to the fact that the protein aggregation solution entered the so-called 'growth phase' where the rate of formation of amyloid species with characteristic β-sheet structures from self-assembly of the peptides increased considerably [54]. In the case of the SSPI solution at pH 2.0, the lag phase appeared to be shorter, as evinced in the substantial increase in fluorescence intensity after heating for 2 h. The increase in ThT fluorescence intensity continued from 2 to about 8 h, suggesting an increasing rate of formation of β-sheet structure as a consequence of protein aggregation. Beyond 12 h of heating, the increase in ThT fluorescence intensity was less dramatic, which may be due to the protein aggregation reaction entering the stationary phase [53]. At this point, the content of peptides in the solution available for self-assembly is lower, which diminishes the vigor of the aggregation reaction. In both protein solutions, the ThT fluorescence intensity following heating from 4 to 24 h was clearly higher than the solution without heating, underscoring the progressive formation of amyloid nanostructures with thermal treatment of the protein solutions at acidic conditions (pH 2.0). A similar increase in the ThT fluorescence intensity of rice glutelin solution (pH 2.0) was noted by Li et al. after heating for 24 h, confirming the formation of amyloid-based nanostructures [8]. It is worthy of note that the impact of heating on formation of amyloid nanostructures was varied between both SSG and SSPI samples, perhaps due to difference in the unfolding transition temperature of both samples resulting in slightly different ThT fluorescence intensity pattern (Figure 6). Indeed, it has been previously noted that the unfolding transition of proteins subjected to aggregation is contingent to the mode of sample preparation, ionic strength of the medium, structure of proteins as well as modification of the proteins [11]. The result from the ThT fluorescence assay of sesame protein aggregation indicated a progressive increase in protein amyloid formation with increase in heating time under acidic condition. When this result is considered in light of the different morphologies revealed by TEM, the ThT result could be an indication that there may be an optimal heating time for obtaining amyloid nanostructures with different features.

3.3. Secondary Structure Analysis of Sesame Proteins and Amyloid Nanostructures

Further insights pertaining to the secondary structure and likely conformational alterations of the protein isolates and amyloid nanostructures were facilitated with the aid of FTIR analysis. Figure 7 depicts the FTIR spectra of the protein samples in the region 1800–1300 cm^{-1}, which encompasses the Amide I band located between 1600 and 1700 cm^{-1} and the Amide II band delineated by IR absorption from 1580 to 1480 cm^{-1}. The full FTIR spectra (4000–400 cm^{-1}) are included in Supplementary Materials (Figure S1). The Amide I band represents the stretching vibration of C=O bond (80%), whereas the Amide II band corresponds to the vibrations emerging from a combination of the C–N bond stretching (30%), N–H bond bending (60%), and C–C bond stretching [55]. The strong IR absorption peaks noticeable in the Amide I and II spectral regions (Figure 7) are a testament to the proteinaceous nature of the sesame samples. The peak position of SSPI (1653 cm^{-1}) and SSG (1652 cm^{-1}) Amide I bands was similar (1654/1655 cm^{-1}) to those obtained by other authors for sesame protein isolate [33]. By dint of its exceptional signal-to-noise ratio and remarkable sensitivity to protein secondary structure, the Amide I (1700–1590 cm^{-1}) band of each sample was used for evaluating the composition of the various secondary structural elements present in the proteins pre and post-aggregation. This was accomplished by deconvolution of the Amide I band into component peaks. Peak position and assignment were in accordance with insights drawn from previous studies. The corresponding peak area as a percent of the total area of all peaks in the Amide I region was used to calculate the composition the four major structural elements, namely β-sheets (1618 cm^{-1}, 1625 cm^{-1}, and 1634 cm^{-1}, 1680 cm^{-1}), random coils (1644 cm^{-1}), α-helix (1651 cm^{-1} and 1658 cm^{-1}), and β-turns (1667 cm^{-1}) [43,55,56].

In all the samples, β-sheets emerged as the most abundant secondary structural element. The difference in the relative composition of the secondary structure of the native proteins, when contrasted with those subjected to aggregation, attests to the fact that the proteins were structurally modified by the aggregation process. Among the various samples, it was noticed that the content of β-sheets increased remarkably after the proteins were subjected to aggregation treatment. Meanwhile, the composition of random coils decreased following the acidic heating of the native proteins. This observation is consistent with the notion that the protein conformation became well-organized and ordered after aggregation. This was more pronounced in the case of SSGAN, which exhibited greater content of β-sheet structures compared to SSPAN. The increased content of β-sheet in SSGAN and conformational re-alignment support the presence of amyloid structures in the sample.

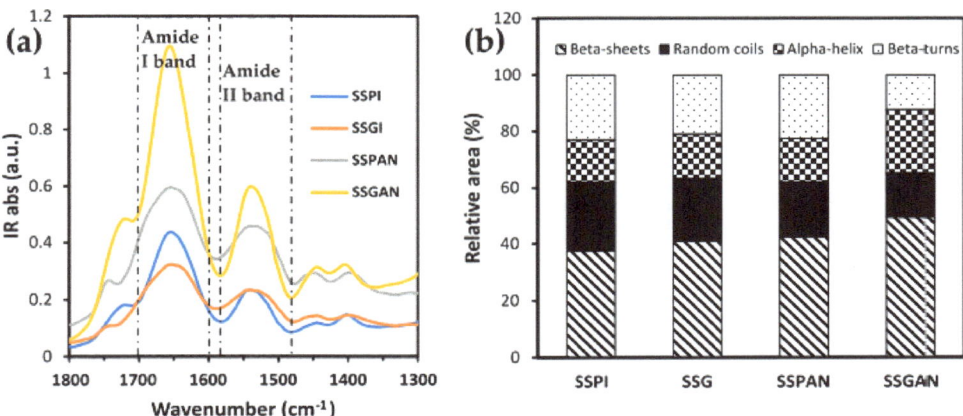

Figure 7. (a) FTIR spectra of sesame proteins (SSPI and SSG) and amyloid-based nanostructures (SSPAN and SSGAN) revealing noticeable peaks in the Amide I and Amide II regions; (b) relative content of secondary structure in the Amide I band obtained from the relative area of corresponding Gaussian components, i.e., size of each resolved peak in percent.

3.4. Solubility and Surface Hydrophobicity

Solubility is a critical attribute of vital importance in the application of proteins in food. The protein solubility could profoundly impact essential aspects such as the foaming, gelation, emulsification, color, texture, and sensory attributes of the protein-incorporated food product. One factor that greatly impacts protein solubility is its hydrophilicity/hydrophobicity balance, which in turn is dependent on the composition of the surface amino acids. In a case where there is a greater number of surface hydrophilic amino acid residues and a higher net charge leading to greater electrostatic repulsion and ionic hydration, the tendency of the protein is one of higher solubility and vice versa [14].

Apparently, protein solubility is pH-dependent. Solubility of both SSPI, SSG, and SSGAN from pH 2.0 to 10.0 is presented in Figure 8a. Limited availability of SSPAN samples meant that solubility could only be performed on SSGAN. The solubility of all protein samples was at least around pH 3.5–5.0, which is consistent with the pH range of the isoelectric point of sesame proteins. In general, there was an increase in solubility toward the alkaline pH range, which is from 6.5 to 10. This is consistent with previous reports. Also, there appears to be an increase in solubility of the proteins, especially SSGAN, at acidic pH (2.0) and alkaline pH (8.0–10.0), which is valuable for food product formulation. SSGAN was significantly more soluble than SSPI and SSG at pH 2.0–3.5 ($p < 0.05$). The increase in SSGAN solubility at low pH can be explained in part by the fact that at low acidic pH, amyloid samples typically present greater positively charged structures compared to their native monomers [4,8,57]. This increased net positive charge engenders greater electrostatic repulsion of the amyloid species, thereby obviating their precipitation in solution. Consequently, the physical stability of the amyloid suspension is enhanced [8]. Also, the presence of hydrolyzed peptides with small molecular weights in the amyloid sample can improve the solubility of SSGAN at low pH. It has been reported that the extraction method or processing could have a considerable impact on the solubility of obtained products such as protein isolate, concentrate, etc. For instance, it was observed that the processing of raw sesame meals by cooking, microwave, or ultrasound improved the protein solubility of the processed sesame meal relative to its raw counterpart [14]. It can, therefore, be inferred from the result herein that modification of SSG via amyloid formation can improve the protein solubility in an acidic milieu.

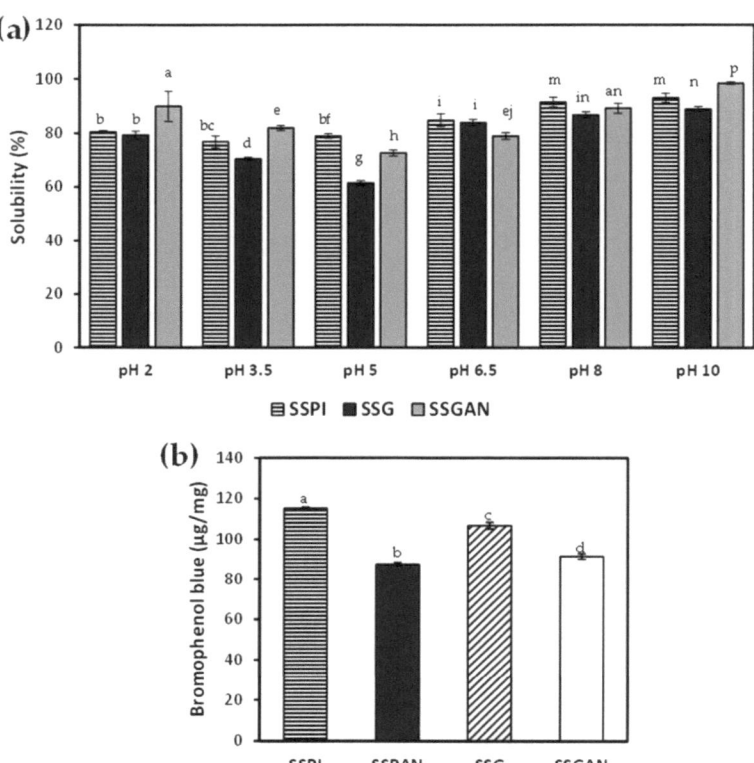

Figure 8. (a) Solubility of sesame protein isolate, globulin, and amyloid fibrils under different pH conditions; (b) Surface hydrophobicity of sesame protein samples. Different lowercase letters (e.g., a, b, c) on the columns represent values that are significantly different from each other ($p < 0.05$).

Understanding the surface hydrophobicity of plant-based proteins and protein amyloid species is important because hydrophobic interactions have a considerable impact on protein conformation, structure, as well as functional characteristics such as the proteins' affinity toward the water–oil interface. To obtain insights with respect to the surface hydrophobicity of the proteins, bromophenol blue dye was used as a probe. That is, the higher the amount of bromophenol blue bound to the protein, the higher the surface hydrophobicity. Data on the surface hydrophobicity are shown in Figure 8b. The results revealed that in both SSPI and SSG, the formation of amyloid-based nanostructures after heating for 24 h at pH 2.0 markedly reduced the surface hydrophobicity. It has been reported that at the initial heating period of the aggregation process (≈2–4 h), surface hydrophobicity typically increases due to protein hydrolysis coupled with partial unfolding of the protein structure [51]. As a result, hydrophobic amino acid residues, which are normally buried inside the globular protein, become exposed to the surface. The exposed hydrophobic patches on the protein surface interact, causing self-assembly and the formation of aggregate species or protofibrils. Interestingly, extending the heating time has proven to not only promote protein aggregation into amyloid nanostructures but also decrease the surface hydrophobicity of the obtained amyloid species. The latter is attributed to the reburial of the hydrophobic regions rich in β-sheets as they re-assembled to form ordered amyloid structures [58]. So, while it is common to see an increase in surface hydrophobicity of some plant proteins post-formation of amyloid species, it is not uncommon to also see the reverse in others such as chicken pea, lentil, and pumpkin seed protein isolates [4]. The result obtained from this study concurred with a similar work by Xu et al., who also noticed that the surface hydrophobicity of oat globulin decreased substantially after heating for 24 h at acidic conditions because of the formation of oat globulin fibrils [10].

3.5. Water Holding and Oil Holding Capacity of Sesame Protein and Amyloid Structures

The water absorption capacity of the protein samples is represented in Figure 9. It can be observed that the water retention ability of the sesame protein amyloid nanostructures was higher than that of the native proteins. It is understood that for proteins subjected to thermal-mediated unfolding under acidic conditions, there is an increase in the reactive sulfhydryl groups, the distribution and number of charges, as well as the hydrophilic and hydrophobic groups. This increases the tendency and potential of the protein functional groups to form strong intermolecular interactions with water molecules [59]. Besides the obvious fibrillar species in the amyloid nanostructures, non-fibrillar and peptides resulting from acid-mediated hydrolysis of the polypeptides are also present as part of the amyloid sample [56]. These peptide species tend to be more hydrophilic and, therefore, contribute toward increasing the water holding capacity of the amyloid nanostructure [59]. Meanwhile, it is understood that protein fibrillation increases the overall charge of the fibrillated protein relative to the native proteins [57], and this increase in charge improves the ability of the amyloid protein sample to interact with water molecules.

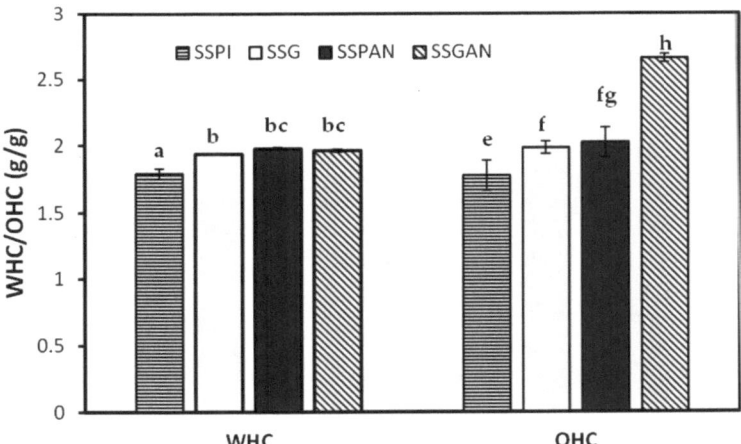

Figure 9. The water holding capacity (WHC) and oil holding capacity (OHC) of sesame seed protein isolate (SSPI), sesame seed globulin (SSG), as well as their respective amyloid nanostructures, SSPAN and SSGAN. Different lowercase letters (e.g., a, b, c) on the columns represent values that are significantly different from each other ($p < 0.05$).

With respect to the oil absorption capacity of the protein samples, the trend was positive following amyloid formation, i.e., the OHC increased markedly for the aggregated proteins relative to their native counterpart. Protein aggregation is enabled by the unfolding of the compact globular proteins due to the thermal treatment that exposes the hydrophobic residues and polypeptide chains, which become converted into various oligomeric and fibrillar species. As such, the surface area for hydrophobic interaction between protein species and monolayer oil is increased in the amyloid protein compared to native protein. This leads to a greater amount of oil binding and retention by the amyloid nanostructure.

As such, the sesame amyloid nanostructures could be valuable in the preparation of food products with high fluid holding requirements.

3.6. Antioxidant Activity of Sesame Protein Amyloid Nanostructures

Oxidation of food can lead to a significant decline in quality, resulting in poor taste, aroma, flavor, texture, and color, as well as enrichment with undesirable and unhealthy metabolites such as free radial and reactive oxygen species. Thus, fortifying and increasing the antioxidant content of food products have become common place as a valuable strategy to mitigate oxidative deterioration, preserve food quality, and enhance shelf-life. Increasingly, consumer preference has been shifting toward the use of natural antioxidants for food application. Plant proteins in their various forms and compositions, including concentrates, hydrolysates, and peptides, have been reported for their antioxidant activity. Interestingly, amyloid fibrils from food proteins are increasingly being recognized for having appreciable radical scavenging properties.

As evidenced in the results from DPPH and ABTS assays (Figure 10), the antioxidant activity of sesame proteins was remarkably increased after the formation of amyloid nanostructures. This finding is consistent with results from other studies. For instance, Mohammadian et al. found that the antioxidant property of amyloid fibrils derived from whey protein isolate and hydrolysate was substantially higher than that of their respective precursor proteins. Meanwhile, Li et al. established that fibrillization of rice glutelin markedly elevated the antioxidant activity of the obtained amyloid nanostructures [60]. As often noted, early events of protein amyloid formation under elevated temperature and acid pH are characterized by partial unfolding and hydrolysis of the native proteins into amyloid-competent peptides. Du et al. [20], in their study, also found that the antioxidant activity increased following enzymatic hydrolysis and unfolding of black sesame seed protein. The antioxidant properties of protein amyloid species have been credited to their peptide building blocks. Wei et al. noted that the antioxidant activity of ovotransferrin amyloid fibril prepared via heating at 90 °C in acidic milieu (pH 2.0) was due to its constituting peptides [61]. Indeed, it has been previously noted that antioxidant property is one of the common features shared by both peptides in amyloid structures and bioactive peptides [62]. Compared to their precursor proteins, the peptides in these structures tend to offer greater solvent accessibility of their active amino acid residues—the main driver for the antioxidant activity [63].

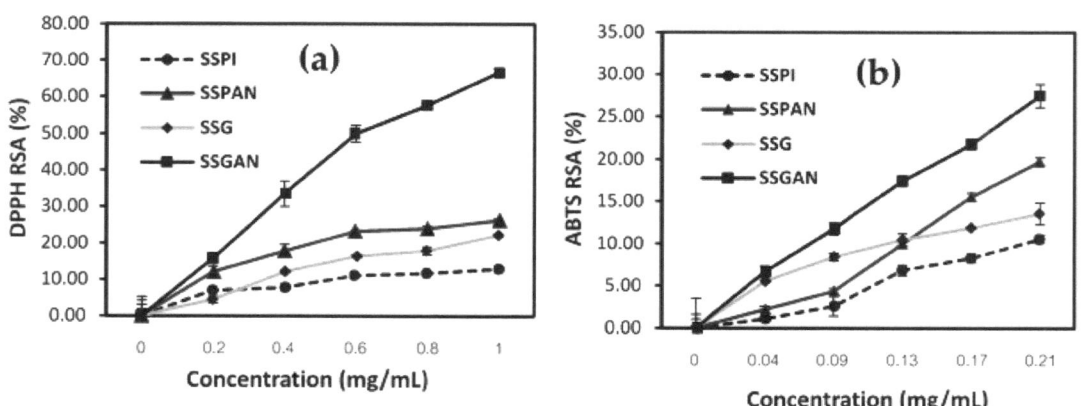

Figure 10. Radical scavenging activity of sesame seed proteins and amyloid nanostructure against the following: (a) DPPH; (b) ABTS radicals.

Amino acid resides such as methionine, cysteine, phenylalanine, tryptophan, tyrosine and histidine are among the most active in terms of antioxidant properties. A good number of these residues, including tyrosine and tryptophan, are not only typically found in amyloid structures where they participate in π–π stacking [64], but also in sesame seed proteins [14]. The antioxidant activity of these amyloid structures could be modulated via a number of mechanisms, such as chelating metal ions, scavenging free radicals and reactive chemical species, or participating in the breakage of free-radical chain reactions. The antioxidant activity of sesame proteins and amyloid nanostructures is a testament that the amyloid nanostructures could be used as ingredients in enhancing food quality by inhibiting oxidation as well as in the development of functional foods.

3.7. Biocompatibility of Sesame Protein Nanofibrils

A fundamental and cogent consideration in the production and utilization of protein amyloids is their safety as food ingredients for nutrition. Unlike native food proteins, which have been part of human diets from time immemorial without many worries besides the occasional incidence of allergenicity, protein amyloids derived from food proteins have been restricted in their application as food ingredients for nutrition and health. The reluctance in use of food protein amyloids is linked to the historical association of amyloids in general to the development of many chronic and neurodegenerative diseases. Notable examples of this include the polypeptide hormone amylin, which is released together with insulin as a response to food. Aggregation of amylin causes the protein to suppress the activity of insulin and glucagon, leading to type 2 diabetes mellitus [65]. Also, β-amyloid peptide and α-synuclein aggregation and amyloid formation have been linked to the development of Alzheimer's disease and Parkinson's disease, respectively. Specifically, with respect to the safety of amyloids derived from food proteins, a consensus is yet to emerge since some studies indicate that they (e.g., egg-white lysozyme) are toxic [66], whereas others deem them (soy, whey, egg-white, kidney, and bean amyloid) as safe [67]. In this context, to obtain clarity as to whether sesame-derived amyloids are safe or potentially harmful, in vitro biocompatibility studies were performed.

Figure 11 represents the results of in vitro cytotoxicity and hemolytic effect of SSPAN and SSGAN. As shown in Figure 11a, following the exposure of murine macrophage (RAW264.7) cells to the sesame amyloids (7.81–1000 µg/mL) for 24 h, there was a noticeable decrease in the cell viability to about 90% following exposure of cells to both amyloid samples at a concentration of 250 µg/mL. At higher concentrations of SSPAN and SSGAN (up to 1000 µg/mL), the cell viability was further reduced by not lower than 88%. A similar decrease could also be seen in the cell viability of human keratinocyte (HaCaT) cells following treatment with the amyloid samples. In both cell lines, it can also be seen that the cells were marginally more viable in the presence of SSGAN compared to SSPAN. Importantly, the viability at all treatment concentrations remained high (greater than 75% in HaCaT cells and 80% in RAW264.7 cells), suggesting that the amyloids were non-toxic to the cells [68]. Furthermore, the results on erythrocyte hemolysis show that for erythrocytes exposed to either SSPAN or SSGAN (250–1000 µg/mL), the percentage hemolysis was less than 2% (Figure 11b), which is below the threshold for materials which do not possess any hemolytic activity (that is hemolysis value <5%) [69]. From these findings, it can be inferred that the amyloids prepared from SSG and SSPI were non-toxic to the treated cells. A similar observation was also made by Lassé and co-researchers, who found that amyloid fibrils obtained from food proteins, namely kidney bean, whey, egg white, and soy protein isolates, were non-toxic to Hec-1a and Caco-2 cell lines in vitro [67]. Further credence on the safety of food amyloid was offered by a recent study in which the authors demonstrated that in vitro digested amyloid fibrils from milk β-lactoglobulin and hen egg-white globulin were non-toxic in both in vitro and in vivo models. In fact, the digested amyloid fibrils promoted cell viability in Caco-2 and HCEC cell lines because they served as nutrients to the cells [1]. Based on the result obtained in the present work, it is apparent that SSPAN and SSGAN are biocompatible and, thus, are potentially safe for use as ingredients for possible advancement of nutrition and health.

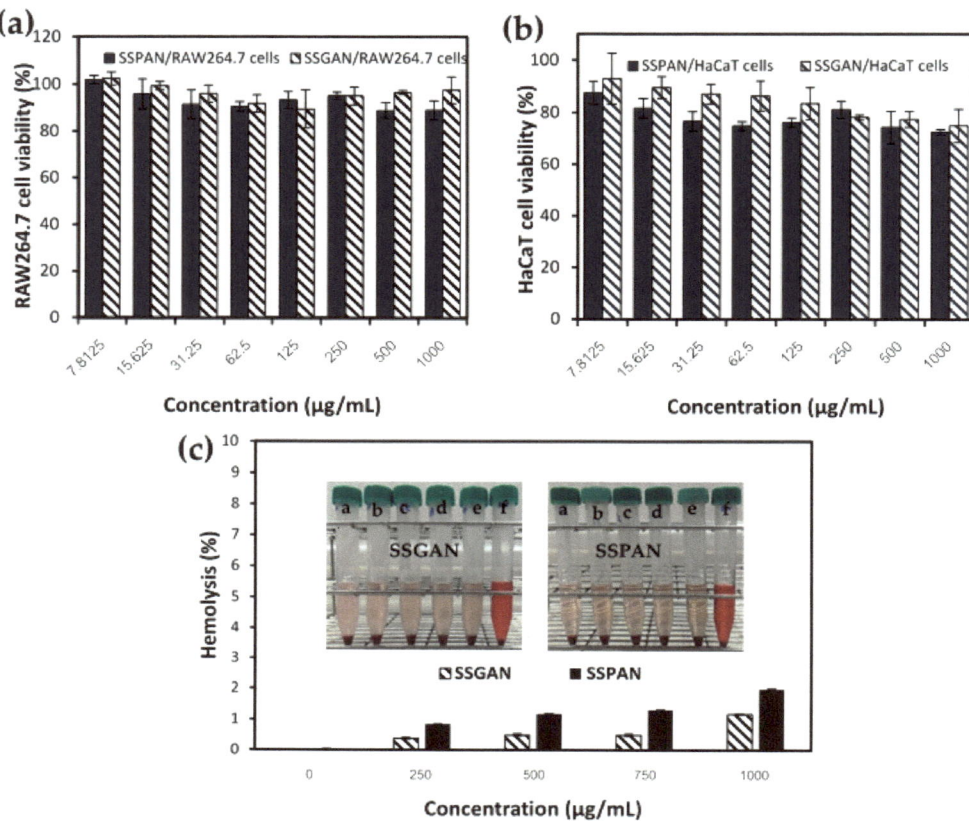

Figure 11. Effect of SSGAN and SSPAN on viability of (**a**) murine macrophage RAW264.7 and (**b**) human keratinocyte HaCaT cells 24 h after treatment; (**c**) Hemolytic effect of SSGAN and SSPAN on rat red blood cells (insert depicts image of the sample treated erythrocyte suspension after centrifugation at 1500 rpm for 10 min. a–e represent samples treated with 0–1000 µg/mL protein amyloid, whereas f connotes sample treated with distilled water only).

4. Conclusions

The self-assembly of plant proteins into amyloid-based structures presents an interesting modification strategy for improving their functionalities and extending their potential application in food and healthcare. Up until now, proteins derived from sesame seeds have yet to be investigated as potential amyloid materials. So, in this work, the viability of defatted sesame seed meal was explored as an inexpensive, abundant, and sustainable source of plant-based proteins and amyloid structures. We demonstrated that sesame seed protein fraction rich in 11S globulin could be facilely prepared from DSSM. Under acid-induced hydrolyses and heating, the sesame seed globulin was converted into fibrillar amyloid nanostructures with high β-sheet content via self-assembly of the hydrolyzed peptides. Importantly, the prepared amyloid nanostructure from sesame seed globulin exhibited improved solubility at low pH, oil retention, and radical scavenging properties. In addition, the sesame protein amyloid preparation demonstrated a good safety profile. In aggregate, these findings indicate that amyloid-based nanostructures derived from sesame seed globulin could be a promising ingredient for application in functional foods and plant-based food products. It also highlights the fact that proteins and amyloid preparation could be a valuable strategy for upcycling and adding value to defatted sesame seed meal

In the future, the sesame seed globulin amyloid nanostructures could be further exploited in the delivery of valuable nutraceutical ingredients, in tailoring the functional properties of foods, such as gelation, fluid holding capacity, interfacial properties, as well as in the development of edible food coatings and scaffolds for plant-based meat analogs.

Supplementary Materials: The following supporting information can be downloaded at https://www.mdpi.com/article/10.3390/foods13142281/s1, Figure S1: FTIR spectra of sesame proteins (SSPI and SSG) and amyloid-based nanostructure (SSPAN and SSGAN) from 4000–400 cm^{-1} highlighting prominent peaks.

Author Contributions: Conceptualization, F.N.E. and R.M.; methodology, F.N.E. and S.S.; validation, F.N.E., R.M., S.S., W.J., T.S. and Y.C.; formal analysis, F.N.E., R.M., S.S., W.J., T.S. and Y.C.; investigation, F.N.E., S.S., W.J. and T.S.; resources, F.N.E., R.M., S.S., W.J., T.S. and Y.C.; data curation, F.N.E.; writing—original draft preparation, F.N.E.; writing—review and editing, F.N.E., R.M., W.J., T.S. and Y.C.; supervision, R.M., W.J., T.S. and Y.C. All authors have read and agreed to the published version of the manuscript.

Funding: This research was partially supported by CMU Proactive Researcher Scheme (2023), Chiang Mai University.

Institutional Review Board Statement: Not applicable.

Informed Consent Statement: Not applicable.

Data Availability Statement: The original contributions presented in the study are included in the article, further inquiries can be directed to the corresponding author.

Acknowledgments: This work was partially supported by CMU Proactive Researcher Scheme (2023), Chiang Mai University.

Conflicts of Interest: The authors declare no conflicts of interest.

References

1. Xu, D.; Zhou, J.; Soon, W.L.; Kutzli, I.; Molière, A.; Diedrich, S.; Radiom, M.; Handschin, S.; Li, B.; Li, L.; et al. Food Amyloid Fibrils Are Safe Nutrition Ingredients Based on In-Vitro and in-Vivo Assessment. *Nat. Commun.* **2023**, *14*, 6806. [CrossRef]
2. Chen, X.; Yi, J.; Wen, Z.; Fan, Y. Ultrasonic Pretreatment and Epigallocatechin Gallate Incorporation Enhance the Formation, Apparent Viscosity, and Antioxidant Activity of Pea Protein Amyloid-like Fibrils. *Food Hydrocoll.* **2024**, *149*, 109630. [CrossRef]
3. Dong, Y.; Lan, T.; Wang, J.; Wang, X.; Xu, Z.; Jiang, L.; Zhang, Y.; Sui, X. Development of Composite Electrospun Films Utilizing Soy Protein Amyloid Fibrils and Pullulan for Food Packaging Applications. *Food Chem. X* **2023**, *20*, 100995. [CrossRef]
4. Li, T.; Zhou, J.; Peydayesh, M.; Yao, Y.; Bagnani, M.; Kutzli, I.; Chen, Z.; Wang, L.; Mezzenga, R. Plant Protein Amyloid Fibrils for Multifunctional Sustainable Materials. *Adv. Sustain. Syst.* **2023**, *7*, 2200414. [CrossRef]
5. Wei, Z.; Dai, S.; Huang, J.; Hu, X.; Ge, C.; Zhang, X.; Yang, K.; Shao, P.; Sun, P.; Xiang, N. Soy Protein Amyloid Fibril Scaffold for Cultivated Meat Application. *ACS Appl. Mater. Interfaces* **2023**, *15*, 15108–15119. [CrossRef]
6. Liu, S.; Sun, N.; Ren, K.; Tan, X.; Li, L.; Wang, Z.; Dai, S.; Tong, X.; Wang, H.; Jiang, L. Utilization of Self-Assembled Soy Protein Nanoparticles as Carriers for Natural Pigments: Examining Non-Interaction Mechanisms and Stability. *Food Hydrocoll.* **2024**, *148*, 109491. [CrossRef]
7. Wang, Y.-R.; Yang, Q.; Jiang, Y.-X.; Chen, H.-Q. Enhanced Solubility, Thermal Stability and Antioxidant Activity of Resveratrol by Complexation with Ovalbumin Amyloid-like Fibrils: Effect of pH. *Food Hydrocoll.* **2024**, *148*, 109463. [CrossRef]
8. Li, T.; Zhou, J.; Wu, Q.; Zhang, X.; Chen, Z.; Wang, L. Modifying Functional Properties of Food Amyloid-Based Nanostructures from Rice Glutelin. *Food Chem.* **2023**, *398*, 133798. [CrossRef]
9. Zhao, Y.; Wang, C.; Lu, W.; Sun, C.; Zhu, X.; Fang, Y. Evolution of Physicochemical and Antioxidant Properties of Whey Protein Isolate during Fibrillization Process. *Food Chem.* **2021**, *357*, 129751. [CrossRef]
10. Xu, J.; Tang, M.; Wang, D.; Xie, Q.; Xu, X. Exploring the Self-Assembly Journey of Oat Globulin Fibrils: From Structural Evolution to Modified Functionality. *Food Hydrocoll.* **2024**, *149*, 109587. [CrossRef]
11. Mykolenko, S.; Soon, W.L.; Mezzenga, R. Production and Characterization of Amaranth Amyloid Fibrils from Food Protein Waste. *Food Hydrocoll.* **2024**, *149*, 109604. [CrossRef]
12. Kutzli, I.; Zhou, J.; Li, T.; Baier, S.K.; Mezzenga, R. Formation and Characterization of Plant-Based Amyloid Fibrils from Hemp Seed Protein. *Food Hydrocoll.* **2023**, *137*, 108307. [CrossRef]
13. Ramachandran, S.; Singh, S.K.; Larroche, C.; Soccol, C.R.; Pandey, A. Oil Cakes and Their Biotechnological Applications—A Review. *Bioresour. Technol.* **2007**, *98*, 2000–2009. [CrossRef] [PubMed]

14. Sá, A.G.A.; Pacheco, M.T.B.; Moreno, Y.M.F.; Carciofi, B.A.M. Cold-Pressed Sesame Seed Meal as a Protein Source: Effect of Processing on the Protein Digestibility, Amino Acid Profile, and Functional Properties. *J. Food Compos. Anal.* **2022**, *111*, 104634. [CrossRef]
15. Sarkis, J.R.; Boussetta, N.; Blouet, C.; Tessaro, I.C.; Marczak, L.D.F.; Vorobiev, E. Effect of Pulsed Electric Fields and High Voltage Electrical Discharges on Polyphenol and Protein Extraction from Sesame Cake. *Innov. Food Sci. Emerg. Technol.* **2015**, *29*, 170–177. [CrossRef]
16. Koysuren, B.; Oztop, M.H.; Mazi, B.G. Sesame Seed as an Alternative Plant Protein Source: A Comprehensive Physicochemical Characterisation Study for Alkaline, Salt and Enzyme-Assisted Extracted Samples. *Int. J. Food Sci. Technol.* **2021**, *56*, 5471–5484. [CrossRef]
17. Saini, C.S.; Sharma, H.K.; Sharma, L. Thermal, Structural and Rheological Characterization of Protein Isolate from Sesame Meal. *Food Meas.* **2018**, *12*, 426–432. [CrossRef]
18. Hu, S.; Gao, H.; Ouyang, L.; Li, X.; Zhu, S.; Wu, Y.; Yuan, L.; Zhou, J. Mechanistic Insights into the Improving Effects of Germination on Physicochemical Properties and Antioxidant Activity of Protein Isolate Derived from Black and White Sesame. *Food Chem.* **2023**, *429*, 136833. [CrossRef]
19. Biswas, A.; Dhar, P.; Ghosh, S. Antihyperlipidemic Effect of Sesame (*Sesamum indicum* L.) Protein Isolate in Rats Fed a Normal and High Cholesterol Diet. *J. Food Sci.* **2010**, *75*, H274–H279. [CrossRef] [PubMed]
20. Du, T.; Huang, J.; Xiong, S.; Zhang, L.; Xu, X.; Xu, Y.; Peng, F.; Huang, T.; Xiao, M.; Xiong, T. Effects of Enzyme Treatment on the Antihypertensive Activity and Protein Structure of Black Sesame Seed (*Sesamum indicum* L.) after Fermentation Pretreatment. *Food Chem.* **2023**, *428*, 136781. [CrossRef]
21. Shu, Z.; Liu, L.; Geng, P.; Liu, J.; Shen, W.; Tu, M. Sesame Cake Hydrolysates Improved Spatial Learning and Memory of Mice. *Food Biosci.* **2019**, *31*, 100440. [CrossRef]
22. Buranachokpaisan, K.; Chalermchat, Y.; Muangrat, R. Economic Evaluation of the Production of Oil Extracted from Pressed Sesame Seed Cake Using Supercritical CO2 in Thailand. *J. Appl. Res. Med. Aromat. Plants* **2022**, *31*, 100410. [CrossRef]
23. Goyeneche, R.; Agüero, M.V.; Roura, S.; Di Scala, K. Application of Citric Acid and Mild Heat Shock to Minimally Processed Sliced Radish: Color Evaluation. *Postharvest Biol. Technol.* **2014**, *93*, 106–113. [CrossRef]
24. Eze, F.N.; Ingkaninan, K.; Prapunpoj, P. Transthyretin Anti-Amyloidogenic and Fibril Disrupting Activities of *Bacopa monnieri* (L.) Wettst (Brahmi) Extract. *Biomolecules* **2019**, *9*, 845. [CrossRef] [PubMed]
25. Jiang, S.; Liu, S.; Zhao, C.; Wu, C. Developing Protocols of Tricine-SDS-PAGE for Separation of Polypeptides in the Mass Range 1-30 kDa with Minigel Electrophoresis System. *Int. J. Electrochem. Sci.* **2016**, *11*, 640–649. [CrossRef]
26. Bühler, J.M.; Dekkers, B.L.; Bruins, M.E.; van der Goot, A.J. Modifying Faba Bean Protein Concentrate Using Dry Heat to Increase Water Holding Capacity. *Foods* **2020**, *9*, 1077. [CrossRef] [PubMed]
27. Hu, S.; Zhu, S.; Luo, J.; Ouyang, L.; Feng, J.; Zhou, J. Effect of Extrusion on Physicochemical Properties and Antioxidant Potential of Protein Isolate Derived from Baijiu Vinasse. *Food Chem.* **2022**, *384*, 132527. [CrossRef] [PubMed]
28. Ernst, O.; Zor, T. Linearization of the Bradford Protein Assay. *J. Vis. Exp.* **2010**, *12*, 1918. [CrossRef]
29. Hadnađev, M.; Dapčević-Hadnađev, T.; Lazaridou, A.; Moschakis, T.; Michaelidou, A.-M.; Popović, S.; Biliaderis, C.G. Hempseed Meal Protein Isolates Prepared by Different Isolation Techniques. Part I. Physicochemical Properties. *Food Hydrocoll.* **2018**, *79*, 526–533. [CrossRef]
30. Hu, J.; Qi, Q.; Zhu, Y.; Wen, C.; Olatunji, O.J.; Jayeoye, T.J.; Eze, F.N. Unveiling the Anticancer, Antimicrobial, Antioxidative Properties, and UPLC-ESI-QTOF-MS/ GC–MS Metabolite Profile of the Lipophilic Extract of Siam Weed (*Chromolaena odorata*). *Arab. J. Chem.* **2023**, *16*, 104834. [CrossRef]
31. Singh, S.; Chidrawar, V.R.; Hermawan, D.; Nwabor, O.F.; Olatunde, O.O.; Jayeoye, T.J.; Samee, W.; Ontong, J.C.; Chittasupho, C. Solvent-Assisted Dechlorophyllization of *Psidium guajava* Leaf Extract: Effects on the Polyphenol Content, Cytocompatibility, Antibacterial, Anti-Inflammatory, and Anticancer Activities. *S. Afr. J. Bot.* **2023**, *158*, 166–179. [CrossRef]
32. Eze, F.N.; Nwabor, O.F. Valorization of *Pichia* Spent Medium via One-Pot Synthesis of Biocompatible Silver Nanoparticles with Potent Antioxidant, Antimicrobial, Tyrosinase Inhibitory and Reusable Catalytic Activities. *Mater. Sci. Eng. C* **2020**, *115*, 111104. [CrossRef]
33. Yang, K.; Xu, T.-R.; Fu, Y.-H.; Cai, M.; Xia, Q.-L.; Guan, R.-F.; Zou, X.-G.; Sun, P.-L. Effects of Ultrasonic Pre-Treatment on Physicochemical Properties of Proteins Extracted from Cold-Pressed Sesame Cake. *Food Res. Int.* **2021**, *139*, 109907. [CrossRef]
34. Gundogan, R.; Can Karaca, A. Physicochemical and Functional Properties of Proteins Isolated from Local Beans of Turkey. *LWT* **2020**, *130*, 109609. [CrossRef]
35. Tanger, C.; Engel, J.; Kulozik, U. Influence of Extraction Conditions on the Conformational Alteration of Pea Protein Extracted from Pea Flour. *Food Hydrocoll.* **2020**, *107*, 105949. [CrossRef]
36. Segla Koffi Dossou, S.; Xu, F.; You, J.; Zhou, R.; Li, D.; Wang, L. Widely Targeted Metabolome Profiling of Different Colored Sesame (*Sesamum indicum* L.) Seeds Provides New Insight into Their Antioxidant Activities. *Food Res. Int.* **2022**, *151*, 110850. [CrossRef]
37. Dossou, S.S.K.; Luo, Z.; Wang, Z.; Zhou, W.; Zhou, R.; Zhang, Y.; Li, D.; Liu, A.; Dossa, K.; You, J.; et al. The Dark Pigment in the Sesame (*Sesamum indicum* L.) Seed Coat: Isolation, Characterization, and Its Potential Precursors. *Front. Nutr.* **2022**, *9*, 858673. [CrossRef]

38. Panzella, L.; Eidenberger, T.; Napolitano, A.; d'Ischia, M. Black Sesame Pigment: DPPH Assay-Guided Purification, Antioxidant/Antinitrosating Properties, and Identification of a Degradative Structural Marker. *J. Agric. Food Chem.* **2012**, *60*, 8895–8901. [CrossRef]
39. Misra, S.; Pandey, P.; Panigrahi, C.; Mishra, H.N. A Comparative Approach on the Spray and Freeze Drying of Probiotic and Gamma-Aminobutyric Acid as a Single Entity: Characterization and Evaluation of Stability in Simulated Gastrointestinal Conditions. *Food Chem. Adv.* **2023**, *3*, 100385. [CrossRef]
40. Sahni, P.; Sharma, S.; Surasani, V.K.R. Influence of Processing and pH on Amino Acid Profile, Morphology, Electrophoretic Pattern, Bioactive Potential and Functional Characteristics of Alfalfa Protein Isolates. *Food Chem.* **2020**, *333*, 127503. [CrossRef]
41. Chen, Y.; Zhu, J.; Zhang, C.; Kong, X.; Hua, Y. Sesame Water-Soluble Proteins Fraction Contains Endopeptidases and Exopeptidases with High Activity: A Natural Source for Plant Proteases. *Food Chem.* **2021**, *353*, 129519. [CrossRef]
42. Orruño, E.; Morgan, M.R.A. Purification and Characterisation of the 7S Globulin Storage Protein from Sesame (*Sesamum indicum* L.). *Food Chem.* **2007**, *100*, 926–934. [CrossRef]
43. Nouska, C.; Deligeorgaki, M.; Kyrkou, C.; Michaelidou, A.-M.; Moschakis, T.; Biliaderis, C.G.; Lazaridou, A. Structural and Physicochemical Properties of Sesame Cake Protein Isolates Obtained by Different Extraction Methods. *Food Hydrocoll.* **2024**, *151*, 109757. [CrossRef]
44. Zhou, J.; Li, T.; Peydayesh, M.; Usuelli, M.; Lutz-Bueno, V.; Teng, J.; Wang, L.; Mezzenga, R. Oat Plant Amyloids for Sustainable Functional Materials. *Adv. Sci.* **2022**, *9*, 2104445. [CrossRef]
45. Afify, A.E.-M.M.R.; Rashed, M.M.; Mahmoud, E.A.; El-Beltagi, H.S. Effect of Gamma Radiation on Protein Profile, Protein Fraction and Solubility's of Three Oil Seeds: Soybean, Peanut and Sesame. *Not. Bot. Horti Agrobot. Cluj-Napoca* **2011**, *39*, 90–98. [CrossRef]
46. Idowu, A.O.; Alashi, A.M.; Nwachukwu, I.D.; Fagbemi, T.N.; Aluko, R.E. Functional Properties of Sesame (*Sesamum indicum* L.) Seed Protein Fractions. *Food Prod. Process. Nutr.* **2021**, *3*, 4. [CrossRef]
47. Tai, S.S.K.; Lee, T.T.T.; Tsai, C.C.Y.; Yiu, T.-J.; Tzen, J.T.C. Expression Pattern and Deposition of Three Storage Proteins, 11S Globulin, 2S Albumin and 7S Globulin in Maturing Sesame Seeds§. *Plant Physiol. Biochem.* **2001**, *39*, 981–992. [CrossRef]
48. Yang, J.; Kornet, R.; Diedericks, C.F.; Yang, Q.; Berton-Carabin, C.C.; Nikiforidis, C.V.; Venema, P.; van der Linden, E.; Sagis, L.M.C. Rethinking Plant Protein Extraction: Albumin—From Side Stream to an Excellent Foaming Ingredient. *Food Struct.* **2022**, *31*, 100254. [CrossRef]
49. Hu, A.; Li, L. Effect Mechanism of Ultrasound Pretreatment on Fibrillation Kinetics, Physicochemical Properties and Structure Characteristics of Soy Protein Isolate Nanofibrils. *Ultrason. Sonochem.* **2021**, *78*, 105741. [CrossRef]
50. Xu, Z.; Wang, Y.; Gao, Y.; Zhang, Y.; Jiang, L.; Sui, X. Structural Insights into Acidic Heating-Induced Amyloid Fibrils Derived from Soy Protein as a Function of Protein Concentration. *Food Hydrocoll.* **2023**, *145*, 109085. [CrossRef]
51. Yu, Z.; Li, N.; Liu, Y.; Zhang, B.; Zhang, M.; Wang, X.; Wang, X. Formation, Structure and Functional Characteristics of Amyloid Fibrils Formed Based on Soy Protein Isolates. *Int. J. Biol. Macromol.* **2024**, *254*, 127956. [CrossRef] [PubMed]
52. Gade Malmos, K.; Blancas-Mejia, L.M.; Weber, B.; Buchner, J.; Ramirez-Alvarado, M.; Naiki, H.; Otzen, D. ThT 101: A Primer on the Use of Thioflavin T to Investigate Amyloid Formation. *Amyloid* **2017**, *24*, 1–16. [CrossRef]
53. Liu, C.; Wu, D.; Wang, P.; McClements, D.J.; Cui, S.; Liu, H.; Leng, F.; Sun, Q.; Dai, L. Study on the Formation Mechanism of Pea Protein Nanofibrils and the Changes of Structural Properties of Fibril under Different pH and Temperature. *Food Hydrocoll.* **2024**, *150*, 109735. [CrossRef]
54. Liu, J.; Tang, C.-H. Heat-Induced Fibril Assembly of Vicilin at pH 2.0: Reaction Kinetics, Influence of Ionic Strength and Protein Concentration, and Molecular Mechanism. *Food Res. Int.* **2013**, *51*, 621–632. [CrossRef]
55. Martínez-Velasco, A.; Lobato-Calleros, C.; Hernández-Rodríguez, B.E.; Román-Guerrero, A.; Alvarez-Ramirez, J.; Vernon-Carter, E.J. High Intensity Ultrasound Treatment of Faba Bean (*Vicia faba* L.) Protein: Effect on Surface Properties, Foaming Ability and Structural Changes. *Ultrason. Sonochem.* **2018**, *44*, 97–105. [CrossRef]
56. Herneke, A.; Lendel, C.; Johansson, D.; Newson, W.; Hedenqvist, M.; Karkehabadi, S.; Jonsson, D.; Langton, M. Protein Nanofibrils for Sustainable Food–Characterization and Comparison of Fibrils from a Broad Range of Plant Protein Isolates. *ACS Food Sci. Technol.* **2021**, *1*, 854–864. [CrossRef]
57. Song, Y.; Li, T.; Zhang, X.; Wang, L. Investigating the Effects of Ion Strength on Amyloid Fibril Formation of Rice Proteins. *Food Biosci.* **2023**, *51*, 102068. [CrossRef]
58. Yang, Q.; Wang, Y.-R.; Du, Y.-N.; Chen, H.-Q. Comparison of the Assembly Behavior and Structural Characteristics of Arachin and Conarachin Amyloid-like Fibrils. *Food Hydrocoll.* **2023**, *138*, 108479. [CrossRef]
59. Farrokhi, F.; Badii, F.; Ehsani, M.R.; Hashemi, M. Functional and Thermal Properties of Nanofibrillated Whey Protein Isolate as Functions of Denaturation Temperature and Solution pH. *Colloids Surf. A Physicochem. Eng. Asp.* **2019**, *583*, 124002. [CrossRef]
60. Mohammadian, M.; Madadlou, A. Characterization of Fibrillated Antioxidant Whey Protein Hydrolysate and Comparison with Fibrillated Protein Solution. *Food Hydrocoll.* **2016**, *52*, 221–230. [CrossRef]
61. Wei, Z.; Huang, Q. Impact of Covalent or Non-Covalent Bound Epigallocatechin-3-Gallate (EGCG) on Assembly, Physicochemical Characteristics and Digestion of Ovotransferrin Fibrils. *Food Hydrocoll.* **2020**, *98*, 105314. [CrossRef]
62. Zhang, M.; Zhao, J.; Zheng, J. Molecular Understanding of a Potential Functional Link between Antimicrobial and Amyloid Peptides. *Soft Matter* **2014**, *10*, 7425–7451. [CrossRef] [PubMed]
63. Elias, R.J.; Kellerby, S.S.; Decker, E.A. Antioxidant Activity of Proteins and Peptides. *Crit. Rev. Food Sci. Nutr.* **2008**, *48*, 430–441. [CrossRef] [PubMed]

64. Cao, Y.; Mezzenga, R. Food Protein Amyloid Fibrils: Origin, Structure, Formation, Characterization, Applications and Health Implications. *Adv. Colloid Interface Sci.* **2019**, *269*, 334–356. [CrossRef] [PubMed]
65. Paul, A.; Kalita, S.; Kalita, S.; Sukumar, P.; Mandal, B. Disaggregation of Amylin Aggregate by Novel Conformationally Restricted Aminobenzoic Acid Containing α/β and α/γ Hybrid Peptidomimetics. *Sci. Rep.* **2017**, *7*, 40095. [CrossRef] [PubMed]
66. Wu, J.W.; Liu, K.-N.; How, S.-C.; Chen, W.-A.; Lai, C.-M.; Liu, H.-S.; Hu, C.-J.; Wang, S.S.-S. Carnosine's Effect on Amyloid Fibril Formation and Induced Cytotoxicity of Lysozyme. *PLoS ONE* **2013**, *8*, e81982. [CrossRef] [PubMed]
67. Lassé, M.; Ulluwishewa, D.; Healy, J.; Thompson, D.; Miller, A.; Roy, N.; Chitcholtan, K.; Gerrard, J.A. Evaluation of Protease Resistance and Toxicity of Amyloid-like Food Fibrils from Whey, Soy, Kidney Bean, and Egg White. *Food Chem.* **2016**, *192*, 491–498. [CrossRef] [PubMed]
68. Markstedt, K.; Mantas, A.; Tournier, I.; Martínez Ávila, H.; Hägg, D.; Gatenholm, P. 3D Bioprinting Human Chondrocytes with Nanocellulose–Alginate Bioink for Cartilage Tissue Engineering Applications. *Biomacromolecules* **2015**, *16*, 1489–1496. [CrossRef]
69. Amin, K.; Dannenfelser, R.-M. In Vitro Hemolysis: Guidance for the Pharmaceutical Scientist. *J. Pharm. Sci.* **2006**, *95*, 1173–1176. [CrossRef]

Disclaimer/Publisher's Note: The statements, opinions and data contained in all publications are solely those of the individual author(s) and contributor(s) and not of MDPI and/or the editor(s). MDPI and/or the editor(s) disclaim responsibility for any injury to people or property resulting from any ideas, methods, instructions or products referred to in the content.

Article

Physical Treatments Modified the Functionality of Carrot Pomace

Jordan Richards [1], Amy Lammert [1], Jack Madden [1], Iksoon Kang [2] and Samir Amin [1,*]

[1] Food Science and Nutrition Department, California Polytechnic State University, San Luis Obispo, CA 93407, USA; jricha@calpoly.edu (J.R.); alammert@calpoly.edu (A.L.)
[2] Animal Science Department, California Polytechnic State University, San Luis Obispo, CA 93407, USA; ikang01@calpoly.edu
* Correspondence: samin02@calpoly.edu

Abstract: This study addressed the critical issue of food waste, particularly focusing on carrot pomace, a by-product of carrot juice production, and its potential reutilization. Carrot pomace, characterized by high dietary fiber content, presents a sustainable opportunity to enhance the functional properties of food products. The effects of physical pretreatments—high shearing (HS), hydraulic pressing (HP), and their combination (HSHP)—alongside two drying methods (freeze-drying and dehydration) on the functional, chemical, and physical properties of carrot pomace were explored. The results indicated significant enhancements in water-holding capacity, fat-binding capacity, and swelling capacity, particularly with freeze-drying. Freeze-dried pomace retained up to 33% more carotenoids and demonstrated an increase of up to 22% in water-holding capacity compared to dehydrated samples. Freeze-dried pomace demonstrated an increase of up to 194% in fat-binding capacity compared to dehydrated samples. Furthermore, HSHP pretreatment notably increased the swelling capacity of both freeze-dried (54%) and dehydrated pomace (35%) compared to pomace without pretreatments. Freeze-drying can enhance the functional properties of dried carrot pomace and preserve more carotenoids. This presents an innovative way for vegetable juice processors to repurpose their processing by-products as functional food ingredients, which can help reduce food waste and improve the dietary fiber content and sustainability of food products.

Keywords: carrot pomace; food waste; functional properties; freeze-drying; dehydration; dietary fibers; carotenoids

1. Introduction

Approximately one-third of all food produced globally is lost or wasted somewhere along the food supply chain [1]. Food loss and waste have been reported to occur throughout the food-processing cycle; this includes everything from in-field harvest to processing and packaging facilities and retail grocery stores. This represents a waste of the water, land, energy, and natural resources used to produce food and is estimated to cause USD 940 billion in economic losses and produce more than 4.4 gigatons of greenhouse gas emissions (CO_2 equivalent) annually [1]. The United States Environmental Protection Agency (EPA) estimates that annual food loss and waste are equivalent to 170 million metric tons of CO_2 equivalent emissions within the U.S. [2]. Reducing food waste within the U.S. presents opportunities to address climate change, conserve resources, and increase food security, productivity, and economic efficiency. According to the United States Department of Agriculture (USDA), 31% of food is wasted, amounting to a total of USD 218 billion, or 1.3% of the country's Gross Domestic Product (GDP).

In 2019, the total production of carrots in the U.S. reached 2.53 million metric tons, which was a 13% increase from the 2018 total [3]. Carrots are the sixth-most consumed fresh vegetable in the U.S. [3]. Per capita consumption of fresh carrots in the U.S. peaked at 6.4 kg in 1997 and then decreased to around 3.8 kg in 2022 [4]. Over the past 35 years, the U.S. carrot industry has changed with the introduction of fresh-cut technology for more value-added carrot products such as pre-cut carrots, baby carrots, and carrot juice, which has increased the amount of carrot waste from carrot processing. When producing carrot juice, a pulp by-product is generated that is equivalent to 50% of the raw material [5]. Carrot by-products are rich in bioactive substances such as carotenoids (especially β-carotene), insoluble and soluble fiber is composed of pectic polysaccharides, hemicellulose and cellulose [5,6].

Using carrot pomace reduces food waste and produces functional ingredients for the food industry [7]. Carrot pomace contains approximately 55% dietary fiber, which could increase water-holding capacity from 17.9 to 23.3 g water/g fiber [7,8]. Dietary fiber can also hold fat particles and play a key functional role in foods [9]. Fat-binding capacity (FBC) and WHC are important for improving product quality, such as juiciness, flavor, and mouthfeel [10].

Drying is the process of removing moisture from a material via natural or unnatural conditions. Drying technologies for fruits and vegetables include hot air drying, microwave drying, vacuum drying, freeze-drying, and heat pump drying [10]. Drying is a frequently used method to reduce volume and weight, therefore reducing the costs of packaging, storage, and transportation. Drying can also affect the flavor and textural properties of fruits and vegetables [11]. Dehydration is, by definition, the removal of water via evaporation from solid or liquid food to obtain a solid product with low water activity to inhibit microbial growth [12]. Drying methods influence food products' density, porosity, and rehydration features. Convective drying can reduce hydrophilic properties due to irreversible cellular rupture, resulting in dense structure and integrity losses by broken and shrunken capillaries, which hinders water absorption and rehydration. Freeze-dried fruits and vegetables are usually characterized by minimal shrinkage and less structural collapse due to their highly porous structures after water removal via sublimation [13–15]. Different drying methods resulted in different porous structures, and freeze-drying produced higher porosity in food structures (80–90%) [13]. Microwave-dried potato and carrot had a porosity of approximately 75%, while vacuum-drying decreased the porosity to 50% in carrot and 25% in potato [13,16].

The objective of this study is to investigate how pretreatments such as high-shear mixing, hydraulic pressing, and a combination of both, followed by drying methods such as dehydration and freeze-drying, affect the functional properties of carrot pomace. Specifically, the study aims to evaluate the impact of these treatments on water-holding capacity, fat-binding capacity, swelling capacity, dietary fiber composition, and carotenoid content. This presents an innovative way for vegetable juice processors to repurpose their processing by-products as functional food ingredients, which can help reduce food waste and improve the dietary fiber content and sustainability of food products.

2. Materials and Methods

Carrot pomace was obtained from Grimmway Family Farms (Arvin, CA, USA). Carrot pomace was placed in 22 kg sealed, food-grade pails and stored in a dark freezer at $-20\ °C$ until further processed. Freezing pomace prior to processing minimized chances of microbial growth and degradation of carotenoids.

2.1. Mechanical Pretreatments of Carrot Pomace

The frozen carrot pomace was thawed overnight in a refrigerated room and pretreated using one of the three methods prior to drying: (1) high-shear (HS) for 5 min @ 15,000 RPM (Yuchengtech AD300L-H High-Shear Mixer, Shanghai, China), (2) hydraulic press (HP) (Hydraulic Wells Juice Press, Samson Brands, Danbury, CT, USA), and (3) the combination of high-shear and hydraulic press (HSHP).

2.2. Mechanical Drying Treatment of Carrot Pomace

Carrot pomace with and without pretreatment was dried using one of two methods: (1) dehydration (D) using a drying oven (Harvest Saver R4 drying oven, Commercial Dehydrator Systems, Inc., Eugene, OR, USA) at 40 °C for 24 h on fan speed 1 (0.13 m/s) and (2) lyophilization, or free drying (FD), using a freeze-dryer (Harvest Right Freeze Dryer, Salt Lake City, UT, USA) at −20 °C and 6.67 Pa for 24 h (Table 1). Non-pretreated and pretreated dried carrot pomace was then ground using a commercial spice grinder (VEVOR 2500 g Electric Grain Mill Grinder, Sacramento, CA, USA) to pass through a 20-mesh sieve (0.85 mm) and stored at −22 °C after placing into gallon-sized plastic bags (Ziplock, SC Johnson & Sons, Inc., Racine, WI, USA) wrapped in aluminum foil (Reynolds Wrap Reynolds, Consumer Products, Lake Forest, IL, USA).

Table 1. Pretreatment and drying methods applied to carrot pomace.

Samples *	Pretreatment	Drying Method
CD	No Pretreatment	Dehydration
CFD	No Pretreatment	Freeze-Drying
HSD	High-shear	Dehydration
HSFD	High-shear	Freeze-Drying
HPD	Hydraulic Press	Dehydration
HPFD	Hydraulic Press	Freeze-Drying
HSHPD	High-shear and Hydraulic Press	Dehydration
HSHPFD	High-shear and Hydraulic Press	Freeze-Drying

* CD = control/dehydration; CFD = control/freeze-drying; HSD = high-shear/dehydrated; HSFD = high-shear/freeze-drying; HPD = hydraulic press/dehydrated; HPFD = hydraulic press/freeze-drying; HSHPD = high-shear and hydraulic press/dehydrated; HSHPFD = high-shear and hydraulic press/freeze-drying.

2.3. Chemical Properties

2.3.1. Total Moisture

Total moisture content was determined for both solid and liquid fractions of the carrot pomace. Approximately 2.50 g of carrot pomace was weighed, recorded, and placed in the Ohaus MB45 Moisture Analyzer (Ohaus Corp., Parsippany, NJ, USA) at 105 °C until no weight change was detected. The moisture content was determined using the following equation:

$$\text{Moisture Content } (\%) = \frac{\text{Dried weight of sample (g)}}{\text{Weight of initial sample (g)}} \times 100$$

2.3.2. Carotenoid Content

Carotenoid contents were determined for carrot pomace samples (Table 1) according to the method described by Amin [17]. One gram of each carrot pomace sample (Table 1) was added to 25 mL of extraction solvent and homogenized for 30 s at 7500 rpm (Senstry Cyclone I.Q. 2 Sentry Microprocessor Digital Homogenizer, SP Industries Inc., Warminster, PA, USA) in 50 mL centrifuge tubes. The centrifuge tubes were centrifuged for 5 min at 6500 rpm and 5 °C (Eppendorf 5810 R Centrifuge, Hauppauge, NY, USA). After centrifuging, the supernatant layer containing hexane and non-polar carotenoids (β-carotene) was transferred to a 25.00 mL volumetric flask. The supernatant volume was adjusted to 25.00 mL with additional hexane. Absorbance values were measured at λmax 450 nm (Shimadzu UV–1900 UV-VIS spectrophotometer, Shimadzu, MD, USA). An extinction coefficient of 2505 for β-carotene was used to calculate the concentration of carotenoids in the samples using Beer's law.

2.3.3. Total Dietary Fiber

Total dietary fiber (TDF), soluble dietary fiber (SDF), and insoluble dietary fiber (IDF) were determined for all pretreated and dried carrot pomace samples (Control, HSD, HSFD, HPD, HPFD, HSHPD, and HSHPFD) using the Megazyme total dietary fiber assay kit (K-TDFR-200A, Neogen, Lansing, MI, USA; Megazyme, Wicklow, Ireland) with modifications of AOAC 991.43 [18] and AACC 32-07.01 [19] (Figure 1). Samples were incubated with 50 mL of heat-stable alpha-amylase (Megazyme cat. no. E-BLAAM) (100 °C, 30 min) and then enzymatically digested with 100 mL protease (Megazyme cat. No. E-BSPRT) (60 °C, 30 min), followed by incubation with 200 mL of amyloglucosidase (Megazyme cat. No. E-AMGDF) (60 °C, 30 min) to remove protein and starch. The samples were filtered, washed (with water, 95% ethanol, and acetone), dried, and weighed to determine insoluble fiber (IDF). Four volumes of 95% ethanol (preheated to 60 °C) were added to the filtrate and the wash water. The precipitates were filtered and washed with 78% ethanol. The residues of soluble dietary fiber (SDF) were dried and weighed. The obtained values were corrected for ash and protein. TDF was determined by summing insoluble IDF and SDF. Fiber ratios were calculated as a ratio of IDF:SDF. Total dietary fiber was calculated using the equation below.

$$\text{Dietary Fiber } (\%) = \frac{\frac{R_1 + R_2}{2} - P - A - B}{m_1} \times 100$$

R_1 = IDF residue weight.
R_2 = SDF residue weight.
m_1 = sample weight.
A = ash weight from R_1.
P = protein weight from R_2.
B = blank.

2.3.4. Amylase Neutral Detergent Fiber

Amylase neutral detergent fiber (aNDF) was determined for control dehydrated (CD) and freeze-dried (CFD) carrot pomace samples. The amounts of 0.45–0.55 g of sample and 0.5 g of sodium sulfite (Na_2SO_3) were weighed and combined. The samples were heated until boiling in 50 mL of neutral detergent solution. An amount of 2 mL of α-amylase was added before the beaker was heated. The sample was boiled for 1 h and filtered using a pretared fritted glass crucible. Fritted crucibles containing aNDF residue were dried at 100 °C for 24 h. The residue weight was then recorded. All samples were analyzed in triplicate.

2.3.5. Acid Detergent Fiber

For sequential analysis of acid detergent fiber (ADF), the crucible containing the aNDF fiber preparation was analyzed sequentially. The crucible was placed on its side in a 600 mL Berzelius beaker, and the sample was boiled in 200 mL of acid detergent solution for 1 h. At the end of boiling, the crucible was removed with tongs, and the solution was gravimetrically transferred and filtered through the fritted crucible. Fritted crucibles containing ADF residue were dried at 100 °C for 24 h. The residue weight was then recorded. All samples were analyzed in triplicate.

2.4. Functional Properties

Functional properties were evaluated for all carrot pomace samples after drying using two methods (Table 1).

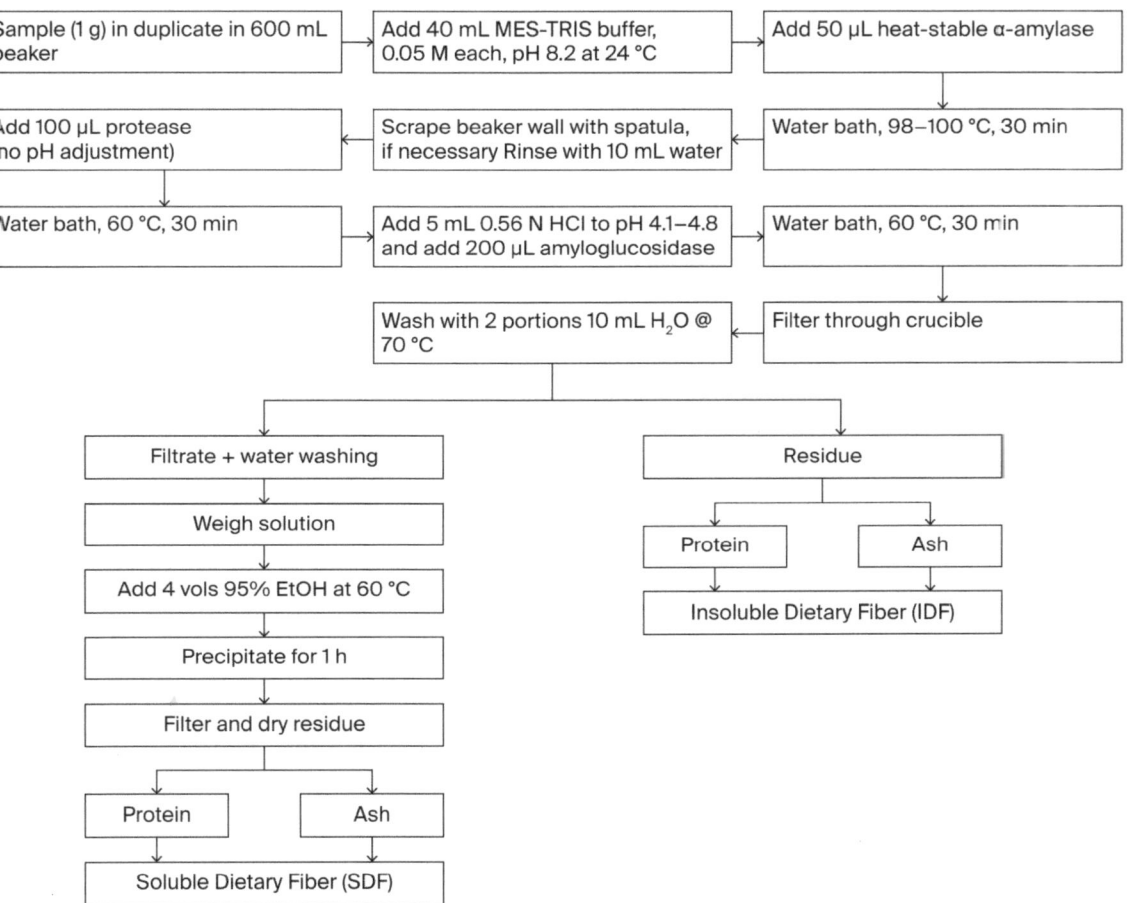

Figure 1. Total dietary fiber process flowchart from Megazyme [20].

2.4.1. Water-Holding Capacity

Water-holding capacity (WHC) was determined according to the method described by Raghavendra et al. [21]. Dried carrot pomace (0.50 g) was added to 15.00 mL of water in a graduated cylinder and mixed. After storing at ambient temperature for 24 h, the supernatant was filtered through a sintered glass crucible under vacuum. The hydrated residue weight was recorded before being dried at 105 °C for 1 h to obtain the residue dry weight. The water-holding capacity was measured as one gram of water held by one gram of pomace and calculated using the equation below.

$$\text{WHC} \left(\frac{\text{g water}}{\text{g dry pomace}} \right) = \frac{(\text{residue hydrated weight} - \text{residue dry weight})}{(\text{residue dry weight})}$$

2.4.2. Fat-Binding Capacity

Fat-binding capacity (FBC) was determined according to Beuchat's method [22] with modification. Canola oil (5.60 g) was added to dehydrated dried carrot pomace (1.00 g) in a 50 mL centrifuge tube. Due to the increased volume of freeze-dried pomace, the weight of the pomace used was reduced from 1.00 g to 0.10 g. Canola oil (5.60 g) was added to freeze-dried pomace in a 50 mL centrifuge tube. Each slurry was vortexed for 30 s, allowed to sit for 30 min at 22 °C, and then centrifuged at 1610× g for 25 min. The supernatant was decanted from the sample, the weight of the decanted sample was determined, and grams of oil retained per gram of sample was calculated. The fat-binding capacity was calculated using the equation below.

$$\text{Fat Binding Capacity} \left(\frac{g}{g}\right) = \frac{\text{Weight of decanted sample}}{\text{Weight of initial sample}}$$

2.4.3. Swelling Capacity

Swelling capacity was determined according to the method of Raghavendra et al. [21]. A total of 25 mL of deionized water was added to 1.00 g of dried carrot pomace in a 50.00 mL graduated cylinder. Graduated cylinders were covered with parafilm to reduce evaporation, and the samples were allowed to sit at 22 °C for 24 h. After 24 h, the volume of the swollen sample was measured. The swelling capacity was expressed as mL of water per 1.00 g of carrot pomace and was calculated using the equation below.

$$\text{Swelling Capacity} \left(\frac{mL}{g}\right) = \frac{\text{Volume occupied by sample}}{\text{Original sample weight}}$$

2.5. Statistical Analysis

Results of chemical and physical properties are reported as mean ± standard deviation. Two-way analysis of variance (ANOVA) was used to determine significant differences between functional properties based on the drying method and pretreatment using JMP Pro version 17 statistics software (Cary, NC, USA). Tukey's post hoc analysis was performed to identify significant differences between treatments at $p \leq 0.05$.

3. Results and Discussion

3.1. Characterization of Carrot Pomace

3.1.1. Total Moisture and Solids Content

Hydraulic pressing (HP) and high-shearing/hydraulic pressing pretreatments (HSHP) significantly ($p \leq 0.05$) reduced the moisture content of carrot pomace compared to the control and high-shear (HS) pretreatment (Table 2). HP increased the total solids content by 157% compared to the control.

Table 2. Impact of physical pretreatment methods on commercially produced carrot pomace's moisture and solids content.

Pretreatments	Moisture Content (%)	Solids Content (%)
Control	94.45 ± 0.07 [a]	5.55 ± 0.70 [c]
HS	94.52 ± 0.38 [a]	5.48 ± 0.38 [c]
HP	85.68 ± 0.79 [c]	14.31 ± 0.79 [a]
HSHP	89.70 ± 0.28 [b]	10.29 ± 0.28 [b]

[a–c] Different letters within columns indicate significant differences at $p \leq 0.05$.

The moisture content of whole carrots has been reported to be in the range of 86–89% [23], while the moisture content of carrot pomace has been reported as approximately 85% [24]. The application of hydraulic and expeller pressing significantly decreased the moisture content of commercially produced carrot mash by 9.10% and 12.56%, respectively [17]. The application of HP decreased carrot pomace's moisture content by 9.29%, while HS had no significant impact, while the combination of HSHP pomace only decreased the moisture content by 5.03%. The difference observed can be attributed to the expeller press being able to apply high shear and compression simultaneously.

HSHP pretreatment could increase the soluble fiber content of carrot pomace. Soluble fibers have demonstrated the capacity to improve viscosity, gel-formation, and emulsification [25]. This could be why HSHP had a 4% higher moisture content than HP. The physical action of water removal could reduce drying time and increase the solids content of dried carrot pomace during processing.

3.1.2. Total Carotenoid Content

The drying method significantly impacted the total carotenoid concentration of dried carrot pomace ($p < 0.05$). Freeze-drying significantly increased the total carotenoid concentration of carrot pomace compared to dehydration (Table 3).

Table 3. Impact of physical treatment methods on the total carotenoid concentration of commercially produced carrot pomace.

Pretreatments	Freeze-Dried (µg/g)	Dehydrated (µg/g)
Control	100.07 ± 4.51 [a]	44.91 ± 3.02 [c]
HS	67.94 ± 3.80 [b]	35.47 ± 1.35 [d]
HP	90.11 ± 2.76 [ab]	35.85 ± 1.19 [d]
HSHP	67.79 ± 1.11 [b]	48.72 ± 3.49 [c]

[a–d] Different letters within the same column indicate significant differences at $p < 0.05$.

Carotenoids are sensitive to quality loss by heat, light, and oxygen. During drying, the isomerization and oxidation of carotenoids can cause thermal degradation and quality degradation in color, flavor, and nutritional quality [26]. Carrot pomace that has not undergone pretreatments and drying contains 144.64 µg/g of carotenoids [27]. Both drying methods decreased carotenoids in dried pomace. Freeze-drying resulted in a 30.8%–53.1% reduction in carotenoids compared to untreated and undried pomace. Dehydration resulted in a 66.3%–75.5% reduction in carotenoids compared to untreated and undried pomace. Freeze-drying can increase the stability of carotenoids by reducing heat exposure and the oxidation rate at low temperatures and pressure [28]. Air-dried purple carrots had a 36.2% decrease in carotenoid content compared to freeze-dried purple carrots, which had a small decrease [29]. Freeze-dried carrot pomace subjected to high-shearing (HS and HSHP) showed a 33% reduction in carotenoid concentration compared to the control ($p \leq 0.05$). Disruptions in cell walls due to shearing can enhance the release of phytochemicals such as carotenoids from the solid matrix. These phytochemicals could be released into the liquid fraction of the material. The bioavailability of carotenoids has been shown to increase with heating or cell wall disruption through chopping or shearing. This is consistent with the previous reports that attributed higher values to the high-shear's ability to break down and release trapped carotenoid crystals within the cells [17,28–30].

3.1.3. Fiber Composition

Drying methods and pretreatment application significantly modified the total dietary fiber (TDF), neutral detergent fiber (aNDF), and acid detergent fiber (ADF) contents ($p \leq 0.05$) of carrot pomace (Tables 4 and 5). Both the pretreatment and drying methods used had a significant effect on the levels of TDF ($p \leq 0.0001$ and 0.0005, respectively). The interaction between the pretreatment and drying method was also significant ($p \leq 0.01$) and impacted TDF levels. Similarly, both the pretreatment and drying methods had a statistically significant effect on IDF (insoluble dietary fiber) levels ($p \leq 0.0001$). The interaction between the pretreatment and drying methods also significantly influenced IDF levels ($p \leq 0.0001$). Furthermore, the pretreatment had a statistically significant effect ($p = 0.0049$) on SDF (soluble dietary fiber) levels. Additionally, the drying method had a significant impact on SDF levels ($p \leq 0.0001$). Finally, the interaction between the pretreatment and drying methods significantly impacted SDF levels ($p < 0.0058$).

Physical pretreatments significantly affected the ratio of insoluble and soluble dietary fiber. Overall, freeze-dried samples showed a decrease in IDF and an increase in SDF in all pretreatments. The higher soluble fiber after physical shearing or pressing is presumed from the chemical interactions between insoluble fractions, hemicellulose, and lignin, which could convert insoluble fibers to soluble fibers [31].

Pretreatments had a more significant impact on TDF (total dietary fiber) compared to the drying methods and the combination of drying methods and pretreatments. The drying method alone had a more significant effect on SDF (soluble dietary fiber) compared to the pretreatment and the combination. On the other hand, the interaction between the drying method and pretreatment had a more significant effect on IDF (insoluble dietary fiber) than either factor alone.

Table 4. Significance of pretreatment and/or drying method on the fiber composition of dried carrot pomace.

	Source	Nparm	DF	Sum of Squares	F Ratio	Prob > F
TDF	Pretreatment	1	1	732.57	27.3662	<0.0001 *
	Drying Method	3	3	739.53	9.2088	0.0005 *
	Interaction	3	3	453.81	5.6510	0.0057 *
IDF	Pretreatment	1	1	175.12	33.1700	<0.0001 *
	Drying Method	3	3	976.41	61.6637	<0.0001 *
	Interaction	3	3	1258.89	79.5034	<0.0001 *
SDF	Pretreatment	1	1	182.07	9.9863	0.0049 *
	Drying Method	3	3	753.88	13.7828	<0.0001 *
	Interaction	3	3	308.19	5.6345	<0.0058 *

* Source with a significant impact ($p < 0.05$) on the fiber content (TDF, IDF, and SDF).

Table 5. Amylase neutral detergent fiber and acid detergent fiber composition of commercially produced dried carrot pomace.

Treatments	Freeze-Dried (g/100 g)	Dehydrated (g/100 g)
Amylase Neutral Detergent Fiber (aNDF)	37.57 ± 1.40 [a]	23.13 ± 1.76 [b]
Acid Detergent Fiber (ADF)	30.60 ± 0.30 [a]	17.82 ± 1.32 [b]

[a,b] Different letters within the same row indicate significant differences at $p \leq 0.05$.

Freeze-drying increased the aNDF and ADF contents compared to dehydration (Table 5). Freeze-dried carrot pomace showed a 32% increase in aNDF and a 43% increase in ADF compared to dehydrated pomace. Amylase neutral detergent fiber accounts for the hemicellulose, cellulose, and lignin present in the product, while ADF accounts for the removal of hemicellulose. In freeze-dried carrot pomace, cellulose and lignin made up 81% and 73% of the fiber content, respectively, compared to dehydrated pomace.

The total dietary fiber (TDF) content was higher in freeze-dried carrot pomaces (71.86 g/100 g) than in dehydrated carrot pomaces (51.84 g/100 g). These results indicated that the content of insoluble dietary fiber was higher in freeze-dried pomace (55.38 g/100 g), whereas the content of soluble fiber was higher in dehydrated pomace (20.70 g/100 g) than in freeze-dried pomace (16.49 g/100 g). Thermal processes have been an important factor in modifying insoluble and soluble fiber ratios and physiochemical properties [32]. The quantity of soluble fiber is generally influenced by the processing temperatures. Higher temperatures can break down glycosidic bonds in polysaccharides, lead to an increase in oligosaccharides, and, therefore, increase the quantity of soluble dietary fiber [33]. This may explain the increase of soluble fiber found in dehydrated pomaces. Various thermal treatments were shown to alter the insoluble and soluble ratio of barley fiber [34].

Pretreatments modified the TDF content in dehydrated and freeze-dried carrot pomaces (Table 6). HSHP pretreatment significantly increased the TDF content of dehydrated pomace from 52 g/100 g to 68.22 g/100 g. Dehydrated HS and HP pomace showed no significant difference in TDF compared to dehydrated control samples. However, HS and HP pretreatments decreased the TDF in freeze-dried carrot pomace. Pretreatments significantly affected the ratio of insoluble and soluble dietary fibers. Freeze-drying decreased the IDF content and increased the SDF content in all pretreatments. The higher soluble fiber in pomace after physical shearing or pressing is presumed from the chemical interactions that might convert the insoluble fractions to soluble fractions [31]. Physical treatments such as ball milling resulted in the redistribution of TDF in grape pomace and grape pomace fiber concentrate, causing an increase in SDF and a decrease in IDF [31].

Table 6. Impact of physical treatment methods on the fiber content of dehydrated and freeze-dried commercially produced carrot pomace (g/100 g).

Sample *	Total Dietary Fiber (TDF)	Insoluble Dietary Fiber (IDF)	Soluble Dietary Fiber (SDF)	Fiber Ratio (IDF:SDF)
CD	51.84 ± 7.08 [c]	29.51 ± 2.66 [c]	22.33 ± 5.30 [cd]	2:1.5
CFD	71.87 ± 2.11 [a]	55.38 ± 2.53 [a]	16.49 ± 1.47 [d]	3.4:1
HSD	54.68 ± 5.07 [bc]	32.20 ± 0.79 [d]	22.47 ± 4.40 [bcd]	2:1.4
HSFD	65.27 ± 9.42 [ab]	29.90 ± 2.12 [c]	35.37 ± 4.30 [a]	2:2.4
HPD	53.45 ± 2.58 [bc]	28.95 ± 3.10 [d]	24.49 ± 3.50 [bc]	2:1.7
HPFD	56.21 ± 1.29 [bc]	27.19 ± 1.01 [d]	29.02 ± 0.48 [ab]	2:1.9
HSHPD	67.17 ± 0.80 [ab]	39.75 ± 2.10 [bc]	27.42 ± 0.50 [b]	2:1.4
HSHPFD	73.29 ± 5.78 [a]	38.19 ± 2.28 [b]	35.10 ± 3.59 [a]	2:1.9

[a-d] Different letters within the same column indicate significant differences at $p \leq 0.05$. * CD = control/dehydration; CFD = control/freeze-drying; HSD = high-shear/dehydrated; HSFD = high-shear/freeze-drying; HPD = hydraulic press/dehydrated; HPFD = hydraulic press/freeze-drying; HSHPD = high-shear and hydraulic press/dehydrated; HSHPFD = high-shear and hydraulic press/freeze-drying.

The IDF/SDF ratio is an important factor as both fractions are complementary in their functional properties. As an acceptable food ingredient, the IDF/SDF ratio should be approximately 2:1 [35]. Carrot pomace dietary fiber could be a high-quality food ingredient due to the physiological effects of soluble and insoluble fibers. The total dietary fiber of carrot pomace has been reported to be 63.6%, with insoluble and soluble fractions of 50.1% and 13.5%, respectively [36].

3.2. Functional Properties of Carrot Pomace

3.2.1. Water-Holding Capacity

When looking at the effect of the drying method, pretreatment or the interaction between the drying method and pretreatment (F = 43.215) had a more significant effect on WHC than the drying method (F = 1.474) and interactions between pretreatments and drying methods (F = 5.880) (Table 7). Freeze-drying significantly increased the water-holding capacity of the FDC and FHS pretreated samples compared to CD, HSCD, HSFD, HPCD, HPFD, HSHPD, and HSHPFD ($p \leq 0.05$), while no significant differences were observed between freeze-drying and dehydration in the HS, HP, and HSHP pretreated samples at either drying method (Figure 2).

Table 7. Significance of pretreatment and/or drying method on the water-holding capacity of dried carrot pomace.

Source	Nparm	DF	Sum of Squares	F Ratio	Prob > F
Pretreatment	1	1	504.284	43.215	<0.0001 *
Drying Method	3	3	5.734	1.474	0.2318
Interaction	3	3	68.612	5.880	0.0020 *

* Source with a significant impact ($p < 0.05$) on the water-holding capacity.

Carrot dietary fiber has a high water-holding capacity, ranging from 17.9 to 23.3 g water/g fiber compared to other vegetable fibers such as coconut, potato, and pea [8,21]. The water-holding capacity of fibers can be impacted by their tissue structure and structure shrinkage. Freeze-drying has been shown to affect the physicochemical and structural characteristics of foods. Removing water via sublimation during freeze-drying produces a highly porous structure and little shrinkage [14,15]. Freeze-dried spinach exhibits high porosity and increased surface area [37]. In contrast, carrot slices after conventional dehydration had a higher shrinkage rate (35.53%) than freeze-dried slices (20.83%) [38]. Hot air–drying has been shown to affect carrot tissue, causing the tissue to exhibit highly dense, less porous, and collapsed structures due to cellular damage from hot air–drying [39].

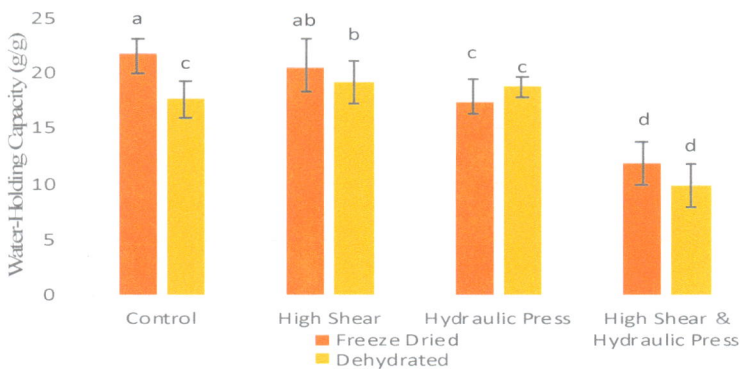

Figure 2. Impact of physical treatment methods on the water-holding capacity of commercially produced carrot pomace. $^{a-d}$ Different letters within the same physical treatment indicate significant differences at $p \leq 0.05$.

The ability of dietary fiber to hold water can be affected by its structure and how effectively it traps water. In the case of freeze-dried treatments, HS and HP pretreatments resulted in significant decreases in water-holding capacity (WHC) of 5.90% and 20.50%, respectively. The reduction in the WHC could be due to the change in insoluble and soluble fiber ratio and the reduction of insoluble to soluble fiber (Figure 2). Physically treated samples (HS, HP, and HSHP) showed a decrease in insoluble fiber but an increase in soluble fiber (Table 6). High-shear treatments could also be responsible for breaking up the fiber chains within carrot pomace. After milling, the WHC of grape pomace decreased from 2.52 g/g to 2.17 g/g [31]. However, an increase was seen in dehydrated HS and HP treatments. Thus, the length and degree of physical shearing can impact fiber degradation.

These results in this study support the findings that carrot pomace has a high water-holding capacity. Coconut pomace produced from coconut milk production is reported to have a high water-retention capacity of 5.4 g/g, which is lower than the WHC of carrot pomace in this study. The high WHC of carrot pomace suggests its potential uses in food applications as a functional ingredient.

3.2.2. Fat-Binding Capacity

When looking at the effects of the drying method, pretreatment, or interaction, the drying method (F = 3058.060) had a more significant effect on FBC than the pretreatment (F = 170.72) and the interaction (F = 132.65) (Table 8). Freeze-drying significantly increased the fat-binding capacity (FBC) of carrot pomace compared to dehydration in all pretreatments ($p < 0.05$) (Figure 3). The FBC increased from 4.00 g/g in CD to 11.78 g/g in FDC, from 3.26 g/g in HSD to 16.6 g/g HSFD, from 2.88 g/g in HPD to 7.82 g/g in HSFD, and from 2.77 g/g in HSHPD to 9.59 g/g in HSHPFD.

Table 8. Significance of pretreatment and/or drying method on the fat-binding capacity of dried carrot pomace.

Source	Nparm	DF	Sum of Squares	F Ratio	Prob > F
Pretreatment	1	1	133.360	170.725	<0.0001 *
Drying Method	3	3	796.255	3058.060	<0.0001 *
Interaction	3	3	103.637	132.675	<0.0001 *

* Source with a significant impact ($p < 0.05$) on the fat-binding capacity.

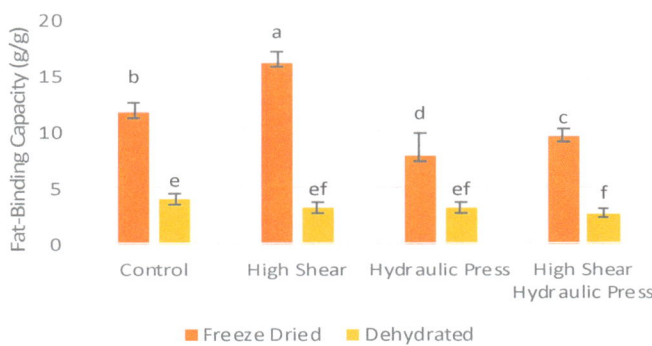

Figure 3. Impact of physical treatment methods on the fat-binding capacity of commercially produced carrot pomace. [a–f] Different letters within the same pretreatment indicate significant differences at $p \leq 0.05$.

Fat-binding capacity can be affected by various factors, such as plant polysaccharides, hydrophobic particle character, particle size, and the ratio of insoluble and soluble fibers [40]. The exposure of plant fibers to higher temperatures for an extended period can alter the physicochemical structure of polysaccharides and the hydrophobic nature of the particles [41]. Dehydration can result in the shrinking and deformation of fiber particles, causing a loss of the original shape, the formation of particle aggregates, and the reduction of space for water or fat to be absorbed [42]. During freeze-drying, intercellular water is frozen and removed as gas by sublimation. Therefore, freeze-dried particles can keep their original shape, showing larger sizes and more porous surfaces. Due to their larger volume and porosity, freeze-dried samples can bind more oil and water than conventionally dehydrated samples. The fat-binding capacity of dehydrated carrot pomace has been reported to be 3.95 ± 0.17 g/g [40], consistent with the values obtained from the present study. The fat-binding capacities of the fibrous residues of coconut fiber and banana powder are 5.30 g oil/g and 2.20 g oil/g, respectively [21]. Freeze-dried carrot pomace contained more total dietary fiber than conventionally dehydrated carrot pomace (Table 3). The composition of fibers plays a crucial role in the fat-binding capacity as well, as observed in date fiber concentrate, which showed a high oil-holding capacity (9.6–9.9 g oil/g). Date fiber concentrate was characterized by high levels of insoluble (81.3–84.7%) and soluble (6.7–7.69%) fibers [43]. Carrot pomace, a rich source of insoluble and soluble fibers, exhibits a higher fat-binding capacity, which could explain the higher FBC in carrot fibers than other fibers [40,44].

3.2.3. Swelling Capacity (SC)

When looking at the effects of the drying method, pretreatment, and their interaction, the drying method (F = 384.312) had a more significant effect on SC than the pretreatment (F = 109.035) and interaction (F = 42.134) (Table 9). The results of the swelling capacity demonstrated that any physical treatment (HS, HP, and HSHP) could significantly increase the swelling capacity of both freeze-dried and dehydrated carrot pomaces (Figure 4). SC increased from 26.25 mL/g to 31.83 mL/g and 35.5 mL/g after HSD and HSHPD pretreatments, respectively, while SC increased from 27.50 mL/g to 43.33 mL/g after FDHP and FDHSHP pretreatments. These increases can be attributed to the enhanced exposure of hydrophilic groups in cellulose and hemicellulose, as well as the increased surface area and surface energy of particles after shearing [45].

Table 9. Significance of pretreatment and/or drying method on the swelling capacity of dried carrot pomace.

Source	Nparm	DF	Sum of Squares	F Ratio	Prob > F
Pretreatment	1	1	950.307	109.035	<0.0001 *
Drying Method	3	3	1116.505	384.312	<0.0001 *
Interaction	3	3	367.224	42.134	<0.0001 *

* Source with a significant impact ($p < 0.05$) on the swelling capacity.

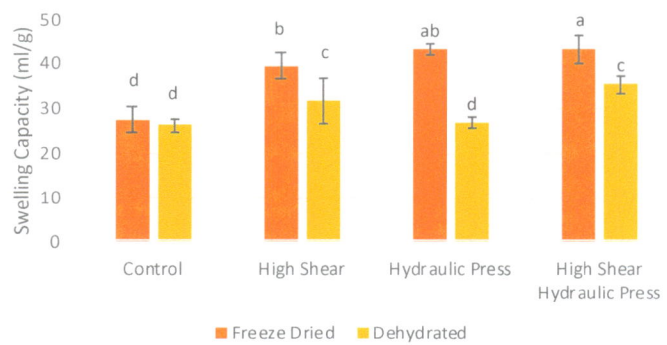

Figure 4. Impact of physical treatment methods on the swelling capacity of commercially produced carrot pomace. [a–d] Different letters within the same pretreatment indicate significant differences at $p \leq 0.05$.

The excessive grinding of vegetables may substantially disrupt the integrity of their dietary fiber chain, adversely affecting their hydration properties [46]. Asparagus pomace samples subjected to superfine grinding showed an initial increase in swelling capacity from 1.45 mL/g to 3.42 mL/g with a particle size reduction from 141.0 μm to 32.7 μm, which decreased thereafter as the particle size was further reduced to 6.1 μm [44]. Excessive micronation can lead to structural damage, reduce the total dietary fiber content, and consequently decrease the water-retention functionality of the fiber [45]. However, pretreatments showed benefits in increasing SC, possibly due to the breakage of long-chain dietary fibers into shorter-chain fibers and the porosity of fiber affecting its binding sites, resulting in enhanced hydration properties [47].

Soluble and insoluble fiber contents directly influence the swelling capacity and functionality of the product. The structural and chemical properties of fiber can play a role in the kinetics of water uptake. Water can be held by the capillary structures of dietary fiber due to surface tension. Additionally, water can interact with the molecular components of fiber through hydrogen bonds. The swelling capacities of orange and lemon fiber concentrates were 6.11 mL/g and 9.19 mL/g, respectively, corresponding to insoluble dietary fiber values of 54.0 g/100 g and 63.9 g/100 g, respectively [48]. The pectin in carrot pomace also has a greater water-holding capacity than cellulose fibers [40]. Pectin-rich citrus fibers have been shown to have a high water-swelling capacity [49]. The increased swelling capacity of pretreated carrot pomace could be due to the change in the insoluble to soluble fiber ratio after various pretreatments (Figure 4). Carrot pomace had a swelling capacity higher than that of coconut fibers [21]. The swelling capacity of carrot mash/peel has been reported to be 29.23 mL/g [17], comparable to both control pretreatments. The high swelling capacity of carrot pomace shows the benefit of dietary fibers with functional properties.

4. Conclusions

Freeze-drying was the most effective method compared to conventional dehydration in improving the functional properties of dried carrot pomace, including the water-holding, fat-binding, and swelling capacities. Freeze drying increased the water-holding capacity by 22% and the fat-binding capacity by 194%, with a greater retention of carotenoids. This is because of the lower temperatures and pressure during freeze-drying. Moreover, freeze-drying shows additional advantages by retaining up to 60% more carotenoid content within the carrot pomace. Physical pretreatments could also influence the functional properties of carrot pomace in combination with drying methods. Combining high-shearing and hydraulic pressing pretreatments increased the swelling capacity of freeze-dried carrot pomace and dehydrated carrot pomace compared to high-shearing or hydraulic pressing alone.

Author Contributions: Conceptualization, J.R. and S.A.; data curation, S.A.; formal analysis, J.R. and S.A.; funding acquisition, A.L. and S.A.; investigation, J.R. and J.M.; methodology, J.R. and S.A.; project administration, S.A.; supervision, S.A.; validation, A.L., I.K., and S.A.; writing—original draft, J.R. and S.A.; writing—review & editing, J.R., A.L., J.M., I.K. and S.A. All authors have read and agreed to the published version of the manuscript.

Funding: Partial funding for this project was made available by the California State University Agricultural Research Institute (ARI) Award No. 19-03-127, Carrot Pomace as a Source of Carotenoids and Nondigestible Oligosaccharides, and 23-01-103, Developing Case Studies for Undergraduates—Valorization of Fruit and Vegetable Byproducts for New Product Development. Carrot pomace was donated by Grimmway Family Farms (Arvin, CA, USA).

Institutional Review Board Statement: Not applicable.

Informed Consent Statement: Not applicable.

Data Availability Statement: The original contributions presented in the study are included in the article, further inquiries can be directed to the corresponding author.

Conflicts of Interest: The authors declare no conflicts of interest. The funders had no role in the design of the study; in the collection, analyses, or interpretation of data; in the writing of the manuscript; or in the decision to publish the results.

References

1. FAO. Global Initiative on Food Loss and Waste Reduction. Available online: https://www.fao.org/3/i7657e/i7657e.pdf (accessed on 1 March 2023).
2. Buzby, J. *Food Waste and Its Links to Greenhouse Gases and Climate Change*; USDA: Washington, DC, USA, 2022. Available online: https://www.usda.gov/media/blog/2022/01/24/food-waste-and-its-links-greenhouse-gases-and-climate-change (accessed on 6 February 2023).
3. Davis, W.; Lucier, G. *Vegetable and Pulses Outlook 2021*; USDA: Washington, DC, USA, 2021.
4. USDA-NASS QuickStats. Available online: https://quickstats.nass.usda.gov/ (accessed on 6 February 2023).
5. Yadav, S.; Pathera, A.K.; Islam, R.U.; Malik, A.K.; Sharma, D.P. Effect of Wheat Bran and Dried Carrot Pomace Addition on Quality Characteristics of Chicken Sausage. *Asian-Australas. J. Anim. Sci.* **2018**, *31*, 729–737. [CrossRef] [PubMed]
6. Elik, A.; Yanık, D.K.; Göğüş, F. Microwave-Assisted Extraction of Carotenoids from Carrot Juice Processing Waste Using Flaxseed Oil as a Solvent. *LWT* **2020**, *123*, 109100. [CrossRef]
7. Singh, B.; Panesar, P.; Nanda, V. Utilization of Carrot Pomace for the Preparation of a Value Added Product. *World J. Dairy Food Sci.* **2006**, *1*, 22–27.
8. Robertson, J.A.; Eastwood, M.A.; Yeoman, M.M. An Investigation into the Physical Properties of Fiber Prepared from Several Carrot Varieties at Different Stages of Development. *J. Sci. Food Agric.* **1979**, *31*, 633–638. [CrossRef]
9. Thebaudin, J.Y.; Lefebvre, A.C.; Harrington, M.; Bourgeois, C.M. Dietary Fibres: Nutritional and Technological Interest. *Trends Food Sci. Technol.* **1997**, *8*, 41–48. [CrossRef]

10. Nagai, T.; Reiji, I.; Hachiro, I.; Nobutaka, S. Preparation and Antioxidant Properties of Water Extract of Propolis. *Food Chem.* **2003**, *80*, 29–33. [CrossRef]
11. Law, C.L.; Khan, M.I.H.; Wellard, R.M.; Mahiuddin, M.; Karim, M.A. Cellular Level Water Distribution and Its Investigation Techniques. In *Intermittent and Nonstationary Drying Technologies: Principles and Applications*; CRC Press: Boca Raton, FL, USA, 2017; pp. 193–215.
12. Berk, Z. *Food Process Engineering and Technology*; Academic Press: Cambridge, MA, USA, 2018.
13. Calín-Sánchez, Á.; Lipan, L.; Cano-Lamadrid, M.; Kharaghani, A.; Masztalerz, K.; Carbonell-Barrachina, Á.A.; Figiel, A. Comparison of Traditional and Novel Drying Techniques and Its Effect on Quality of Fruits, Vegetables and Aromatic Herbs. *Foods* **2020**, *9*, 1261. [CrossRef] [PubMed]
14. Meda, L.; Ratti, C. Rehydration of Freeze-Dried Strawberries at Varying Temperatures. *J. Food Process Eng.* **2005**, *28*, 233–246. [CrossRef]
15. Jia, Y.; Khalifa, I.; Hu, L.; Zhu, W.; Li, J.; Li, K.; Li, C. Influence of Three Different Drying Techniques on Persimmon Chips' Characteristics: A Comparison Study among Hot-Air, Combined Hot-Air-Microwave, and Vacuum-Freeze Drying Techniques. *Food Bioprod. Process.* **2019**, *118*, 67–76. [CrossRef]
16. Krokida, M.K.; Philippopoulos, C. Rehydration of Dehydrated Foods. *Dry. Technol.* **2005**, *23*, 799–830. [CrossRef]
17. Amin, S.; Duval, A.; Jung, S.; Kang, I. Valorization of Baby Carrot Processing Waste. *J. Culin. Sci. Technol.* **2021**, *21*, 1–17. [CrossRef]
18. AOAC. *Official Methods of Analysis of AOAC International*; AOAC International: Rockville, MD, USA, 2012; ISBN 978-0-935584-83-7.
19. AACC. Soluble, Insoluble, and Total Dietary Fiber in Foods and Food Products. In *AACC Approved Methods of Analysis*; Method 32-07.01; Cereals & Grains Association: St. Paul, MN, USA, 1991.
20. Total Dietary Fiber Assay Kit. Available online: https://www.megazyme.com/total-dietary-fiber-assay-kit (accessed on 30 December 2023).
21. Raghavendra, S.N.; Rastogi, N.K.; Raghavarao, K.S.M.S.; Tharanathan, R.N. Dietary Fiber from Coconut Residue: Effects of Different Treatments and Particle Size on the Hydration Properties. *Eur. Food Res. Technol.* **2004**, *218*, 563–567. [CrossRef]
22. Beuchat, L.R. Functional and Electrophoretic Characteristics of Succinylated Peanut Flour Protein. *J. Agric. Food Chem.* **1977**, *25*, 258–261. [CrossRef]
23. Gopalan, C.; Ramasastry, B.V.; Balasubramanian, S.C. *Nutritive Valueof Indian Foods*; National Institute of Nutrition: Hyderabad, India, 1991.
24. Sharma, S.; Sagar, N.A.; Pareek, S.; Yahia, E.M.; Lobo, M.G. Fruit and Vegetable Waste: Bioactive Compounds, Their Extraction, and Possible Utilization. *Compr. Rev. Food Sci. Food Saf.* **2018**, *17*, 512–531.
25. Spotti, M.J.; Campanella, O.H. Functional Modifications by Physical Treatments of Dietary Fibers Used in Food Formulations. *Sens. Sci. Consum. Percept. Food Phys. Mater. Sci.* **2017**, *15*, 70–78. [CrossRef]
26. Schultz, A.K.; Barrett, D.M.; Dungan, S.R. Effect of Acidification on Carrot (*Daucus carota*) Juice Cloud Stability. *J. Agric. Food Chem.* **2014**, *62*, 11528–11535. [CrossRef] [PubMed]
27. Honda, M.; Takasu, S.; Nakagawa, K.; Tsuda, T. Differences in Bioavailability and Tissue Accumulation Efficiency of (All-E)- and (Z)-Carotenoids: A Comparative Study. *Food Chem.* **2021**, *361*, 130119. [CrossRef] [PubMed]
28. de la Rosa, L.; Alvarez-Parrilla, E.; Gonzalez-Agular, G.A. (Eds.) *Fruit and Vegetable Phytochemicals: Chemistry, Nutritional Value and Stability*; Wiley Blackwell: Ames, IA, USA, 2010.
29. Macura, R.; Michalczyk, M.; Fiutak, G.; Maciejaszek, I. Effect of freeze-drying and air-drying on the content of carotenoids and anthocyanins in stored purple carrot. *Acta Sci. Pol. Technol. Aliment.* **2019**, *18*, 135–142. [PubMed]
30. Gärtner, C.; Stahl, W.; Sies, H. Lycopene Is More Bioavailable from Tomato Paste than from Fresh Tomatoes. *Am. J. Clin. Nutr.* **1997**, *66*, 116–122. [CrossRef] [PubMed]
31. Bender, A.B.B.; Speroni, C.S.; Moro, K.I.B.; Morisso, F.D.P.; Santos, D.R.; Silva, L.P.; Penna, N.G. EfFfects of Micronization on Dietary Fiber Composition, Physicochemical Properties, Phenolic Compounds, and Antioxidant Capacity of Grape Pomace and Its Dietary Fiber Concentrate. *LWT* **2020**, *117*, 108652. [CrossRef]
32. Zhou, X.L.; Qian, Y.F.; Zhou, Y.M.; Zhang, R. Effect of Enzymatic Extraction Treatment on Physicochemical Properties, Microstructure and Nutrient Composition of Tartary Buckwheat Bran: A New Source of Antioxidant Dietary Fiber. *Adv. Mater. Res.* **2012**, *396*, 2052–2059. [CrossRef]
33. Wang, X.; Xu, Y.; Liang, D.; Yan, X.; Shi, H.; Sun, Y. Extrusion-Assisted Enzymatic Hydrolysis Extraction Process of Rice Bran Dietary Fiber. In Proceedings of the 2015 ASABE Annual International Meeting, New Orleans, IL, USA, 26–29 July 2015; American Society of Agricultural and Biological Engineers: St. Joseph, MI, USA, 2015; p. 1.
34. Bader Ul Ain, H.; Saeed, F.; Khan, M.A.; Niaz, B.; Rohi, M.; Nasir, M.A.; Tufail, T.; Anbreen, F.; Anjum, F.M. Modification of Barley Dietary Fiber through Thermal Treatments. *Food Sci. Nutr.* **2019**, *7*, 1816–1820. [CrossRef] [PubMed]
35. de Moraes Crizel, T.; Jablonski, A.; de Oliveira Rios, A.; Rech, R.; Flôres, S.H. Dietary Fiber from Orange Byproducts as a Potential Fat Replacer. *LWT—Food Sci. Technol.* **2013**, *53*, 9–14. [CrossRef]
36. Chau, C.-F.; Chen, C.-H.; Lee, M.-H. Comparison of the Characteristics, Functional Properties, and in Vitro Hypoglycemic Effects of Various Carrot Insoluble Fiber-Rich Fractions. *LWT—Food Sci. Technol.* **2004**, *37*, 155–160. [CrossRef]
37. King, V.A.E.; Liu, C.F.; Liu, Y.J. Chlorophyll Stability in Spinach Dehydrated by Freeze-Drying and Controlled Low-Temperature Vacuum Dehydration. *Food Res. Int.* **2001**, *34*, 167–175. [CrossRef]

38. Rajkumar, G.; Shanmugam, S.; Galvao, M.D.S.; Leite Neta, M.T.S.; Dutra Sandes, R.D.; Mujumdar, A.S.; Narain, N. Comparative Evaluation of Physical Properties and Aroma Profile of Carrot Slices Subjected to Hot Air and Freeze Drying. *Dry. Technol.* **2017**, *35*, 699–708. [CrossRef]
39. Nahimana, H.; Zhang, M. Shrinkage and Color Change during Microwave Vacuum Drying of Carrot. *Dry. Technol.* **2011**, *29*, 836–847. [CrossRef]
40. Sharoba, A.; Farrag, M.; Abd El-Salam, A. Utilization of some fruits and vegetables wastes as a source of dietary fibers in cake making. *J. Food Dairy Sci.* **2013**, *4*, 433–453. [CrossRef]
41. Fernández-López, J.; Sendra-Nadal, E.; Navarro, C.; Sayas, E.; Viuda-Martos, M.; Alvarez, J.A.P. Storage Stability of a High Dietary Fibre Powder from Orange By-products. *Int. J. Food Sci. Technol.* **2009**, *44*, 748–756. [CrossRef]
42. Fu, W.; Zhao, G.; Liu, J. Effect of Preparation Methods on Physiochemical and Functional Properties of Yeast β-Glucan. *LWT* **2022**, *160*, 113284. [CrossRef]
43. Elleuch, M.; Besbes, S.; Roiseux, O.; Blecker, C.; Deroanne, C.; Drira, N.E.; Attia, H. Date Flesh: Chemical Composition and Characteristics of the Dietary Fibre. *Food Chem.* **2008**, *111*, 676–682. [CrossRef]
44. Gao, W.; Chen, F.; Zhang, L.; Meng, Q. Effects of Superfine Grinding on Asparagus Pomace. Part I: Changes on Physicochemical and Functional Properties. *J. Food Sci.* **2020**, *85*, 1827–1833. [CrossRef] [PubMed]
45. He, S.; Tang, M.; Sun, H.; Ye, Y.; Cao, X.; Wang, J. Potential of Water Dropwort (*Oenanthe javanica* DC.) Powder as an Ingredient in Beverage: Functional, Thermal, Dissolution and Dispersion Properties after Superfine Grinding. *Powder Technol.* **2019**, *353*, 516–525. [CrossRef]
46. Zhao, Y.; Wu, X.; Wang, Y.; Jing, R.; Yue, F. Comparing Physicochemical Properties of Hawthorn Superfine and Fine Powders. *J. Food Process. Preserv.* **2017**, *41*, e12834. [CrossRef]
47. Meng, X.; Liu, F.; Xiao, Y.; Cao, J.; Wang, M.; Duan, X. Alterations in Physicochemical and Functional Properties of Buckwheat Straw Insoluble Dietary Fiber by Alkaline Hydrogen Peroxide Treatment. *Food Chem.* **2019**, *3*, 100029. [CrossRef] [PubMed]
48. Figuerola, F.; Hurtado, M.L.; Estevez, A.M.; Chiffelle, I.; Fernando, A. Fibre Concentrates from Apple Pomace and Citrus Peel as Potential Fibre Sources for Food Enrichment. *Food Chem.* **2005**, *91*, 395–401. [CrossRef]
49. Huang, J.; Liao, J.; Qi, J.; Jiang, W.; Yang, X. Structural and Physicochemical Properties of Pectin-Rich Dietary Fiber Prepared from Citrus Peel. *Food Hydrocoll.* **2021**, *110*, 106140. [CrossRef]

Disclaimer/Publisher's Note: The statements, opinions and data contained in all publications are solely those of the individual author(s) and contributor(s) and not of MDPI and/or the editor(s). MDPI and/or the editor(s) disclaim responsibility for any injury to people or property resulting from any ideas, methods, instructions or products referred to in the content.

Article

Green Extraction of Natural Colorants from Food Residues: Colorimetric Characterization and Nanostructuring for Enhanced Stability

Victoria Baggi Mendonça Lauria [1,2] and Luciano Paulino Silva [1,2,*]

1. Embrapa Recursos Genéticos e Biotecnologia, Laboratório de Nanobiotecnologia (LNANO), Parque Estação Biológica, Final W5 Norte, Brasília 70770-917, DF, Brazil; victoriabaggi@gmail.com
2. Programa de Pós-Graduação em Nanociência e Nanobiotecnologia, Universidade de Brasília, Campus Universitário Darcy Ribeiro, Brasília 70910-900, DF, Brazil
* Correspondence: luciano.paulino@embrapa.br

Abstract: Food residues are a promising resource for obtaining natural pigments, which may replace artificial dyes in the industry. However, their use still presents challenges due to the lack of suitable sources and the low stability of these natural compounds when exposed to environmental variations. In this scenario, the present study aims to identify different food residues (such as peels, stalks, and leaves) as potential candidates for obtaining natural colorants through eco-friendly extractions, identify the colorimetric profile of natural pigments using the RGB color model, and develop alternatives using nanotechnology (e.g., liposomes, micelles, and polymeric nanoparticles) to increase their stability. The results showed that extractive solution and residue concentration influenced the RGB color profile of the pigments. Furthermore, the external leaves of *Brassica oleracea* L. *var. capitata f. rubra*, the peels of *Cucurbita maxima*, *Cucurbita maxima x Cucurbita moschata*, and *Beta vulgaris* L. proved to be excellent resources for obtaining natural pigments. Finally, the use of nanotechnology proved to be a viable alternative for increasing the stability of natural colorants over storage time.

Keywords: food residues; green extraction; natural colorants; food colorants; RGB color model; nanotechnology

1. Introduction

Colors are one of the most important sensory attributes in foods. They directly affect final product consumption since they serve as key criteria for judging the quality and even taste of foods by consumers [1]. Given the importance of colors in influencing eating behaviors, the industry uses artificial colorings (e.g., tartrazine, erythrosine, brilliant blue, and others) to intensify or preserve the initial colors of food during processing and storage stages [2,3]. This guarantee of product color quality control is carried out through color models, such as CIELab and RGB. The latter, for example, is capable of measuring the color of any object based on the sum of the variations in intensity between red, green, and blue [4]. The main advantages of using artificial dyes are their good stability to temperature, light, pH, and storage time, as well as their good coloring properties and low-cost production [2].

Regardless of the clear importance of artificial dyes, many have witnessed concerns about the potential adverse impacts of their production process on the environment and the possible risks to human health caused by their excessive intake [5,6]. On the other hand, there is a wide variety of natural pigments, such as anthocyanins, betanins, and carotenoids, which can replace artificial colorants. These biopigments are chemical compounds (metabolites) present in virtually all living organisms. Due to their rich structural diversity, they usually differ among themselves in terms of antioxidant, antineoplastic, anti-inflammatory and antimicrobial properties [7–9].

In this scenario, by-products from industrial processes, specifically vegetable residues, such as peels, pomace, and seeds, among others, have been considered promising sources for developing natural food colorants due to their low cost and abundance in industrial, commercial, and household sectors, as well as the presence of health-beneficial phytopigments [10]. Furthermore, the reuse of food residues minimizes environmental pollution since their decomposition results in gas emissions that contribute to climate change [11].

Despite the environmental benefits associated with the use of food residues, obtaining these natural colorants generally requires organic solvents. Due to this fact, there is an increasing interest in technology development aimed at using alternative and nontoxic (green) solvents due to global environmental and climate imbalances [12]. That is why several studies have aimed at finding suitable sources for obtaining natural colorants from the use of sustainable technological innovations in correspondence with green chemistry principles [13,14].

Another challenge that needs to be overcome is the instability of natural compounds when subjected to environmental variations. In this sense, the use of nanotechnological strategies to minimize this degradation process appears as an emerging alternative, given that studies in the literature have already demonstrated the efficiency of nanosystems in protecting biopigments [15]. Nanotechnology refers to a field of technology responsible for manipulating materials, devices, and systems on a nanometric scale (10^{-9} m) [16]. Currently, in the food industry, nanotechnology has been used in several ways, such as in the nanoencapsulation of functional substances, aiming to improve the nutritional quality of food, as well as in the application of nanoparticles in packaging to increase the shelf life of foods [17,18].

In this context, this study innovates by developing an eco-friendly method to extract natural pigments from food residues and produce lipid and polymeric nanosystems to enhance the stability of these compounds. This pioneering approach opens new avenues for the food, pharmaceutical, and cosmetic industries, enabling the development of sustainable products.

2. Materials and Methods

2.1. Materials

In total, eight food residues were chosen as raw materials for the extraction of natural dyes, such as (i) hybrid tetsukabuto pumpkin peels (*Cucurbita maxima x Cucurbita moschata*); (ii) moranga pumpkin (*Cucurbita maxima*) peels; (iii) sweet potato peels (*Ipomoea potatoes*); (iv) beetroot peels (*Beta vulgaris* L.); (v) carrot peels (*Daucus carota* subsp. *sativus*); (vi) chayote shells (*Sechium edule* Sw); (vii) cabbage stalks (*Brassica oleracea* L. *var. acephala*); and (viii) purple cabbage external leaves (*Brassica oleracea* L. *var. capitata f. rubra*). These food residues were obtained through a donation made by the SESI industrial kitchen in the Federal District (DF) and/or by the team members of the Laboratory of Nanobiotechnology (LNANO) at Embrapa Genetic Resources and Biotechnology. Sodium hypochlorite (NaClO) (Agistereli Ltda, Itapevi, SP, Brazil) and sodium bicarbonate ($NaHCO_3$) (Kitano, São Bernardo do Campo, SP, Brazil) were purchased at a local supermarket. Ethanol (Itajá, Goianésia, GO, Brazil) and ultrapure water were used to prepare the extraction solutions. Nonionic surfactant Tween 80 (Sigma-Aldrich, St. Louis, MO, USA), soy lecithin (Saint Charbel, Viçosa, MG, Brazil), glacial acetic acid (Merck, Darmstadt, Germany), chitosan from shrimp shells, ≥75% deacetylated (Sigma-Aldrich, St. Louis, MO, USA), and sodium tripolyphosphate (Sigma-Aldrich, St. Louis, MO, USA) were used for the development of nanosystems containing natural pigments.

2.2. Sanitization Process

Firstly, all vegetables were subjected to pre-washing with running water, aiming to eliminate the surface dirt. Then, they went through a sanitization process with 2.5% NaClO, remaining submerged for 15 min. Subsequently, the residues were immersed in a $NaHCO_3$ solution at a concentration of 10 mg/mL for 15 min. Finally, they were rinsed with distilled water for 2 min, dried with a paper towel, and stored at −20 °C in plastic packaging until the extraction assays.

2.3. Extraction of Natural Pigments

2.3.1. Solvent Screening

The extraction of natural pigments was carried out using ultrapure water and ethanol 25% and 96% as solvents. The food residues were individually weighed (0.5 g) and cut into small pieces to increase the contact surface and optimize the extraction process. The pigment extractions were conducted in a 1:10 solids-to-solvent ratio. The samples were sonicated for 30 min at 40 kHz in an ultrasound bath and then magnetic stirred for 30 min. Finally, the food residues were separated from the extraction solvent by centrifugation at $1500 \times g$ for 10 min. The supernatants (natural pigments) were collected and stored at 2–8 °C for colorimetric evaluation. From the RGB analyses, the best solvents and food residues were chosen. From there, extractions varying the residue mass concentrations were carried out.

2.3.2. Variation of Residue Mass Concentrations

The extraction of natural pigments was performed in triplicate using five different mass concentrations of food residues (3.3%, 5%, 10%, 20%, and 30%). The extractions were carried out in triplicate, following the same methodology described in Section 2.3.1. The evaluation of RGB color profiles was carried out the day after the extraction and storage process.

2.4. Colorimetric Characterization

The colorimetric characterization of natural pigments was carried out in triplicate and determined by the RGB color space model. A volume of 200 µL of each pigment was added to a 96-well microplate. The microplate was placed on top of an acrylic box illuminated by a LightPad containing an LED light with a cool white color temperature of 6500 Kelvin (K). The images were acquired with an adapted iPod-based BiO Assay system using the Experimental Assistant app (n3D BioSciences Inc., Houston, TX, USA). The digital images were analyzed through the image processing software ImageJ (version 1.52a), which measured the light intensity in the RGB spectrums.

2.5. Color Profile Statistical Analysis

The mean and standard deviation data of R, G, and B color intensities were subjected to analysis of variance (ANOVA) at a significance level of 5% ($p < 0.05$) with the Paleontological Statistics software package—PAST (version 2.7c). The results are presented as the arithmetic mean ± standard deviation of the mean.

2.6. Development of Lipid and Polymeric Nanosystems

Using the best solvents and residue concentrations pre-established in previous trials, new extractions were conducted to obtain an adequate volume of natural pigment to produce micelles, liposomes, and polymeric chitosan nanoparticles. The nanostructured natural pigments were stored at 2–8 °C.

2.6.1. Micelles

Micelles were produced from natural pigment obtained from pumpkin peels (*Cucurbita maxima*). A total volume of 10 mL of pigment was added to 0.15 µL of the surfactant Tween 80. The sample was stirred until complete dissolution. Then, an aliquot was submitted as a rotary evaporator with an immersion bath at 50 °C. After the solvent evaporated, the sample was hydrated with ultrapure water. An empty control, without the presence of natural pigment, was also produced under the same conditions.

2.6.2. Liposomes

Liposomes were produced by the lipid film hydration method from natural pigment obtained from the peels of hybrid pumpkin (*Cucurbita maxima* x *Cucurbita moschata*) and red beet (*Beta vulgaris* L.). A total volume of 4 mL of pigments was added to 0.020 g of soy lecithin. The samples were stirred until complete dissolution. Then, an aliquot was submitted as a rotary evaporator with an immersion bath at 50 °C. After the solvent evaporation, the films were hydrated with ultrapure water. Empty controls, without the presence of natural pigments, were also produced under the same conditions.

2.6.3. Chitosan Polymeric Nanoparticles

Polymeric chitosan nanoparticles were produced from natural pigment obtained from the external leaves of purple cabbage (*Brassica oleracea* L. *var. capitata f. rubra*). A total volume of 2 mL of pigments was submitted as a rotary evaporator in 4 cycles (2 cycles of 1 h at 40 °C, followed by 2 cycles of 1 h at 50 °C). After the solvent evaporation, 2.0 mL of water acidified (0.1 mol/L) with glacial acetic acid was added. Subsequently, the solution was added to 0.004 g of medium molecular mass chitosan, remaining under stirring for 1 h. Then, a 1 mg/mL sodium tripolyphosphate solution was dripped gradually into the acidified solution. The formulation remained under constant stirring for 5 min after the end of dripping. Finally, it was subjected to the breaking of its particles using an ultraturrax at 21,500 rpm. An empty control, without the presence of natural pigment, was also produced under the same conditions.

2.7. Characterization of Nanostructured Pigments

The hydrodynamic diameter (HD) and the polydispersity index (PdI) of the nanosystems and their respective empty controls were evaluated using the dynamic light scattering (DLS) technique. The Zeta potential (ZP) was determined from its electrophoretic mobilities. The samples were diluted in a ratio of 1:10 in ultrapure water. The analyses were carried out on ZetaSizer Nano ZS equipment (Malvern Panalytical, Worcestershire, UK) by DLS and ZP. DLS analyses were performed at an angle of 173° using a He-Ne laser (4 mW) at a wavelength of 633 nm. Three measurements were taken of each sample at 25 °C in automatic run mode. The results are presented as the arithmetic mean ± standard deviation of the mean.

2.8. Evaluation of Colorimetric Stability

Colorimetric stability tests were carried out with natural pigments and their nanosystems stored at 2–8 °C protected from light for 26 days. Stability assessment occurred through RGB analysis carried out weekly, in triplicate, according to the methodology described in Section 2.4. The results are presented as the arithmetic mean ± standard deviation of the mean.

3. Results

3.1. Influence of Solvent on Natural Pigment Extraction

In total, eight food residues were chosen to extract natural pigments, as seen in Figure 1.

To identify the optimal solvent for extracting the natural pigments, ultrapure water and ethanol 25% and 96% were used for extraction. Ethanol 96% showed to be the most suitable solvent for the extraction of natural pigments from residues of moranga pumpkin (*Cucurbita maxima*), carrot (*Daucus carota* subsp. *sativus*), hybrid tetsukabuto pumpkin (*Cucurbita maxima* x *Cucurbita moschata*), chayote (*Sechium edule* Sw), and cabbage (*Brassica oleracea* L. *var. acephala*) from RGB color profiles. It is possible to observe that the choice of extractive solution directly influenced the intensities of red, green, and blue. For example, the use of 25% or 96% ethanol solvents compared to water resulted in natural pigments with more pronounced green intensities, which are extracted from the peel of pumpkin, sweet potato, carrot, chayote, and cabbage stalks (Figure 2).

Figure 1. Food residues. (**A**) Tetsukabuto hybrid pumpkin peels; (**B**) moranga pumpkin peels; (**C**) sweet potato peels; (**D**) red beet peels; (**E**) carrot peels; (**F**) chayote shells; (**G**) cabbage stalks; (**H**) purple cabbage external leaves.

On the other hand, ethanol 25% proved to be a more promising alternative compared to ultrapure water and ethanol 96% for extracting the natural pigment obtained from the external leaves of purple cabbage (*Brassica oleracea* L. *var. capitata f. rubra*). A similar result can be found in a study carried out by Song et al. (2011) [19], where 25% ethanol was considered the best solvent for obtaining natural pigment from purple cabbage, compared to water, acetone, petroleum ether, and ethanol in higher concentrations.

Furthermore, it is possible to observe that both ultrapure water and ethanol at concentrations of 25% were good extractive solvents for obtaining natural pigments from red beet peels (*Beta vulgaris* L.) due to the similarities in the average RGB color profile (Figure 2). Similar results can be found in studies carried out by Zin et al. (2020) [20] and Kushwana et al. (2017) [21], who demonstrated the efficiency of ethanol and water solvents in extracting natural pigments obtained from the peels and bagasse of beetroot, respectively. Finally, none of the solvents proved to be efficient in extracting natural pigment from the peels of sweet potatoes (*Ipomoea potatoes*).

Although all residues present phytopigments of varied hues and tonal scales, the external leaves of purple cabbage and peels of moranga pumpkin, tetsukabuto hybrid pumpkin, and beetroot were considered sources of the greatest potential to continue this study owing to obtaining more intense dyes at the end of the extraction process. These more intense dyes have a wider range of applications and can be used to create new colors. Additionally, intense colors are more stable over time under favorable conditions. In addition, it was decided to use ultrapure water and ethanol 25% to extract the natural pigments present in beetroot peels and external leaves of purple cabbage, respectively. Finally, ethanol 96% was the solvent chosen to extract the natural pigments present in the peels of moranga pumpkin and tetsukabuto hybrid pumpkin, varying the residue mass concentrations.

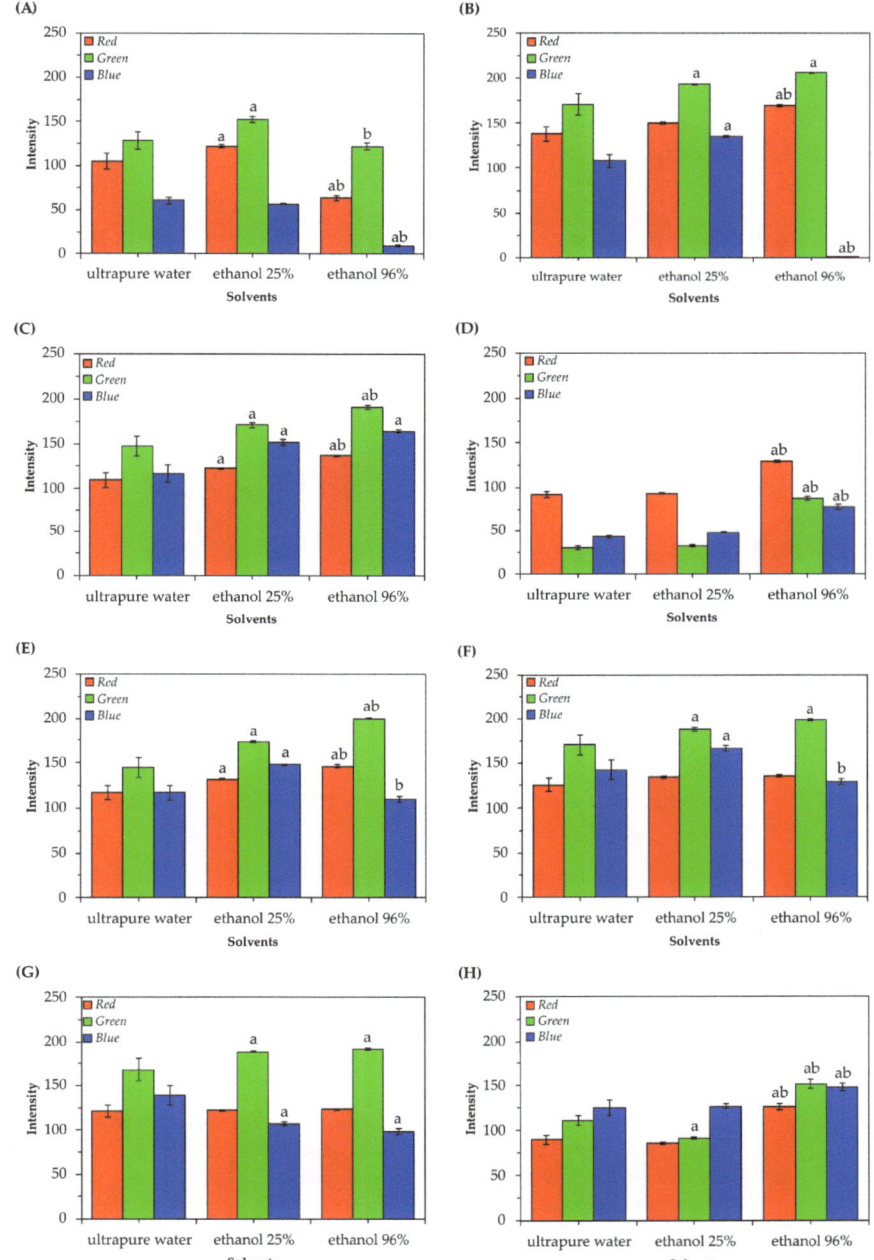

Figure 2. Average values (± standard deviation of the mean) of R′G′B′ of natural pigments obtained in different extractive solutions. (**A**) Tetsukabuto hybrid pumpkin peels; (**B**) moranga pumpkin peels; (**C**) sweet potato peels; (**D**) beetroot peels; (**E**) carrot peels; (**F**) chayote shells; (**G**) cabbage stalks; (**H**) purple cabbage external leaves. In each figure, columns with different letters indicate statistically significant differences at the 5% level by the Tukey Test, with "a" to water and "b" to ethanol 25% of the corresponding color.

3.2. Influence of Residue Mass Concentrations on Natural Pigment Extraction

Once the best food residues and their respective extractive solutions were identified, assays were carried out to verify the influence of the mass concentrations of residues in the extraction of natural pigments. It is possible to observe that the triplicates of the natural pigments presented different RGB color profiles, demonstrating that this variable affects the extraction process (Figure 3).

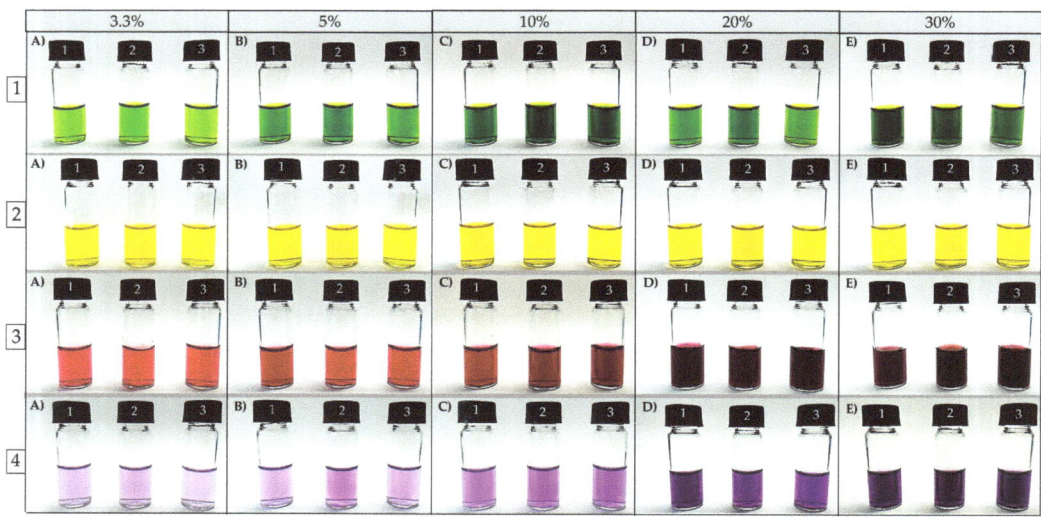

Figure 3. Natural pigments extracted from different concentrations of food residues (3.3%; 5%; 10%; 20; and 30%). (**1 A–E**) Pigments from tetsukabuto hybrid pumpkin peels; (**2 A–E**) pigments from moranga pumpkin peels; (**3 A–E**) pigments from beetroot peels; (**4 A–E**) pigments from purple cabbage external leaves.

As seen in Figure 4, it is possible to notice that triplicates of natural pigments extracted from peels of tetsukabuto hybrid pumpkin (*Cucurbita maxima x Cucurbita moschata*) present high variations in red intensity for concentrations of 10%, 20%, and 30% of residue. The triplicates of natural pigments extracted from moranga pumpkin (*Cucurbita maxima*) peels showed small variations in the red and green intensities at all concentrations studied; only an average variation was observed in the blue intensity between the triplicates at the concentrations of 3.3%, 20%, and 30%. Furthermore, it is possible to observe that natural pigments extracted from beetroot peels (*Beta vulgaris* L.) showed greater heterogeneity in color patterns between concentrations of 5%, 10%, 20%, and 30%. Finally, small variations were observed in the intensities of red, green, and blue between triplicates at concentrations of 3.3%, 5%, and 10% for pigments extracted from the external leaves of purple cabbage (*Brassica oleracea* L. *var. capitata f. rubra*). Considering the 30% residue concentration yielded more intense colorants, it was selected for the nanostructuring process.

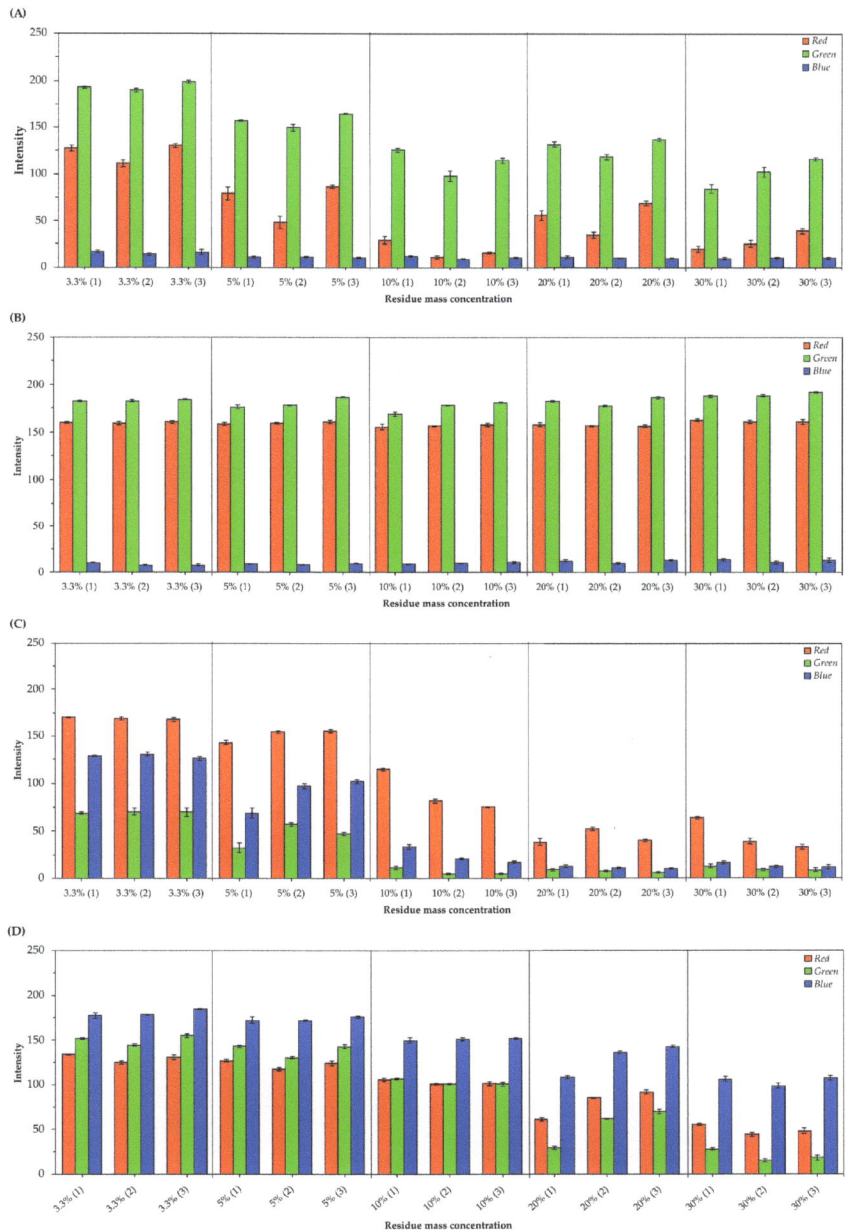

Figure 4. RGB average values (±standard deviation of the mean) of natural pigments triplicates extracted from different concentrations of waste (3.3%; 5%; 10%; 20%; and 30%). (**A**) Pigments obtained from tetsukabuto hybrid pumpkin peels; (**B**) pigments obtained from moranga pumpkin peels; (**C**) pigments obtained from beetroot peels; (**D**) pigments obtained from purple cabbage external leaves.

3.3. Hydrodynamic Diameter (HD), Polydispersity Index (PdI), and Zeta Potential (ZP) of Nanostructured Natural Pigments

The HD and PdI of the nanosystems were determined by DLS. This technique evaluates the size distribution (diameter) of particles and molecules in a liquid medium. In general, the movement of particles causes the incident light to be scattered with different intensities. It is from the analysis of these intensity fluctuations that the size of the particles is determined. The PdI offers information regarding the degree of sample dispersity, which can vary from 0 to 1. Values close to 1 indicate a highly polydisperse sample, while values close to 0 indicate homogeneity (monodisperse). The surface Zeta potential (ZP) of the particles was determined from their electrophoretic mobilities. The ZP values, represented in mV, indicate the colloidal stability of the suspension. ZP values greater than ±30 mV are indicative of colloidal dispersion results; ZP values lower than ±30 mV are indicative of colloidal instability [22]. A code designation was given to each nanostructure produced, as shown in Table 1.

Table 1. Nanosystems containing natural pigments and their respective designations.

Food Residue	Nanosystem	Sample Designation
Tetsukabuto hybrid pumpkin peels	Micelles	MPCA
Moranga pumpkin peels	Liposomes	LPCA
Beetroot peels	Liposomes	LPCB
Purple cabbage external leaves	Chitosan polymeric nanoparticles	NPQPFR

The nanosystems containing natural pigments presented different sizes (HD), varying between 194.3 ± 19.0 nm and 645.2 ± 37.4 nm (Table 2). The average HDs exhibited by liposomes (194.3 ± 19.0 nm for LPCA and 273.0 ± 37.6 nm for LPCB) suggest structures consisting of only a phospholipid bilayer with a large internal aqueous compartment. The MPCA has an average HD of 645.2 ± 37.4 nm, providing the formation of structures with different dimensions already reported in the literature for the encapsulation of natural pigments [23].

Table 2. Average HD, PdI, and ZP (±standard deviation of the mean) of the nanosystems containing natural pigments extracted from different food residues and their respective empty controls.

Sample	HD Z-Average (nm)	PdI	ZP (mV)
Empty micelle control	604.1 ± 329.7	0.520 ± 0.039	-7.5 ± 1.8
MPCA	645.2 ± 37.4	0.802 ± 0.076	-45.0 ± 0.7
Empty liposome control	255.8 ± 44.4	0.524 ± 0.132	-46.3 ± 1.0
LPCA	194.3 ± 19.0	0.752 ± 0.009	-63.6 ± 0.1
LPCB	273.0 ± 37.6	0.478 ± 0.043	-35.9 ± 0.9
Empty control of polymeric chitosan nanoparticles	1022.7 ± 66.9	0.861 ± 0.046	43.3 ± 8.3
NPQPFR	228.8 ± 40.7	0.482 ± 0.060	24.4 ± 1.0

Furthermore, size heterogeneity was observed for all nanosystems since the samples exhibited PdI $\geq 0.478 \pm 0.043$ (LPCB), as seen in Table 2. This high value may be caused by the presence of agglomerates or aggregates. An alternative to reducing this polydispersity is applying membrane separation methods, such as extrusion or high-pressure homogenization.

The nanosystems showed different colloidal stabilities. The NPQPFR showed incipient instability (24.4 ± 1.0 mV). In contrast, the LPCB, MPCA, and LPCA showed moderate (-35.9 ± 0.9 mV), good (-45.0 ± 0.7 mV), and excellent (-63.6 ± 0.1 mV) stability, respectively.

The MPCA and LPCA showed smaller average sizes and greater stability compared to their controls. When comparing the NPQPFR with its respective control, it is noted that the presence of the natural pigment obtained from the external leaves of purple cabbage enabled the formation of particles with smaller average sizes and homogeneous distribution. Finally, the presence of natural pigment extracted from beetroot peels resulted in the formation of liposomes with larger sizes and more monodispersity compared to their respective control. This increase in the nanosystem size can be explained by the fact that betalains (the main group of pigments present in beetroot peels) are relatively large molecules, which may explain the observed particulate growth.

3.4. Colorimetric Stability of Natural Pigments and Nanosystems

Significant variations ($p < 0.05$) were observed in the RGB profiles of natural pigments, as well as their respective nanosystems. However, a more pronounced RGB color variation was observed in the pigment from hybrid tetsukabuto pumpkin and beetroot peels (Figure 5A,C).

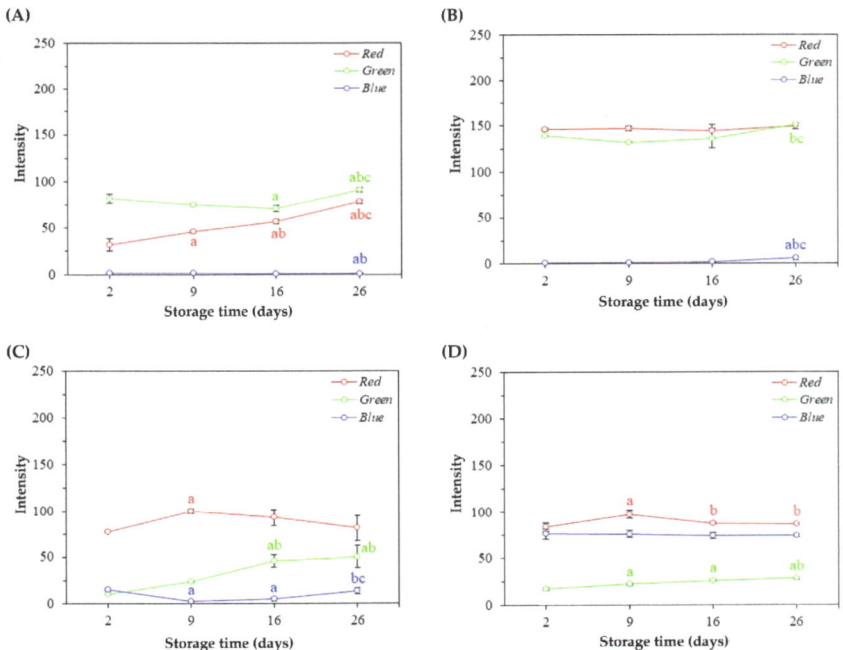

Figure 5. RGB average values (±standard deviation of the mean) of natural pigments during the storage time. (**A**) Pigments obtained from tetsukabuto hybrid pumpkin peels; (**B**) pigments obtained from moranga pumpkin peels; (**C**) pigments obtained from beetroot peels; (**D**) pigments obtained from purple cabbage external leaves. In each figure, different letters indicate statistically significant differences at the 5% level using the Tukey Test, with "a" in relation to 2 days, "b" in relation to 9 days, and "c" in relation to 16 days.

The natural pigment extracted from tetsukabuto pumpkin peels showed significant changes ($p < 0.05$) in red intensity over time, indicating that this color is unstable even under favorable storage conditions (Figure 5A). Significant variations in the LPCA colorimetric profile were also observed, as shown in Figure 6A. However, from the 16th day of storage, variations in RGB values are not significant, signaling the color stability of this nanosystem.

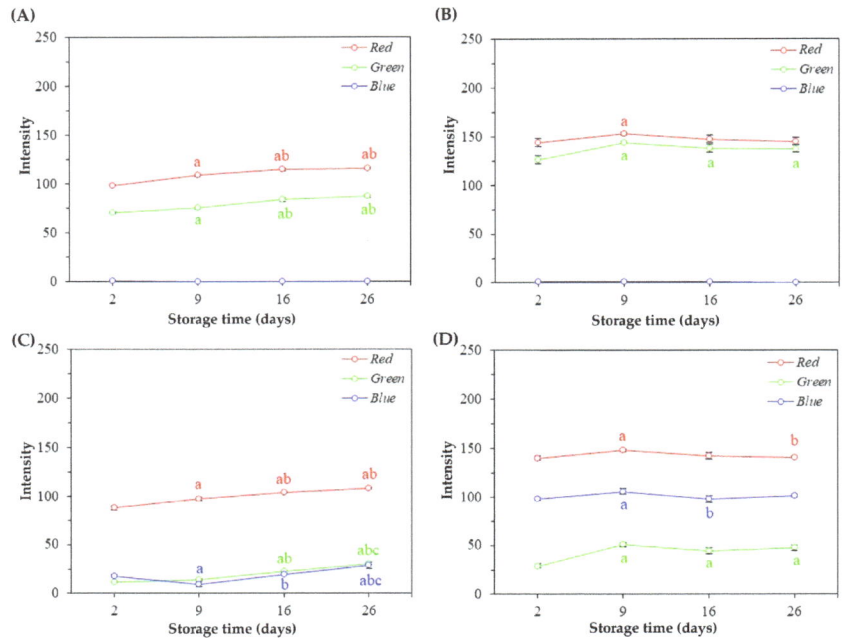

Figure 6. RGB average values (±standard deviation of the mean) of nanosystems during the storage time. (**A**) LPCA; (**B**) MPCA; (**C**) LPCB; (**D**) NPQPFR. In each figure, different letters indicate statistically significant differences at the 5% level using the Tukey Test, with "a" in relation to 2 days, "b" in relation to 9 days, and "c" in relation to 16 days.

Regarding the natural pigment obtained from moranga pumpkin peels, it is possible to observe a color difference only after the 26th day of storage (Figure 5B). About the micelles containing the pigment extracted from the same plant residue (MPCA), a significant difference was observed in the green color on the 9th, 16th, and 26th day of storage compared to the 2nd day, as well as in the red color on the 9th day compared to the 2nd day (Figure 6B). However, the RGB color profile of the nanostructured pigment did not show a significant difference between the 9th, 16th, and 26th days of storage, indicating colorimetric stability.

About the natural pigment extracted from beetroot peels, it is possible to notice significant variations in the green and blue intensities over time, signaling changes in the colorimetric profile, as shown in Figure 5C. In addition, the LPCB showed a significant variation in red, green, and blue intensities from the 9th day of storage, as seen in Figure 6C. Finally, no significant changes in blue intensity were observed in the natural pigment obtained from the purple cabbage's external leaves (Figure 5D). On the other hand, significant differences in the intensities of red and green are observed in the first few days. Similar results were found in the nanostructured pigment (NPQPFR), as no significant changes in RGB intensities were observed between the 16th and 26th days of storage, signaling color stability (Figure 6D).

4. Discussion

The food industry commonly uses artificial colorings in its products, seeking to ensure more attractive sensory characteristics. However, the demand for healthier alternatives is growing in the market, and an alternative to meet this demand is the use of natural dyes. The exploitation of food waste as a source of natural dyes opens doors for innovation in the food industry, promoting sustainability and the valorization of resources that would previously be discarded. However, to ensure the viability of using food waste as a source of natural dyes, it is necessary to consider the application of appropriate and eco-friendly extractive methods. In this study, the eco-friendly extraction method of natural pigments using green solvents was explored.

The ultrasonic bath treatment followed by magnetic stirring has proved to be a simple, fast, and efficient green technological method for extracting natural pigments. The ultrasound technique has been widely applied mainly for the extraction of bioactive compounds due to its several advantages, such as its short time of operation, accuracy, and non-destructive method [24]. Higher extraction rates are commonly obtained using this technology due to the formation of acoustic cavitations in liquid media, thus generating greater penetration of the solvent into the plant matrix and release of intracellular content [25]. In the study conducted by Sharma and Bhat [26], ultrasound-assisted and microwave-assisted extractions were performed to obtain carotenoids. These extraction methods were compared with the conventional extraction method, and ultrasound-assisted extraction proved to be a more efficient technique due to its higher yields. The combination of the ultrasound technique with other extraction methods has recently become a focus in studies aimed at obtaining natural dyes due to the higher yields achieved [27].

The color variations observed during the extraction of natural pigments can be explained by the chemical structure of the potential classes of pigments extracted: (i) anthocyanins, present in the external leaves of purple cabbage and sweet potato peels; (ii) betalains, present in beetroot peels; (iii) carotenoids, present in pumpkin and carrot peels; and (iv) chlorophylls, present in cabbage stalks, chayote peels, and tetsukabuto hybrid pumpkin. The polarity of these natural compounds and the interactions between these pigments with carbohydrates, proteins, and other components in foods can influence the extraction process [28]. In addition, variations in the concentration of secondary metabolites affect the pigment hues. The content of secondary metabolites produced by a plant depends on several factors, such as temperature, humidity, light, altitude, macro- and micronutrients present in the soil, atmospheric composition, seasonality, plant genome, and mechanical stimuli, among others [29]. A promising alternative to avoid this variability in the color profile is the use of screening tests and phytochemical analyses to identify the chemical constituents' values of interest. Furthermore, dilution and/or concentration processes can be adopted to achieve greater similarity in color between natural pigments.

Regarding the use of green solvents for the extraction of natural dyes, ethanol has proved to be an effective solvent, as it is possible to obtain attractive natural colorants from different food residues. In this way, it is possible to replace organic solvents with green alternatives to obtain dyes from natural sources, as can also be seen in the study by Li et al. (2013) [14].

The selection of the nanocarrier system for encapsulation was guided by the physicochemical properties of each dye. For instance, dyes extracted from moranga pumpkin peels are lipophilic and are thus frequently incorporated into micelles due to their affinity for hydrophobic environments. This affinity stems from the molecular structure of these compounds, which typically possess long non-polar hydrocarbon chains.

Micelles are aggregates of surfactant molecules that possess hydrophilic and hydrophobic moieties. The hydrophobic portions of the surfactant molecules cluster together at the core of the micelle, forming a hydrophobic environment [30]. The micelles produced in this study exhibit structures with different dimensions compared to those reported in the literature. This variation is influenced by some factors, including the chemical structure and concentration of the surfactant; the presence of counterion; and temperature variations during micelle formation [30].

The dye extracted from the external leaves of purple cabbage exhibits greater stability in an acidic environment. Therefore, chitosan polymeric nanoparticles emerge as a promising alternative to ensure this stability since chitosan, a natural biopolymer, is easily solubilized at acidic pH [31]. Among the methods for developing chitosan nanoparticles, ionic gelation stands out. This technique relies on the electrostatic interaction between the positively charged amino groups of chitosan and the negatively charged groups of a polyanionic substance, such as sodium tripolyphosphate (TPP), under acidic pH [31]. These nanoparticles can encapsulate and protect the dyes, reducing their degradation and increasing their resistance to photodegradation and other adverse conditions.

Dyes extracted from beetroot peels and tetsukabuto hybrid pumpkin peels are highly hydrophilic; therefore, a strategy for improving the stability of these compounds is their incorporation into liposomes. Liposomes are small, spherical vesicular structures formed by one or more phospholipid bilayers [32]. These nanosystems are composed of a non-polar tail consisting of fatty acid chains (hydrophobic region) and a polar head (hydrophilic region) formed by a phosphate group and a base (choline, glycerol, etc.). In the presence of water, these structures align, originating one or more concentric lipid bilayers, which are separated by an aqueous compartment [33]. Due to these characteristics, liposomes carry hydrophilic or lipophilic compounds.

Among the three nanotechnological strategies explored to enhance the stability of natural pigments, all nanosystems exhibited promising results. Compared to their unencapsulated counterparts, the pigments incorporated into these nanosystems displayed less color variation, indicating successful protection, particularly for pumpkin-derived pigments. The observed color changes in the LPCB and NPQPFR nanosystems can potentially be attributed to their high water content. Since these pigments are susceptible to nucleophilic attack by water molecules, the presence of water within the nanosystems could lead to alterations in color intensity [34]. Therefore, the application of water removal strategies, such as freeze-drying and spray-drying, could be valuable tools to further improve colorimetric stability [35].

5. Conclusions

Food residues (e.g., peels, stalks, and leaves) are promising sources to obtain natural pigments with varied color profiles. The extraction method is a simple and economical technique, proving to be a technological process with potential application on an industrial scale for obtaining natural pigments from different sources. In addition, the hydroethanolic extraction was very effective when performed with an ultrasound bath followed by magnetic stirring. This green solvent is an alternative to the use of organic solvents since there is a minimization of the negative impacts on both the environment and human health.

It is worth highlighting that studies of physicochemical characteristics and toxicity tests need to be explored to evaluate whether these natural colorants could be implemented in the food industrial sector at the stages of development, processing, and storage of foods, as well as in other industrial sectors, such as textiles, cosmetics, and inks, among others.

Furthermore, liposomes, micelles, and chitosan nanoparticles offer a compelling solution to overcome the challenge of rapid degradation observed with natural pigments during storage. Liposomes, formed by phospholipid bilayers, can encapsulate pigments, enhancing their stability and controlled release. Micelles also provide another encapsulation strategy, improving the water solubility and bioavailability of natural dyes. In addition, chitosan nanoparticles offer an attractive option to encapsulate stable dyes in an acidic environment. Their cationic nature allows for electrostatic interactions with the often negatively charged pigments, promoting efficient encapsulation and improved stability.

This research opens avenues for the further optimization of natural pigment extraction techniques, paving the way to explore new sources of natural dyes and aiming to meet the current market demand for more sustainable products.

Author Contributions: Conceptualization, V.B.M.L. and L.P.S.; methodology, V.B.M.L.; formal analysis, V.B.M.L. and L.P.S.; investigation, V.B.M.L. and Luciano Paulino.; resources, L.P.S.; data curation, L.P.S.; writing—original draft preparation, V.B.M.L. and L.P.S.; writing—review and editing, L.P.S.; visualization, L.P.S.; supervision, L.P.S.; project administration, L.P.S.; funding acquisition, L.P.S. All authors have read and agreed to the published version of the manuscript.

Funding: This study was funding in part by the Coordenação de Aperfeiçoamento de Pessoal de Nível Superior—Brazil (CAPES)—Finance Code 001 and 23038.019088/2009-58; FAP-DF (193.001.392/2016); CNPq (311825/2021-4, 307853/2018-7, 408857/2016-1, 306413/2014-0, and 563802/2010-3); and Embrapa (23.17.00.069.00.02, 13.17.00.037.00.00, 21.14.03.001.03.05, 13.14.03.010.00.02, 12.16.04.010.00.06, 22.16.05.016.00.04, and 11.13.06.001.06.03).

Institutional Review Board Statement: Not applicable.

Informed Consent Statement: Not applicable.

Data Availability Statement: The data presented in this study are available on request from the corresponding author due to intellectual property subjects.

Conflicts of Interest: Victoria Baggi Mendonça Lauria developed her master's degree at the Universidade de Brasília in partnership with the company Embrapa Recursos Genéticos e Bictecnologia. She participated in the conceptualization, methodology, formal analysis, investigation, and writing—original draft preparation. Luciano Paulino Silva is an employee of the company Embrapa Recursos Genéticos e Biotecnologia and is a collaborating professor at the Universidade de Brasília. He participated in the conceptualization, formal analysis, investigation, resources, data curation, writing—original draft preparation, writing—review and editing, visualization, supervision, project administration, and funding acquisition. The authors declare that the research was conducted in the absence of any commercial or financial relationships that could be construed as potential conflicts of interest. The authors declare that this study received funding from Coordenação de Aperfeiçoamento de Pessoal de Nível Superior—Brazil (CAPES)—Finance Code 001 and 23038.019088/2009-58; FAP-DF (193.001.392/2016); CNPq (311825/2021-4, 307853/2018-7, 408857/2016-1, 306413/2014-0, and 563802/2010-3); and Embrapa (23.17.00.069.00.02, 13.17.00.037.00.00, 21.14.03.001.03.05, 13.14.03.010.00.02, 12.16.04.010.00.06, 22.16.05.016.00.04, and 11.13.06.001.06.03). The funders were not involved in the study design, collection, analysis, interpretation of data, the writing of this article, or the decision to submit it for publication. The involvement of the funders had no impact on the objectivity and authenticity of the study.

References

1. Clydesdale, F.M. Color perception and food quality. *J. Food Qual.* **1991**, *14*, 61–74. [CrossRef]
2. Bobbio, P.A.; Bobbio, F.O. *Química do Processamento de Alimentos*, 3rd ed.; Varela: São Paulo, Brazil, 1992.
3. Constant, P.B.L.; Stringheta, P.C.; Sandi, D. Corantes alimentícios. *Bol. Cent. Pesqui. Process. Aliment.* **2002**, *20*, 203–220. [CrossRef]
4. Ibraheen, N.A.; Hasan, M.M.; Khan, R.Z.; Mishra, P.K. Understanding color models: A review. *J. Sci. Technol.* **2012**, *2*, 265–275.
5. Carocho, M.; Barreiro, M.F.; Morales, P.; Ferreira, I.C.F.R. Adding molecules to food, pros and cons: A review of synthetic and natural food additives. *Compr. Rev. Food Sci. Food Saf.* **2014**, *13*, 377–399. [CrossRef]
6. Ramesh, M.; Muthuraman, A. Flavoring and coloring agents: Health risks and potential problems. In *Natural and Artificial Flavoring Agents and Food Dyes*; Grumezescu, A.M., Holban, A.M., Eds.; Academic Press: Waltham, MA, USA, 2018; pp. 1–28.
7. Delgado-Vargas, F.; Jiménez, A.R.; Paredes-López, O. Natural pigments: Carotenoids, anthocyanins, and betalains—Characteristics, biosynthesis, processing and stability. *Crit. Rev. Food Sci. Nutr.* **2000**, *40*, 173–189. [CrossRef]
8. Delgado-Vargas, F.; Paredes-López, O.P. *Natural Colorants for Food and Nutraceutical Uses*, 1st ed.; CRC Press: New York, NY, USA, 2002.
9. He, J.; Giusti, M.M. Anthocyanins: Natural colorants with health-promoting properties. *Food. Sci. Technol.* **2010**, *1*, 163–187. [CrossRef]
10. Baiano, A. Recovery of biomolecules from food wastes—A review. *Molecules* **2014**, *19*, 14821–14842. [CrossRef]
11. Hall, K.D.; Guo, J.; Dore, M.; Chow, C.C. The progressive increase of food waste in America and its environmental impact. *PLoS ONE* **2009**, *4*, e7940. [CrossRef]
12. Clark, J.H.; Macquarrie, D. *Handbook of Green Chemistry and Technology*, 1st ed.; Blackwell Science: London, UK, 2002.
13. Li, Y.; Fabiano-Tixier, A.S.F.; Tomao, V.; Cravotto, G.; Chemat, F. Green ultrasound-assisted extraction of carotenoids based on the bio-refinery concept using sunflower oil as an alternative solvent. *Ultrason. Sonochem.* **2013**, *20*, 12–18. [CrossRef]
14. Cardoso-Ugarte, G.A.; Sosa-Morales, M.E.; Ballard, T.; Liceaga, A.; San Martín-González, M.F. Microwave-assisted extraction of betalains from red beet (*Beta vulgaris*). *LWT-Food Sci. Technol.* **2014**, *59*, 276–282. [CrossRef]
15. Ravanfar, R.; Tamaddon, A.M.; Niakousari, M.; Moein, M.R. Preservation of anthocyanins in solid lipid nanoparticles: Optimization of a microemulsion dilution method using the Placket–Burman and Box–Behnken designs. *Food Chem.* **2016**, *199*, 573–580. [CrossRef]
16. Binns, C. *Introduction to Nanoscience and Nanotechnology*, 2nd ed.; John Wiley & Sons: Hoboken, NJ, USA, 2021; pp. 13–32.
17. Ghorbanzade, T.; Jafari, S.M.; Akhavan, S.; Hadavi, R. Nano-encapsulation of fish oil in nano-liposomes and its application in fortification of yogurt. *Food Chem.* **2017**, *216*, 146–152. [CrossRef]
18. Sarwar, M.S.; Niazi, M.B.K.; Jahan, Z.; Ahmad, T.; Hussain, A. Preparation and characterization of PVA/nanocellulose/Ag nanocomposite films for antimicrobial food packaging. *Carbohydr. Polym.* **2018**, *184*, 453–464. [CrossRef]

19. Song, X.Q.; Ye, L.; Yang, X.B. Comparative study of three methods for extraction of purple cabbage pigment. *Food Sci.* **2011**, *32*, 74–77.
20. Zin, M.M.; Márki, E.; Bánvölgyi, S. Conventional extraction of betalain compounds from beetroot peels with aqueous ethanol solvent. *Acta Aliment.* **2020**, *49*, 163–169. [CrossRef]
21. Kushwaha, R.; Kumar, V.; Vyas, G.; Kaur, J. Optimization of different variable for eco-friendly extraction of betalains and phytochemicals from beetroot pomace. *Waste Biomass Valorization* **2018**, *9*, 1485–1494. [CrossRef]
22. *ASTM Standard D4187, 1982*; Zeta Potential of Colloids in Water and Waste Water. American Society for Testing and Materials: West Conshohocken, PA, USA, 1985; pp. 4182–4187.
23. De Paz, E.; Martín, A.; Mateos, E.; Cocero, M.J. Production of water-soluble β-carotene micellar formulations by novel emulsion techniques. *Chem. Eng. Process. Process Intensif.* **2013**, *74*, 90–96. [CrossRef]
24. McClements, D.J. Advances in the application of ultrasound in food analysis and processing. *Trends Food Sci. Technol.* **1995**, *6*, 293–299. [CrossRef]
25. Corbin, C.; Fidel, T.; Leclerc, E.A.; Barakzoy, E.; Sagot, N.; Falguiéres, A.; Hano, C. Development and validation of an efficient ultrasound assisted extraction of phenolic compounds from flax (*Linum usitatissimum* L.) seeds. *Ultrason. Sonochem.* **2015**, *26*, 176–185. [CrossRef]
26. Sharma, M.; Bhat, R. Extraction of carotenoids from pumpkin peel and pulp: Comparison between innovative green extraction technologies (ultrasonic and microwave-assisted extractions using corn oil). *Foods* **2021**, *10*, 787. [CrossRef]
27. Wizi, J.; Wang, L.; Hou, X.; Tao, Y.; Ma, B.; Yang, Y. Ultrasound-microwave assisted extraction of natural colorants from sorghum husk with different solvents. *Ind. Crop Prod.* **2018**, *120*, 203–213. [CrossRef]
28. Andreo, D.; Jorge, N. Antioxidantes naturais: Técnicas de extração. *Bol. Cent. Pesqui. Process. Aliment* **2006**, *24*, 319–336. [CrossRef]
29. Gobbo-Neto, L.; Lopes, N.P. Plantas medicinais: Fatores de influência no conteúdo de metabólitos secundários. *Química Nova* **2007**, *30*, 374–381. [CrossRef]
30. Wennerström, H.; Lindman, B. Micelles. Physical chemistry of surfactant association. *Phys. Rep.* **1979**, *52*, 3–32. [CrossRef]
31. Sabliov, C.M.; Astete, C.E. Polymeric nanoparticles for food applications. In *Nanotechnology and Functional Foods: Effective Delivery of Bioactive Ingredients*; John Wiley & Sons: Hoboken, NJ, USA, 2015; pp. 272–296.
32. Akbarzadeh, A.; Rezaei-Sadabady, R.; Davaran, S.; Joo, S.W.; Zarghami, N.; Hanifehpour, Y.; Nejati-Koshki, K. Liposome: Classification, preparation, and applications. *Nanoscale Res. Lett.* **2013**, *8*, 102. [CrossRef]
33. Shailesh, S.; Neelam, S.; Sandeep, K.; Gupta, G.D. Liposomes: A review. *J. Pharm. Res.* **2009**, *2*, 1163–1167.
34. Schiozer, A.L.; Barata, L.E.S. Estabilidade de corantes e pigmentos de origem vegetal. *Rev. Fitos* **2013**, *2*, 6–24. [CrossRef]
35. De Souza, V.B.; Thomazini, M.; de Carvalho Balieiro, J.C.; Fávaro-Trindade, C.S. Effect of spray drying on the physicochemical properties and color stability of the powdered pigment obtained from vinification byproducts of the Bordo grape (*Vitis labrusca*). *Food Bioprod. Process.* **2015**, *93*, 39–50. [CrossRef]

Disclaimer/Publisher's Note: The statements, opinions and data contained in all publications are solely those of the individual author(s) and contributor(s) and not of MDPI and/or the editor(s). MDPI and/or the editor(s) disclaim responsibility for any injury to people or property resulting from any ideas, methods, instructions or products referred to in the content.

Article

Broccoli, Amaranth, and Red Beet Microgreen Juices: The Influence of Cold-Pressing on the Phytochemical Composition and the Antioxidant and Sensory Properties

Spasoje D. Belošević [1,†], Danijel D. Milinčić [2,†], Uroš M. Gašić [3], Aleksandar Ž. Kostić [2], Ana S. Salević-Jelić [1], Jovana M. Marković [1], Verica B. Đorđević [4], Steva M. Lević [1], Mirjana B. Pešić [2,*] and Viktor A. Nedović [1,*]

[1] Food Biotechnology Laboratory, Department of Food Technology and Biochemistry, Faculty of Agriculture, University of Belgrade, Nemanjina 6, 11080 Belgrade, Serbia; sbelosevic@agrif.bg.ac.rs (S.D.B.); ana.salevic@agrif.bg.ac.rs (A.S.S.-J.); jovana.markovic@agrif.bg.ac.rs (J.M.M.); slevic@agrif.bg.ac.rs (S.M.L.)

[2] Food Chemistry and Biochemistry Laboratory, Department of Food Technology and Biochemistry, Faculty of Agriculture, University of Belgrade, Nemanjina 6, 11080 Belgrade, Serbia; danijel.milincic@agrif.bg.ac.rs (D.D.M.); akostic@agrif.bg.ac.rs (A.Ž.K.)

[3] Department of Plant Physiology, Institute for Biological Research Siniša Stanković-National Institute of Serbia, University of Belgrade, Bulevar Despota Stefana 142, 11060 Belgrade, Serbia; uros.gasic@ibiss.bg.ac.rs

[4] Department of Chemical Engineering, Faculty of Technology and Metallurgy, University of Belgrade, Karnegijeva 4, 11000 Belgrade, Serbia; vmanojlovic@tmf.bg.ac.rs

* Correspondence: mpesic@agrif.bg.ac.rs (M.B.P.); vnedovic@agrif.bg.ac.rs (V.A.N.); Tel.: +381-11-441-3315 (M.B.P.); +381-441-3154 (V.A.N.)

† These authors equally contributed to this work.

Citation: Belošević, S.D.; Milinčić, D.D.; Gašić, U.M.; Kostić, A.Ž.; Salević-Jelić, A.S.; Marković, J.M.; Đorđević, V.B.; Lević, S.M.; Pešić, M.B.; Nedović, V.A. Broccoli, Amaranth, and Red Beet Microgreen Juices: The Influence of Cold-Pressing on the Phytochemical Composition and the Antioxidant and Sensory Properties. *Foods* **2024**, *13*, 757. https://doi.org/10.3390/foods13050757

Academic Editors: Gianluca Nardone, Rosaria Viscecchia and Francesco Bimbo

Received: 25 January 2024
Revised: 19 February 2024
Accepted: 27 February 2024
Published: 29 February 2024

Copyright: © 2024 by the authors. Licensee MDPI, Basel, Switzerland. This article is an open access article distributed under the terms and conditions of the Creative Commons Attribution (CC BY) license (https://creativecommons.org/licenses/by/4.0/).

Abstract: The aim of this study was to analyze in detail the phytochemical composition of amaranth (AMJ), red beet (RBJ), and broccoli (BCJ) microgreens and cold-pressed juices and to evaluate the antioxidant and sensory properties of the juices. The results showed the presence of various phenolic compounds in all samples, namely betalains in amaranth and red beet microgreens, while glucosinolates were only detected in broccoli microgreens. Phenolic acids and derivatives dominated in amaranth and broccoli microgreens, while apigenin *C*-glycosides were most abundant in red beet microgreens. Cold-pressing of microgreens into juice significantly altered the profiles of bioactive compounds. Various isothiocyanates were detected in BCJ, while more phenolic acid aglycones and their derivatives with organic acids (quinic acid and malic acid) were identified in all juices. Microgreen juices exhibited good antioxidant properties, especially ABTS$^{\bullet+}$ scavenging activity and ferric reducing antioxidant power. Microgreen juices had mild acidity, low sugar content, and good sensory acceptability and quality with the typical flavors of the respective microgreen species. Cold-pressed microgreen juices from AMJ, RBJ, and BCJ represent a rich source of bioactive compounds and can be characterized as novel functional products.

Keywords: broccoli microgreens; amaranth microgreens; red beet microgreens; microgreen juices; antioxidant activity; apigenin *C*-glycosides

1. Introduction

Microgreens are recognized as new crops and potential foods of the future [1]. They represent a novel and promising source of highly valuable bioactive compounds with health-promoting effects [2–6]. The most commonly grown and studied microgreens are from the Brassicaceae and Amaranthaceae families with crops such as broccoli, cabbage, kale, argula, red beet, chard, amaranth, etc. [1]. So far, the aforementioned microgreen species have been mostly consumed in raw form or as culinary ingredients in dishes due to their high content of bioactive compounds and specific flavor [7]. Previous studies have shown that broccoli, amaranth and red beet microgreens are high in bioactive compounds such as vitamins, glucosinolates, isothiocyanates, phenolic compounds and betalains and have good antioxidant properties [4,8]. However, there are few studies that provide insight into the biocompound profiles of these microgreens and their correlation with antioxidant properties [9–14]. In addition, microgreens can be successfully used for healthy beverage production [15–18] or incorporated into various bakery/confectionary products [19]. Some studies have preliminarily analyzed microgreen juices from broccoli [15], Alternanthera sessilis [16], and wheatgrass [17], as well as functional microgreen/fruit juices [20]. However, in general, microgreen juices have only become attractive in recent years and have not been extensively studied. Various processed, treated, or stored wheatgrass and wheat sprout juices have shown good scavenging activity of $ABTS^+$, DPPH, and oxygen radicals [17,18,21]. These antioxidant assays have not yet been performed on juices from other microgreens. Recent and rare in vitro and in vivo studies conducted with cold-pressed broccoli microgreen and sprout juices have shown that the juices may have anticancer [22] and antiobesity effects [15] and protective properties against oxidative stress-related diseases [23]. However, considering previous research on microgreen juices [15], their health effects, and the lack of their bioactive compounds profile, further research is thus needed. On the other hand, microgreen juices from amaranth and red beet have not been analyzed to our knowledge. Finally, sensory analysis is often a key parameter for the acceptance of new products based on microgreens, mainly due to their specific flavor attributes. Tests for consumers' sensory perception and acceptance of microgreens have been conducted most frequently [24–28], while microgreen juices have hardly been tested. To date, only microgreen juices of Alternanthera sessilis and Brassica juncea have been the subject of sensory evaluation [16], showing good overall acceptability, using the hedonic test. To our knowledge, sensory evaluations have not yet been performed on the juices from broccoli, amaranth, and red beet microgreens. However, these microgreen varieties have unique sensory attributes such as astringency, bitterness, and sourness which most likely contribute to the overall acceptability of their products by consumers [24,25].

In view of the aforementioned, the aim of this study was to prepare cold-pressed juices from broccoli, amaranth, and red beet microgreens and to analyze in detail their phytochemical composition, antioxidant properties, and sensory acceptability. In addition, the profiles of bioactive compounds of raw broccoli, amaranth, and red beet microgreens were also analyzed to better follow the migration of individual glucosinolates, phenolic compounds, and betalains from the microgreens to juices and to explain their transformation during the juices' production process.

2. Materials and Methods

2.1. Microgreen Sample

Samples of broccoli (*Brassica oleracea* var. *italica*), red beet (*Beta vulgaris*), and amaranth (*Amaranthus tricolour* L.) microgreens were obtained from a local company (Plantica) from Belgrade, Serbia. Briefly, the microgreens were grown in a controlled environment, including vertical cultivation in the growing channels. They were grown under artificial light and at room temperature (20 °C). The humidity in the room (85%) and the air temperature was ensured by fans. The year in which the microgreens of broccoli, amaranth, and red beet were produced was 2023. All microgreens used in this study were harvested 12 days after germination and when the first pair of true leaves and the fully expanded embryonic leaves (cotyledons) had developed.

2.2. Preparation of Cold-Pressed Microgreen Juices

Selected microgreens were cut with scissors a few centimeters above the ground, then weighed and washed to remove impurities. The cleaned microgreens were pressed using a super-slow juicer (Angel juicer 8500, Angel Co., Ltd., Busan, Republic of Korea), and the obtained juices from broccoli (BCJ), amaranth (AMJ), and red beet (RBJ) (cold-pressed microgreen juices) were collected in plastic flasks (Figure 1). Part of the prepared juices was stored in the refrigerator for sensory evaluation. The other part of the squeezed juices was centrifuged at 9000× g rpm for 12 min to remove solid fractions, and the collected supernatants were stored at −20 °C for further spectrophotometric and chromatographic analyses.

Figure 1. Schematic representation of cold-pressed juices preparation from amaranth, broccoli, and red beet microgreens.

2.3. Preparation of Microgreens for Chromatographic Analysis

The microgreen species were cut and finely grounded using liquid nitrogen. These microgreen powders were extracted with 80% methanol (+0.1% HCl) (1:10 w/v) for 1 h with constant stirring on a mechanical shaker (Thys 2, MLW Labortechnik GmbH, Seelbach, Germany) [29]. After that, the samples were centrifuged at 4000× g for 10 min. Collected supernatants of broccoli (BC), amaranth (AM), and red beet (RB) microgreens were filtered through 0.22 µm filters and used for further characterization of their bioactive compounds by UHPLC Q-ToF MS. This extraction solvent was most commonly used for the extraction of bioactive compounds from plant materials [12,14,29–32], while Pintać et al. [30] showed that this solvent provided the highest yield of phenolic compounds. For additional characterization of highly sensitive glucosinolates (GLSs), the broccoli sample was extracted with boiled 70% methanol for 1 h on a thermoshaker (70 °C), because the activity of the enzyme myrosinase was inhibited in boiled methanol [10]. The obtained supernatant (BC1) was filtered and used for additional characterization of GLSs.

2.4. Preparation of Cold-Pressed Microgreen Juices for Chromatographic Analysis

Microgreen juices were passed through an SPE cartridge (CLEAN-UP[R], C18 Extraction columns, Unendcapped-PKG50, UCT, Bristol, UK) before chromatographic analysis to remove sugars and other colloidal impurities. The SPE cartridge was conditioned by washing with 5 mL of acidified methanol (methanol containing 0.1% HCl) and milliQ water, respectively. After that, the samples were passed through the cartridge and washed with 5 mL of milliQ water. Adsorbed bioactive compounds were eluted with 1 mL of acidified methanol (methanol containing 0.1% HCl), filtered through 0.45 µm syringe filters, and analyzed by UHPLC Q-ToF MS.

2.5. UHPLC Q-ToF MS of Microgreens and Cold-Pressed Microgreen Juices

The phytochemical profiles of the microgreen extracts and prepared juices were analyzed using the Agilent 1290 Infinity ultra-high-performance liquid chromatography (UHPLC) system coupled with quadrupole time-of-flight mass spectrometry (6530C Q-ToF-MS) (Agilent Technologies, Inc., Santa Clara, CA, USA), according to the method described in detail previously by Kostić and Milinčić [33]. The QToF-MS system was equipped with a dual Agilent Jet Stream electrospray ionization (ESI) source that operated in both positive

(ESI$^+$) and negative (ESI$^-$) ionization modes. The operating parameters for ESI were the same as previously reported by Kostić and Milinčić [33]. Agilent Mass Hunter software ver. 10.0 was used for instrument control, data acquisition, and analysis.

Individual glucosinolates, phenolic compounds, and betalains were identified based on their monoisotopic mass and MS fragmentation. In addition, data already published in the literature were also used for the identification of glucosinolates [9,34–38], phenolic compounds [10,31,39,40], and betalains [41–44]. Phenolic compounds were quantified by direct comparison with available standards. However, as specific phenolic derivatives were detected for which no specific standards exist, the amounts of the individual phenolic derivatives were quantified using available standards (sinapic acid for phenolic acid derivatives and apigenin for flavonoid derivatives), and expressed as mg/100 g fresh weight of microgreens (FW) or mg/100 mL juice. Table S1 shows a phenolic compounds used for quantification, together with their equation parameters and correlation coefficient (r^2). The relative content of individual betalains in amaranth and red beet microgreens and juices (%) was calculated as the ratio of the area of each individual compound and the area of total compounds detected. The exact masses of the compounds were calculated using ChemDraw software (version 12.0, CambridgeSoft, Cambridge, MA, USA).

2.6. Proximate Compositions of Cold-Pressed Microgreen Juices

The pH of the microgreen juices was determined with a digital pH meter, while the total soluble solids (°Bx) were measured with a refractometer (ATC 0–32 Brix, Huixia Supply Co., Ltd., Fuzhou, China). The dry weight of the juices was determined gravimetrically by drying the samples at 105 °C to constant mass. Juice yield (%) was calculated as the ratio between the mass of obtained juice ($m1$) and the mass of fresh microgreens ($m2$) and calculated according to the following equation:

$$Yield\ of\ juice\ (\%) = \frac{m1}{m2} \times 100 \qquad (1)$$

2.7. Total Phenolics, Flavonoids, Betalains, and Chlorophyll Content in Microgreen Juices

The total phenolic content (TPC) and flavonoid content (TFC) of the microgreen juices were determined using the Folin–Ciocalteu colorimetric assay and the aluminum chloride assay [29]. Briefly, TPC was determined after reacting diluted juices (0.5 mL) with Folin–Ciocalteu reagent (2.5 mL) and 7.5% Na_2CO_3 (2.5 mL). For TFC, the juices (2 mL) were mixed and incubated with 5% $NaNO_2$ (0.15 mL), 10% $AlCl_3$ (0.15 mL), 1 M NaOH (1 mL), and milliQwater (1.2 mL), respectively. After incubation, the absorbance of the mixtures was measured at 760 nm (TPC) and 510 nm (TFC), using a UV-Vis spectrophotometer (model HALO DB-20S, Dynamica Scientific Ltd., Livingston, UK). Results for TPC and TFC were expressed as mg of gallic acid (mgGAE/100 mL) and quercetin (mgQE/100 mL) equivalents per 100 mL juice.

Total betacyanins and betaxanthins were determined according to the previously described method of Stintzing and Schieber [45]. Total betalains are the sum of total betacyanins and betaxanthins. The absorbance of the appropriately diluted microgreen juices was measured at 485 nm, 536 nm, and 650 nm. The results were expressed in mg per 100 mL juice and calculated according to the following equation:

$$Betaxanthins(Betacyanins)\left(\frac{mg}{100mL}\right) = \frac{A \times DF \times MW \times 100}{\varepsilon \times i} \qquad (2)$$

where A = $A_{485} - A_{650}$ (betaxanthins) and A = $A_{536} - A_{650}$ (betacyanins). DF—dilution factor; MW—molecular weight (339 g/mol for betaxanthins and 550 g/mol for betacyanins); ε—molar extinction (48,000 for betaxanthins and 60,000 for betacyanins); i—the path length (cm).

Total chlorophyll a and chlorophyll b were calculated using Equations (2) and (3), respectively, as previously described by Ali and Popović [17]. Briefly, the microgreen juices were appropriately diluted with 80% acetone, while the absorbance of the prepared samples was measured at two wavelengths (645 nm and 663 nm).

$$Chlorophyll\ a\ \left(\frac{mg}{100mL}\right) = (12.71 \times A_{663} - 2.59 \times A_{645}) \times DF/10 \quad (3)$$

$$Chlorophyll\ b\ \left(\frac{mg}{100mL}\right) = (12.71 \times A_{645} - 2.59 \times A_{663}) \times DF/10 \quad (4)$$

where DF is the dilution factor, and A refers to the absorbances recorded at 645 nm and 663 nm. The results were expressed as mg per 100 mL of juice.

2.8. Antioxidant Properties of Cold-Pressed Microgreen Juices

The antioxidant properties of the microgreen juices were analyzed using three assays: DPPH radical scavenging activity assay (DPPH• scavenging activity), ABTS radical cation scavenging activity assay (ABTS•+ scavenging activity), and ferric reducing antioxidant power assay (FRAP assay) [29,46]. Briefly, diluted juice (0.1 mL and 0.03 mL) was mixed with 1.9 mL of DPPH• and 3 mL of ABTS•+ working solution, respectively. After incubation of the two reaction mixtures, the absorbance of the samples was measured at 515 nm (DPPH•) and 734 nm (ABTS•+). For the FRAP assay, the juice (0.1 mL) was mixed with 0.3 mL of milliQ water and 3 mL of FRAP reagents. The mixture was then incubated (37 °C, 40 min), and absorbance was recorded at 593 nm. Trolox was used as the standard for all antioxidant assays, and results were expressed as mg Trolox equivalent (TE) per 100 mL juice.

2.9. Sensory Properties of Cold-Pressed Microgreen Juices

Sensory evaluation (overall quality and consumer acceptance) of the microgreen juices was performed by trained evaluators/selected consumers at the Faculty of Agriculture, University of Belgrade. In addition, the sensory testing conditions were conducted in a controlled environment in accordance with ISO standards for sensory analysis, including ISO 8589:2007 [47] and ISO 11136:2014 [48]. The samples of microgreen juices were prepared as follows: (1) the selected microgreens were cut, weighed, washed, and cold-pressed using a super-slow juicer (Angel Juicer 8500, Angel Co., Ltd., Busan, Republic of Korea), (2) the obtained cold-pressed juices of microgreens (broccoli, amaranth, and red beet) were placed and served in a transparent glass to ensure good transparency of microgreen juices and coded with random three-digit numbers in an amount of 20 mL. The microgreen juices were tasted in the sensory room under the known conditions, such as controlled ventilation, with white light and separated from the sample preparation room and away from inappropriate odors and noise. Tap water and unsalted crackers were used as taste neutralizers after tasting each sample. The temperature of the juices was 8 °C, as for commercially available juices. The conditions of the sensory tests ensured that each evaluator/consumer could objectively award an appropriate score. All sensory tests with participants were conducted in accordance with the Code of Professional Ethics of the University of Belgrade [49]. Before sensory evaluation, all participants gave informed consent via the statement that they were aware that their responses were confidential, they agreed to participate in this study, their responses could be used, they could withdraw from the study at any time, and that there would be no release of participant data without their knowledge. The products tested were safe for consumption.

2.9.1. Overall Quality Evaluation

The overall quality of the microgreen juices was determined using the quality assessment method, taking into account the following quality criteria: appearance, odor, texture, and flavor. The quality of the juices was assessed using category scales ranging from 0 (unsatisfactory quality) to 5 (excellent quality). This evaluation was performed using a 5-level quality scoring system described in [50].

As the sensory attributes have different effects on the overall quality of the juices, the following importance coefficients (CI) were used: 1 (appearance), 6 (odor), 5 (texture), and 8 (taste). To calculate the overall quality score for each evaluator, individual scores given to the selected sensory attributes were first multiplied by the corresponding CI, and then the sum of corrected score values was divided by the sum of CI.

The quality rating method was used to determine the overall quality of cold-pressed microgreen juices. Sensory evaluation using quality rating method was performed by trained evaluators who are employed at the Faculty of Agriculture and are well-versed in the process of producing cold-pressed beverages and vegetable/plant-based juices, and they also used a guide for sensory evaluation of quality microgreen juices. The 10 trained evaluators participated in this sensory testing (ISO 8586:2023) [51]. Before performing the quality rating method, all evaluators attended training sessions for three weeks for 2 h each and tasted the juices. The microgreen juices were presented to the evaluators monadically in random order.

2.9.2. Consumer Acceptance Evaluation

Sensory acceptability of the microgreen juices by consumers was assessed using the 9-point hedonic scale (1–4: dislike; 5: neither like nor dislike; 6–9: like). The evaluators did not compare the samples with each other, but rated the sample according to individual sensory attributes and overall acceptability. The aim of the hedonic test is not the comparison of products or to range products by evaluator, but rather evaluation of the product in terms of overall acceptability and the similarity of the selected sensory attributes [52,53]. To ensure an absolutely independent evaluation, consumers did not assign scores to compare the results with each other or between samples. The first sensory characteristic that the evaluator rated was color, i.e., the appearance of the product: for example, "How much do you like the color of the product?", and so on for each of the attributes. If the evaluator felt the need to describe any sensations related to the product or its attributes, they could write comments in the comment field. The numerical data obtained for the hedonic test were expressed as mean values in radar diagram. These data were statistically processed using a one-way ANOVA (Duncan's post hoc test), which allowed a comparison of the juices with each other.

The prepared cold-pressed microgreen juices were served to consumers in clear plastic glasses, to help them perceive the color and appearance of the juices. The hedonic test was carried out on 74 consumers who regularly consume vegetable beverages. Criteria for the selection of consumers were as follows: (1) they must be consumers of these or similar products, and (2) they must have no health problems, in particular, no problems with oral perception or dysfunctional senses, or dental disfunction. The sensory panel included both men (57%) and women (43%) who consume different types of vegetable beverages, and the average age of evaluators was 29 years.

2.10. Statistical Analysis

Results for proximate compositions, spectrophotometric assays, chromatographic quantification, consumer acceptance evaluation (radar chart), and quality ranking were expressed as mean values ± standard deviation (n = 3). Significant differences between means were determined by one-way ANOVA, using Duncan's post hoc test (IBM SPSS ver 25 statistical software, SPSS Inc., Chicago, IL, USA). The correlation analysis was carried out by calculating Pearson's correlation coefficient ($p < 0.05$).

3. Results and Discussion

3.1. UHPLC Q-ToF MS Profile of Bioactive Compounds of Broccoli, Amaranth, and Red Beet Microgreens

The major classes of bioactive compounds in broccoli, amaranth, and red beet microgreens were identified by UHPLC Q-ToF MS, taking into account the exact m/z masses of the molecular ions, typical MS fragments and available data in the literature [9,10,37–39,41,43,44]

Glucosinolates (GLSs) were detected only in broccoli microgreens, as expected. A total of seven GLSs compounds were detected in different broccoli microgreen extracts (Table 1).

Table 1. Characterization of glucosinolates in broccoli microgreens by UHPLC-QToF-MS. Target compounds, expected retention time (RT), molecular formula, calculated mass, exact mass, and MS fragments are presented.

RT	Formula	Calculated Mass	mDa	Compound Name	m/z Exact Mass	Major MS Fragments (Base Peak)	BC	BC1	Ref
1.72	$C_{11}H_{18}NO_9S_2^-$	372.0423	−5.00	Gluconapin	372.0473	130(100), **195**, **259**, **275**, 241, 291, 139	-	+	
6.71	$C_{15}H_{20}NO_9S_2^-$	422.0579	−2.60	Gluconasturtiin	422.0605	205(100), 247, 164, **259**, **275**, 180, 226, 244, 342	+	-	
6.14	$C_{16}H_{19}N_2O_9S_2^-$	447.0532	−4.15	Glucobrassicin	447.0574	130(100), **259**, **205**, 447, **275**, 165, **195**	+	+	
1.77	$C_{16}H_{19}N_2O_{10}S_2^-$	463.0481	−4.35	4-Hydroxy-glucobrassicin	463.0525	169(100), 160, 221, **259**, **275**, **195**, 463, **205**, 285, 383, 267, 241, 186, 176	+	+	[9,37,38]
2.56	$C_{16}H_{19}N_2O_{10}S_2^-$	463.0481	−4.35	5-Hydroxy-glucobrassicin	463.0525	169(100), 160, 221, **259**, 267, **275**, **195**, **205**, 285, 463	+	+	
7.13	$C_{17}H_{21}N_2O_{10}S_2^-$	477.0638	−4.54	Neo-glucobrassicin	477.0683	477(100), **259**, **275**, 284, 235, **195**, 241, 145	+	+	
7.88	$C_{17}H_{21}N_2O_{10}S_2^-$	477.0638	−4.54	4-Methoxy-glucobrassicin	477.0683	167(100), **259**, **205**, 241, **275**, 282, 285, **195**, 315, 447	+	+	

Abbreviations: BC—broccoli microgreens extracted with acidified 80% methanol at room temperature; BC1—broccoli microgreens extracted with 70% boiling methanol, at thermoshaker. "+"—detected glucosinolates.

The identification was performed in two different extracts of broccoli microgreens to obtain a better insight into the profiles of GLSs, as they are very unstable and rapidly hydrolyze to different degradation products [2,8]. Glucobrassicin and its various substituted hydroxy and methoxy derivatives were detected in both extracts, which is consistent with other studies that analyzed broccoli microgreens [11,54,55]. However, gluconapin and gluconasturtiin were detected in different extracts, depending on the extraction conditions and the extractant used. The absence of gluconapin in BC extract is probably due to its increased sensitivity to myrosinase activity. On the other hand, the presence of gluconasturtiin in the BC extract may be due to the different polarity of the extraction solvents used and the better tendency of acidified 80% methanol to extract this compound. Interestingly, glucoraphanin was not detected, which has been reported as the dominant GLS in broccoli microgreens in most studies [2,11,54]. Its absence in the extracts may be due to its lower stability, its rapid conversion into sulforaphane or its tendency to form conjugates with other compounds of the microgreens [56].

Phenolic compounds (PCs) were identified and quantified in the extracts of all microgreens studied. However, the phenolic compounds found in the analyzed microgreen species differed significantly (Table 2).

Table 2. Characterization and quantification (mg/100 g) of phenolic compounds detected in amaranth, red beet, and broccoli microgreens by UHPLC-QToF-MS. Target compounds, expected retention time (RT), molecular formula, calculated mass, exact mass, and MS fragments are presented.

No	Compounds Name	RT	Formula	Calculated Mass	m/z Exact Mass	mDa	MS Fragments (% of Base Peaks)	Samples (mg/100 g)		
								AM	BC	RB
						Phenolic acid and derivatives				
1	Dihydroxy-benzoic acid hexoside isomer I [b]	3.83	$C_{13}H_{15}O_9^-$	315.07216	315.07504	−2.88	**108.0228 (100)**, 109.0298 (41), 110.0330 (4), 152.0132 (58), 153.0199 (19), 154.0198 (2)	10.08 ± 0.08 [B]	104.14 ± 1.14 [A]	/
2	Vanillic acid hexoside isomer I [b]	4.45	$C_{14}H_{17}O_9^-$	329.08781	329.08943	−1.63	**108.021 (100)**, 109.0263 (8), 113.0218 (4), 123.0449 (35), 124.0473 (3), 125.0240 (8), 152.0109 (76), 153.0136 (10), 167.0364 (34), 169.019 (3)	4.23 ± 0.03 [B]	11.25 ± 0.06 [A]	/
3	Hydroxy-benzoic acid dihexoside [b]	5.18	$C_{24}H_{19}O_{10}^-$	467.09837	467.10054	−2.17	**137.0246 (100)**, 138.0278 (9), 299.0771 (2), 431.1188 (5)	25.08 ± 0.10	/	/
4	Vanillic acid hexoside isomer II [b]	5.18	$C_{14}H_{17}O_9^-$	329.08781	329.08927	−1.46	**108.0229 (100)**, 109.0265 (8), 122.0367 (3), 123.0464 (39), 124.0504 (4), 152.0129 (61), 153.0172 (4), 167.0369 (32), 168.0409 (4)	7.34 ± 0.04 [B]	38.53 ± 0.54 [A]	4.41 ± 0.12 [C]
5	Hydroxy-benzoic acid pentosyl hexoside isomer I [b]	5.52	$C_{18}H_{23}O_{12}^-$	431.11950	431.12157	−2.07	**137.0246 (100)**, 138.0287 (8)	95.15 ± 0.37 [A]	/	2.58 ± 0.08 [B]
6	Carboxy-vanillic acid [b]	5.59	$C_9H_7O_6^-$	211.02430	211.02751	−3.21	108.0235 (39), 109.0275 (2), 121.0287 (2), **122.0386 (100)**, 123.0453 (35), 124.0465 (3)	/	28.99 ± 0.22	/
7	Dihydroxy-benzoic acid hexoside isomer II [b]	5.65	$C_{13}H_{15}O_9^-$	315.07216	315.07845	−6.29	108.0222 (8), **109.0305 (100)**, 110.0335 (14), 152.0114 (12), 153.0208 (61)	10.33 ± 0.12 [B]	19.23 ± 0.20 [A]	3.83 ± 0.03 [C]
8	Vanillic acid pentosyl hexoside [b]	5.79	$C_{19}H_{25}O_{13}^-$	461.13007	461.13684	−6.77	108.0226 (5), 123.0448 (8), 152.0113 (18), 153.0166 (2), **167.0356 (100)**, 168.0383 (10)	123.40 ± 0.83	/	/
9	Sinapoyl syringic acid [b]	5.92	$C_{20}H_{19}O_9^-$	403.10290	403.10103	1.87	138.0306 (51), 153.0543 (31), 154.0590 (12), 161.0362 (27), 182.0204 (34), 189.0316 (89), 190.0316 (8), **197.0445 (100)**, 198.0494 (13), 203.0441 (7), 204.0558 (15)	/	7.93 ± 0.05	/
10	Dihydroxy-benzoic acid pentosyl hexoside [b]	6.05	$C_{18}H_{23}O_{13}^-$	447.11442	447.11850	−4.08	101.0249 (4), 108.022 (14), 109.0298 (15), 151.0397 (3), **152.0118 (100)**, 153.0169 (13), 154.0189 (1), 161.0464 (3), 315.0738 (2), 447.1157 (37)	65.56 ± 0.98 [A]	/	4.55 ± 0.15 [B]
11	Dihydroxy-benzoic acid pentoside [b]	6.05	$C_{12}H_{13}O_8^-$	285.06159	285.06344	−1.85	**108.0221 (100)**, 109.0282 (19), 152.0118 (43), 153.0171 (9)	2.65 ± 0.06 [B]	/	59.79 ± 1.89 [A]
12	Syringic acid hexoside [b]	6.32	$C_{15}H_{19}O_{10}^-$	359.09837	359.10153	−3.15	101.0242 (59), 113.0246 (64), 121.0289 (38), 137.0261 (45), 138.0326 (71), 152.0488 (52), 153.0559 (56), 166.0279 (18), 181.0140 (41), 182.0234 (56), 196.0406 (22), **197.0449 (87)**, 211.0609 (20), 239.0567 (34), 359.1005 (100)	/	54.42 ± 0.98 [A]	17.16 ± 0.26 [B]
13	Coumaric acid pentoside [b]	6.32	$C_{14}H_{15}O_7^-$	295.08180	295.08967	−7.87	108.0206 (22), 127.1141 (13), 149.0220 (24), 151.1095 (19), 152.0415 (13), **163.0405 (100)**	/	15.80 ± 0.24	/
14	Hydroxy-benzoic acid hexoside [b]	6.45	$C_{13}H_{15}O_8^-$	299.07724	299.08261	−5.37	108.0823 (8), 121.024 (11), 122.0373 (61), 123.0458 (51), **137.0252 (100)**, 138.027 (10)	9.44 ± 0.04 [C]	10.02 ± 0.13 [B]	25.76 ± 0.54 [A]
15	Dihydroxy-benzoic acid dipentoside [b]	6.66	$C_{17}H_{21}O_{12}^-$	417.10385	417.10765	−3.80	108.022 (18), 109.0298 (23), 110.0326 (1), 151.0402 (4), **152.0121 (100)**, 153.0168 (13), 285.0628 (2), 417.1050 (25)	/	/	252.35 ± 2.32

Table 2. Cont.

No	Compounds Name	RT	Formula	Calculated Mass	m/z Exact Mass	mDa	MS Fragments (% of Base Peaks)	Samples (mg/100 g)		
								AM	BC	RB
16	Caffeic acid hexoside [b]	6.66	$C_{15}H_{17}O_9^-$	341.08730	341.09286	−5.56	134.0356 (5), **135.0466 (100)**, 136.0467 (9), 137.0575 (11), 145.0860 (5), 161.0255 (6), 164.0495 (6), 178.0266 (3), 179.0352 (61)	/	16.47 ± 0.08	/
17	Feruloyl quinic acid [b]	6.86	$C_{17}H_{19}O_9^-$	367.10290	367.10878	−5.88	111.0471 (4), 117.0343 (15), 120.9974 (9), **134.0387 (100)**, 135.0427 (14), 146.0620 (11), 149.0617 (8), 155.0341 (5), 173.0472 (5), 190.0531 (6), 191.0589 (6), **193.0521 (49)**, 194.0527 (6)	/	21.67 ± 0.24	/
18	Sinapic acid dihexoside [b]	6.80	$C_{23}H_{31}O_{15}^-$	547.16630	547.17128	−4.98	101.0242 (22), 113.0239 (10), 119.0346 (12), 149.0263 (9), 164.0477 (26), 179.0676 (13), 190.0262 (18), 205.0523 (80), 206.0563 (27), 221.0789 (40), **223.0616 (69)**, **247.0621 (100)**	/	35.71 ± 0.48	/
19	Ferulic acid hexoside [b]	7.07	$C_{16}H_{19}O_9^-$	355.10346	355.10802	−4.56	**111.009 (100)**, 112.0128 (7), 132.0233 (6), 134.0375 (85), 135.0408 (8), 149.0616 (20), 154.9994 (12), 160.0170 (73), 161.0204 (8), 175.0402 (59), 176.0450 (8), 178.0277 (44), 179.0308 (5), **193.0514 (28)**, 194.0534 (4)	39.70 ± 0.14 [A]	7.80 ± 0.13 [B]	/
20	Sinapic acid hexoside [b,***]	7.14	$C_{17}H_{21}O_{10}^-$	385.11402	385.11743	−3.41	101.0262(2), 113.0255(2), 119.0222(2), 147.0115(2), 149.0268(4), 164.0499 (10), 175.0056 (12), **190.0296 (100)**, 191.0334 (13), 192.03511(2), **205.05315 (77)**, 206.0571 (13), 207.0594 (2), 223.0633 (3)	/	214.46 ± 2.22	/
21	Feruloyl isocitric acid [b]	8.14	$C_{16}H_{15}O_{10}^-$	367.06707	367.07166	−4.59	**111.0092 (100)**, 112.0128 (7), 129.0199 (3), 134.0379 (2), 154.9992 (12), 173.0099 (3)	175.78 ± 1.10 [A]	/	6.59 ± 0.09 [B]
22	Sinapoyl malic acid [b,***]	8.35	$C_{15}H_{15}O_9^-$	339.07216	339.07690	−4.74	115.0054 (34), 116.0085 (2), 121.0311 (8), 132.0237 (2), **133.0161 (29)**, 134.0199 (2), 147.0469 (7), **149.0263 (100)**, 150.0297 (10), 164.0499 (79), 165.0532 (9), 193.0168 (2), 208.0406 (3), **223.064 (6)**	/	159.12 ± 1.90	/
23	Sinapic acid [a]	8.35	$C_{11}H_{11}O_5^-$	223.06120	223.06408	−2.89	104.0281 (4), 117.0361 (3), **121.0309 (100)**, 122.0342 (10), 132.0250 (1), 135.0460 (3), 149.0257 (62), 150.0288 (6), 163.0413 (2), 165.0227 (4), 193.0166 (9)	/	526.06 ± 2.29	/
24	Benzoyl malic acid [b]	8.48	$C_{11}H_9O_6^-$	237.03990	237.04732	−7.42	103.4469 (3), 115.0112 (3), **121.0295 (100)**, 122.0361 (8)	53.27 ± 1.27 [A]	4.45 ± 0.12 [B]	/
25	Disinapoyl-dihexoside [b,***]	8.75	$C_{34}H_{41}O_{19}^-$	753.22420	753.23231	−8.11	119.0359 (3), 164.0496 (4), 179.0659 (3), 190.0294 (7), **205.0529 (100)**, 206.0565 (14), 208.0398 (3), 223.0637 (66), 224.0672 (8), 247.0642 (9), 265.0760 (4), 289.0751 (3), 529.1625 (42), 530.1661 (14), 531.1661 (3)	/	122.51 ± 1.95	/
26	Trisanapoyl-dihexoside [b,***]	9.29	$C_{45}H_{51}O_{23}^-$	959.28210	959.28907	−6.97	**205.0525 (75)**, 206.0564 (7), 223.0637 (31), 247.0641 (14), 265.0763 (7), 289.0759 (7), 511.1509 (28), 512.1537 (8), 529.1607 (30), 530.1613 (9), **735.2217 (100)**, 736.2243 (46), 737.2255 (13), 959.2905 (14)	/	130.53 ± 1.96	/
			Σ					622.02	1529.08	377.03

Table 2. Cont.

No	Compounds Name	RT	Formula	Calculated Mass	m/z Exact Mass	mDa	MS Fragments (% of Base Peaks)	Samples (mg/100 g)		
								AM	BC	RB
	Apigenin C-glycosides ****									
27	2″-Hexosyl vitexin [c]	7.82	$C_{27}H_{31}O_{15}^+$	595.16630	595.17351	−7.21	271.0608 (32), 283.0597 (17), 295.06144 (13), **313.0709 (100)**, 314.0746 (25), 337.0715 (20), 367.0819 (10), 379.0809 (9), 397.0928 (28), 398.0949 (9), 415.1031 (48), 416.1061 (14), 433.1138 (87), 434.1180 (26), 435.1191 (7)	/	/	65.09 ± 0.31
28	2″-Pentosyl vitexin [c]	7.95	$C_{26}H_{29}O_{14}^+$	565.15570	565.16177	−6.07	283.0599 (16), 295.0603 (6), **313.0714 (100)**, 314.0753 (24), 337.0711 (16), 343.0821 (7), 367.0818 (14), 379.0818 (11), 397.0923 (42), 398.097 (12), 415.1031 (68), 416.1074 (19), 433.1138 (90), 434.1172 (28), 565.1550 (14)	/	/	87.04 ± 1.02
29	2″-Hexosyl-6″-malonyl vitexin [c]	8.22	$C_{30}H_{33}O_{18}^+$	681.16670	681.17274	−6.04	271.0606 (19), 283.0604 (7), 295.0610 (21), **313.0712 (100)**, 314.0739 (24), 337.0706 (13), 345.1099 (7), 439.1031 (9), 457.1122 (8), 475.1238 (13), 483.0918 (7), 501.1043 (22), 502.1063 (8), 519.1149 (60), 520.1182 (23)	/	/	67.35 ± 2.12
30	2″-Hexosyl-6″-acetyl vitexin [c,*]	8.27	$C_{29}H_{31}O_{16}^-$	635.16120	635.16454	−3.34	101.0257 (5), 175.0376 (6), 193.0516 (8), **293.0464 (100)**, 311.0613 (15), 337.0938 (7), 413.0880 (21), 431.0900 (5), **455.0981 (88)**, 473.1113 (12), 575.1456 (9)	/	/	6.97 ± 0.06
31	2″-Pentosyl-6″-malonyl vitexin [c]	8.36	$C_{29}H_{31}O_{17}^+$	651.15610	651.15914	−3.04	283.0600 (9), 295.0611 (19), **313.0713 (100)**, 314.0744 (23), 337.0707 (13), 379.0820 (8), 439.1029 (11), 445.1139 (8), 457.1149 (10), 475.1246 (7), 483.0936 (7), 501.1035 (20), **519.1146 (41)**, 520.1178 (15), 651.1564 (11)	/	/	76.12 ± 0.79
32	2″-Hexosyl cytisoside [c]	8.49	$C_{28}H_{33}O_{15}^+$	609.18190	609.18544	−3.54	285.0764 (34), 297.0764 (15), 309.0762 (14), **327.0874 (100)**, 351.0868 (18), 381.0977 (9), 393.0975 (8), 411.1082 (28), 429.1191 (47), **447.1296 (88)**	/	/	170.18 ± 1.63
33	2″-Pentosyl cytisoside [c]	8.69	$C_{27}H_{31}O_{14}^+$	579.17140	579.17423	−2.83	297.0767 (15), **327.0872 (100)**, 328.0908 (25), 351.0863 (15), 357.0970 (9), 381.0975 (14), 393.0975 (10), 411.1088 (41), 412.1115 (13), 429.1193 (70), 430.1223 (23), **447.1296 (94)**, 448.1334 (31), 579.1716 (18), 580.1764 (7)	/	/	184.04 ± 1.83
34	2″-Hexosyl-6″-malonyl cytisoside [c,**]	8.83	$C_{31}H_{35}O_{18}^+$	695.18230	695.18826	−5.96	285.0763 (24), 297.0759 (9), 309.0767 (20), **327.0869 (100)**, 328.0904 (25), 351.0871 (12), 393.0976 (8), 453.1185 (10), 471.1286 (10), 489.1415 (15), 497.1094 (7), 515.1202 (24), 516.1235 (9), **533.1299 (76)**, 534.1330 (28)	/	/	64.77 ± 1.35
35	2″-Hexosyl-6″-acetyl cytisoside [c]	8.96	$C_{30}H_{35}O_{16}^+$	651.19250	651.19507	−2.57	297.0764 (6), **327.0871 (7)**, 369.0976 (22), 370.1010 (7), 381.0966 (4), 393.0958 (4), 411.1094 (12), 423.1086 (4), 429.1180 (10), 430.1229 (4), 471.1295 (14), 472.1314 (5), **489.1406 (100)**, 490.1438 (36), 491.1459 (8)	/	/	194.21 ± 1.98

Table 2. Cont.

No	Compounds Name	RT	Formula	Calculated Mass	m/z Exact Mass	mDa	MS Fragments (% of Base Peaks)	Samples (mg/100 g)		
								AM	BC	RB
36	2″-Pentosyl-6″-acetyl cytisoside [c,*]	9.08	$C_{29}H_{31}O_{15}^-$	619.16684	619.17055	−3.70	101.0255 (18), 113.0246 (26), 131.0344 (12), 283.0643 (12), **307.0619 (100)**, 308.0676 (17), 325.0731 (62), 326.0740 (15), 337.0726 (44), 349.0726 (22), 367.0843 (33), 409.0936 (17), 427.1057 (16), 469.1162 (13), 619.1704 (16)	/	/	35.15 ± 0.93
			Σ					/	/	950.92
					Other detected flavonoids					
37	Kaempferol-3-O-(6″-hexosyl)hexoside-7-O-hexoside with HCOOH [c,***]	6.53	$C_{34}H_{41}O_{23}^-$	817.20390	817.20689	−2.99	**284.0375 (3)**, 285.0391 (4), 288.6129 (2), 299.0519 (2), 446.0897 (4), **447.0986 (48)**, 448.1022 (16), 489.0970 (3), 609.1533 (100), 610.1576 (39), 611.1627 (9), 612.1502 (2), 771.2062 (5), 772.2040 (3)	/	6.50 ± 0.08	/
38	Kaempferol-3-O-sinapoyl-trihexoside-7-O-hexoside [c,***]	7.04	$C_{50}H_{59}O_{30}^-$	1139.30910	1139.32320	−14.10	**1139.318 (100)**, 815.2109 (52), **977.2654 (43)**, 609.1536 (9), **284.0357 (8)**	/	26.27 ± 0.25	/
39	Kaempferol-3-O-sinapoyl-dihexoside-7-O-hexoside [c,***]	7.07	$C_{44}H_{51}O_{25}^+$	979.27190	979.27923	−7.33	127.0423 (4), **207.0669 (95)**, 208.0701 (16), 225.0765 (4), **287.0564 (55)**, 288.0608 (11), 291.0869 (3), 351.1114 (31), 352.1174 (7), **369.1216 (100)**, 370.1249 (26), 371.1262 (6), **449.1127 (49)**	/	24.84 ± 0.72	/
40	Quercetin 3-O-(6″-rhamnosyl)hexoside [c]	8.02	$C_{27}H_{29}O_{16}^-$	609.14611	609.14934	−3.24	151.0039 (3), 178.9985 (4), 255.0319 (3), 271.0253 (7), **300.0283 (100)**, 301.0355 (73)	24.37 ± 0.36 [A]	2.01 ± 0.02 [C]	3.06 ± 0.07 [B]
			Σ					24.37	59.63	3.06
			ΣΣ					646.38	1588.71	1331.0

Abbreviations: "/"—nonidentified phenolic compounds; AM—amaranth microgreens; BC—broccoli microgreens; RB—red beet microgreens; "*"—compounds detected only in negative ionization mode; "**"—compounds detected only in positive ionization mode; "***"—previously detected compounds in broccoli microgreens and reported by Liu and Shi [10]; "****"—apigenin C glycosides detected in accordance to previously reported date by Isayenkova and Wray [39] and da Silva and Morelli [40]. Compound quantities expressed using available standards [a]; compounds expressed as sinapic acid equivalents [b]; compounds expressed as apigenin equivalents [c]. Means with the same uppercase letter in the same raw are significantly different according to Duncan's test, ($p < 0.05$), (mean ± S.D.; $n = 3$).

The highest total detected PC content is found in broccoli (1588.71 mg/100 g FW), followed by red beet (1331.01 mg/100 g FW), and the lowest is found in amaranth (646.38 mg/100 g FW) microgreen extracts. The high total PC content in amaranth and broccoli extracts is mainly due to various derivatives of phenolic acids (>95% of the total PC content), and in red beet microgreen extract, it is due to apigenin C-glycosides (>70% of the total PC content). Various hydroxybenzoic, dihydroxybenzoic, and vanillic acid glycosides were detected in all analyzed microgreen extracts, but their content varied and depended strongly on the microgreen species. For example, pentosyl hexoside glycosides of hydroxybenzoic, dihydrobenzoic, and vanillic acid were dominant in amaranth microgreens along with feruloyl isocitric (175.78 mg/100 g FW) and benzoyl malic acid. On the other hand, dipentosyl, and pentosyl glycosides of dihydroxybenzoic acid (252.35 and 59.79 mg/100 g FW) were the most abundant phenolic acid derivatives in red beet microgreens. Similarly to our results, Wojdyło and Nowicka [14] identified phenolic acids as the predominant class of phenolic compounds in amaranth microgreens. In addition, sinapic acid (526.06 mg/100 g FW) and its various derivatives (sinapic acid hexoside, sinapoyl malic acid, disinapoyl-dihexoside, and trisinapoyl-dihexoside) were the dominant compounds in broccoli microgreens. Identical sinapoyl derivatives were previously discovered and reported by Liu and Shi [10] in the analysis of differentially grown broccoli microgreens. In contrast to phenolic acids, the detected flavonoids can be characterized as specific markers for broccoli, amaranth, and red beet microgreens. Several apigenin C-glycosides, i.e., vitexin and cytisoside (3′-methyl vitexin) derivatives (10 compounds) were, to our knowledge, detected for the first time in red beet microgreens. Quantification confirmed that the total amount of cytisoside derivatives was significantly higher compared to vitexin derivatives, primarily contributed by compounds **32** (194.21 mg/100 g), **33** (184.04 mg/100 g), and **35** (170.18 mg/100 g) (Table 2). These compounds have a common MS base at 327 m/z, which is a typical fragment of the cytisoside molecule (m/z 447), obtained after cross-ring cleavage of the sugar unit. Compound 35 (m/z 651) was identified as 2″-hexosyl-6″-acetyl cytisoside. The key MS fragment (Y_0^+) for its identification was at 489 m/z ([M-acetyl residue-$^{0.2}X_8$ scission]$^+$), followed by a fragment at 327 m/z ([Y_0^+-hexosyl residue (162Da)]$^+$). Compounds **32** and **33** were recognized as 2″-hexosyl cytisoside (m/z 609) and 2″-pentosyl cytisoside (m/z 579), respectively, with MS secondary peak at 447 m/z obtained by the loss of hexosyl (-162 Da) or pentosyl (-132 Da) sugar units, respectively. In addition, both compounds yielded a fragment followed by another loss of 18 Da (H_2O), indicating that the secondary sugar is linked to the primary sugar by an interglycosidic 1–2 bond. A typical MS^2 fragment repeated in all cytisoside derivatives was found at 351 m/z, and its proposed structure is shown in Figure 2b. To date, only a few studies have confirmed the presence of several vitexin derivatives (compounds **27**, **28**, **29**, and **32** in Table 2) in mature red beet (*Beta vulgaris* L.) stalks and leaves [39,40,57]. Moreover, the cited studies have confirmed and characterized vitexin derivatives as potentially very useful compounds for human health, due to their antioxidant, anticancer and anti-inflammatory activities [40,57,58]. In contrast, apigenin C-glycoside was not found in broccoli and amaranth microgreens. Macromolecular kaempferol (compound **37**) and kaempferol-sinapoyl derivatives (compounds **38** and **39**) were identified only in broccoli microgreens and accounted for about 4% of the total quantified phenolics. These kaempferol derivatives were previously identified by Liu and Shi [10] and represent typical compounds found in *Brassica* microgreens and vegetables. Quercetin 3-O-(6″-rhamnosyl)-hexoside was found in all three analyzed microgreen species (amaranth, broccoli, and red beet). However, the highest content of this compound was found in amaranth microgreens (24.37 mg/100 g).

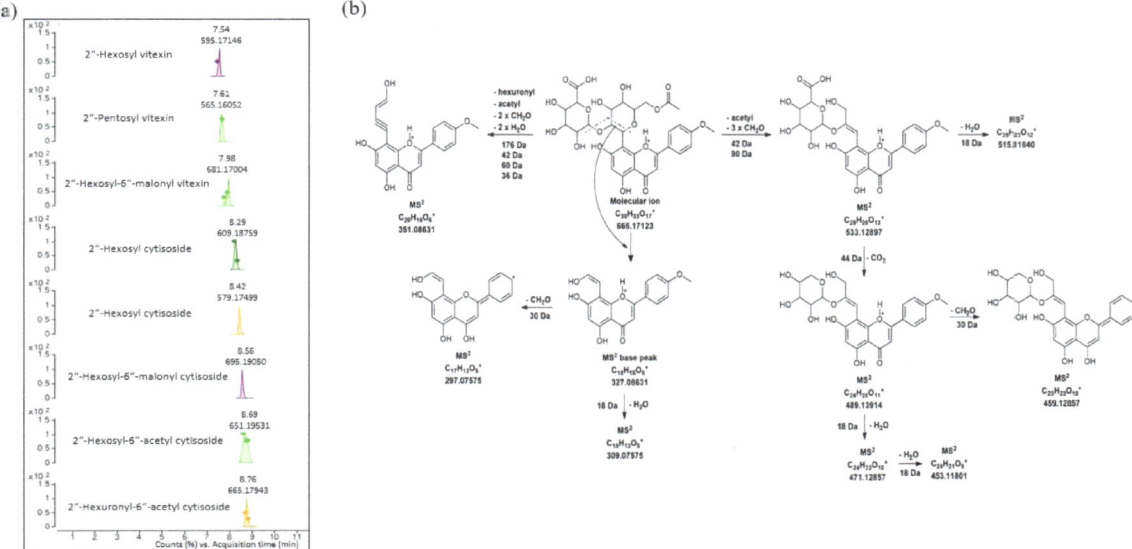

Figure 2. Chromatograms of the dominant cytisoside and vitexin derivatives in red beet cold-pressed juice, with retention time and exact mass (**a**); proposed fragmentation pathway of 2″-Hexuronyl-6″-acetyl cytisoside (**b**).

Based on a detailed analysis of betalains, the dominant presence of (iso)amarnthin (73.56%) and (iso)betanin (49.20%) was confirmed in amaranth and red beet microgreens, respectively (Table 3).

Both compounds produced a MS base peak at 389 m/z (betanidin aglycon), which resulted from the loss of glucosyl-glucuronyl residue ([M+H-176 Da-162 Da]$^+$) for (iso)amaranthin and glucosyl unit ([M+H-162 Da]$^+$) for (iso)betanin [41]. Other detected compounds are decorboxy derivatives of amaranthin (m/z 683) and betanin (m/z 507), with characteristic fragments at 345 m/z ([betanidin aglycone-44 Da (CO$_2$)]$^+$), 299 m/z, and 150 m/z. These betalain derivatives were previously identified and analyzed in detail in mature red beet root [43,44] or leaves/sprouts of various *Amaranthus* species [59]. However, to our knowledge, this is the first study to analyze betalain profiles in amaranth and red beet microgreens.

Table 3. Characterization and relative content (%) of betalains in amaranth and red beet microgreen samples by UHPLC Q-ToF-MS. Target compounds, expected retention time (RT), molecular formula, calculated mass, exact mass, and MS fragments are presented.

No	Compounds Name	RT	Formula	Calculated Mass	m/z Exact Mass	mDa	MS Fragments (% of Base Peaks)	Samples (%) AM	Samples (%) RB	Ref
				Betalains						
41	(Iso)Amaranthin	5.05	$C_{30}H_{35}N_2O_{19}^+$	727.18285	727.18431	−1.45	150.0552(1), **389.0982 (100)**, 390.1014 (28), 391.1044 (5), **551.1509 (5)**, 552.1541 (2), 727.1837 (21)	73.56	0.17	
42	(Iso)betanin	5.72	$C_{24}H_{27}N_2O_{13}^+$	551.15130	551.15259	−1.29	150.0549 (2), 343.0931 (2), **389.0987 (100)**, 390.1021 (29), 391.1041 (5), 551.1522 (4)	5.22	49.20	
43	17-Decarboxy-(iso)amaranthin	5.85	$C_{29}H_{35}N_2O_{17}^+$	683.19302	683.19430	−1.27	150.056 (1), **345.1084(100)**, 346.1116 (27), 347.1135 (4), **507.1618 (9)**, 508.1649 (4), 683.1930 (31)	17.35	-	
44	(2, 15 or 17)-Decarboxy-(iso)betanin	5.93	$C_{23}H_{27}N_2O_{11}^+$	507.16150	507.16233	−0.83	106.0660 (2), 150.0549 (2), 299.1035 (1), 301.1186 (1), **345.1089 (100)**, 346.1124 (25), 347.1145 (4), 507.1617 (5)	1.66	23.29	[41,43,44]
45	(2, 15 or 17)-Decarboxy(iso)betanin	6.40	$C_{23}H_{27}N_2O_{11}^+$	507.16150	507.16406	−2.56	150.0548 (2), 299.1030 (2), 301.1176 (1), **345.1088 (100)**, 346.1126 (25), 347.1144 (4), 507.1622 (2)	2.22	23.55	
46	(2, 15 or 17)-Decarboxy-(iso)betanidin	7.01	$C_{17}H_{17}N_2O_6^+$	345.10811	345.11267	−4.56	100.0392 (27), 106.0643 (37), 132.0449 (53), 144.0302 (47), **150.0541 (99)**, 151.0626 (62), 152.0708 (36), 202.0881 (34), 209.0726 (36), 227.0862 (35), 253.0849 (61), 255.1138 (65), 281.0767 (49), 299.1034 (43), **345.1061 (100)**	-	3.79	
				Total (%)				100	100	

Abbreviations: AM—amaranth microgreens; RB—red beet microgreens.

3.2. Proximate Composition of Microgreen Juices

Results for yield and general physicochemical parameters (moisture, dry weight, pH, and °Brix) of microgreen juices (BCJ, AMJ, and RBJ) are shown in Table 4. The yield of the microgreen juices varied from 53.4% (AMJ) to 70.2% (BCJ), which was a very high yield for the microgreen juices obtained by mechanical pressing. All microgreen juices had high moisture content (>98%) and low dry matter content (from 1.64 to 2.0%). The pH values of the microgreen juices ranged from 5.96 (BCJ) to 6.52 (AMJ), which is consistent with previously published results for the same microgreen species [24]. The values for total soluble solids (°Bx) are similar for all analyzed microgreen juices ranging from 1.8 to 2.0 °Bx. It can be concluded that these microgreen juices have a mild acidity and low sugar content, which places them among the low-calorie beverages.

Table 4. Proximate compositions of broccoli, red beet, and amaranth microgreen juices.

Microgreen Juices	Family and Species	Yield of Juices (%)	Percentage of Dry Weight (%)	Percentage of Moisture (%)	pH Values	°Brix
BCJ	*Brassica oleracea* var. *italica*	70.20 ± 0.15 [a]	1.84 ± 0.05 [b]	98.26 ± 0.05 [b]	5.96 ± 0.01 [c]	2.00 ± 0.01 [a]
RBJ	*Beta vulgaris*	62.00 ± 0.10 [b]	2.00 ± 0.01 [a]	98.00 ± 0.01 [c]	6.44 ± 0.01 [b]	2.00 ± 0.01 [a]
AMJ	*Amaranthus tricolour* L.	53.40 ± 0.15 [c]	1.64 ± 0.01 [c]	98.46 ± 0.01 [a]	6.52 ± 0.01 [a]	1.80 ± 0.01 [b]

Abbreviations: AMJ—cold-pressed amaranth microgreen juice; BCJ—cold-pressed broccoli microgreen juice; RBJ—cold-pressed red beet microgreen juice. Results were presented as mean values ± standard deviation. Different small letters in the same column denote a significant difference according to Duncan's test, $p < 0.05$.

3.3. UHPLC Q-ToF MS Profile of Microgreen Juices

In contrast to broccoli microgreens, GLSs were not detected in broccoli microgreen juice. During pressing of microgreens to juice, mechanical damage to the plant tissue occurs with the release of GLSs which are rapidly hydrolyzed to various isothiocyanates by the action of enzyme myrosinase [2]. Similarly to our results, Bello and Maldini [56] reported the complete absence of GLSs (glucoraphane) in broccoli sprout juices, obtained by mechanical pressing of raw and microwave-treated sprouts. However, various degradation products of GLSs (isothiocyanates) have also shown strong biological activity [8,15,22,23], and likely contribute to the antioxidant potential of BCJ along with phenolic compounds.

The prepared juices from amaranth, broccoli, and red beet microgreens (BCJ, AMJ, and RBJ) had high content of phenolic compounds, namely, 49.84 mg/100 mL, 362.37 mg/100 mL, and 342.02 mg/100 mL juice, respectively. As for microgreens (Table 2), phenolic acids and their derivatives were identified and quantified dominantly in BCJ and AMJ, while various apigenin C-glycosides were found mainly in RBJ. Based on the results presented in Tables 2 and 5, the differences among phenolic acid derivatives detected in microgreens and their juices can be clearly observed. This is probably due to the different migration of individual phenolic compounds from the plant tissue into the juice or their transformation by enzyme action during juice production.

Table 5. Characterization and quantification (mg/100 mL juices) of phenolic compounds detected in amaranth, red beet, and broccoli microgreen juices by UHPLC-QToF-MS. Target compounds, expected retention time (RT), molecular formula, calculated mass, exact mass, and MS fragments are presented.

No	Compounds Name	RT	Formula	Calculated Mass	m/z Exact Mass	mDa	MS Fragments (% of Base Peaks)	Samples (mg/100 mL) AMJ	BCJ	RBJ
							Phenolic acid and derivatives			
1a	Hydroxy-benzoic acid hexoside isomer I [b]	1.68	$C_{13}H_{15}O_8^-$	299.07724	299.08054	−3.30	**137.0248 (100)**, 138.0301 (10)	/	3.47 ± 0.03	/
2a	Shikimic quinic acid hexoside [b]	2.15	$C_{22}H_{23}O_{15}^-$	527.10370	527.11099	−7.29	143.0015 (3), 167.0342 (4), 173.0272 (2), **191.0555 (100)**, 192.0593 (10), 193.0580 (3), 353.0837 (4)	0.92 ± 0.01	/	/
3a	Dihydroxy-benzoic acid isomer I [b]	2.35	$C_7H_5O_4^-$	153.01933	153.02154	−2.21	**108.0229 (100)**, 109.0294 (81), 110.0311 (6)	/	0.72 ± 0.02	/
4a	Hydroxy-benzoic acid dihexoside [b]	3.46	$C_{24}H_{19}O_{10}^-$	467.09837	467.09935	−0.98	**137.0255 (100)**, 138.0284 (9), 299.0774 (2)	2.40 ± 0.02	/	/
5a	Hydroxy-benzoic acid [b]	3.91	$C_7H_5O_3^-$	137.02390	137.02614	−2.24	/	14.97 ± 0.19 [A]	5.15 ± 0.01 [C]	9.81 ± 0.02 [B]
6a	Dihydroxy-benzoic acid isomer III [b]	4.23	$C_7H_5O_4^-$	153.01933	153.02021	−0.88	106.9976 (65), 107.0293 (52), 107.053(20), **108.0218 (100)**, 122.9839 (14), 123.0203 (34), 135.0194 (11), 135.0538 (12)	/	0.99 ± 0.03	/
7a	Vanillic acid pentosyl hexoside [b]	4.43	$C_{19}H_{25}O_{13}^-$	461.13007	461.13222	−2.16	108.0226 (5), 123.0461(7), 152.0122 (18), 153.0161 (3), **167.0374 (100)**, 168.0382 (10)	6.86 ± 0.03	/	/

Table 5. Cont.

No	Compounds Name	RT	Formula	Calculated Mass	m/z Exact Mass	mDa	MS Fragments (% of Base Peaks)	Samples (mg/100 mL)		
								AMJ	BCJ	RBJ
8a	Dihydroxy-benzoic acid pentoside [b]	4.71	$C_{12}H_{13}O_8^-$	285.06159	285.06303	−1.44	**108.0231 (100)**, 109.0291 (22), 110.0312 (2), 152.0117 (47), 153.0164 (9), 154.0176 (2)	/	/	7.33 ± 0.03
9a	Dihydroxy-benzoic acid pentosyl hexoside [b]	5.32	$C_{18}H_{23}O_{13}^-$	447.11390	447.11985	−5.95	101.02230(3), 108.0229 (13), 109.0289 (14), 136.0394 (11), 151.0374 (3), **152.0114 (100)**, 153.0161 (14), 161.0453 (3), 163.0387 (6), 315.0666 (3), 447.1152 (46)	13.83 ± 0.07	/	/
10a	Benzoic acid derivative (like as carboxy benzoic acid) [b]	5.66	$C_8H_5O_4^-$	165.01880	165.02126	−2.46	105.0153(52), 105.0395 (58), 108.0156 (13), 120.0197(38), **121.0306 (100)**, 122.0288 (11), 123.9880 (27), 124.0190(39), 135.0394 (17), 147.8908 (8), 151.9801(24), 152.0114 (32)	/	48.02 ± 0.02	/
11a	Hydroxy-benzoic acid hexoside isomer II [b]	5.78	$C_{13}H_{15}O_8^-$	299.07724	299.08174	−4.50	**137.0252 (100)**, 138.0307 (9)	/	/	1.06 ± 0.04
12a	Coumaroyl-quinic acid isomer I [b]	5.78	$C_{16}H_{17}O_8^-$	337.09289	337.09853	−5.64	111.0452 (5), **119.0519 (100)**, 120.0542 (11), 163.0406 (50), 164.0437 (7), 173.0448 (4), 191.0564 (60)	/	55.84 ± 0.14	/
13a	Benzoic acid [b]	5.80	$C_7H_5O_2^-$	121.02900	121.03002	−1.02	/	3.68 ± 0.02 [B]	14.80 ± 0.02 [A]	0.46 ± 0.01 [C]
14a	5-O-Caffeoyl-quinic acid isomer I [b]	6.19	$C_{16}H_{17}O_9^-$	353.08781	353.08964	−1.83	135.0452 (1), 161.0242 (2), 173.0454 (1), **191.0554 (100)**	/	1.61 ± 0.02	/
15a	Coumaric acid hexoside [b]	6.27	$C_{15}H_{17}O_8^-$	325.09230	325.09815	−5.85	117.0354 (6), **119.0513 (100)**, 120.0544 (11), **163.0398 (24)**, 164.0436 (3)	2.63 ± 0.03	/	/
16a	Carboxy hydroxybenzoic acid [b]	6.53	$C_8H_5O_5^-$	181.01370	181.01425	−0.55	107.0304(15), 107.0612 (12), 117.0185 (9), **119.0235 (100)**, 120.0294 (16), 134.0376(30), 135.0487 (14), **137.0287 (26)**	/	0.66 ± 0.01	/
17a	Sinapic acid hexoside [b]	6.59	$C_{17}H_{21}O_{10}^-$	385.11402	385.11652	−2.50	149.0249(21), 164.0481 (56), 165.0516 (8), 175.0042 (12), 179.0701(14), **190.0274 (100)**, 191.0325 (25), **205.0510 (99)**, 206.0569(45), 207.0492 (9), 217.0156 (11), 221.0806(14), 223.0620 (12)	/	18.15 ± 0.02	/
18a	Ferulic acid hexoside [b]	6.67	$C_{16}H_{19}O_9^-$	355.10346	355.10148	1.98	111.0102(50), 112.0133 (16), 113.0147 (18), **134.0379 (100)**, 135.0424(12), 149.0610 (28), 150.0672 (3), 154.9760 (5), 155.0063 (6), 157.0035 (3), 178.0270 (62), 179.0308 (9), **193.0504 (37)**, 194.0542 (6)	1.41 ± 0.01	/	/
19a	Hydroxy-benzoyl malic acid [b]	6.84	$C_{11}H_9O_7^-$	253.03480	253.03892	−4.12	102.9829 (2), 103.0087 (2), 114.0580 (2), **121.0305(100)**, 122.0332(10), 123.0058 (2), 123.0383(2), 130.0424 (2)	2.02 ± 0.02	/	/
20a	Coumaroyl-quinic acid isomer II [b]	6.88	$C_{16}H_{17}O_8^-$	337.09289	337.09441	−1.52	109.0311(2), 111.0446 (18), 112.0467 (2), 119.0508 (44), 120.0531 (5), 137.0257 (11), 138.0320 (1), 155.0348 (6), **163.0402 (26)**, 164.0441(4), **173.0455 (100)**, 174.0484(10), 191.0549 (3)	/	1.16 ± 0.02	/
21a	Sinapic acid [a]	7.88	$C_{11}H_{11}O_5^-$	223.06120	223.06222	−1.03	105.0352(1), **121.0308 (100)**, 122.0339 (9), 134.0359 (1), 135.0456 (13), 136.0548 (1), 148.0172 (5), 149.0248 (50), 150.0277 (5), 163.0396 (13), 164.0469 (5), 165.0197 (27), 166.0219 (3), 193.0142(60), 194.0177 (7)	/	60.00 ± 0.54	/
22a	Sinapoyl malic acid [b]	7.94	$C_{15}H_{15}O_9^-$	339.07216	339.07540	−3.24	115.0047 (47), 116.0085 (2), 117.0301 (1), 121.0313 (8), 132.0226 (2), **133.0156 (43)**, 134.0193 (2), 147.0462 (7), **149.0248 (100)**, 150.0291 (11), 164.0480 (86), 165.0519 (10), 179.0716 (2), 208.0385 (2), 223.0620 (7)	/	134.18 ± 0.04	/

Table 5. Cont.

No	Compounds Name	RT	Formula	Calculated Mass	m/z Exact Mass	mDa	MS Fragments (% of Base Peaks)	Samples (mg/100 mL)		
								AMJ	BCJ	RBJ
23a	Benzoylmalic acid [b]	7.94	$C_{11}H_9O_6^-$	237.03990	237.04514	−5.24	114.9839 (2), 115.0099 (3), **121.0310 (100)**, 122.0333 (10)	/	2.09 ± 0.02	/
24a	Dihydroxy-benzoic acid dihexoside [b]	8.15	$C_{21}H_{19}O_{13}^-$	479.08260	479.08873	−6.13	108.0228(20), 109.0346 (21), 137.0257 (61), **152.0122 (67)**, 153.0151(18), **435.0914 (100)**	/	8.21 ± 0.08	/
25a	Hydroxyferulic acid [b]	11.25	$C_{10}H_9O_5^-$	209.04500	209.04494	0.06	**105.0353 (100)**, 107.0146(58), 121.0291 (23), 123.0439 (44), 125.0253 (13), 131.0141 (16), 149.02305 (77), 150.0276(13), 151.0024 (70), 165.0555 (18), 167.0333 (19), 191.0347(10), **193.0143 (63)**, 209.0157 (12)	/	2.53 ± 0.04	/
			Σ					48.74	357.59	18.66
					Apigenin C-glycosides					
26a	2″-Hexosyl vitexin [c]	7.54	$C_{27}H_{31}O_{15}^+$	595.16630	595.17146	−5.16	271.0586(29), 283.0586 (17), 295.0580 (13), **313.0710 (100)**, 337.0691(18), 367.0794 (10), 379.0794 (8), 397.0905(29), 415.1016 (45), **433.1133 (88)**	/	/	28.81 ± 0.05
27a	2″-Pentosyl vitexin [c]	7.61	$C_{26}H_{29}O_{14}^+$	565.15570	565.16052	−4.82	283.0596(16), **313.0724 (100)**, 337.0699 (17), 343.0802 (9), 367.0806 (16), 379.0806 (11), 397.0921 (41), 415.1041(74), **433.1144 (100)**	/	/	34.45 ± 0.05
28a	2″-Hexosyl-6″-acetyl vitexin [c]	7.81	$C_{29}H_{33}O_{16}^+$	637.17690	637.18029	−3.39	283.0596(4), 295.0573 (4), **313.0710 (10)**, 337.0694 (3), 355.0789 (28), 367.0793 (3), 397.0898 (6), 415.1022 (10), 457.1121 (12), **475.1226 (100)**	/	/	4.34 ± 0.04
29a	2″-Hexosyl-6″-malonyl vitexin [c]	7.95	$C_{30}H_{33}O_{18}^+$	681.16670	681.17004	−3.34	271.0583(18), 283.0590 (8), 295.0587 (21), **313.0712 (100)**, 337.0684 (13), 379.0797 (7), 439.1005 (9), 457.1118 (9), 475.1212(14), 483.0908 (8), 501.1012 (23), **519.1140 (60)**	/	/	26.43 ± 0.05
30a	2″-Hexosyl cytisoside [c]	8.29	$C_{28}H_{33}O_{15}^+$	609.18190	609.18759	−5.69	285.0753(32), 297.0745 (16), 309.0745 (13), **327.0844 (100)**, 351.0852 (18), 381.0953 (8), 393.0952 (8), 411.1073 (28), 429.1186(50), **447.1272 (91)**	/	/	48.94 ± 0.12
31a	2″-Rhamnosyl cytisoside [c]	8.35	$C_{28}H_{33}O_{14}^+$	593.18700	593.19311	−6.11	297.0738(12), **327.0850 (57)**, 351.0832 (12), 357.0946 (7), 381.0956 (11), 393.0956 (8), 411.1058 (34), 429.1168 (58), **447.1273 (100)**	/	/	5.48 ± 0.06
32a	2″-Pentosyl cytisoside [c]	8.42	$C_{27}H_{31}O_{14}^+$	579.17140	579.17499	−3.59	297.0746(14), **327.0847 (100)**, 351.0851 (15), 357.0956 (9), 381.0957 (14), 393.0954 (9), 411.1078 (43), 429.1174(71), **447.1276 (95)**	/	/	46.00 ± 0.23
33a	2″-Hexosyl-6″-malonyl cytisoside [c]	8.56	$C_{31}H_{35}O_{18}^+$	695.18230	695.19050	−8.20	285.0741(24), 297.0736 (9), 309.0747(21), **327.0859 (100)**, 351.0840 (12), 393.0956 (8), 453.1166 (10), 471.1282 (12), **489.1373 (17)**, 497.1060 (8), 515.1173 (27), **533.1284 (77)**	/	/	17.90 ± 0.07
34a	Cytisoside (3′-Methyl vitexin) [c]	8.62	$C_{22}H_{23}O_{10}^+$	447.12910	447.13521	−6.11	135.0459 (8), 297.0737 (51), 309.0717 (8), **327.0846 (100)**, 337.1007 (14), 351.0832(22), 357.0948 (14), 365.1001 (10), 381.0924(11), 393.0937 (15), 411.1024 (31), 429.1197 (16)	/	/	4.71 ± 0.05
35a	2″-Hexosyl-6″-acetyl cytisoside [c]	8.69	$C_{30}H_{35}O_{16}^+$	651.19250	651.19531	−2.81	297.0739(3), 309.0738 (3), **327.0852 (8)**, 351.0839 (2), 369.0966 (27), 393.095(2), 411.1064 (7), 429.1165 (9), 471.1271 (10), **489.1382 (100)**	/	/	67.44 ± 0.34
36a	2″-Hexuronyl-6″-acetyl cytisoside [c]	8.76	$C_{30}H_{33}O_{17}^+$	665.17180	665.17943	−7.63	297.0742(10), 309.0736 (18), **327.0865 (100)**, 351.0839 (12), 453.1161(11), 459.1265 (11), 471.1254(12), 489.1366(10), 515.1169 (21), **533.1280 (49)**	/	/	18.49 ± 0.08

Table 5. *Cont.*

No	Compounds Name	RT	Formula	Calculated Mass	m/z Exact Mass	mDa	MS Fragments (% of Base Peaks)	Samples (mg/100 mL)		
								AMJ	BCJ	RBJ
37a	6″-Acetyl cytisoside [c]	9.30	$C_{24}H_{25}O_{11}^+$	489.13914	489.14587	−6.73	297.0740(33), 309.074 (21), **327.0846 (95)**, 351.0846 (13), **369.0954 (100)**, 381.0946(18), 393.0946 (15), 411.1052 (34), 429.1160(45), 471.1267 (16)	/	/	9.39 ± 0.04
38a	2″-Malonyl-6″-acetyl-cytisoside [c]	9.57	$C_{27}H_{27}O_{14}^+$	575.14010	575.14610	−6.00	127.0370(7), 129.1006 (7), 297.0701 (12), 309.0736(51), **327.0849 (66)**, 351.0842 (19), **369.0949 (100)**, 375.0937 (8), 393.0966 (13), 453.1132(9), 471.1253 (9)	/	/	1.40 ± 0.01
39a	Apigenin [a]	10.44	$C_{15}H_9O_5^-$	269.04500	269.05007	−5.07	136.9884(53), 139.0059 (53), 141.0708 (26), 143.0506(19), 167.0342 (31), 169.0656 (44), 171.0446(35), 179.0495 (19), 195.0448 (50), 197.0606(25), 223.0392 (52), 241.0492 (43), 251.0359(19), **269.0453 (100)**	1.06 ± 0.02 [B]	/	9.39 ± 0.05 [A]
				Σ				1.06	/	323.17
						Other flavonoids				
40a	Kaempferol-3-O-sinapoyl-dihexoside-7-O-hexoside [c]	6.85	$C_{44}H_{49}O_{25}^-$	977.25630	977.26510	−8.80	**815.2079 (100)**, 816.2111(54), 977.2594 (20), 609.1468 (13), **284.0332 (9)**, 446.085 (3)	/	4.77 ± 0.07	/
41a	Chalcan-flavan 3-ol dimer [c]	7.68	$C_{27}H_{31}O_{14}^-$	579.17140	579.17180	−0.40	116.0382(13), 117.0445 (1), 125.0248 (10), 151.0035 (2), 167.0345 (28), 179.0413 (1), 201.1035 (3), 203.0823 (31), 204.0835 (4), **245.0924(100)**, 246.0951(19), 247.0961 (2), 271.0607 (4), **289.0706 (47)**	/	/	0.19 ± 0.01
42a	Europetin [c]	9.37	$C_{16}H_{11}O_8^-$	331.04594	331.04602	−0.08	110.0017(37), 111.0082 (6), 121.0299(14), 137.9962(20), 139.0037 (26), 140.0085 (4), **165.9906 (100)**, 166.9962(24), 181.0143 (11), 193.9856 (6), 243.0284 (5), 271.0239 (8), 287.0173 (5), 316.0210 32), 317.0235 (7)	0.04 ± 0.001	/	/
				Σ				0.04	4.77	0.19
				ΣΣ				49.84	362.37	342.02
						Other detected compounds				
43a	Tuberonic acid	9.63	$C_{12}H_{17}O_4^-$	225.11270	225.11193	0.77	109.0414(11), 109.0694 (11), **110.0387(100)**, 111.0439(16), 123.0416 (14), 123.0707 (10), 135.0837(10), 136.0548 (36), 161.0720 (8), 161.1000(9), 163.1125 (25), 179.1085 (8), 181.1220(24), 207.1026 (88), 208.1046 (13)	+	+	-
44a	Methyl jasmonate	11.53	$C_{13}H_{19}O_3^-$	223.13340	223.13275	0.65	120.0274(27), 121.0237 (23), 123.1018 (28), 141.8757(21), 142.0382 (32), 142.0750 (25), **143.0676 (100)**, 143.1133 (57), 143.1413(26), 151.0361 (38), 168.8724 (26), 205.8271(26), 205.8642 (22), 214.9475 (21)	+	+	+

Abbreviations: AMJ—cold-pressed amaranth microgreen juice; BCJ—cold-pressed broccoli microgreen juice; RBJ—cold-pressed red beet microgreen juice. Compound quantities expressed using available standards [a]; Compounds expressed as sinapic acid equivalents [b]; compounds expressed as apigenin equivalents [c]. "/"—nonidentified phenolic compounds; "+"—other detected compounds. Means with the same uppercase letter in the same raw are significantly different according to Duncan's test, ($p < 0.05$), (mean ± S.D.; $n = 3$).

In addition, more phenolic acids (aglycones) and their derivatives with organic acids (quinic acid and malic acid) were identified in the juices of microgreens (Table 5), probably due to the increased contact between these molecules and their tendency to interact with each other. In amaranth microgreens and its juice, a high content of pentosyl hexoside glycosides of vanillic acid and dihydroxybenzoic acid was confirmed. However, some compounds that were dominantly detected in amaranth microgreens, such as feruloyl isocitric acid, benzoyl malic acid, and some glycosides of vanillic and hydroxybenzoic acids, were not detected in AMJ. It can be assumed that these compounds were degraded or transformed during juice production. This is partially supported by the fact that the high content of hydroxybenzoic (14.97 mg/100 mL) and benzoic (3.68 mg/100 mL) acid was detected only in amaranth juice (AMJ). A wide variety of phenolic acids derivatives were identified in BCJ. Commonly present sinapic acid derivatives (sinapic acid, sinapoyl malic acid, and sinapic acid hexoside) were detected in high amounts in both broccoli microgreens and juice (Tables 2 and 5). However, the characteristic di- and trisinapoyl glycosides previously detected in broccoli microgreens [10], were not found in BCJ, so it can be assumed that these macromolecules were retained in the residues generated during the pressing of broccoli juice. In addition, high levels of hydroxybenzoic acid and dihydroxybenzoic acid along with benzoic acid derivatives were also detected in BCJ, while their glycoside forms were present only in trace amounts. It is worth mentioning that kaempferol-3-O-sinapoyl-dihexoside-7-O-hexoside was the only flavonoid detected in BCJ, with a content of 4.77 mg/100 mL. In contrast to AMJ and BCJ, low levels of phenolic acid derivatives (about 5.5% of the total amount of PCs) were detected in red beet microgreen juice (Table 5). Among the individual phenolic acids, dihydroxybenzoic acid pentosyl hexoside (similar to that in BR) and hydroxybenzoic acid were the dominant compounds in this juice, with contents of 7.33 mg/100 mL and 9.81 mg/100 mL, respectively. Other phenolic acid derivatives were found in small or trace amounts. Interestingly, dihydroxybenzoic acid dipentoside, the predominant compound in red beet microgreens, was not detected in the juice. This compound was probably retained in the solid waste of the microgreens or converted to other phenolic acid derivatives during juice production. However, the juice of red beet microgreens is a good source of several apigenin C-glycosides (Table 5), which appear to be highly soluble and readily transferred from the microgreens to the juice. A total of 13 vitexin and cytisoside derivatives were identified, accounting for 94.4% of the total amount of PCs in RBJ. Cytisoside derivatives dominated, especially $2''$-hexosyl-$6''$-acetyl cytisioside (67.44 mg/100 mL), followed by hexosyl cytisoside (48.94 mg/100 mL) and pentosyl cytisoside (46.00 mg/100 mL), just as in red beet microgreens. In addition, $2''$-hexuronyl-$6''$-acetyl cytisoside (compound 36, Table 5) was also confirmed at a high level (18.49 mg/mL), but only in RBJ. The proposed fragmentation pathway for this compound is shown in Figure 2b. Among the identified vitexin derivatives, special attention should be paid to $2''$-hexosyl-$6''$-malonyl vitexin, $2''$-hexosyl vitexin and $2''$-pentosyl vitexin, whose individual contents exceeded 25 mg/100 mL. Several studies have already identified these vitexin derivatives in red beet leaves and pointed out their numerous health benefits [39,40,57,58]. Other apigenin C-glycosides were present in significantly lower amounts. Chromatograms extracted on the accurate masses of dominant cytisoside and vitexin derivatives are shown in Figure 2a. Finally, to the best of our knowledge, this study marked the first time that detailed profiles of the phenolic compounds of microgreen juices were obtained, making comparisons with other studies difficult.

In addition to the phenolic compounds, several characteristic betalains were identified in RBJ and AMJ (Table 6).

The good water solubility of betalains contributes to their effective and rapid leaching (migration) from the tissues to juice. Amaranthin and isoamaranthine were detected mainly in AMJ. In contrast, the most abundant betalains in red beet juice were betaine and isobetanin. Similarly to our results, Sawicki and Martinez-Villaluenga [43] reported a high total content of betalains in fresh red beet juice, with betanin dominating. In both microgreen juices (AMJ and RBJ), betalamic acid was identified as the third dominant compound. Betalamic acid is a precursor in the synthesis of betalains, which explains its presence in these microgreen juices. On the other hand, betanin can form complexes with feruloyl residues, which explains the presence of $6'$-O-feruloyl-betanin in RBJ. This compound was previously found in dried red beet [44] and other red beet products [43]. The relative content of decarboxy derivatives of amaranthine, betanin, and neobetanin was low in both juices (<3% of the total betalains). This means that the betalains remained as carriers of the red color in these cold-pressed microgreen juices. A higher content of decarboxy derivatives is characteristic of thermally treated juices, resulting in a reduction in betalains, an increased content of decarboxy derivatives of betalains, and the appearance of a brown color [60]. Betaxanthins (two compounds) were found only in red beet juice (RBJ), but their relative content was low (<4% of the total betalains and betaxanthins).

Table 6. Characterization and relative content (%) of betalains in amaranth (AMJ) and red beet (RBJ) microgreen juices by UHPLC-QToF-MS. Target compounds, expected retention time (RT), molecular formula, calculated mass, exact mass, and MS fragments are presented.

No	Compounds Name	RT	Formula	Calculated Mass	m/z Exact Mass	mDa	MS Fragments (% of Base Peaks)	Samples (%) AMJ	Samples (%) RBJ	Ref
				Betalains and betaxanthins						
45a	Amaranthin	2.68	$C_{30}H_{35}N_2O_{19}^+$	727.18340	727.18497	−1.57	389.0989 (100), 551.1524 (7)	78.22	-	
46a	Betalamic acid	3.77	$C_9H_{10}NO_5^+$	212.05590	212.05630	−0.40	102.0344 (3), 106.0293 (9), **120.0454 (100)**, 121.0475 (12), 122.0468 (2), 130.0292 (3), 138.0547 (1), 148.0389 (8), 149.0394 (1), 166.0469 (1)	7.51	8.29	
47a	γ-Aminobutyric acid-betaxanthin	3.16	$C_{12}H_{15}N_2O_6^+$	283.09300	283.09369	−0.69	102.0341(2), 116.0698 (4), 119.0361(3), **136.0610(100)**, 137.0632(9), 148.0400 (60), 149.0426 (8), 212.0448(2), 237.0866(3), 239.0570 (4), 248.0540 (2), 266.0677 (3), 283.0931 (33), 284.0950 (8)	-	0.86	
48a	Isoamaranhthin	4.91	$C_{30}H_{35}N_2O_{19}^+$	727.18340	727.18378	−0.38	389.0985 (100), 551.1515 (6)	8.78	-	
49a	Betanin	5.11	$C_{24}H_{27}N_2O_{13}^+$	551.15130	551.15207	−0.77	150.0540 (2), 343.0909 (2), 345.1058 (1), **389.0978(100)**, 390.0990(32), 551.1503 (5)	3.91	70.54	
50a	Isobetanin	5.59	$C_{24}H_{27}N_2O_{13}^+$	551.15130	551.15703	−5.73	150.0531 (2), 343.0895 (1), **389.0959 (100)**, 390.0999 (25), 391.1010(5), 551.1478 (4), 552.1526 (2)	-	7.49	
51a	Decarboxy-dehydro-(iso)amaranthin	5.86	$C_{29}H_{33}N_2O_{17}^+$	681.17790	681.18521	−7.31	297.0847 (8), 299.0987 (3), 343.0913 (100), 505.1446 (28)	1.58	-	[41,43,44]
52a	(2 or 17)-Decarboxy(iso)-betanin	6.13	$C_{23}H_{27}N_2O_{11}^+$	507.16150	507.16866	−7.16	150.0535 (2), 151.0613 (1), 299.0993 (2), **345.1059(100)**, 346.1116 (27), 347.1114 (4), 507.1575 (3)	-	1.33	
53a	(2 or 17)-Decarboxy-neobetanin	6.13	$C_{23}H_{25}N_2O_{11}^+$	505.14580	505.15438	−8.58	253.0948 (3), 255.0956 (4), 269.0894 (5), 281.0887 (2), 297.0851 (20), 298.0888 (5), 299.0987 (3), **343.0899 (100)**, 344.0943 (24), 345.1067 (49), 346.1086 (8), 505.1434 (10)	-	1.53	
54a	Isoleucine-betaxanthin	6.94	$C_{15}H_{21}N_2O_6^+$	325.14000	325.14250	−2.50	104.0494(12), 106.0621 (17), 119.0612(10), 132.0511(14), 133.0753 (35), 147.0868 (14), 148.0480(12), 150.0540 (16), 173.0704 (14), 189.1365(35), **191.081 (100)**, 192.0843(17), 205.1273(11), 233.1240 (17), 325.1360 (10)	-	2.88	
55a	6′-O-Feruloyl-betanin	7.41	$C_{34}H_{35}N_2O_{16}^+$	727.19870	727.20406	−5.36	389.0975 (100)	-	7.07	
				Total (%)				100	100	

Abbreviations: AMJ—cold-pressed amaranth microgreen juice; RBJ—cold-pressed red beet microgreen juice.

3.4. Total Phenolic, Flavonoid, Betalain, and Chlorophyll Content in Cold-Pressed Microgreen Juices

The spectrophotometrically determined total phenolic and flavonoid content of the microgreen juices is shown in Figure 3a.

The highest TPC value was obtained for broccoli juice (92.92 mg GAE/100 mL), followed by red beet (71.26 mg GAE/100 mL) and amaranth (50.86 mg GAE/100 mL) microgreen juices, which showed the same trend as the results of chromatographic analysis.

The obtained results are in the range of TPC values previously reported by other authors for differently processed *Alternanthera sessilis* [16] and wheatgrass [18] microgreen juices. The TFC values for amaranth, red beet and broccoli microgreen juices were 45.94, 46.35 and 61.56 mg QE/100 mL, respectively. As shown in Figure 3a, the TFC values for RBJ and AMJ were not significantly different. These TFC results do not agree with the results of chromatographic analysis, which showed the dominant presence of flavonoids in RBJ, a low level in BCJ and traces in AMJ. However, the interpretation of these results should take into account the limitation of the spectrophotometric TFC method. Finally, the obtained variations in TFC values can be explained by the following facts: (1) the most abundant apigenin derivatives in RBJ do not contribute to the absorbance at 510 nm, and (2) phenolic acids show considerable absorbance at the same wavelength [61].

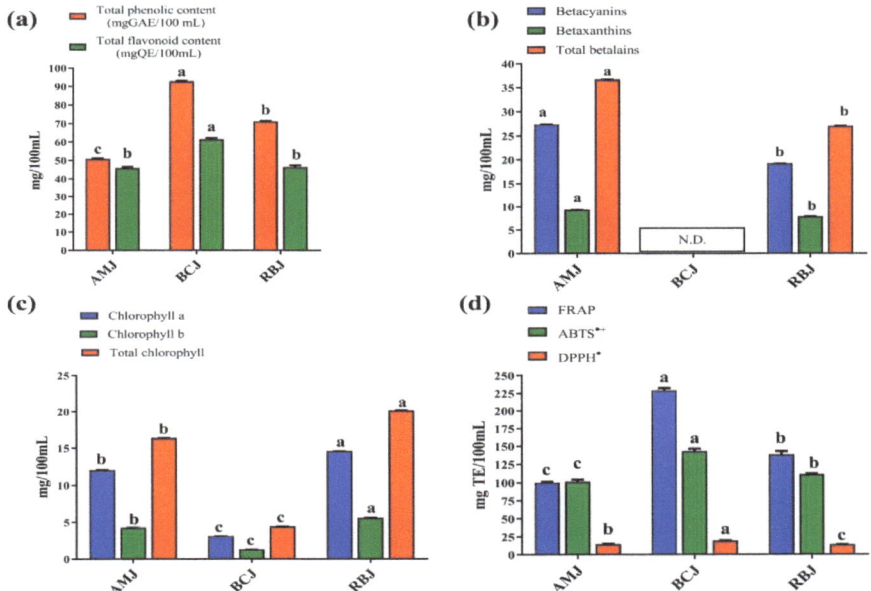

Figure 3. Total phenolic and flavonoid content (**a**), total betalain and betaxanthin content (**b**), total chlorophylls content (**c**), and antioxidant properties (**d**) of microgreen juices. Different lowercase letters denote significant differences between the microgreen juices, according to Duncan's test ($p < 0.05$). Abbreviations: AMJ—cold-pressed amaranth microgreen juice; BCJ—cold-pressed broccoli microgreen juice; RBJ—cold-pressed red beet microgreen juice.

Pigments are responsible for the appealing and attractive color of microgreens. The intense color of amaranth and red beet juices comes from betalain pigments, which have various health benefits. Amaranth microgreen juice had significantly higher levels of total betalains (36.73 mg/100 mL), betacyanins (27.39 mg/100 mL), and betaxanthins (9.34 mg/100 mL), compared to red beet microgreen juice (Figure 3b). However, the obtained results suggest that the prepared cold-pressed microgreen juices are a good source of betalains. Moreover, betacyanins were predominantly detected in both juices (RBJ and AMJ), almost three times more than betaxanthins. Betalains were not detected in broccoli juice. These results are in agreement with the chromatographic profiles (Table 6). On the other hand, the green color of microgreens comes from chlorophylls, and the green hue of plant samples is often directly proportional to the amount of these pigments. The highest levels of total chlorophyll (20.19 mg/100 mL), chlorophyll *a* (14.60 mg/100 mL), and chlorophyll *b* (5.60 mg/100 mL) were found in red beet microgreen juice, while the lowest levels were found in broccoli microgreen juice (Figure 3c). Chlorophylls are unstable pigments that are readily converted to pheophytin (dark green/brown color) at an acidic pH, and this conversion may occur primarily in cold-pressed broccoli microgreen juices.

3.5. Antioxidant Properties of the Cold-Pressed Microgreen Juices

The results of the antioxidant assays for the microgreen juices are shown in Figure 3d. All microgreen juices prepared had good ABTS$^{\bullet+}$ scavenging activity. The highest activity against ABTS$^{\bullet+}$ was shown by broccoli microgreen juice (143.38 mg TE/100 mL), to which phenolic compounds and isothiocyanates probably contribute. On the other hand, the obtained values of DPPH$^{\bullet}$ scavenging activity for all microgreen juices were significantly lower (about 7-fold) compared to the ABTS$^{\bullet+}$ values. The prepared microgreen juices mostly contain hydrophilic biocompounds, which apparently have a greater affinity to interact with more polar ABTS radical cations than with the hydrophobic DPPH radicals. Similar observations and results were reported by Skoczylas and Korus [18], who analyzed the ABTS$^{\bullet+}$ and DPPH$^{\bullet}$ scavenging activity of fresh and frozen wheatgrass juice. Moreover, all microgreen juices had a high tendency to reduce the $[Fe^{3+}\text{-}(TPTZ)_2]^{3+}$ complex, indicating their good reduction capacity. The highest FRAP value was obtained for BCJ, followed by RBJ and AMJ. Correlation analysis revealed that FRAP of the microgreen juices had a strong positive correlation with TPC ($r = 0.98$), TFC ($r = 0.96$) and total phenolic acids ($r = 0.93$).

3.6. Sensory Properties of Cold-Pressed Microgreen Juices

The results for the sensory quality evaluation of microgreen juices produced are shown in Figure 4a.

The mean quality scores for odor, texture, and overall quality did not significantly differ for all analyzed microgreen juices. Amaranth microgreen juice had the highest score for texture (4.7) due to its completely clear appearance and absence of colloidal cloudiness. All analyzed juices had a typical odor, characteristic of the respective microgreens and plant species. On the other hand, the mean value for the appearance of broccoli juice was significantly lower ($p < 0.05$) compared to the juices from red beet and amaranth microgreens. According to the comments of evaluators, the low score for the appearance of broccoli juice was due to high colloidal turbidity and rapid precipitation of particles. Amaranth and red beet microgreen juices received excellent scores for appearance due to their light purple/red color, which is characteristic of these microgreen species. Evaluators provided critical comments and low scores for taste, ranging from 4.0 (BCJ) to 2.9 (RBJ). The main deficiencies of amaranth and red beet microgreen juices were a slight astringency and an earthy and stale flavor, while broccoli microgreen juice had a flavor typical of plants from the *Brassicaceae* family, with herbaceous, grassy, and sulfurous notes. The overall quality of the microgreen juices was rated 3.5 to 4.5 points, indicating good acceptability and sensory quality of these juices.

The results for overall acceptability of the microgreen juices ranged from 5.0 (neither like nor dislike) to 6.4 (slightly like) (Figure 4b). The broccoli microgreen juice received the lowest rating for overall acceptability. The evaluators pointed out the bitter and astringent taste of broccoli juice, which was probably due to the phenolic acids and GLSs metabolites (isothiocyanates) that were predominantly present. In addition, the score of overall acceptability showed a strong positive correlation with the results for taste ($r = 0.771$) and odor ($r = 0.611$). This indicates that these sensory attributes had the greatest influence on the overall acceptance of the microgreen juices by consumers. The high overall acceptance and good appearance of the amaranth microgreen juice was probably mainly influenced by its intense color. Similar observations were reported by other authors [24,25], who studied the sensory properties of broccoli, amaranth and red beet microgreens.

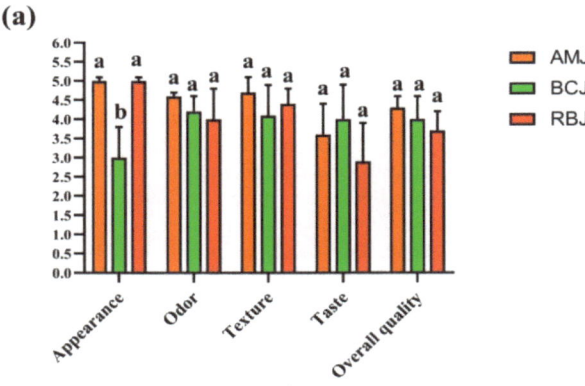

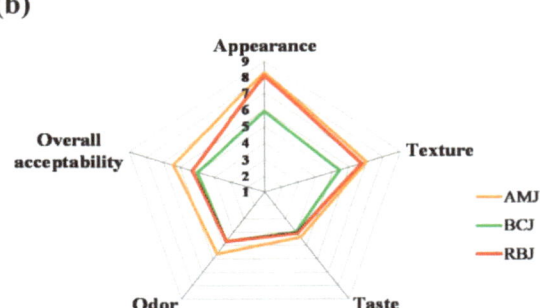

Figure 4. (**a**) Sensory quality scores; (**b**) sensory radar chart for the likeability testing of amaranth, red beet, and broccoli microgreen juices. Different lowercase letters denote a significant difference between the microgreen juices, separately evaluated for each sensory attribute, according to Duncan's test ($p < 0.05$). Abbreviations: AMJ—cold-pressed amaranth microgreen juice; BCJ—cold-pressed broccoli microgreen juice; RBJ—cold-pressed red beet microgreen juice.

4. Conclusions

In summary, the prepared juices from amaranth, broccoli, and red beet microgreens had a high content of phenolic compounds, but this was significantly lower than in the corresponding extracts of microgreens. In addition, the composition of the bioactive compounds in the juices differed significantly from their profile in the starting materials. Various glucosinolates have been detected in broccoli microgreens. However, they were not found in broccoli microgreen juice, as they are degraded to isothiocyanates. Further, more phenolic acid aglycones and their derivatives with organic acids (quinic acid and malic acid) were detected in the juices compared to extracts of microgreens. This is probably due to the differential migration of individual phenolic compounds from the plant tissue into the juice or their transformation due to enzyme action and increased contact between the molecules during the production process. Phenolic acids, especially sinapic acid and its derivatives (sinapoyl malic acid and sinapic acid hexoside), were predominantly found in broccoli microgreen juice, while pentosyl hexoside glycosides of vanillic acid and dihydroxybenzoic acid were most frequently detected in amaranth microgreen juices. Low levels of phenolic acid derivatives were found in red beet microgreen juice. However, red beet microgreens and juices were a good source of apigenin C-glycosides (vitexin and cytosioside derivatives), which apparently readily transfer from the microgreens to the juice. Betalains were detected in both red beet and amaranth microgreens and juices, with betanin and decarboxy betanin dominating in red beet microgreens/juice, and amaranthin dominating in amaranth microgreens/juice. Interestingly, betalamic acid and some betaxanthins were detected, although at low levels, only in red beet and amaranth juices.

All microgreen juices exhibited good ABTS$^{\bullet+}$ scavenging activity and ferric reducing antioxidant power (FRAP). Considering the previous characterizations, it can be concluded that phenolic acid derivatives/isothiocyanate, phenolic acid derivatives/amaranthin, and apigenin-C-glycoside/betanin have the greatest influence on the functional properties of broccoli, amaranth, and red microgreen juices, respectively. In addition, these microgreen juices showed good overall quality and overall acceptability. Cold-pressing resulted in a high yield of microgreen juices characterized by mild acidity and low total soluble solids, which means they can be recommended as novel functional and low-calorie beverages. However, the specific flavor and astringency were the main drawbacks of the prepared microgreen juices. Therefore, it would be desirable to produce and characterize microgreen/fruit-based juices in future studies to mask the flavor and improve their sensory properties.

Supplementary Materials: The following supporting information can be downloaded at: https://www.mdpi.com/article/10.3390/foods13050757/s1, Table S1: Equation parameters and correlation coefficient (R^2) of the phenolic standards used for quantification.

Author Contributions: Conceptualization: S.D.B., D.D.M., S.M.L. and V.A.N.; Methodology: S.D.B., D.D.M. and U.M.G.; Investigation: S.D.B., D.D.M. and M.B.P.; Validation: D.D.M. and U.M.G.; Formal analysis: S.D.B., D.D.M., U.M.G., A.Ž.K., A.S.S.-J., J.M.M., V.B.Đ. and S.M.L.; Data curation: S.D.B. and D.D.M.; Writing—original draft preparation: S.D.B. and D.D.M.; Supervision: U.M.G., A.Ž.K., A.S.S.-J., S.M.L., M.B.P. and V.A.N.; Writing—review and editing: A.S.S.-J., D.D.M., M.B.P. and V.A.N.; Funding acquisition: M.B.P. and V.A.N.; Resources: M.B.P. and V.A.N. All authors have read and agreed to the published version of the manuscript.

Funding: This work was supported by the Ministry of Science, Technological Development and Innovation of the Republic of Serbia, Grant No. 451-03-47/2023-01/200116 and 451-03-65/2024-03/200116 and the Science Fund of the Republic of Serbia, #GRANT No. 7744714, FUNPRO and #GRANT No. 7751519, MultiPromis.

Institutional Review Board Statement: All sensory tests with participants were conducted in accordance with the Code of Professional Ethics of the University of Belgrade [49]. The study follows the Serbian Law on the Protection of Personal Data (Official Gazzet of the Republic of Serbia, 87/2018).

Informed Consent Statement: Informed consent was obtained from all subjects involved in the study.

Data Availability Statement: The authors declare that the data supporting the findings of this study are available within the paper. Should any raw data files be needed in another format, they are available from the corresponding author upon reasonable request.

Conflicts of Interest: The authors declare no competing financial interests.

References

1. Ebert, A.W. Sprouts and Microgreens-Novel Food Sources for Healthy Diets. *Plants* **2022**, *11*, 571. [CrossRef]
2. Le, T.N.; Chiu, C.H.; Hsieh, P.C. Bioactive Compounds and Bioactivities of *Brassica oleracea* L. var. Italica Sprouts and Microgreens: An Updated Overview from a Nutraceutical Perspective. *Plants* **2020**, *9*, 946. [CrossRef] [PubMed]
3. Zhang, Y.; Xiao, Z.; Ager, E.; Kong, L.; Tan, L. Nutritional quality and health benefits of microgreens, a crop of modern agriculture. *J. Future Foods* **2021**, *1*, 58–66. [CrossRef]
4. Bhaswant, M.; Shanmugam, D.K.; Miyazawa, T.; Abe, C.; Miyazawa, T. Microgreens—A Comprehensive Review of Bioactive Molecules and Health Benefits. *Molecules* **2023**, *28*, 867. [CrossRef] [PubMed]
5. Sharma, S.; Shree, B.; Sharma, D.; Kumar, S.; Kumar, V.; Sharma, R.; Saini, R. Vegetable microgreens: The gleam of next generation super foods, their genetic enhancement, health benefits and processing approaches. *Food Res. Int.* **2022**, *155*, 111038. [CrossRef]
6. Fuente, B.; López-García, G.; Máñez, V.; Alegría, A.; Barberá, R.; Cilla, A. Antiproliferative Effect of Bioaccessible Fractions of Four Brassicaceae Microgreens on Human Colon Cancer Cells Linked to Their Phytochemical Composition. *Antioxidants* **2020**, *9*, 368. [CrossRef]
7. Renna, M.; Di Gioia, F.; Leoni, B.; Mininni, C.; Santamaria, P. Culinary Assessment of Self-Produced Microgreens as Basic Ingredients in Sweet and Savory Dishes. *J. Culin. Sci. Technol.* **2017**, *15*, 126–142. [CrossRef]

8. Alloggia, F.P.; Bafumo, R.F.; Ramirez, D.A.; Maza, M.A.; Camargo, A.B. Brassicaceae microgreens: A novel and promissory source of sustainable bioactive compounds. *Curr. Res. Food Sci.* **2023**, *6*, 100480. [CrossRef] [PubMed]
9. Castellaneta, A.; Losito, I.; Cisternino, G.; Leoni, B.; Santamaria, P.; Calvano, C.D.; Bianco, G.; Cataldi, T.R.I. All Ion Fragmentation Analysis Enhances the Untargeted Profiling of Glucosinolates in Brassica Microgreens by Liquid Chromatography and High-Resolution Mass Spectrometry. *J. Am. Soc. Mass Spectrom.* **2022**, *33*, 2108–2119. [CrossRef]
10. Liu, Z.; Shi, J.; Wan, J.; Pham, Q.; Zhang, Z.; Sun, J.; Yu, L.; Luo, Y.; Wang, T.T.Y.; Chen, P. Profiling of Polyphenols and Glucosinolates in Kale and Broccoli Microgreens Grown under Chamber and Windowsill Conditions by Ultrahigh-Performance Liquid Chromatography High-Resolution Mass Spectrometry. *ACS Food Sci. Technol.* **2022**, *2*, 101–113. [CrossRef]
11. Demir, K.; Sarıkamış, G.; Çakırer Seyrek, G. Effect of LED lights on the growth, nutritional quality and glucosinolate content of broccoli, cabbage and radish microgreens. *Food Chem.* **2023**, *401*, 134088. [CrossRef] [PubMed]
12. Rocchetti, G.; Tomas, M.; Zhang, L.; Zengin, G.; Lucini, L.; Capanoglu, E. Red beet (*Beta vulgaris*) and amaranth (*Amaranthus* sp.) microgreens: Effect of storage and in vitro gastrointestinal digestion on the untargeted metabolomic profile. *Food Chem.* **2020**, *332*, 127415. [CrossRef] [PubMed]
13. Kyriacou, M.C.; El-Nakhel, C.; Pannico, A.; Graziani, G.; Soteriou, G.A.; Giordano, M.; Zarrelli, A.; Ritieni, A.; De Pascale, S.; Rouphael, Y. Genotype-Specific Modulatory Effects of Select Spectral Bandwidths on the Nutritive and Phytochemical Composition of Microgreens. *Front. Plant Sci.* **2019**, *10*, 1501. [CrossRef] [PubMed]
14. Wojdyło, A.; Nowicka, P.; Tkacz, K.; Turkiewicz, I.P. Sprouts vs. Microgreens as Novel Functional Foods: Variation of Nutritional and Phytochemical Profiles and Their In Vitro Bioactive Properties. *Molecules* **2020**, *25*, 4648. [CrossRef] [PubMed]
15. Li, X.; Tian, S.; Wang, Y.; Liu, J.; Wang, J.; Lu, Y. Broccoli microgreens juice reduces body weight by enhancing insulin sensitivity and modulating gut microbiota in high-fat diet-induced C57BL/6J obese mice. *Eur. J. Nutr.* **2021**, *60*, 3829–3839. [CrossRef] [PubMed]
16. Senthilnathan, K.; Muthusamy, S. Process optimization & kinetic modeling study for fresh microgreen (*Alternanthera sessilis*) juice treated under thermosonication. *Prep. Biochem. Biotechnol.* **2022**, *52*, 433–442.
17. Ali, N.; Popović, V.; Koutchma, T.; Warriner, K.; Zhu, Y. Effect of thermal, high hydrostatic pressure, and ultraviolet-C processing on the microbial inactivation, vitamins, chlorophyll, antioxidants, enzyme activity, and color of wheatgrass juice. *J. Food Process. Eng.* **2019**, *43*, e13036. [CrossRef]
18. Skoczylas, Ł.; Korus, A.; Tabaszewska, M.; Gędoś, K.; Szczepańska, E. Evaluation of the quality of fresh and frozen wheatgrass juices depending on the time of grass harvest: SKOCZYLAS et al. *J. Food Process. Preserv.* **2017**, *42*, e13401. [CrossRef]
19. Klopsch, R.; Baldermann, S.; Voss, A.; Rohn, S.; Schreiner, M.; Neugart, S. Bread Enriched With Legume Microgreens and Leaves—Ontogenetic and Baking-Driven Changes in the Profile of Secondary Plant Metabolites. *Front. Chem.* **2018**, *6*, 322. [CrossRef]
20. Sharma, P.; Sharma, A.; Rasane, P.; Dey, A.; Choudhury, A.; Singh, J.; Kaur, S.; Dhawan, K.; Kaur, D. Optimization of a process for microgreen and fruit-based functional beverage. *An. Acad. Bras Cienc.* **2020**, *92*, e20190596. [CrossRef] [PubMed]
21. Manzoor, M.F.; Hussain, A.; Goksen, G.; Ali, M.; Khalil, A.A.; Zeng, X.-A.; Jambrak, A.R.; Lorenzo, J.M. Probing the impact of sustainable emerging sonication and DBD plasma technologies on the quality of wheat sprouts juice. *Ultrason. Sonochem.* **2023**, *92*, 106257. [CrossRef]
22. Ferruzza, S.; Natella, F.; Ranaldi, G.; Murgia, C.; Rossi, C.; Trošt, K.; Mattivi, F.; Nardini, M.; Maldini, M.; Giusti, A.M.; et al. Nutraceutical Improvement Increases the Protective Activity of Broccoli Sprout Juice in a Human Intestinal Cell Model of Gut Inflammation. *Pharmaceuticals* **2016**, *9*, 48. [CrossRef]
23. Chartoumpekis, D.V.; Ziros, P.G.; Chen, J.G.; Groopman, J.D.; Kensler, T.W.; Sykiotis, G.P. Broccoli sprout beverage is safe for thyroid hormonal and autoimmune status: Results of a 12-week randomized trial. *Food Chem. Toxicol.* **2019**, *126*, 1–6. [CrossRef]
24. Xiao, Z.; Lester, G.E.; Park, E.; Saftner, R.A.; Luo, Y.; Wang, Q. Evaluation and correlation of sensory attributes and chemical compositions of emerging fresh produce: Microgreens. *Postharvest Biol. Technol.* **2015**, *110*, 140–148. [CrossRef]
25. Michell, K.; Isweiri, H.; Newman, S.; Bunning, M.; Bellows, L.; Dinges, M.; Grabos, L.; Rao, S.; Foster, M.; Heuberger, A.; et al. Microgreens: Consumer sensory perception and acceptance of an emerging functional food crop. *J. Food Sci.* **2020**, *85*, 926–935. [CrossRef] [PubMed]
26. Cano-Lamadrid, M.; Martínez Zamora, L.; Castillejo, N.; Cattaneo, C.; Pagliarini, E.; Artés-Hernández, F. How does the phytochemical composition of sprouts and microgreens from Brassica vegetables affect the sensory profile and consumer acceptability? *Postharvest Biol. Technol.* **2023**, *203*, 112411. [CrossRef]
27. Caracciolo, F.; El-Nakhel, C.; Raimondo, M.; Kyriacou, M.C.; Cembalo, L.; De Pascale, S.; Rouphael, Y. Sensory Attributes and Consumer Acceptability of 12 Microgreens Species. *Agronomy* **2020**, *10*, 1043. [CrossRef]
28. Chen, H.; Tong, X.; Tan, L.; Kong, L. Consumers' acceptability and perceptions toward the consumption of hydroponically and soil grown broccoli microgreens. *J. Agric. Food Res.* **2020**, *2*, 100051. [CrossRef]
29. Kolarević, T.; Milinčić, D.D.; Vujović, T.; Gašić, U.M.; Prokić, L.; Kostić, A.Ž.; Cerović, R.; Stanojevic, S.P.; Tešić, Ž.L.; Pešić, M.B. Phenolic compounds and antioxidant properties of field-grown and in vitro leaves, and calluses in blackberry and blueberry. *Horticulturae* **2021**, *7*, 420. [CrossRef]
30. Pintać, D.; Majkić, T.; Torović, L.; Orčić, D.; Beara, I.; Simin, N.; Mimica–Dukić, N.; Lesjak, M. Solvent selection for efficient extraction of bioactive compounds from grape pomace. *Ind. Crops Prod.* **2018**, *111*, 379–390. [CrossRef]

31. Milinčić, D.D.; Stanisavljević, N.S.; Kostić, A.Ž.; Soković Bajić, S.; Kojić, M.O.; Gašić, U.M.; Barać, M.B.; Stanojević, S.P.; Lj Tešić, Ž.; Pešić, M.B. Phenolic compounds and biopotential of grape pomace extracts from Prokupac red grape variety. *LWT* **2021**, *138*, 110739. [CrossRef]
32. Acharya, J.; Gautam, S.; Neupane, P.; Niroula, A. Pigments, ascorbic acid, and total polyphenols content and antioxidant capacities of beet (*Beta vulgaris*) microgreens during growth. *Int. J. Food Prop.* **2021**, *24*, 1175–1186. [CrossRef]
33. Kostić, A.Ž.; Milinčić, D.D.; Špirović Trifunović, B.; Nedić, N.; Gašić, U.M.; Tešić, Ž.L.; Stanojević, S.P.; Pešić, M.B. Monofloral Corn Poppy Bee-Collected Pollen—A Detailed Insight into Its Phytochemical Composition and Antioxidant Properties. *Antioxidants* **2023**, *12*, 1424. [CrossRef] [PubMed]
34. Li, Z.; Zheng, S.; Liu, Y.; Fang, Z.; Yang, L.; Zhuang, M.; Zhang, Y.; Lv, H.; Wang, Y.; Xu, D. Characterization of glucosinolates in 80 broccoli genotypes and different organs using UHPLC-Triple-TOF-MS method. *Food Chem.* **2021**, *334*, 127519. [CrossRef] [PubMed]
35. Dong, M.; Tian, Z.; Ma, Y.; Yang, Z.; Ma, Z.; Wang, X.; Li, Y.; Jiang, H. Rapid screening and characterization of glucosinolates in 25 Brassicaceae tissues by UHPLC-Q-exactive orbitrap-MS. *Food Chem.* **2021**, *365*, 130493. [CrossRef] [PubMed]
36. Clarke, D.B. Glucosinolates, structures and analysis in food. *Anal. Methods Anal. Methods* **2010**, *2*, 310. [CrossRef]
37. Cataldi, T.R.; Rubino, A.; Lelario, F.; Bufo, S.A. Naturally occurring glucosinolates in plant extracts of rocket salad (*Eruca sativa* L.) identified by liquid chromatography coupled with negative ion electrospray ionization and quadrupole ion-trap mass spectrometry. *Rapid Commun. Mass Spectrom.* **2007**, *21*, 2374–2388. [CrossRef] [PubMed]
38. Guo, Q.; Sun, Y.; Tang, Q.; Zhang, H.; Cheng, Z. Isolation, identification, biological estimation, and profiling of glucosinolates in *Isatis indigotica* roots. *J. Liq. Chromatogr. Relat. Technol.* **2020**, *43*, 645–656. [CrossRef]
39. Isayenkova, J.; Wray, V.; Nimtz, M.; Strack, D.; Vogt, T. Cloning and functional characterisation of two regioselective flavonoid glucosyltransferases from Beta vulgaris. *Phytochemistry* **2006**, *67*, 1598–1612. [CrossRef]
40. da Silva, L.G.S.; Morelli, A.P.; Pavan, I.C.B.; Tavares, M.R.; Pestana, N.F.; Rostagno, M.A.; Simabuco, F.M.; Bezerra, R.M.N. Protective effects of beet (*Beta vulgaris*) leaves extract against oxidative stress in endothelial cells in vitro. *Phytother. Res.* **2020**, *34*, 1385–1396. [CrossRef]
41. Xie, G.-R.; Chen, H.-J. Comprehensive Betalain Profiling of Djulis (*Chenopodium formosanum*) Cultivars Using HPLC-Q-Orbitrap High-Resolution Mass Spectrometry. *J. Agric. Food Chem.* **2021**, *69*, 15699–15715. [CrossRef] [PubMed]
42. Herbach, K.M.; Stintzing, F.C.; Carle, R. Identification of heat-induced degradation products from purified betanin, phyllocactin and hylocerenin by high-performance liquid chromatography/electrospray ionization mass spectrometry. *Rapid Commun. Mass Spectrom.* **2005**, *19*, 2603–2616. [CrossRef]
43. Sawicki, T.; Martinez-Villaluenga, C.; Frias, J.; Wiczkowski, W.; Peñas, E.; Bączek, N.; Zieliński, H. The effect of processing and in vitro digestion on the betalain profile and ACE inhibition activity of red beetroot products. *J. Funct. Foods* **2019**, *55*, 229–237. [CrossRef]
44. Nemzer, B.; Pietrzkowski, Z.; Spórna, A.; Stalica, P.; Thresher, W.; Michałowski, T.; Wybraniec, S. Betalainic and nutritional profiles of pigment-enriched red beet root (*Beta vulgaris* L.) dried extracts. *Food Chem.* **2011**, *127*, 42–53. [CrossRef]
45. Stintzing, F.; Schieber, A.; Carle, R. Evaluation of colour properties and chemical quality parameters of cactus juices. *Eur. Food Res. Technol.* **2003**, *216*, 303–311. [CrossRef]
46. Benzie, I.F.F.; Strain, J.J. The Ferric Reducing Ability of Plasma (FRAP) as a Measure of "Antioxidant Power": The FRAP Assay. *Anal. Biochem.* **1996**, *239*, 70–76. [CrossRef]
47. *ISO 8589:2007*; Sensory Analysis: General Guidance for Design of Test Rooms. ISO: Genèva, Switzerland, 2007. Available online: https://www.iso.org/standard/76667.html (accessed on 3 January 2024).
48. *ISO 11136:2014*; Sensory Analysis—Methodology—General Guidance for Conducting Hedonic Tests with Consumers in a Controlled Area. ISO: Genèva, Switzerland, 2014. Available online: https://www.iso.org/standard/50125.html (accessed on 3 January 2024).
49. Senate of the University of Belgrade. The Code of Professional Ethics of the University of Belgrade. *Off. Gaz. Repub. Serb.* **2016**, *189*, 16.
50. Tomic, N.; Djekic, I.; Hofland, G.; Smigic, N.; Udovicki, B.; Rajkovic, A. Comparison of Supercritical CO_2-Drying, Freeze-Drying and Frying on Sensory Properties of Beetroot. *Foods* **2020**, *9*, 1201. [CrossRef]
51. *ISO 8586:2023*; Sensory Analysis—Selection and Training of Sensory Assessors. ISO: Genèva, Switzerland, 2023. Available online: https://www.iso.org/standard/36385.html (accessed on 3 January 2024).
52. Petrović, M.; Veljović, S.; Tomić, N.; Zlatanović, S.; Tosti, T.; Vukosavljević, P.; Gorjanović, S. Formulation of Novel Liqueurs from Juice Industry Waste: Consumer Acceptance, Phenolic Profile and Preliminary Monitoring of Antioxidant Activity and Colour Changes during Storage. *Food Technol. Biotechnol.* **2021**, *59*, 282–294. [CrossRef]
53. Tomic, N.; Dojnov, B.; Miocinovic, J.; Tomasevic, I.; Smigic, N.; Djekic, I.; Vujcic, Z. Enrichment of yoghurt with insoluble dietary fiber from triticale–A sensory perspective. *LWT* **2017**, *80*, 59–66. [CrossRef]
54. Zeng, W.; Yang, J.; Yan, G.; Zhu, Z. CaSO(4) Increases Yield and Alters the Nutritional Contents in Broccoli (*Brassica oleracea* L. Var. *italica*) Microgreens under NaCl Stress. *Foods* **2022**, *11*, 3485.
55. Di Bella, M.C.; Toscano, S.; Arena, D.; Moreno, D.A.; Romano, D.; Branca, F. Effects of Growing Cycle and Genotype on the Morphometric Properties and Glucosinolates Amount and Profile of Sprouts, Microgreens and Baby Leaves of Broccoli (*Brassica oleracea* L. var. *italica* Plenck) and Kale (*B. oleracea* L. var. *acephala* DC.). *Agronomy* **2021**, *11*, 1685. [CrossRef]

56. Bello, C.; Maldini, M.; Baima, S.; Scaccini, C.; Natella, F. Glucoraphanin and sulforaphane evolution during juice preparation from broccoli sprouts. *Food Chem.* **2018**, *268*, 249–256. [CrossRef]
57. Lorizola, I.; Furlan, C.P.; Portovedo, M.; Milanski, M.; Botelho, P.; Bezerra, R.; Sumere, B.; Rostagno, M.; Capitani, C. Beet Stalks and Leaves (*Beta vulgaris* L.) Protect Against High-Fat Diet-Induced Oxidative Damage in the Liver in Mice. *Nutrients* **2018**, *10*, 872. [CrossRef] [PubMed]
58. Ninfali, P.; Antonini, E.; Frati, A.; Scarpa, E.S. C-Glycosyl Flavonoids from Beta vulgaris Cicla and Betalains from Beta vulgaris rubra: Antioxidant, Anticancer and Antiinflammatory Activities-A Review. *Phytother. Res.* **2017**, *31*, 871–884. [CrossRef] [PubMed]
59. Li, H.; Deng, Z.; Liu, R.; Zhu, H.; Draves, J.; Marcone, M.; Sun, Y.; Tsao, R. Characterization of phenolics, betacyanins and antioxidant activities of the seed, leaf, sprout, flower and stalk extracts of three *Amaranthus* species. *J. Food Compos. Anal.* **2015**, *37*, 75–81. [CrossRef]
60. Celli, G.B.; Brooks, M.S.-L. Impact of extraction and processing conditions on betalains and comparison of properties with anthocyanins—A current review. *Food Res. Int.* **2017**, *100*, 501–509. [CrossRef]
61. Pękal, A.; Pyrzynska, K. Evaluation of aluminium complexation reaction for flavonoid content assay. *Food Anal. Methods* **2014**, *7*, 1776–1782. [CrossRef]

Disclaimer/Publisher's Note: The statements, opinions and data contained in all publications are solely those of the individual author(s) and contributor(s) and not of MDPI and/or the editor(s). MDPI and/or the editor(s) disclaim responsibility for any injury to people or property resulting from any ideas, methods, instructions or products referred to in the content.

Article

Optimization of Bioactive Compound Extraction from Saffron Petals Using Ultrasound-Assisted Acidified Ethanol Solvent: Adding Value to Food Waste

Nikoo Jabbari [1], Mohammad Goli [1,*] and Sharifeh Shahi [2]

[1] Department of Food Science and Technology, Laser and Biophotonics in Biotechnologies Research Center, Isfahan (Khorasgan) Branch, Islamic Azad University, Isfahan 81551-39998, Iran; n.jabbari@khuisf.ac.ir
[2] Department of Medical Engineering, Laser and Biophotonics in Biotechnologies Research Center, Isfahan (Khorasgan) Branch, Islamic Azad University, Isfahan 81551-39998, Iran; shahilaser@khuisf.ac.ir
* Correspondence: mgolifood@yahoo.com or m.goli@khuisf.ac.ir; Tel.: +98-9132252910; Fax: +98-3135354060

Abstract: The saffron industry produces large by-products, including petals with potential bioactive compounds, which are cheap and abundant, making them an attractive alternative to expensive stigmas for extracting bioactive components. This study aimed to optimize the extraction conditions of bioactive compounds from vacuum-dried saffron petals using an ultrasound-assisted acidified ethanol solvent. Three factors were considered: ethanol concentration (0–96%), citric acid concentration in the final solvent (0–1%), and ultrasound power (0–400 watt). This study examined the effects of these factors on parameters like maximum antioxidant activity, total anthocyanin content, total phenolic content, and the total flavonoid content of the extraction. This study found that saffron petal extract's antioxidant activity increases with higher ethanol concentration, citric acid dose, and ultrasound power, but that an increased water content leads to non-antioxidant compounds. Increasing the dosage of citric acid improved the extraction of cyanidin-3-glucoside at different ultrasound power levels. The highest extraction was achieved with 400 watts of ultrasound power and 1% citric acid. Ethanol concentration did not affect anthocyanin extraction. Higher ethanol concentration and greater citric acid concentration doses resulted in the maximum extraction of total phenolic content, with a noticeable drop in extraction at higher purity levels. This study found that increasing the proportion of citric acid in the final solvent did not affect flavonoid extraction at high ethanol concentration levels, and the highest efficiency was observed at 200 watts of ultrasound power. The optimum values of the independent parameters for extracting bioactive compounds from saffron petals included 96% ethanol concentration, 0.67% citric acid concentration, and 216 watts of ultrasound power, resulting in a desirability value of 0.82. This ultrasound-assisted acidified ethanolic extract can be used in the food industry as a natural antioxidant and pigment source.

Keywords: vacuum-dried saffron petal; bioactive compounds; response surface methodology; acidified-ethanol solvent; ultrasound treatment; optimal extraction conditions

Citation: Jabbari, N.; Goli, M.; Shahi, S. Optimization of Bioactive Compound Extraction from Saffron Petals Using Ultrasound-Assisted Acidified Ethanol Solvent: Adding Value to Food Waste. *Foods* **2024**, *13*, 542. https://doi.org/10.3390/foods13040542

Academic Editors: Rosaria Viscecchia, Francesco Bimbo and Gianluca Nardone

Received: 6 January 2024
Revised: 2 February 2024
Accepted: 6 February 2024
Published: 9 February 2024

Copyright: © 2024 by the authors. Licensee MDPI, Basel, Switzerland. This article is an open access article distributed under the terms and conditions of the Creative Commons Attribution (CC BY) license (https://creativecommons.org/licenses/by/4.0/).

1. Introduction

Studies have shown that medicinal plants contain compounds with significant antioxidant activity, antiradical potential, metal-chelating ability, anticancer properties, and antifungal effects. These compounds, such as phenolics, flavonoids, and carotenoids, can scavenge free radicals, protect against oxidative stress, and potentially prevent chronic diseases. The metal chelating properties of these compounds have been studied for their potential to combat metal-induced toxicity and oxidative stress-related disorders. Additionally, certain compounds derived from medicinal plants have been found to have anticancer properties, exhibiting cytotoxic effects on cancer cells and contributing to apoptosis and cell cycle arrest. Oxidative stress, caused by factors like psychological stress, toxins, industrial lifestyle, infections, drugs, smoking, and obesity, is linked to neurodegenerative disorders like Alzheimer's and Parkinson's. Reactive oxygen species (ROS) and free radicals (FR) are linked to various diseases like inflammation, cardiovascular disease, cancer, and diabetes through lipid peroxidation in cellular membranes. To combat these issues, the body needs to provide a constant supply of antioxidants through dietary supplementation. Natural products have been used to prevent and treat many diseases, including cancer, making them potential candidates for developing anti-cancer drugs [1].

The *Crocus sativus*, a perennial plant cultivated in Iran, accounts for 90% of global saffron production. Its flowers, also known as saffron, are valuable spices and medicinal plants with health-enhancing properties for reducing blood pressure and alleviating depression symptoms, as well as antioxidant, antiradical, metal-chelating, anticancer, and antifungal properties [2]. The saffron industry generates large by-products, including large quantities of petals with potential bioactive compounds. Approximately 98.5% of the saffron-flower material is eventually discarded as waste, since only 15 g of spice can be produced from 1 kg of flowers during the production process. This leads to a large number of by-products. Numerous phytochemical components, including flavonoids, anthocyanins, carotenoids, phenolic acids, monoterpenoids, alkaloids, glycosides, and saponins, are present in saffron by-products [3]. These petals are cheap and abundant, making them an attractive alternative to expensive stigmas for extracting bioactive components, especially in the extraction of bioactive components [4].

Bioactive compounds can be recovered from natural products or agro-industrial by-products using conventional methods like Soxhlet, heat reflux, and maceration. However, these methods have low extraction efficiency due to prolonged heat treatment, thermo-sensitive chemical degradation, and significant energy, time, and solvent usage. The grinding procedure has also been used for phenolic compound extraction, but it is ineffective due to high levels of impurities. Therefore, purifying methods are needed to eliminate undesirable components [5]. Innovative extraction technologies like ultrasound, microwave, pulsed electric field, high voltage electrical discharge, supercritical fluid extraction, and pressurized liquid extraction have been developed to protect the stability of bioactive compounds. These techniques offer excellent extraction efficiency, superior final product quality at lower temperatures, quicker kinetics, and reduced energy/solvent consumption. Ultrasounds and other assisted procedures are widely used to maintain sensitive chemicals, offering key benefits such as high extraction efficiency and reduced energy/solvent usage [6]. Acoustic cavitation is a mechanism where sudden bubbles in a liquid grow and collapse, causing high-speed solvent jets to approach the solid surface, thereby accelerating mass transfer and diffusion by increasing the contact surface area between the solid matrix and solvent phase [7,8]. Ultrasound-assisted extraction (UAE) is a green energy-saving technology that offers benefits like time and energy efficiency, lower extraction temperature, less degradation, higher yield, component selectivity, and reproducibility compared to conventional methods [9–11]. A study investigated the extraction of lipophilic bioactive compounds from saffron petals using hydroethanolic solvent. Conventional methods were time-consuming, less selective, solvent-intensive, and energy inefficient [5]. According to Ferarsa et al. [5], the cavitation mechanism in ultrasound-assisted extraction consisted of two main steps: (i) dissolving the target com-pounds by causing superficial tissue damage, also known as rinsing or rapid extraction, and (ii) diffusion of the desired chemicals into the extraction medium, also known as slow extraction. When cavitation bubbles collapse at the solid matrix's surface, the mechanical force causes a breakdown of the plant cell walls, which encourages a rise in the amount of flavonoids during ultrasonication. This enhances mass transfer and increases the solvent–plant material contact surface area [12].

For the first time, this study aimed to optimize the extraction conditions of bioactive compounds from vacuum-dried saffron petals using ultrasound-assisted acidified ethanol solvent as three factors: factor A (ethanol concentration: 0–100% v/v), factor B (citric acid concentration in the final solvent: 0–1% w/v), and factor C (ultrasound power: 0–400 watt) using the response surface methodology (RSM), central composite design (CCD) with an alpha of 2, one repeat in axial and factorial points, and three central points. The study investigated the effects of these factors on various parameters, including maximum antioxidant activity (measured by DPPH radical scavenging activity and ferric-reducing antioxidant power assay), total anthocyanin content, total phenolic content, and total flavonoid content of the extraction.

2. Materials and Method

2.1. Raw Materials and Sample Extract Preparation

The saffron petals were harvested (Vezvan, Isfahan, Iran) in November 2020 and subjected to vacuum-drying (GT 2A, Leybold Heraeus, Koln, Germany) at an absolute pressure of 150 Pa (1.50 mbar). The final solvents recommended by the response surface method were combined with 3 g saffron petal powder in a 1 to 10 ratio. The resulting samples underwent ultrasound-assisted extraction at a temperature of 60 °C, 3 min of extraction time, and a sonication frequency of 28 kHz. The sonication was accomplished with an ultrasonic transducer and a power supply (HAMEG 8150, Mainhausen, Germany). Sonication enhances bioactive compound extraction by facilitating mass transfer from pomace to solvent and improving fluid dynamics along the extraction column [13]. Subsequently, the liquid part of the extract was evaporated under vacuum at 40 °C using an evaporator. The resulting extract was transferred to glass plates and dried until it attained a non-liquid state, while any remaining solvent was drained using a hot water bath maintained at 45 to 50 °C. The dried plates were then sealed, shielded from light with aluminum foil, placed inside a 4-layer plastic cover, and stored in a freezer at −18 °C for further chemical analysis [12]. The experimental study's entire supply of chemicals came from Merck Co. (Darmstadt, Germany) or Sigma-Aldrich Co. (St. Louis, MO, USA).

2.2. Antioxidant Activity Assessment

2.2.1. DPPH Radical Scavenging Activity

One mL of 0.135 mM DPPH (i.e., 2,2-Diphenyl-1-picrylhydrazyl) prepared in methanol was combined with 1.0 mL of aqueous extract containing 0.2–0.8 mg/mL. The reaction mixture was completely vortexed and kept in the dark at room temperature for 30 min. At 517 nm, the absorbance was determined spectrophotometrically. The scavenging ability of the plant extract was calculated as the following [13–16].

$$\text{DPPH scavenging activity (\%)} = [(\text{Abs}_{control} - \text{Abs}_{sample})/(\text{Abs}_{control})] \times 100$$

where $\text{Abs}_{control}$ indicates the absorbance of DPPH + methanol and Abs_{sample} indicates the absorbance of DPPH radical + sample (i.e., extract or standard).

2.2.2. Ferric Reducing Antioxidant Power (FRAP) Assay

A 5 µL aqueous extract sample, with concentrations ranging from 10 to 1000 g/mL, was mixed with 180 µL of ferric-TPTZ reagent. The ferric-TPTZ reagent was prepared by combining a 300 mM acetate buffer with pH 3.6, 10 mM TPTZ in 40 mM HCl, and 20 mM $FeCl_3 \cdot 6H_2O$ in a 10:1:1 ($v/v/v$) ratio. The mixture was then incubated at 37 °C for 5 min. Absorbance at 593 nm was measured using a Thermo Varioskan Flash Microplate Reader (Thermo Scientific, Waltham, MA, USA). The standard curve for $FeSO_4$ was linear within the range of 0.15 to 5.00 mM $FeSO_4$. The results were reported in terms of $FeSO_4$ levels determined using the established standard curves. Each sample was analyzed three times [13–16].

2.3. Total Monomeric Anthocyanin Pigment Determination

The pH-dependent reversible color change of monomeric anthocyanin pigments was observed, where the oxonium form exhibited a colorful appearance at pH 1.0, while the hemiketal form, which was colorless, predominated at pH 4.5. The difference in absorbance of the pigments at 520 nm was found to be associated with the concentration of the pigments. The results were reported in terms of Cyanidin-3-glucoside. It should be noted that the measurements excluded degraded anthocyanins in their polymeric form, which were resistant to color changes at both pH 4.5 and pH 1.0. The absorbance of a test fraction (1.0 mL of aqueous extract containing 0.1–1 mg/mL) diluted with pH 1.0 buffer (0.025 M potassium chloride) and pH 4.5 buffer (0.4 M sodium acetate) was measured at 520 nm and 700 nm. The diluted test samples were compared to a blank cell filled with distilled water. The absorbance measurements were conducted within 20–50 min after preparation. To calculate the concentration of anthocyanin pigments, expressed as Cyanidin-3-glucoside equivalents, the following formula was employed [12,13,15,16]:

Monomeric anthocyanin (mg/kg dry extraction) = $(A \times MW \times DF \times 1000)/(\varepsilon \times L)$

where A = $[(A_{520nm} - A_{700nm})$ in pH 1.0] − $[(A_{520nm} - A_{700nm})$ in pH 4.5]; MW (molecular weight) = 484.83 g/mol for Cyanidin-3-glucoside; DF = dilution factor; L = path length in cm; ε = 29,000 molar extinction coefficient in L × mol^{-1} × cm^{-1}, for Cyanidin-3-glucoside; and 1000 = factor for conversion from g to mg.

2.4. Total Phenolic Content (TPC) Determination

The total phenolic content (TPC) of the crude extract was determined using the Folin–Ciocalteu method, employing colorimetric measurement. To prepare the sample, one gram of the extract was diluted with water at a dilution factor of 200. Next, triplicate aliquots of 1.0 mL of the diluted extract were transferred into separate test tubes using a 1 mL transfer pipette. These aliquots were then thoroughly mixed with 5.0 mL of Folin–Ciocalteu reagent, which had been previously diluted 1:10 with distilled water. After shaking for 3 min, 4.0 mL of sodium carbonate solution (7.5% w/v) was added and mixed thoroughly. The mixtures were then allowed to stand for 30 min in the dark before measuring the absorbance in a single beam UV–vis spectrophotometer (Ocean optics, Orlando, FL, USA) at 765 nm against the blank of methanol pure solvent. On a dry basis, TPC values were expressed as mg gallic acid equivalent (GAE)/kg of extract. To reach the Lambert–Beer linear zone, each solution was diluted with the extraction solvent [13,15–18].

2.5. Total Flavonoid Content Determination

The flavonoid concentration was ascertained using the aluminum chloride colorimetric technique. A total of 1.5 mL of 96% ethanol was added to 500 µL of the extract. Until a final volume of 5 mL was attained, the additional reagents were added in the same manner and quantity as the hydrophilic extracts. After 30 min of darkness for the mixes, the optical density at 415 nm was determined. As a standard, quercetin (QE), a flavanol present in high amounts as an O-glycoside in both fruits and vegetables, was employed. A calibration curve was obtained. The flavonoid content was measured in mg of quercetin per gram of freeze-dried saffron petals [15–19].

2.6. Experimental Design and Statistical Analysis

Response surface methodology (RSM) is a statistical approach that uses a series of well-prepared tests to determine the optimal response to multiple causal variables [20]. Independent variables including antioxidant activity-DPPH radical scavenging activity (%), antioxidant activity-FRAP assay (mg Fe^{+2}/g vacuum-dried saffron petal), total anthocyanin content (mg Cyanidin-3-glucoside/g vacuum-dried saffron petal), total phenol content (mg Gallic acid/g vacuum-dried saffron petal), and total flavonoid content (mg Quercetin/g vacuum-dried saffron petal) (Table 1). The program Design-Expert 8.1.3 (State-Ease Inc., Minneapolis, MN, USA) was utilized to calculate the linear and quadratic polynomial models coefficients and to perform optimization. Statistical significance was determined based on p values less than 0.05. The data was collected in triplicate and averaged to obtain the findings. The regression coefficients (β) were determined by fitting the experimental data to second-order and third-order polynomial models. The response surface analysis employed generalized first and second-order polynomial models.

$$Y = \beta_0 + \beta_1 A + \beta_2 B + \beta_3 C + \beta_{11} A^2 + \beta_{22} B^2 + \beta_{33} C^2 + \beta_{12} AB + \beta_{13} AC + \beta_{23} BC$$

The coefficients of the polynomial model were represented by β_0 (i.e., constant coefficient), β_1, β_2, and β_3 (i.e., linear coefficients), β_{11}, β_{22}, and β_{33} (i.e., quadratic coefficients), and finally β_{12}, β_{13}, and β_{23} (i.e., interactive coefficients) [21]. The variable A represented the ethanol concentration in percentage, B represented the citric acid concentration in the final solvent as a percentage, and C represented the ultrasound power in watts. The dependent variables or responses, denoted as Y, included the mean values of antioxidant activity based on DPPH and FRAP, total anthocyanin content, total phenol content, and total flavonoid content. Experimental models incorporating linear relationships and quadratic terms were developed based on the acquired data to predict the responses. Subsequently, these models underwent statistical analysis to determine the best-fitting model. The model with the highest R^2 value was considered a good fit from a statistical perspective. The relative error is a measure of the deviation between the predicted value and the actual value, expressed as a percentage of the true value. It provides information about the quality and reliability of the data obtained [22,23]. To calculate the relative error for each response value, use the following formula:

Relative error = [(Predicted value − Actual value)/Actual value] × 100

Table 1. The coded and uncoded levels of the independent variables, the composite central design (CCD), and the experimental data for the response surface methodology (RSM) in the extraction of bioactive compounds from vacuum-dried saffron petals.

Independent Variables	Code	Symbol				
		$-\alpha$	-1	0	1	$+\alpha$
Ethanol concentration (%)	A	0	24	48	72	96
Citric acid concentration concentration in final solvent (%)	B	0	0.25	0.5	0.75	1
Ultrasound power (watt)	C	0	100	200	300	400

Run	Ethanol Concentration (%)	Citric Acid Concentration in Final Solvent (%)	Ultrasound Power (Watt)	Antioxidant Activity (DPPH-Assay) *	Antioxidant Activity (FRAP Assay) **	Total Anthocyanin Content ***	Total Phenol Content ****	Total Flavonoid Content *****
1	72	0.75	300	31.152	85.57	5.444	15.556	55.77
2	48	0.5	0	34.388	39.965	4.447	21.268	44.322
3	24	0.25	100	25.149	34.496	4.872	20.696	48.131
4	48	0.5	200	27.539	106.345	4.038	15.6	49.281
5	72	0.75	100	27.734	144.625	4.278	22.084	58.203
6	48	0.5	400	24.208	67.202	4.708	21.901	43.514
7	24	0.75	100	22.155	54.708	4.528	24.252	61.309
8	24	0.25	300	18.457	71.961	3.133	11.779	42.18
9	48	0.5	200	23.695	52.812	4.609	22.127	54.392
10	48	0	200	37.89	92.541	3.276	13.082	41.152
11	48	0.5	200	23.096	57.766	3.327	22.876	51.787
12	72	0.25	300	27.734	77.523	4.812	18.23	51.867
13	72	0.25	100	20.188	56.691	5.122	23.148	64.063
14	96	0.5	200	16.503	62.702	1.838	8.503	37.755
15	48	1	200	24.218	120.996	4.607	12.348	39.521
16	24	0.75	300	18.477	55.828	5.063	19.506	64.936
17	0	0.5	200	24.208	50.377	4.554	28.423	71.856

* DPPH radical scavenging activity (%), ** Ferric reducing antioxidant power (FRAP) assay (mg Fe^{+2}/g vacuum-dried saffron petal), *** (mg Cyanidin-3-glucoside/g vacuum-dried saffron petal), **** (mg Gallic acid/g vacuum-dried saffron petal), ***** (mg Quercetin/g vacuum-dried saffron petal).

3. Results and Discussion

3.1. Fitting the Response Surface Models for Antioxidant Activity Based on DPPH Radical Scavenging Activity

ANOVA was used to evaluate the first order polynomial models' significance (Table 2). For every term in the models, a high F-value and a low p-value would suggest a more significant impact on the response variable. The antioxidant activity coefficients and matching p values are listed in Table 2. The values of antioxidant activity R^2, R^2-adj, adeq accuracy, and coefficients of variation (CV) were, in that order, 0.85, 0.77, 8.53, and 7.76%. The first polynomial model was therefore demonstrated to be more fitting for the antioxidant activity assay than the other models. The ethanol concentration (A), citric acid concentration in the final solvent (B), and ultrasonic power (C) variables did not show significant linear coefficients ($p > 0.01$). The quadratic and cubic coefficients were not significant ($p > 0.05$), while the AB-interactive coefficients were significant ($p < 0.05$). The fit of the model was evaluated using a lack-of-fit test ($p > 0.05$), which shows how well the model can predict variation. The resulting extract exhibits diminishing and rising trends in antioxidant activity at low and high doses of citric acid, respectively, as the purity of ethanol in the final solvent increases. Moreover, increasing the proportion of citric acid in the final solvent demonstrated a reduction in the antioxidant activity of the extract at low and high levels of ethanol concentration, respectively (Figure 1a). The extraction of organic compounds from plant material is closely correlated with the compatibility of solvent ingredients. If the extracted components are well-polarized at the same polarity as the solvent, extraction will be easy; if not, extraction will be challenging. The low water-soluble phenolic content in saffron petals may explain the results, as only models with lower reliability were obtained, and mathematical models of low reliability should not be included. The rate of DPPH free radical scavenging was decreased when saffron petal extraction was performed with more water. This might be explained by the high impurity content of water extraction, which raises the amount of non-phenolic compounds and reduces the antioxidant property [24]. The antioxidant activity of an extract is strongly correlated with the solvent used, due to the individual components' varying polarity and potential for antioxidant activity. Antioxidant extraction depends critically on the plant materials' antioxidant compounds' solubility in the extraction solvent. Plant substances exhibit a correlation between their antioxidant activity and the amount of phenolic chemicals present. Based on the molecular structure of benzoic and cinnamic acids, phenolic acids are free radical receptors. The presence of side chains and phenolic rings in the molecular structure increases the ability to withstand free radical degradation [25]. When compared to phenolic extracts, anthocyanin extracts had the highest antiradical activity across all solvents examined [26]. This suggests that increasing the water-to-alcohol ratio lowers the amount of anthocyanin extracted and, consequently, the antioxidant qualities of the extract. This is in line with the results of our investigation. When compared to non-polar solvents, extraction with highly polar ones produced a high extract yield but a low phenolic and flavonoid concentration. Strong antioxidant molecules are extracted from polar solvents, as indicated by the polarity-dependent increase in overall antioxidant activity [27,28]. Acidic extraction can facilitate the release of phytochemicals by breaking the plant's cell walls. The performance of extraction depends on the solvent and its concentration. The saffron petal extract obtained from different solvents has different chemical compounds with varying antioxidant activity. Water extraction decreases antioxidant activity with higher citric acid concentration, while alcohol extraction with higher citric acid concentration provides higher activity levels (Figure 1a).

Table 2. The analysis of variance (ANOVA) of antioxidant activity in terms of DPPH radical scavenging activity (first-order fitted polynomial model) and antioxidant activity in terms of ferric-reducing antioxidant power (FRAP) assay (first-order fitted polynomial model) in the extraction of bioactive compounds from vacuum-dried saffron petals.

Antioxidant Activity in Terms of DPPH Radical Scavenging Activity (%)						
Source	Coefficient of Final Equation in Terms of Coded Factors	Sum of Squares	df	Mean Square	F-Value	Prob > F	
Model	44.71	103.57	3	34.52	9.58	0.0163	significant
A-Ethanol concentration (%)	−0.406	0.23	1	0.23	0.064	0.8106	
B-Citric acid concentration in final solvent (%)	−41.723	0.43	1	0.43	0.12	0.7438	
AB	0.841	46.71	1	46.71	12.96	0.0155	
Residual	-	18.03	5	3.61	-	-	
Lack of Fit	-	6.4	3	2.13	0.37	0.7884	not significant
Pure Error	-	11.63	2	5.81	-	-	
Cor Total	-	121.59	8	-	-	-	

R^2: 0.85, Adj-R^2: 0.77, Adeq precision: 8.53, C.V.: 7.76%

Antioxidant activity (based on DPPH radical scavenging assay) (%) = 44.71 − 0.406 (A) − 41.723 (B) + 0.841 (AB)

Antioxidant Activity in Terms of FRAP Assay (mg Fe^{+2}/g Vacuum-Dried Saffron Petal)						
Source	Coefficient of Final Equation in Terms of Coded Factors	Sum of Squares	df	Mean Square	F-Value	Prob > F	
Model	17.769	1474.94	3	491.65	10.24	0.0059	significant
A-Ethanol concentration (%)	0.256	471.48	1	471.48	9.82	0.0165	
B-Citric acid concentration in final solvent (%)	28.99	213.79	1	213.79	4.45	0.0728	
C-Ultrasound power (watt)	0.061	395.92	1	395.92	8.25	0.0239	
Residual	-	336.06	7	48.01	-	-	
Lack of Fit	-	323.78	6	53.96	4.4	0.3497	not significant
Pure Error	-	12.27	1	12.27	-	-	
Cor Total	-	1811	10	-	-	-	

R^2: 0.81, Adj-R^2: 0.73, Adeq precision: 9.30, C.V.: 12.33%

Antioxidant activity (based on FRAP assay) = 17.769 + 0.256(A) + 28.99 (B) + 0.061(C)

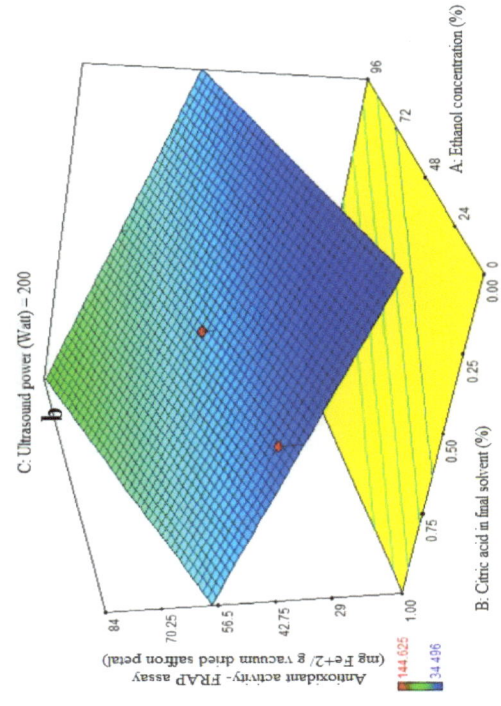

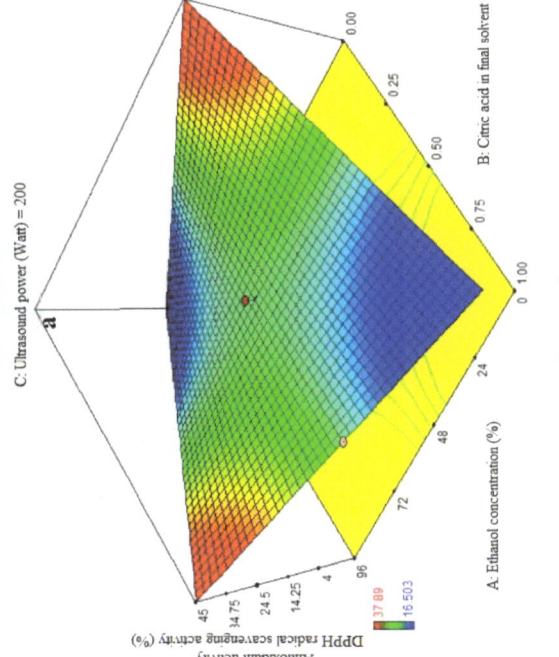

Figure 1. Cont.

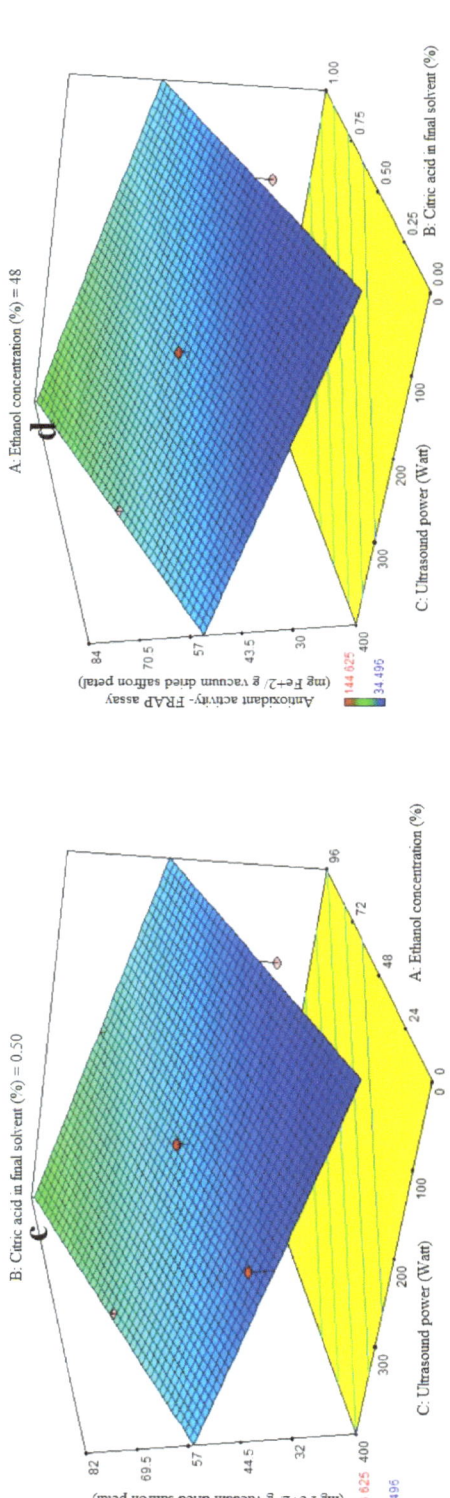

Figure 1. Impact of ethanol concentration (%), citric acid concentration in the final solvent (%), and ultrasound power (watt) on antioxidant activity, as assessed by DPPH radical scavenging activity (%) (**a**) and ferric-reducing antioxidant power (FRAP) (mg Fe^{+2}/g vacuum-dried saffron petal) (**b**–**d**) in extracts obtained from the vacuum-dried saffron petals.

3.2. Fitting the Response Surface Models for Antioxidant Activity Based on FRAP Assay

The first-order polynomial models' significance was assessed using ANOVA (Table 2). Table 2 lists the antioxidant activity coefficients and the p values that correlate to them. For antioxidant activity, the corresponding R^2, R^2-adj, adeq accuracy, and CV values were 0.81, 0.73, 9.30, and 12.33%. Significant linear coefficients were found for the ethanol concentration (A) and ultrasonic power (C) variables ($p < 0.05$). Antioxidant activity was not significantly impacted by the citric acid concentration in the final solvent (B), the AB-, AC-, and BC-interactive terms, or the quadratic and cubic terms of A, B, and C ($p > 0.05$). To assess the model's fitness, which measures how well it can predict variance, a lack-of-fit test ($p > 0.05$) was run. The study found that the antioxidant activity of the extract from saffron petals increased with increasing ethanol concentration, citric acid dose, and ultrasound power. The antioxidant activity of the extract from saffron petals at all levels of citric acid concentration was observed to increase ultrasound power (Figure 1d) and the purity percentage of ethanol (Figure 1b). This trend was observed at all levels of ultrasound power, with increasing ethanol concentration percentage (Figure 1c) or citric acid dose (Figure 1d). Increased water utilization increases the extraction of compounds that lack antioxidant properties and are thus inappropriate for phenolic compound extraction. High ethanol concentrations in the solvent can modify physical properties such as density, dynamic viscosity, and dielectric constant, resulting in increased antioxidant chemical solubility. The acidified ethanolic extract showed the highest radical scavenging effectiveness. This suggests that its chemical composition varies not just numerically, but also qualitatively. Furthermore, the phenols in the ethanol extract may have been endowed with chemical properties that improve radical scavenging efficacy [29,30]. In general, low water levels, high citric acid percentages, and high ultrasonic power indicated the optimum rate of iron reduction power in this study. Previous research has shown that the extraction method has an impact on the phytochemical content and antioxidant potential of dried saffron petal extracts. Ultrasound-assisted extraction (UAE) with water or low methanol percentages often produced vitamin C and phenolic component results comparable to maceration, allowing for a 50% time reduction in extraction processes. The dried saffron petals had lower levels of total phenolics and anthocyanins but higher levels of antioxidant activity when compared to the spice [31]. The greater power and longer duration of UAE may have contributed to the saffron petal extract's notable improvement in its antioxidant qualities and phenolic component profile. In general, the acoustic cavitation phenomenon caused by ultrasonic waves results in strong shear stresses inside the structure and has the ability to rupture cell walls, allowing a solvent to enter the plant material and release the intracellular content into the medium [4].

3.3. Fitting the Response Surface Models for Total Anthocyanin Content

Table 3 shows the ANOVA used to assess the significance of the first-order polynomial model. A high F-value and a low p-value for each term in the models would reflect a more substantial influence on the relevant response variable. Table 3 displays the anthocyanin content coefficient values along with their corresponding p values. The anthocyanin content R^2, R^2-adj, adeq precision, and CV values were 0.84, 0.78, 12.09, and 6.97%, respectively. As a result, the linear model outperformed the other models in terms of anthocyanin content. The linear coefficient for citric acid concentration in the final solvent (B) and the BC-interactive were both significant ($p < 0.05$). A lack-of-fit test ($p > 0.05$) was used to assess the model's fitness, which demonstrated the model's ability to reliably forecast variance. Figure 2a demonstrates that increasing the citric acid dosage enhanced the quantity of cyanidin-3-glucoside extraction at high and low levels of ultrasound power, respectively. The quantity of cyanidin-3-glucoside extracted increased and decreased with increasing ultrasonic power utilized in the extraction process at high and low levels of citric acid consumption in the final solvent. The treatment (400 watt ultrasound power and 1% citric acid concentration) resulted in the greatest quantity of anthocyanin extraction (Figure 2a). According to this study, increasing the purity of ethanol has no effect on anthocyanin extraction and is not a suitable solvent for anthocyanin extraction, while citric acid does increase the rate of anthocyanin extraction to some extent. In the absence of ultrasound, it was suggested that acylated anthocyanins be extracted with a gently acidic solution to prevent hydrolysis. When a moderately acidic medium is used, more and more diverse anthocyanins are extracted. Acid is required to obtain the form of a flavilium cation that is red and stable in acidic solution. Furthermore, at pH = 1.8, anthocyanins have the highest stability [5]. The anthocyanins were distorted by the presence of hydrochloric acid in the solvent, preventing them from hydrolyzing under acidic conditions during solvent evaporation in a vacuum rotary evaporator at 40 °C or deacylation with solvent aliphatic compounds. Increased acidity can improve extraction to some extent, but excessive acidification causes double bond loss and, as a result, anthocyanin loss [32,33]. This is due to the fact that anthocyanin is more stable in acidic solutions and may attach to free radicals in the body such as vitamin C, vitamin E, and beta-carotene. Using acids to aid phytochemical extraction aids in the digestion of plant cell walls, resulting in increased anthocyanin release. The findings agreed with those of a prior work by Li et al. [34], who employed the microwave-assisted extraction of anthocyanins from grape peels with citric acid concentration [28,35]. Ultrasound is a non-thermal food-processing technology that offers an alternative to conventional techniques and can enhance food quality by promoting or damaging enzyme activities at the molecular level. It has been applied to various food by-products, such as fruits, vegetables, and juices, to increase the extraction of polyphenols, improve sensory characteristics, and enhance health benefits due to higher nutraceutical contents. The ultrasonic extraction process of bioactive compounds is a combination of several physical mechanisms such as fragmentation, erosion, the sonocapillary effect, sonoporation, local shear stress, and detexturation [7]. The increase in bioactive components with citric acid treatment could be attributed to its leaching capacity on covalent bonds of the biomass structure, leading to the removal of bioactive compounds from epithelial cells. Further treatment with US could result in the sonolytic cavitation of the biomass structure, improving the availability of phenolic compounds and antioxidant activities. This method has been reported in other fruit by-products, making it an emerging method for agro-allied organizations [36,37].

3.4. Fitting the Response Surface Models for Total Phenol Content (TPC)

The ANOVA was used to evaluate the significance of the first-order polynomial model (Table 3). For each term in the models, a high F-value and a lower p-value would suggest a more significant influence on the response variable. TPC coefficient values are included in Table 3 along with their matching p values. R^2, R^2-adj, adeq accuracy, and the CV for the TPC were 0.96, 0.92, 15.63, and 7.02%, respectively (Table 3). Consequently, it was demonstrated that the second-order polynomial model fit the TPC better than the other models. Table 3, Figure 2b shows that the quadratic coefficient of ethanol concentration (A), the linear coefficient for ultrasound (C), the AB-interactive and acid citric in the final solvent (B), and the ultrasound power (C) were all significant ($p < 0.05$). Table 3, Figure 2c,d, and the AC-interactive, BC-interactive, and cubic coefficients were all found to be non-significant ($p > 0.05$). The model's fitness was assessed using a lack-of-fit test ($p > 0.05$), which showed that the model could correctly predict the variance. Higher and lower ethanol concentrations in the final solvent yielded the maximum extraction of total phenolic content at low and high amounts of citric acid concentration, respectively. There was a noticeable drop in the extraction of the total phenol content at higher ethanol concentration levels and concurrently at greater citric acid doses (Figure 2b). The findings for total phenol content agree exactly with the information in Table 3. Prior studies have demonstrated that anthocyanins from eggplant peels may be extracted more successfully using organic solvents that have undergone acidification [5,13,32]. The extraction success of phenolic compounds is influenced by various factors including the chemical composition, extraction method, sample size, storage conditions, and presence of interfering substances [38].

Table 3. The analysis of variance (ANOVA) of the total anthocyanin content (first-order fitted polynomial model), total phenol content (second-order fitted polynomial model), and total flavonoid content (second-order fitted polynomial model) in the extraction of bioactive compounds from vacuum-dried saffron petals.

Total Anthocyanin Content (mg Cyanidin-3-Glucoside/g Vacuum-Dried Saffron Petal)							
Source	Coefficient of Final Equation in Terms of Coded Factors	Sum of Squares	df	Mean Square	F-Value	Prob > F	
Model	6.73	4.5	3	1.5	15.55	0.0007	significant
B-Citric acid concentration in final solvent (%)	−4.022	1.88	1	1.88	19.44	0.0017	
C-Ultrasound power (watt)	−0.016	0.36	1	0.36	3.72	0.0859	
BC	0.027	3.11	1	3.11	32.19	0.0003	
Residual	-	0.87	9	0.097	-	-	
Lack of Fit	-	0.71	8	0.088	0.54	0.789	not significant
Pure Error	-	0.16	1	0.16	-	-	
Cor Total	-	5.37	12	-	-	-	

R^2: 0.84, Adj-R^2: 0.78, Adeq precision: 12.09, C.V.: 6.97%

Total anthocyanin content (mg Cyanidin − 3 − glucoside/g freeze − dried saffron petal) = 6.73 − 4.022 (B) − 0.016 (C) + 0.027(BC)

Total Phenol Content (mg Gallic Acid/g Vacuum-Dried Saffron Petal)							
Source	Coefficient of Final Equation in Terms of Coded Factors	Sum of Squares	df	Mean Square	F-Value	Prob > F	
Model	8.71	302.31	7	43.19	22.63	0.0007	significant
A-Ethanol concentration (%)	−0.096	0.67	1	0.67	0.35	0.5749	
B-Citric acid concentration in final solvent (%)	55.69	2.31	1	2.31	1.21	0.3136	
C-Ultrasound power (watt)	0.046	89.9	1	89.9	47.1	0.0005	
AB	−0.313	28.2	1	28.2	14.78	0.0085	
A^2	0.0028	21.89	1	21.89	11.47	0.0147	
B^2	−39.15	95.78	1	95.78	50.18	0.0004	
C^2	−0.0002	33.14	1	33.14	17.36	0.0059	

Table 3. Cont.

Source	Coefficient of Final Equation in Terms of Coded Factors	Sum of Squares	df	Mean Square	F-Value	Prob > F	
Total Flavonoid Content (mg Quercetin/g Vacuum-Dried Saffron Petal)							
Residual	-	11.45	6	1.91	-	-	
Lack of Fit	-	11.17	5	2.23	7.97	0.2625	not significant
Pure Error	-	0.28	1	0.28	-	-	
Cor Total	-	313.76	13	-	-	-	

R^2: 0.96, Adj-R^2: 0.92, Adeq precision: 15.63, C.V.: 7.02%

Total phenol content (mg Gallic acid/g freeze – dried saffron petal) = 8.71 − 0.096 (A) + 55.69 (B) + 0.046 (C) − 0.313 (AB) + 0.0028 $(A)^2$ − 39.15 $(B)^2$ − 0.0002 $(C)^2$

Source	Coefficient	Sum of Squares	df	Mean Square	F-Value	Prob > F	
Model	39.863	1026.86	6	171.14	30.62	0.0001	significant
A-Ethanol concentration (%)	−0.548	0.28	1	0.28	0.049	0.8304	
B-Citric acid concentration in final solvent (%)	49.048	289.07	1	289.07	51.71	0.0002	
C-Ultrasound power (watt)	0.075	3.39	1	3.39	0.61	0.4619	
AB	−0.565	73.46	1	73.46	13.14	0.0085	
A^2	0.0088	218.66	1	218.66	39.12	0.0004	
C^2	−0.0002	81.95	1	81.95	14.66	0.0065	
Residual	-	39.13	7	5.59	-	-	
Lack of Fit	-	26.07	5	5.21	0.8	0.6378	not significant
Pure Error	-	13.06	2	6.53	-	-	
Cor Total	-	1065.99	13	-	-	-	

R^2: 0.96, Adj-R^2: 0.93, Adeq precision: 18.36, C.V.: 4.48%

Total flavonoid content(mg Quercetin//g freeze – dried saffron petal) = 39.863 − 0.548 (A) + 49.048 (B) + 0.075 (C) − 0.565 (AB) + 0.0088 $(A)^2$ − 0.0002 $(C)^2$

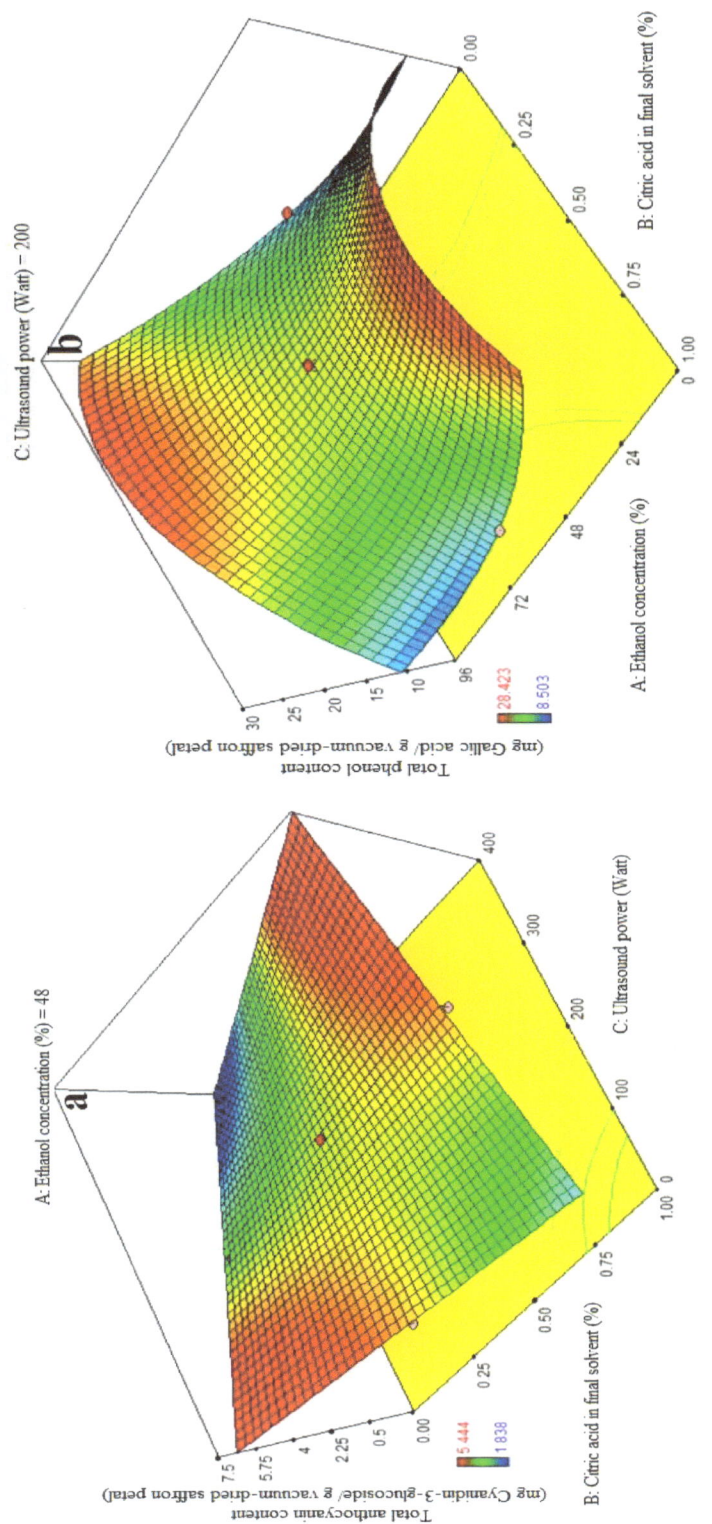

Figure 2. *Cont.*

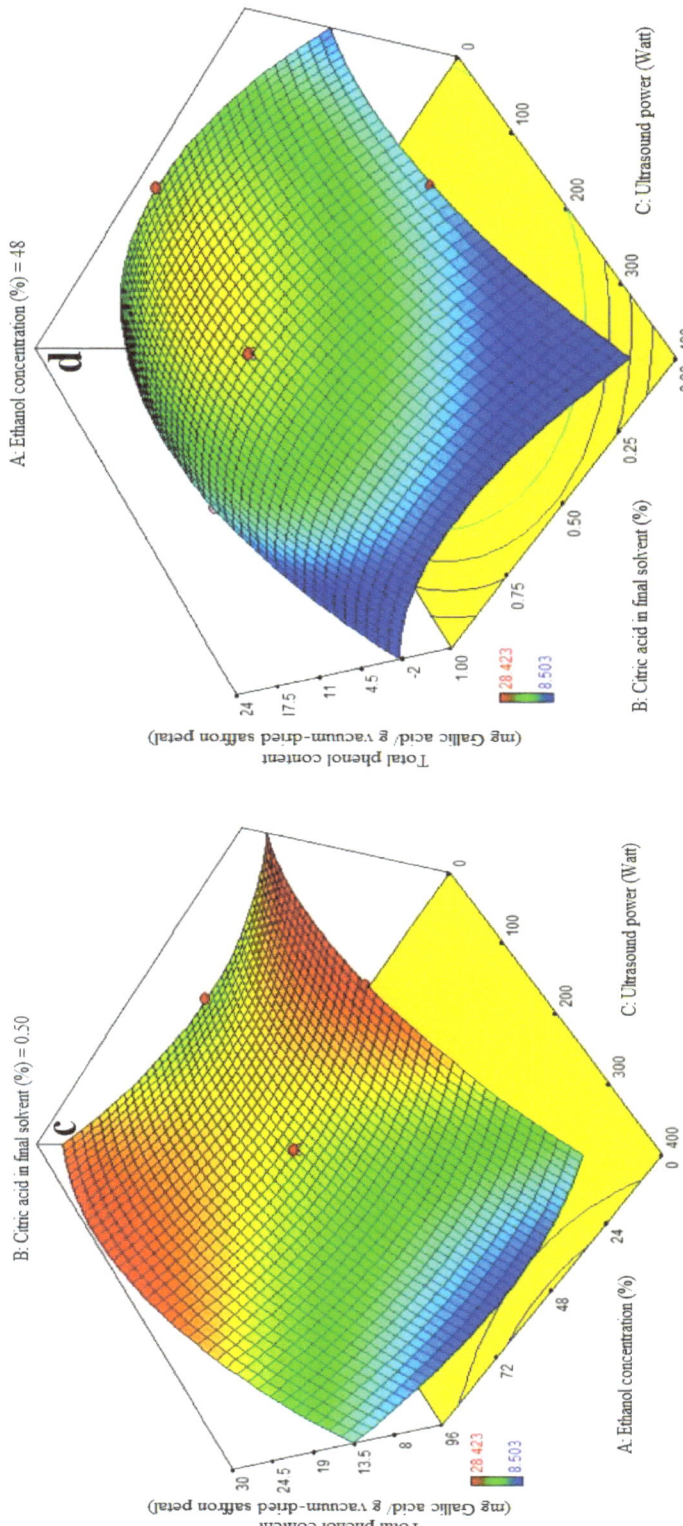

Figure 2. Impact of ethanol concentration (%), citric acid concentration in the final solvent (%), and ultrasound power (watt) on total anthocyanin content (mg Cyanidin-3-glucoside/g vacuum-dried saffron petal) (**a**) and total phenol content (mg Gallic acid/g vacuum-dried saffron petal) (**b**–**d**) in extracts obtained from the vacuum-dried saffron petals.

The phenolic extract, a complex mixture of selectively soluble phenols in various solvents, requires an increase in the solvent's polarity to enhance its solubility [39]. Despite its great extraction efficiency, the water solvent extracted fewer phenolic compounds than the alcohol solvent, as it is accordance with this study (Figure 2b). Actually, more extractable particles have been dissolved by the water solvent; however, not all of these substances are phenolic. Water as the only solvent can affect the identification and measurement of phenols in mint leaves, affecting total phenol content, flavonoids, and antioxidants. Acetone and 75% ethanol are considered the best solvents for phenol extraction [40,41]. This investigation also found that significant amounts of water had no effect on the extraction of phenolic compounds in comparison to alcohol. Low levels of citric acid concentration enhance phenolic compound extraction at high ethanol concentrations, but excessive concentrations limit the extraction of phenolic compounds. The first molecules to be identified are diphenhydroquinone and p-benzoquinone, which exhibit a significant acceleration of phenol oxidation at pH values below 4. Consequently, phenols oxidize more quickly at low pH values [42]. Solvent concentration affected both the total phenolic content and the extraction of total monomeric anthocyanins. Because of this, the solvent's surface tension or dielectric constant may have an impact on how well eggplant peel solubilizes in certain solvents [12]. The US-assisted approach is one of the newly developed strategies of valorization by pertinent agro-allied organizations in light of the improvement in the bioactive yield of fruit waste [37]. Methanolic extracts yielded the highest phenolic compound, followed by water and ethanolic extracts [43]. The number of phenolic compounds increased with increasing citric acid concentration, while extraction using water presented a lower number of compounds. However, when using a lower citric acid concentration, the obtained phenolic compounds were higher. The results were consistent with previous studies on phenolic compounds in plants, which are mainly water-soluble forms. Organic solvents like methanol and ethanol are commonly used for extracting phenolic compounds, with ethanol being the safer option. Antioxidants in plants are typically polar substances, and their activity depends on hydroxyl categories and functional groups [28].

3.5. Fitting the Response Surface Models for Total Flavonoid Content (TFC)

ANOVA was used to assess the significance of the second-order, or quadratic, polynomial models (Table 3). For every term in the models, a high F-value and a low p-value would suggest a more significant impact on the response variable. The TFC coefficient values and the accompanying p values are listed in Table 3. R^2, R^2-adj, adeq accuracy, CV, and extraction yield were, in that order, 0.96, 0.93, 18.36, and 4.48%. Consequently, it was demonstrated that the quadratic model outperformed the other models in terms of TFC. Citric acid's significant linear coefficient ($p < 0.05$) was discovered for the final solvent (B) variable. The cubic coefficient was not significant ($p > 0.05$), but the AB-interactive, quadratic, and citric acid concentration in the final solvent (A) were all significant ($p < 0.05$). The model's fitness was assessed using a lack-of-fit test ($p > 0.05$), which showed that the model could consistently predict variation. While 96% ethanol exhibited the maximum flavonoid extraction, Figure 3a demonstrates that at the highest levels of ethanol concentration, increasing the proportion of citric acid in the final solvent had no effect on the flavonoid extraction ($p > 0.05$). We saw a considerable increase in flavonoid extraction at lower ethanol concentration levels when the final solvent contained a higher proportion of citric acid. As can be seen in Figure 3b, the extraction process's ultrasonic power (200 watts) and the end solvent ethanol concentration have a quadratic influence on flavonoid extraction (Table 3). The extraction process's lowest and maximum levels of ethanol concentration fall within this range. The solvent with the highest flavonoid extraction efficiency was found to be the final one. Figure 3c illustrates that an increasing trend in flavonoid extraction is seen at all ultrasound power levels when the percentage of citric acid concentration in the final solvent increases. At all different levels of citric acid concentration in the final solvent, the highest flavonoid extraction efficiency was observed at 200 watts of ultrasound power. Kaempferols and anthocyanins are the primary chemicals with a high phenolic content, according to the earlier study, which also showed that saffron flower waste includes naturally occurring flavonoid and phenolic compounds. Factors such as origin, altitude, growth circumstances, and picking times can be blamed for variations in these chemicals [44]. This may help to explain why binary solvent systems generate higher yields than those of monosolvent systems. Previous studies have also demonstrated that using organic solvents with acidification improves the extraction of flavonoids from eggplant peels [32,45].

Ultrasound power plays a crucial role in the extraction yields of bioactive compounds like total phenolics and flavonoids. It enhances extraction efficiency by creating cavitation in the extraction medium, generating microscopic bubbles that collapse, promoting the release and diffusion of compounds. However, excessive ultrasound power can lead to adverse effects, such as the mechanical degradation or alteration of sensitive compounds like flavonoids. Flavonoids are heat-sensitive and susceptible to degradation under harsh extraction conditions, and the intense energy generated by high ultrasound power can break down or modify their molecular structure, resulting in reduced extraction yields. Additionally, the extraction process is influenced by the solubility of the target compounds in the solvent. The optimal ultrasound power for extraction may vary depending on the compound, solvent, and plant matrix being studied. Researchers conduct preliminary studies to determine the appropriate range of ultrasound power that maximizes extraction efficiency without significant degradation, ensuring the highest overall yield of bioactive compounds while preserving their chemical integrity [18,19].

3.6. Optimization of the Solvent Formulation and Survey of Actual and Predicted Data

When the weight and significance values for five responses were determined to be equal, the numerical optimization approach was employed to optimize the extraction conditions (Table 4). In this research, the level of optimization was to be maximized for the antioxidant activity in terms of DPPH radical scavenging activity (18.477–31.152%) and antioxidant activity in terms of ferric-reducing antioxidant power assay (34.496–85.57 mg Fe^{+2}/g vacuum-dried saffron petal), total anthocyanin content (3.133–5.444 mg Cyanidin-3-glucoside/g vacuum-dried saffron petal), total phenol content (11.779–28.423 mg Gallic acid/g vacuum-dried saffron petal), and total flavonoid content (41.152–71.856 mg Quercetin/g vacuum-dried saffron petal) of the saffron petal extract samples (Table 4). The best optimal formula for maximum bioactive compound extraction from saffron petals includes ethanol concentration (96%), citric acid concentration in the final solvent (0.67%), and ultrasound power (216 watts) as the predicted results, whose desirability values were equal to 0.82 (Table 4). Furthermore, the disparity between predicted numbers (State-Ease Inc., Minneapolis, MN, USA) and actual (performed in the laboratory) data was minor.

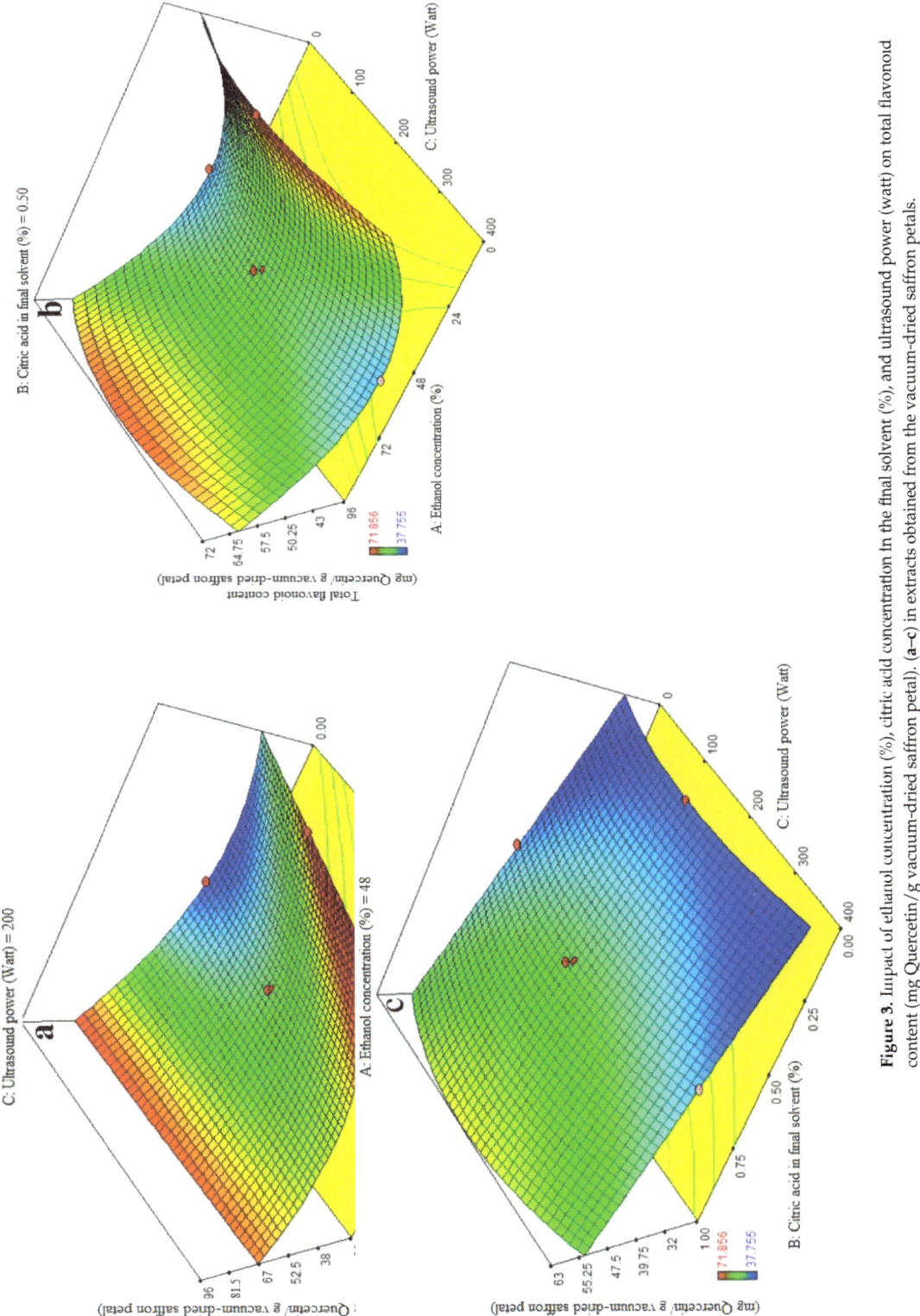

Figure 3. Impact of ethanol concentration (%), citric acid concentration in the final solvent (%), and ultrasound power (watt) on total flavonoid content (mg Quercetin/g vacuum-dried saffron petal). (a–c) in extracts obtained from the vacuum-dried saffron petals.

Table 4. Limitations, optimal formula, and data validation for bioactive compound extraction from vacuum-dried saffron petals and antioxidant activity assessment (DPPH radical scavenging and FRAP assay), total anthocyanin content, total phenol content, and total flavonoid content.

Name	Goal	Lower Limit	Upper Limit	Weight	Weight	Importance
A-Ethanol concentration (%)	is in range	0	96	1	1	3
B-Citric acid concentration in final solvent (%)	is in range	0	1	1	1	3
C-Ultrasound power (watt)	is in range	0	400	1	1	3
Antioxidant activity-DPPH radical scavenging activity (%)	maximize	18.477	31.152	Y	1	3
Antioxidant activity-FRAP assay (mg Fe^{+2}/g vacuum-dried saffron petal)	maximize	34.496	85.57	1	1	3
Total anthocyanin content (mg Cyanidin-3-glucoside/ g vacuum-dried saffron petal)	maximize	3.133	5.444	1	Y	3
Total phenol content (mg Gallic acid/g vacuum-dried saffron petal)	maximize	11.779	28.423	1	1	3
Total flavonoid content (mg Quercetin/g vacuum-dried saffron petal)	maximize	41.152	71.856	1	1	3
Optimal formula for maximum bioactive extraction from saffron petals						
Ethanol concentration (96%) and citric acid concentration in final solvent (0.67%) and ultrasound power (216 watts) with a desirability of 82%						
Parameters	**Predicted Value**		**Actual Value**		**Relative Errors (%)**	
Antioxidant activity- DPPH radical scavenging activity (%)	31.152		31.93		2.44	
Antioxidant activity- FRAP assay (mg Fe^{+2}/g vacuum-dried saffron petal)	74.629		76.644		2.63	
Total anthocyanin content (mg Cyanidin-3-glucoside/ g vacuum-dried saffron petal)	4.613		4.774		3.37	
Total phenol content (mg Gallic acid/g vacuum-dried saffron petal)	24.155		25.17		4.03	
Total flavonoid content (mg Quercetin/g vacuum-dried saffron petal)	70.661		71.721		1.48	

4. Conclusions

This study demonstrated that saffron petals, which are cost-effective and abundant by-products of the saffron industry, contain bioactive compounds that can be efficiently extracted using an ultrasound-assisted acidified ethanol solvent. The optimization of extraction conditions, considering factors such as ethanol concentration, citric acid concentration, and ultrasound power, resulted in the highest antioxidant activity, total anthocyanin content, total phenolic content, and total flavonoid content of the saffron petal extract. The optimum values of the independent parameters, i.e., 96% ethanol concentration, 0.67% citric acid concentration, and 216 watts of ultrasound power, yielded a model desirability value of 0.82, as well as antioxidant activity, including DPPH radical scavenging activity (31.15%), antioxidant activity-FRAP assay (74.63 mg Fe^{+2}/g vacuum-dried saffron petal), total anthocyanin content (4.61 mg Cyanindin-3-glucoside/g vacuum-dried saffron petal), total phenol content (24.16 mg Gallic acid/g vacuum-dried saffron petal), and total flavonoid content (70.66 mg Quercetin/g vacuum-dried saffron petal). This ultrasound-assisted acidified ethanolic extract holds potential for utilization as a natural source of antioxidants and pigments in the food industry. The minor disparity between predicted and actual data further validates the effectiveness of the optimization approach employed in this research.

Author Contributions: Conceptualization, N.J., M.G. and S.S.; Formal analysis, N.J., M.G. and S.S.; Investigation, N.J., M.G. and S.S.; Methodology, N.J., M.G. and S.S.; Project administration, N.J., M.G. and S.S.; Software, N.J., M.G. and S.S.; Supervision, M.G. All authors have read and agreed to the published version of the manuscript.

Funding: This research received no external funding.

Data Availability Statement: The data presented in this study are available on request from the corresponding author. The data are not publicly available due to privacy.

Acknowledgments: The authors would like to express their gratitude to the Islamic Azad University, Isfahan (Khorasgan) Branch, for their scientific and important contributions to this research.

Conflicts of Interest: The authors declare no conflicts of interest.

Abbreviations

Ultrasound-assisted extraction (**UAE**), Response surface methodology (**RSM**), Central composite design (**CCD**), coefficient of variation (**CV**), 2,2-Diphenyl-1-picrylhydrazyl (**DPPH**), Ferric-reducing antioxidant power (**FRAP**), 2,4,6-Tripyndyl-s-triazine (**TPTZ**), Total phenol content (**TPC**), Total flavonoid content (**TFC**), Defatted Hom Nin rice bran (**DHRB**).

References

1. Rahaiee, S.; Moini, S.; Hashemi, M.; Shojaosadati, S.A. Evaluation of antioxidant activities of bioactive compounds and various extracts obtained from saffron (*Crocus sativus* L.): A review. *J. Food Sci. Technol.* **2015**, *52*, 1881–1888. [CrossRef]
2. Ahmadian-Kouchaksaraie, Z.; Niazmand, R. Supercritical carbon dioxide extraction of antioxidants from *Crocus sativus* petals of saffron industry residues: Optimization using response surface methodology. *J. Supercrit. Fluids* **2017**, *121*, 19–31. [CrossRef]
3. Wu, Y.; Gong, Y.; Sun, J.; Zhang, Y.; Luo, Z.; Nishanbaev, S.Z.; Usmanov, D.; Song, X.; Zou, L.; Benito, M.J. Bioactive components and biological activities of *Crocus sativus* L. byproducts: A Comprehensive Review. *J. Agric. Food Chem.* **2023**, *71*, 19189–19206. [CrossRef] [PubMed]
4. Hashemi Gahruie, H.; Parastouei, K.; Mokhtarian, M.; Rostami, H.; Niakousari, M.; Mohsenpour, Z. Application of innovative processing methods for the extraction of bioactive compounds from saffron (*Crocus sativus*) petals. *J. Appl. Res. Med. Aromat. Plants* **2020**, *19*, 100264. [CrossRef]
5. Ferarsa, S.; Zhang, W.; Moulai-Mostefa, N.; Ding, L.; Jaffrin, M.Y.; Grimi, N. Recovery of anthocyanins and other phenolic compounds from purple eggplant peels and pulps using ultrasonic-assisted extraction. *Food Bioprod. Process.* **2018**, *109*, 19–28. [CrossRef]
6. Barba, F.J.; Zhu, Z.; Koubaa, M.; SantAna, A.S.; Orlien, V. Green alternative methods for the extraction of antioxidant bioactive compounds from winery wastes and by-products: A review. *Trends Food Sci. Technol.* **2016**, *49*, 96–109. [CrossRef]
7. Xie, Q.; Tang, Y.; Wu, X.; Luo, Q.; Zhang, W.; Liu, H.; Fung, Y.; Yue, X.; Ju, Y. Combined ultrasound and low temperature pretreatment improve the content of anthocyanins, phenols and volatile substance of Merlot red wine. *Ultrason. Sonochem.* **2023**, *100*, 106636. [CrossRef] [PubMed]
8. Saleh, I.A.; Vinatoru, M.; Mason, T.J.; Abdel-Azim, N.S.; Aboutabl, E.A.; Hammouda, F.M. A possible general mechanism for ultrasound-assisted extraction (UAE) suggested from the results of UAE of chlorogenic acid from *Cynara scolymus* L. (artichoke) leaves. *Ultrason. Sonochem.* **2016**, *31*, 330–336. [CrossRef] [PubMed]
9. Manouchehri, R.; Saharkhiz, M.J.; Karami, A.; Niakousari, M. Extraction of essential oils from damask rose using green and conventional techniques: Microwave and ohmic assisted hydrodistillation versus hydrodistillation. *Sustain. Chem. Pharm.* **2018**, *8*, 76–81. [CrossRef]
10. Maric, M.; Grassino, A.N.; Zhu, Z.; Barba, F.J.; Brncic, M.; Brncic, S.R. An overview of the traditional and innovative approaches for pectin extraction from plant food wastes and by-products: Ultrasound-, microwaves-, and enzyme-assisted extraction. *Trends Food Sci. Technol.* **2018**, *76*, 28–37. [CrossRef]
11. Chemat, F.; Rombaut, N.; Meullemiestre, A.; Turk, M.; Perino, S.; Fabiano-Tixier, A.S.; Abert-Vian, M. Review of green food processing techniques. Preservation, transformation, and extraction. *Innov. Food Sci. Emerg. Technol.* **2017**, *41*, 357–377. [CrossRef]
12. Ozgur, M.U.; Çimen, E. Ultrasound-assisted extraction of anthocyanins from red rose petals and new spectrophotometric methods for the determination of total monomeric anthocyanins. *J. AOAC Int.* **2018**, *101*, 967–980. [CrossRef]
13. Shahabi Mohammadabadi, S.; Goli, M.; Naji Tabasi, S. Optimization of bioactive compound extraction from eggplant peel by response surface methodology: Ultrasound-assisted solvent qualitative and quantitative effect. *Foods* **2022**, *11*, 3263. [CrossRef]
14. Munteanu, I.G.; Apetrei, C. Analytical methods used in determining antioxidant activity: A review. *Int. J. Mol. Sci.* **2021**, *22*, 3380. [CrossRef]
15. Yancheshmeh, B.S.; Panahi, Y.; Allahdad, Z.; Abdolshahi, A.; Zamani, Z. Optimization of ultrasound-assisted extraction of bioactive compounds from *Achillea kellalensis* using response surface methodology. *J. Appl. Res. Med. Aromat. Plants* **2022**, *28*, 100355. [CrossRef]
16. Kobus, Z.; Pecyna, A.; Buczaj, A.; Krzywicka, M.; Przywara, A.; Nadulski, R. Optimization of the ultrasound-assisted extraction of bioactive compounds from *Cannabis sativa* L. leaves and inflorescences using response surface methodology. *Appl. Sci.* **2022**, *12*, 6747. [CrossRef]
17. AOAC. Estimation of total phenolic content using the Folin-C assay. *J. AOAC Int.* **2015**, *98*, 1109–1110. [CrossRef]
18. Matic, P.; Sabljic, M.; Jakobek, L. Validation of spectrophotometric methods for the determination of total polyphenol and total flavonoid content. *J. AOAC Int.* **2017**, *100*, 1795–1803. [CrossRef]
19. Maestre-Hernandez, A.B.; Vicente-Lopez, J.J.; Pérez-Llamas, F.; Candela-Castillo, M.E.; García-Conesa, M.T.; Frutos, M.J.; Cano, A.; Hernandez-Ruiz, J.; Arnao, M.B. Antioxidant activity, total phenolic and flavonoid contents in floral saffron bio-residues. *Processes* **2023**, *11*, 1400. [CrossRef]
20. Jalalizand, F.; Goli, M. Optimization of microencapsulation of selenium with gum Arabian/Persian mixtures by solvent evaporation method using response surface methodology (RSM): Soybean oil fortification and oxidation indices. *J. Food Meas. Charact.* **2021**, *15*, 495–507. [CrossRef]

21. Zaghian, N.; Goli, M. Optimization of the production conditions of primary (W1/O) and double (W1/O/W2) nanoemulsions containing vitamin B12 in skim milk using ultrasound wave by response surface methodology. *J. Food Meas. Charact.* **2020**, *14*, 3216–3226. [CrossRef]
22. Sazesh, B.; Goli, M. Quinoa as a wheat substitute to improve the textural properties and minimize the carcinogenic acrylamide content of the biscuit. *J. Food Process. Preserv.* **2020**, *44*, e14563. [CrossRef]
23. Maghamian, N.; Goli, M.; Najarian, A. Ultrasound-assisted preparation of double nano-emulsions loaded with glycyrrhizic acid in the internal aqueous phase and skim milk as the external aqueous phase. *LWT—Food Sci. Technol.* **2021**, *141*, 108–109. [CrossRef]
24. Agustin, A.R.; Falka, S.; Ju, Y. Influence of extracting solvents on its antioxidant properties of bawang Dayak (*Eleutherine palmifolia* L. Merr). *Int. J. Chem. Petrochem. Technol.* **2016**, *2*, 1–10.
25. Moure, A.; Cruz, J.M.; Franco, D.; Dominguez, J.M.; Sineiro, J.; Dominguez, H.; Nunez, M.J.; Parajo, J.C. Natural antioxidants from residual sources. *Food Chem.* **2001**, *72*, 145–171. [CrossRef]
26. Boulekbache-Makhlouf, L.; Medouni, L.; Medouni-Adrar, S.; Arkoub, L.; Madani, K. Effect of solvents extraction on phenolic content and antioxidant activity of the byproduct of eggplant. *Ind. Crops Prod.* **2013**, *49*, 668–674. [CrossRef]
27. Nawaz, H.; Shad, M.A.; Rehman, N.; Andaleeb, H.; Ullah, N. Effect of solvent polarity on extraction yield and antioxidant properties of phytochemicals from bean (*Phaseolus vulgaris*) seeds. *Braz. J. Pharm. Sci.* **2020**, *56*, e17129. [CrossRef]
28. Halee, A.; Supavititpatana, P.; Ruttarattanamongkol, K.; Jittrepotch, N.; Rojsuntornkitti, K.; Kongbangkerd, T. Effects of solvent types and citric acid concentrations on the extraction of antioxidants from the black rice bran of *Oryza sativa* L. CV. Hom Nin. *J. Microbiol. Biotechnol. Food Sci.* **2018**, *8*, 765. [CrossRef]
29. Alothman, M.; Bhat, R.; Karim, A. Antioxidant capacity and phenolic content of selected tropical fruits from Malaysia, extracted with different solvents. *Food Chem.* **2009**, *115*, 785–788. [CrossRef]
30. Todaro, A.; Cimino, F.; Rapisarda, P.; Catalano, A.E.; Barbagallo, R.N.; Spagna, G. Recovery of anthocyanins from eggplant peel. *Food Chem.* **2009**, *114*, 434–439. [CrossRef]
31. Stelluti, S.; Caser, M.; Demasi, S.; Scariot, V. Sustainable processing of floral bio-residues of saffron (*Crocus sativus* L.) for valuable biorefinery products. *Plants* **2021**, *10*, 523. [CrossRef]
32. Hosseini, S.; Gharachorloo, M.; Ghiassi-Tarzi, B.; Ghavami, M. Evaluation of the organic acids ability for extraction of anthocyanins and phenolic compounds from different sources and their degradation kinetics during cold storage. *Pol. J. Food Nutr. Sci.* **2016**, *66*, 261–270. [CrossRef]
33. Das, A.B.; Goud, V.; Das, C. Extraction of phenolic compounds and anthocyanin from black and purple rice bran (*Oryza sativa* L.) using ultrasound: A comparative analysis and phytochemical profiling. *Ind. Crops Prod.* **2017**, *95*, 332–341. [CrossRef]
34. Li, Y.; Han, L.; Ma, R.; Xu, X.; Zhao, C.; Wang, Z.; Chen, F.; Hu, X. Effect of energy density and citric acid concentration on anthocyanins yield and solution temperature of grape peel in microwave-assisted extraction process. *J. Food Eng.* **2012**, *109*, 274–280. [CrossRef]
35. Andrade, T.A.; Hamerski, F.; Fetzer, D.E.L.; Roda-Serrat, M.C.; Corazza, M.L.; Norddahl, B.; Errico, M. Ultrasound-assisted pressurized liquid extraction of anthocyanins from *Aronia melanocarpa* pomace. *Sep. Purif. Technol.* **2021**, *276*, 119290. [CrossRef]
36. Oladunjoye, A.O.; Olawuyi, I.K.; Afolabi, T.A. Synergistic effect of ultrasound and citric acid treatment on functional, structural and storage properties of hog plum (*Spondias mombin* L.) bagasse. *Food Sci. Technol. Int.* **2023**, 10820132231176579. [CrossRef]
37. Buvaneshwaran, M.; Radhakrishnan, M.; Natarajan, V. Influence of ultrasound-assisted extraction techniques on the valorization of agro-based industrial organic waste—A review. *J. Food Process Eng.* **2022**, *46*, e14012. [CrossRef]
38. Naczk, M.; Shahidi, F. Extraction and analysis of phenolics in food. *J. Chromatogr. A* **2004**, *1054*, 95–111. [CrossRef]
39. Naczk, M.; Shahidi, F. Phenolics in cereals, fruits and vegetables: Occurrence, extraction and analysis. *J. Pharm. Biomed. Anal.* **2006**, *41*, 1523–1542. [CrossRef]
40. Chirinos, R.; Rogez, H.; Camposa, D.; Pedreschi, R.; Larondelle, Y. Optimization of extraction conditions of antioxidant phenolic compounds from mashua (*Tropaeolum tuberosum* Ruiz & Pavon) tubers. *Sep. Purif. Technol.* **2007**, *55*, 217–225.
41. Fatiha, B.; Khodir, M.; Farid, D.; Tiziri, R.; Karima, B.; Sonia, O.; Mohamed, C. Optimisation of solvent extraction of antioxidants (phenolic compounds) from Algerian mint (*Mentha spicata* L.). *Pharmacogn. Commun.* **2012**, *4*, 72–86.
42. Ukrainczyk, L.; McBride, M.B. Oxidation of phenolics in acidic aqueous suspensions of manganese oxides. *Clays Clay Miner.* **1992**, *40*, 157–166. [CrossRef]
43. Jirum, J.; Srihanam, P. Oxidants and antioxidants: Sources and mechanism. *Acad. J. Kalasin Rajabhat Univ.* **2011**, *1*, 59–70.
44. Lachguer, K.; El Merzougui, S.; Boudadi, I.; Laktib, A.; Ben El Caid, M.; Ramdan, B.; Boubaker, H.; Serghini, M.A. Major phytochemical compounds, in vitro antioxidant, antibacterial, and antifungal activities of six aqueous and organic extracts of *Crocus sativus* L. flower waste. *Waste Biomass Valorization* **2023**, *14*, 1571–1587. [CrossRef]
45. Gadioli Tarone, A.; Keven Silva, E.; Dias de Freitas Queiroz Barros, H.; Bau Betim Cazarin, C.; Roberto Marostica, M., Jr. High-intensity ultrasound-assisted recovery of anthocyanins from jabuticaba by-products using green solvents: Effects of ultrasound intensity and solvent composition on the extraction of phenolic compounds. *Food Res. Int.* **2021**, *140*, 110048. [CrossRef]

Disclaimer/Publisher's Note: The statements, opinions and data contained in all publications are solely those of the individual author(s) and contributor(s) and not of MDPI and/or the editor(s). MDPI and/or the editor(s) disclaim responsibility for any injury to people or property resulting from any ideas, methods, instructions or products referred to in the content.

Article

Towards Sustainable Protein Sources: The Thermal and Rheological Properties of Alternative Proteins

Kaitlyn Burghardt [1], Tierney Craven [1], Nabil A. Sardar [2] and Joshua M. Pearce [3,*]

[1] Department of Chemical & Biochemical Engineering, Western University, London, ON N6A 5B9, Canada
[2] BeeHex, LLC, Columbus, OH 43230, USA
[3] Department of Electrical & Computer Engineering and Ivey Business School, Western University, London, ON N6A 5B9, Canada
* Correspondence: joshua.pearce@uwo.ca

Abstract: Reducing meat consumption reduces carbon emissions and other environmental harms. Unfortunately, commercial plant-based meat substitutes have not seen widespread adoption. In order to enable more flexible processing methods, this paper analyzes the characteristics of commercially available spirulina, soy, pea, and brown rice protein isolates to provide data for nonmeat protein processing that can lead to cost reductions. The thermal and rheological properties, as well as viscosity, density, and particle size distribution, were analyzed for further study into alternative protein-based food processing. The differential scanning calorimetry analysis produced dry amorphous-shaped curves and paste curves with a more distinct endothermic peak. The extracted linear temperature ranges for processing within food production were 70–90 °C for spirulina, 87–116 °C for soy protein, 67–77 °C for pea protein, and 87–97 °C for brown rice protein. The viscosity analysis determined that each protein material was shear-thinning and that viscosity increased with decreased water concentration, with rice being an exception to the latter trend. The obtained viscosity range for spirulina was 15,100–78,000 cP, 3200–80,000 cP for soy protein, 1400–32,700 cP for pea protein, and 600–3500 cP for brown rice protein. The results indicate that extrusion is a viable method for the further processing of protein isolates, as this technique has a large temperature operating range and variable screw speed. The data provided here can be used to make single or multi-component protein substitutes.

Keywords: rice protein; spirulina protein; pea protein; soy protein; plant-based diet; thermal properties; rheological properties; protein processing

1. Introduction

Although the UN reports that the annual population growth rate has been dropping (and is expected to keep dropping), the total population of the planet is expected to surpass 10 billion people in the 2050s [1]. This increasing population prompts the need for more sustainable sources of food [2] and, perhaps, the most challenging need: a low-cost source of abundant protein [3]. It is now widely accepted that meat production uses more energy and has a greater negative environmental impact than plant-based meat alternatives [4]. The average American meat-based diet demands more energy, land, and water resources in comparison to an ovo-lacto vegetarian diet, which underscores the need to shift from current meat-based food systems to sustainably meet the dietary demands of the increasing global population [5]. In addition to poor protein conversion efficiency, meat-based diets are unhealthy [6–8], whereas vegetarian diets are associated with a long list of health benefits, including lower rates of death from ischemic heart disease, lower cholesterol levels, lower blood pressure, and lower rates of hypertension and type 2 diabetes [9]. There are also public concerns about animal welfare [10,11] and serious public health issues related to animal diseases and animal husbandry practices that result in antibiotic-resistant bacteria [12,13]. Finally, agriculture is a major contributor to greenhouse gas (GHG) emissions; specifically, methane accounted for 35% of total food system GHG emissions in 2015 [14]. Livestock is a leading source of global methane emissions, and in 2010, 23% of global temperature warming was attributed to livestock emissions [15].

For these reasons, a more efficient plant-based protein diet may be used to mitigate agriculture's contribution to GHG emissions and other environmental hazards [16–18]. This relates to UN Sustainable Development Goal 1, as the cost of animal-based protein has increased globally but can be particularly inaccessible because of costs in the developing world. Consumers have been shifting towards this as the plant-based market [19] has grown from USD 4.8 billion in 2018 to USD 7.4 billion in 2020 [20]. Although clinically vegan/vegetarian diets have proven healthy for humans [21] and the planet [22], converting to a plant-based diet is challenging for some [23,24]. In order to overcome this challenge, a range of plant-based meat substitutes is under development [25–27] and have been commercialized (Beyond Meat, Impossible Burger, etc.) [28]. Unfortunately, these meat substitutes are often more expensive than the meat they aim to replace. For example, at Walmart Canada, 340 g of Beyond Meat Ground Beef is CAD 7.97 (2.3 cents/g), while multiple 450 g farm ground beef options range from CAD 4.97–7.47 (1.1 cent/g to 1.6 cents/g), which substantially restricts their uptake [29,30]. Yet, alternative protein on the market are generally substantially less costly than meat protein, as shown in Table 1. The protein cost was calculated by dividing the purchase cost by the protein content. Note that the purchase cost has been converted from the reported USD/lb to USD/kg.

Table 1. Protein cost (per gram), ranging from bulk costs to retail purchase costs.

Protein	Protein/Mass g/kg	Purchase Cost USD/kg	Protein Cost USD/kg	Sources
Boneless/Skinless Chicken Breast	310	6.28	0.0203	[31,32]
Beef Patties	230	13.69	0.0595	[33,34]
Pasture Raised Pork Shoulder	231	19.93	0.0863	[35,36]
Ground Lamb	166	15.71	0.0946	[37,38]
Alibaba Wholesale Spirulina Isolate Powder	600	8.50	0.0142	[39]
Spirulina Powder	667	95.29	0.143	[40,41]
Retail Spirulina Powder	667 [1]	99.98	0.150	[40,42]
Soybeans	433	0.52	0.00120	[43,44]
Alibaba Wholesale Soy Protein Isolate Powder	900	3.25	0.00361	[45]
Bulk Food Store Wholesale Soy Protein Isolate Powder	900	20.68	0.0230	[46]
Retail Soy Protein Isolate Powder	833	25.66	0.0308	[47]
Green Pea	54.2	0.31	0.00572	[48,49]
Alibaba Wholesale Pea Protein Isolate Powder	800	3.00	0.00375	[50]
Bulk Food Store Wholesale Pea Protein Isolate Powder	800	20.68	0.0259	[51]
Retail Pea Protein Isolate Powder	727	45.19	0.0622	[52]
Crude Rice Bran	134	0.18	0.00134	[53,54]
Alibaba Wholesale Brown Rice Protein Isolate Powder	850	2.50	0.00294	[55]
Bulk Food Store Wholesale Brown Rice Protein Isolate Powder	785	17.34	0.0221	[56]
Retail Brown Rice Protein Isolate Powder	809	43.21	0.0534	[57]

[1] Protein concentration not listed on product site, taken from USDA.

As can be seen in Table 1, the wholesale price of plant protein ranges from 0.00294 USD/g protein (rice)–0.0142 USD/g protein (spirulina), while meat purchase cost has an overall more expensive range of 0.0203 USD/g protein (chicken)–0.0946 USD/g protein (lamb) (Table 1). The most inexpensive plant protein cost is derived from the protein concentration of the raw source material without considering any protein isolation method costs. Raw green peas are the exception to this trend due to their low protein content in comparison to other plant sources. Overall, soy is the most inexpensive plant protein, followed by pea, rice, and spirulina, which are the most expensive. In order to make these meat alternatives more appealing, they normally undergo substantial processing. In order to make meat substitutes more accessible, the basic thermal and rheological properties are needed for alternative sources of protein and their processing, such as extrusion [58]. This paper analyzes the characteristics of spirulina, soy, pea, and brown rice proteins; these compounds naturally grow in various parts of the world and are commercially available. The thermal and rheological properties, viscosity, density, and particle size distribution are analyzed for feasibility comparisons that can be used for further study on alternative protein-based food processing and production. This study is important, as it analyzes more sustainable sources of alternative protein, which can be used to reduce the environmental footprint of the agriculture industry. Specifically, this work provides a means of using economic screw extruders to process a wide range of plant-based proteins into low-cost edible food.

2. Materials and Methods

2.1. Materials

The alternative protein materials were sourced from Healthy Planet Canada and include spirulina [42], soy protein isolate [47], pea protein isolate [52], and rice protein isolate [57].

2.2. Density

The densities of all the materials were determined to provide a full characterization by (1) massing an empty one dram (3.69669 cm^3) vial with a digital scale, (2) filling the vial with the material, and packing it down to flatten the top and record the mass; (3) the bulk density (d) was given by the equation below, where m is mass and V is volume:

$$d\left[\frac{g}{cm^3}\right] = \frac{m_{\text{full vial}} - m_{\text{empty vial}}}{V_{\text{vial}}} \quad (1)$$

2.3. Particle Size

The particle size was quantified to provide the limits for material processing when using the materials (e.g., 3D printing resolution is limited by the maximum particle size). By using a digital microscope (Celestron, Torrance, CA, USA), a micrometer calibration slide (Walfront, Lewes, DE, USA) was imaged and imported into the open source ImageJ package (version 1.53) [59], and a line was drawn on the length of one division to provide the scale for all photos (938 pixels/mm). All five materials were then imaged three times using the same magnification of 200×.

It should be noted that when placing the protein particles on the microscope slides, they often clumped together. To ensure even distribution, after the particles were placed onto the slide, they were brushed off slightly. The images and the scale were imported into ImageJ and analyzed. Firstly, the color threshold of the particles was adjusted to differentiate them from the background. Depending on the image, a threshold method was chosen. The Shanbhag method [60] was the most accurate, which selected only the particles and no additional area in the background. For some images, the Shanbhag method did not work, so other methods, such as the Intermodes method [61], were used. The particle size distribution was determined for each protein sample, and the mean particle size was summarized.

2.4. Differential Scanning Calorimetry (DSC) Analysis

Differential scanning calorimetry was conducted using a Mettler Toledo (Columbus, OH, USA) DSC 3 and the DSC STARe System to analyze the thermal stability of the materials by obtaining endothermic peak temperatures. The DSC samples were both wet and dry. The dry samples were unchanged from the retail purchase. The wet (paste) samples were prepared by measuring 100–150 mg of each material into a clean aluminum boat and adding 50% by weight deionized water with a clean glass pipette. The paste was thoroughly mixed for 1 min. The samples were prepared freshly immediately before each paste DSC run. A total of 1.56–4.64 mg of each powder sample and 3.98–6.38 mg of each paste sample were weighed accurately and placed into an aluminum DSC pan and hermetically sealed. An empty pan was also hermetically sealed to be used as a reference. Both pans were placed into the module for the experiment. All samples were heated from 20–150 °C at a heating rate 5 °C/min. Nitrogen was used as a purge gas at flow rates of 30 or 50 mL/min, depending on the run, to clear out the materials for the next run. All DSC runs were analyzed to find the denaturation temperature (endothermic peak temperature) and the stable temperature processing ranges.

2.5. Viscosity Measurements

A Brookfield (MA, USA) DV-II+ with an S62 spindle viscometer was used to test the materials. A total of 600 g of the sample was measured and mixed with water to reach the desired concentration, referred to as Composition A. This sample slurry was transferred to a 600 mL glass beaker to undergo testing with the viscometer. The spindle was set 1.1 inches above the bottom of the beaker. The viscometer was set at RPM values of 5, 15, 25, 40, and 80. After testing concluded for composition A, additional water was added to reach composition B in the interest of saving sample material. The same experiment was duplicated with the new composition B. This method was repeated for all samples. A summary of compositions tested is shown in Table 2.

Table 2. Viscometer sample water concentrations.

Protein	Composition A (% Water)	Composition B (% Water)
Spirulina	65.5	71.9
Soy	78.5	83.9
Pea	78.6	81.1
Brown Rice	69.1	58.9

3. Results

Each of the material tests for each of the alternative proteins was provided separately below. It should be pointed out that the results provide the processing windows and types for each of the materials, and these are evaluated independently. The primary purpose of this study is to identify the processing windows (e.g., temperature and viscosities) that can be used to reduce the costs of alternative proteins. The following properties are quantified: density, particle size, DSC, and viscosity.

3.1. Density

The density of the alternative proteins is shown in Table 3.

Table 3. Dry density of alternative proteins.

Protein	Density (g/cm^3)
Spirulina	0.49
Soy	0.68
Pea	0.76
Brown Rice	0.54

3.2. Particle Size

Figure 1 shows the micrographs, and Figure 2 shows the histogram of particle size distribution for each protein powder.

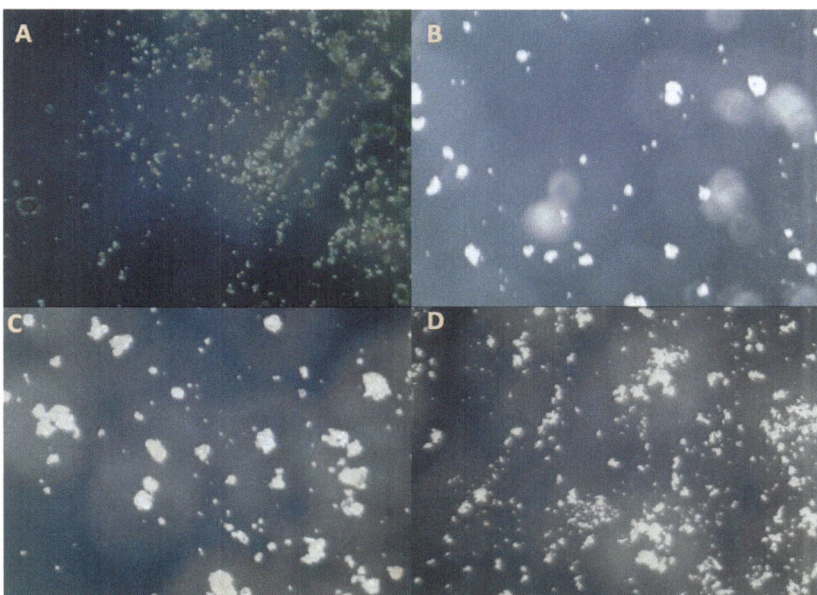

Figure 1. Digital microscope image with a magnification of 200× for (**A**) spirulina, (**B**) soy protein, (**C**) pea protein, and (**D**) brown rice protein.

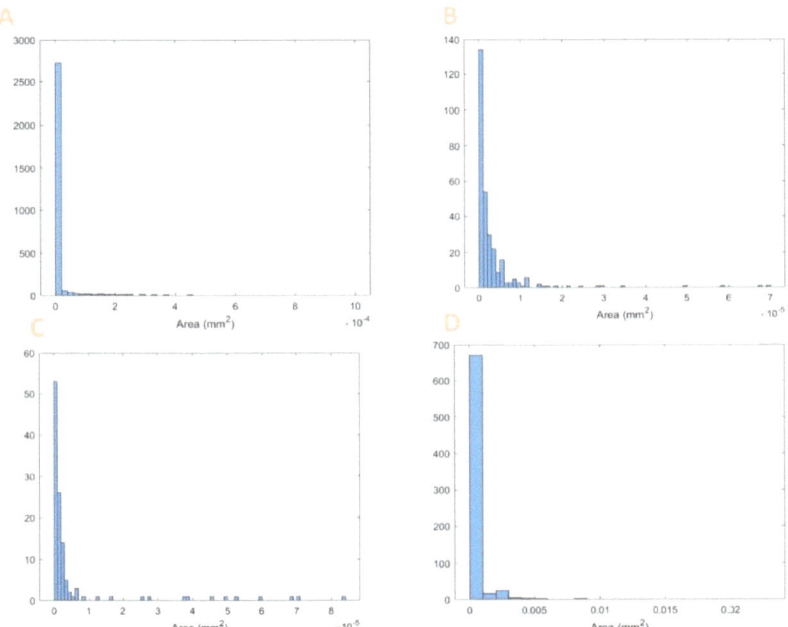

Figure 2. Histograms of particle size distribution for (**A**) spirulina, (**B**) soy protein, (**C**) pea protein, and (**D**) brown rice protein.

Table 4 summarizes the particle size and provides the literature values for comparison.

Table 4. Particle mean area, the mean diameter, and the literature diameter values for the protein powders.

Protein	Mean Area (μm²)	Mean Diameter (μm)	Literature Diameter Values (μm)
Spirulina	28	6.0	<125 [62]
Soy	3.8	2.2	0.1–100 [63]
Pea	6.6	2.9	1.7–270 [64]
Brown Rice	340	21	23–150 [65]

As seen from Table 4, the diameters found fit the range of known values for that particle. Although some are the minimum accepted values, they still fit within the range. The smaller particle sizes do enable some advantages in processing higher-resolution features.

3.3. DSC

The DSC results for each protein source are shown in Figures 3–6.

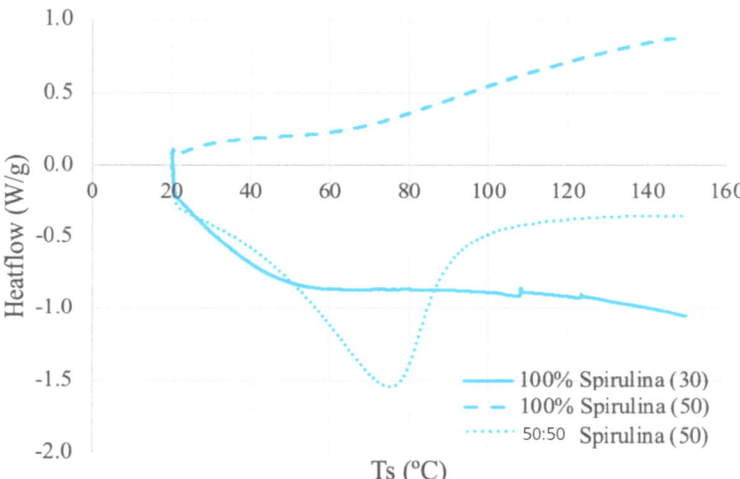

Figure 3. Spirulina DSC curves. The values (30) and (50) denote the nitrogen flow rate in mL/min.

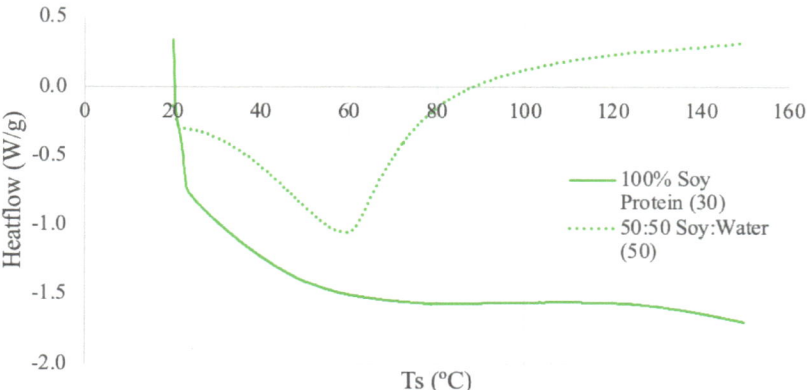

Figure 4. Soy Protein DSC curves. The values (30) and (50) denote the nitrogen flow rate in mL/min.

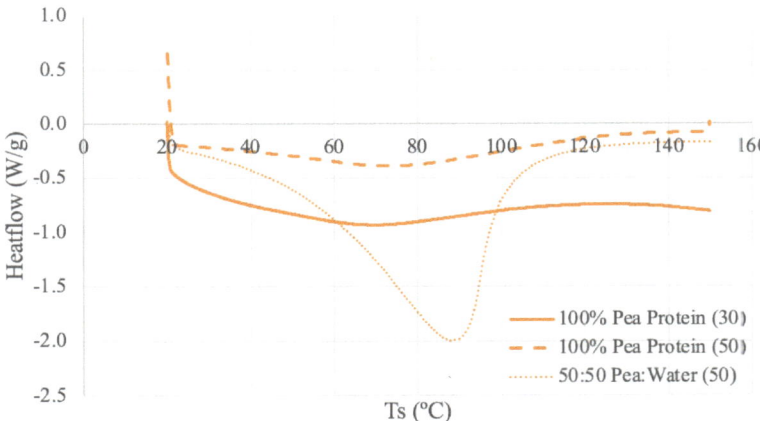

Figure 5. Pea Protein DSC Curves. The values (30) and (50) denote the nitrogen flow rate in mL/min.

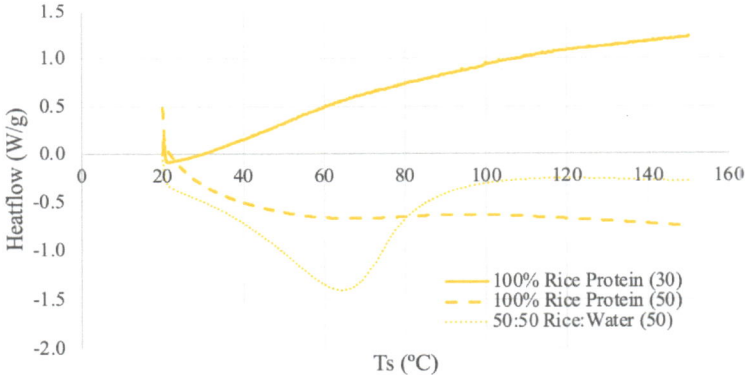

Figure 6. Brown Rice Protein DSC Curves. The values (30) and (50) denote the nitrogen flow rate in mL/min.

3.3.1. Spirulina DSC Results

The spirulina protein samples were run under three different conditions: A 100% spirulina powder using 30 mL/min N_2 and 50 mL/min N_2, and a 50–50 ratio of spirulina and deionized water paste using 50 mL/min N_2. Despite having an overall different curve shape, both non-50:50 powder samples exhibit a relatively linear trend line between 70–90 °C. Thermal deviations are evident in the spirulina (30) curve at around 107 °C and again at around 125 °C. This activity may be interpreted as possible thermal transition temperatures. Alternatively, the slight spikes may be an electrical effect and a result of static electricity discharge, although no significant power disturbance was observed [66]. A DSC study using 4 mg of spirulina protein isolated from spray-dried algal powder and dissolved in a 20 mL buffer solution displayed a DSC curve with a similar stability range of 73–83 °C [67]. The same curve also illustrated an endothermic melting phase change at 108.7 °C, like the spirulina (30) [67]. The DSC analysis in this study used a heating rate of 10 °C/min over a temperature range of 30–180 °C [67]. There are noticeable differences between the literature and obtained results, which can most likely be attributed to the differences in sample preparation and source material. Moreover, the different shapes of the powder curves may be attributed to increased heat transfer through an increased purge gas flow rate [68].

The spirulina and water paste curve looks like a more typical DSC curve with an obvious baseline and clear endothermic peak around 76 °C. In comparison, the dry samples lack the obvious thermal deviation present in the paste, demonstrating the amorphous shape of the dry curves [66]. Further, the difference in curve shape between the paste and powder samples suggests that a lower water content results in a lower denaturation enthalpy (ΔHd), assuming the peak area is equal to the enthalpy of denaturation [69]. This same result was reflected in a study using water and soybean protein [70]. Increasing heat promotes denaturation, and proteins unfold and lose their structure during denaturation [70]. Dehydration was determined to increase destabilization in the unfolded state [70]. Compared to the folded state, the unfolded state permits more contact between the compound and water [70]. Therefore, the unfolded state should have more internal bonding and be more compact at low water levels, which results in a lower ΔHd [70]. Moreover, the paste curve's endothermic peak (76 °C) is located within the assumed stable temperature range of the dry samples (70–90 °C). This discrepancy may be attributed to the decrease in water content. The lack of water may result in a rise in denaturation temperature (Td), which is also due to the destabilization of the unfolded states [70].

3.3.2. Soy Protein DSC Results

The soy protein samples were run under two different conditions: dry powder using 30 mL/min N_2 and a 50–50 ratio of soy protein and deionized water using 50 mL/min N_2. The dry sample thermogram does not exhibit distinct peaks or deviations, but a plateaued temperature range is exhibited from 87–116 °C. The dry sample curve also has a large endothermic start-up hook, which may be attributed to calibration between the sample and reference pans [71]. This large start-up sloping baseline can hinder the detection of weak transitions [71]. The DSC study using soybean protein extracted from soybean flour exhibits similar-looking curves, as obtained for low water contents (1% and 5% water content) [70]. Little thermal activity was observed for the soy globular protein samples at low water contents, and only after 150 °C were any deviations visible, which is outside the temperature range of Figure 4 [70]. Another soybean protein DSC study using silver pans and a heating rate of 5 °C/min from 25 to 200 °C also found similar results [72]. Soybean protein isolates with low water content (11%) did not exhibit an endothermic peak until past 180 °C [72].

The half soy protein and half deionized water sample exhibit an endothermic peak at 59 °C. The thermal activity of the paste sample is more distinct than that of the dry sample, likely due to the increased water content [70]. Further, the slope of the curve increases after the peak, which suggests a shifted baseline [71]. Baseline shifts can be caused by changes in specific heat, which is evidence that the sample has gone through a transition, such as melting [71]. A different DSC study using soybean protein extracted from soybean flour and mixed with distilled water at a 50–50 ratio demonstrated an endothermic peak of approximately 110 °C [73]. The DSC conditions for this study were a heating rate of 10 °C/min between 20 to 130 °C, a nitrogen flow rate of 50 mL/min, and aluminum pans [73]. Despite the similar water content and DSC conditions, the endothermic peak temperatures are not similar, which can likely be attributed to the differences in sample source and preparation. In the literature, the water-protein paste was stored for 24 h at 4 °C in polyethylene bags to promote the even distribution of water, rather than just mixing followed by testing [73]. Another DSC study comparing corn starch and soy protein ratios with 80% water content and a heating rate of 5 °C/min from 25 to 150 °C also found different results [74]. Despite having a water content of 80%, no endothermic peak was observed for the pure soy protein and water sample, possibly due to the previous heat treatment of the soy in the manufacturing stages [74].

3.3.3. Pea DSC Results

The pea protein samples were run under three different conditions: pure powder using 30 mL/min N_2 and 50 mL/min N_2 and a 50–50 ratio of pea protein and deionized water paste using 50 mL/min N$_2$. Both dry samples have similar curve shapes and an indication of thermal activity through a broad endothermic peak from 67–77 °C. Pea protein (50) has a longer endothermic start-up hook, potentially due to a larger baseline adjustment requirement because of the increased purge flow rate [71]. The difference in heat flow levels is likely attributed to the varying nitrogen flow rate for the dry sample.

When compared to the literature, a DSC study using low-denatured pea protein isolate and a heating rate of 5 °C/min from 20 to 110 °C obtained similar results [75]. The thermogram obtained in the literature has only one broad endothermic peak at around 78.5 °C [75]. Another DSC study using field peas at a pea starch-to-water ratio of 1:2 and a heating rate of 10 °C from 30–100 °C also obtained similar results of endothermic peaks from 75.5–89.9 °C [76]. Some of the literature reports both weaker and stronger peaks, possibly due to the denaturation of different components in peas, such as legumin and vicilin [77]. The multiple peaks are not present in the observed data, which may be due to the endothermic start-up hooks, which can diminish the detectability of weaker peaks [71]. Further, previous heat treatment on the sample source, such as spray-drying, may have had thermal effects on the sample, which would influence the shape of the DSC curve [77].

The pea protein and water DSC run produced an endothermic peak at around 89 °C. This value aligns with the high protein sample in the literature despite having a different pea-to-water ratio of 1:2 versus 1:1 [76]. The peak is more distinct than the dry peaks, which, again, suggests that increased water content results in an increase in denaturation enthalpy [70]. Unlike other samples, the paste curve appears to have a higher thermal stability than the powder curves, as the peak is at a higher temperature.

3.3.4. Brown Rice Protein DSC Results

The rice protein samples were run under three different DSC conditions: dry rice protein powder using 30 mL/min N$_2$ and 50 mL/min N$_2$, and a 50–50 ratio of rice protein and deionized water using 50 mL/min N$_2$. The rice protein (30) thermogram appears to calibrate at the beginning of the run and then have a consistently increasing slope. It is difficult to identify any meaningful temperature trends for this range. The rice protein (50) thermogram has a similar smooth and amorphous shape to the other dry samples but with a different slope. A stable, plateaued temperature range can be observed between 87–97 °C for the rice protein (50) curve. There may be a glass transition occurring around 100 °C as the slope of the curve decreases following this point [66]. A study using rice protein derived from long-grain rice combined with soy protein produced curves very similar to rice protein (50) [78]. In this study, a sample with a 1:0.1 ratio of rice protein-to-soy protein produced a very smooth and amorphous curve [78]. However, 2 mg of the sample was mixed with 10 µL of distilled water and allowed to reach full hydration [78]. Therefore, it may not be a completely accurate comparison; however, this finding in the literature does demonstrate the possibility of amorphous curves and highlights the significance of sample preparation and the source [78].

The paste curve reaches a steady baseline and has a more distinct endothermic peak at 65 °C. Again, the water content of the paste sample appears to increase the denaturation enthalpy [70]. A DSC study using three rice protein concentrates and a heating rate of 5 °C between 5 °C and 100 °C in an aluminum pan obtained peak temperatures of 63.6–70.2 °C [65]. These values align with the peak value of the paste; however, the literature states the samples were powders [65]. Regardless, obtaining a peak temperature within the range of the literature confirms that 65 °C is a reasonable endothermic peak temperature for rice protein.

3.4. Viscosity

Viscosity (in cP) at various RPM values was graphed (Figures 7–10). Viscosity curves often compare viscosity (units of cP or mPa·S) to shear rate (units of s^{-1}). For a concentric cylinder viscometer, the shear rate is equal to 1.7 times that of the RPM value of the outer cylinder [79]. For the purposes of this analysis, the shear rate is taken as being proportional to RPM; when RPM increases, so does the shear rate [79].

3.4.1. Spirulina Viscosity Results

The spirulina sample viscosity results were obtained with a water concentration of 71.9% (Figure 7). The curve shape is exponentially decreasing; as RPM increases, viscosity decreases. Therefore, the sample exhibits shear-thinning behavior since the RPM is proportional to the shear rate [79]. At 5 RPM, the viscosity value is 78,000 cP, and this decreases to 15,100 at 80 RPM. The spirulina sample was also tested at 65% water; however, the viscosity was too high and could not be measured. This result is expected, as the increased solid concentration correlates with an increase in viscosity because of stronger intermolecular bonds [80]. Conversely, an increase in water—the solvent—decreases the viscosity of the solution.

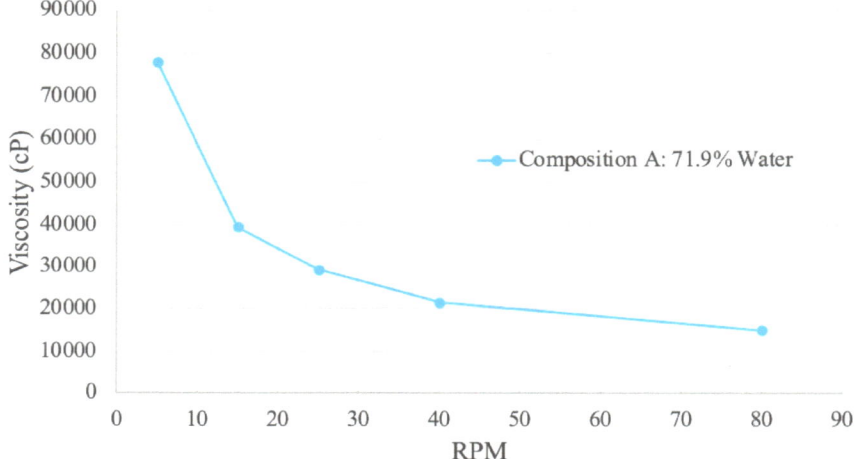

Figure 7. Spirulina Viscosity Curve.

Similar trend results were obtained in a study using a 2% spirulina concentration in a juice sample [81]. In this study, the viscosity versus shear rate curve had a decreasing slope, with viscosity values ranging from 3–5 mPa·s [81]. However, similar viscosity values were not obtained, which is attributed to the different solutions and lower spirulina concentration. Further, the addition of spirulina in the juice sample significantly increased the viscosity of the juice [81]. This observation from the literature aligns with the result of the 65% water sample having a greatly increased viscosity.

3.4.2. Soy Protein Viscosity Results

The soy protein samples were tested in a viscometer using two different concentrations (Figure 8). Composition A has 78.5% water, and composition B has 83.9% water. At all RPM values, composition A is more viscous than composition B, most likely due to the increase in intermolecular forces and friction with a higher solute concentration [80].

At 5 RPM, compositions A and B have viscosities of 80,000 cP and 39,000 cP, respectively. These values continually decrease until 80 RPM, where composition A has a viscosity of 5000 cP and composition B has a viscosity of 3200 cP. The decreasing exponential curve shape is characteristic of certain pseudo-plastic polymer proteins, including soy protein isolate [82,83]. The downward trend indicates shear-thinning behavior since viscosity decreases with increasing RPM [82]. Further, composition A has a steeper slope than composition B. This observation suggests that concentration influences shear-thinning behavior, as a higher concentration has more entanglement [82].

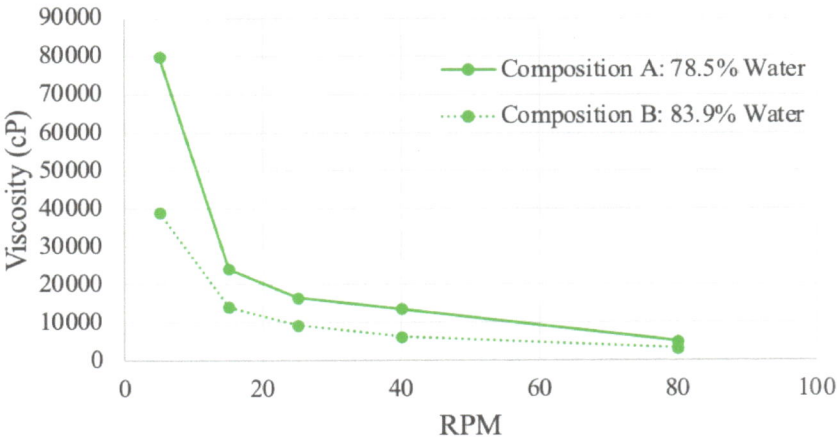

Figure 8. Soy Protein Viscosity Curve.

A study investigating the response of viscosity to shear rate using various soy protein and sodium bisulfate (NaHSO$_3$) solution concentrations also resulted in a decreasing exponential curve shape [84]. This study illustrated shear-thinning viscosity that decreased with an increase in the NaHSO$_3$ solvent [84]. The viscosity of the soy protein-NaHSO$_3$ solution was attributed to the various forces within soy protein, such as disulfide bonds, hydrophobic forces, and electrostatic interactions [84]. The influence of these forces on viscosity would decrease with a lower solid concentration, which was exhibited in the study and in the obtained results [84].

3.4.3. Pea Protein Viscosity Results

Pea protein samples were tested in a viscometer with two different concentrations (Figure 9). Composition A has 78.6% water, and composition B has 81.1% water. At all RPM, composition A was found to be more viscous. This result again demonstrates that decreased water content tends to result in increased viscosity.

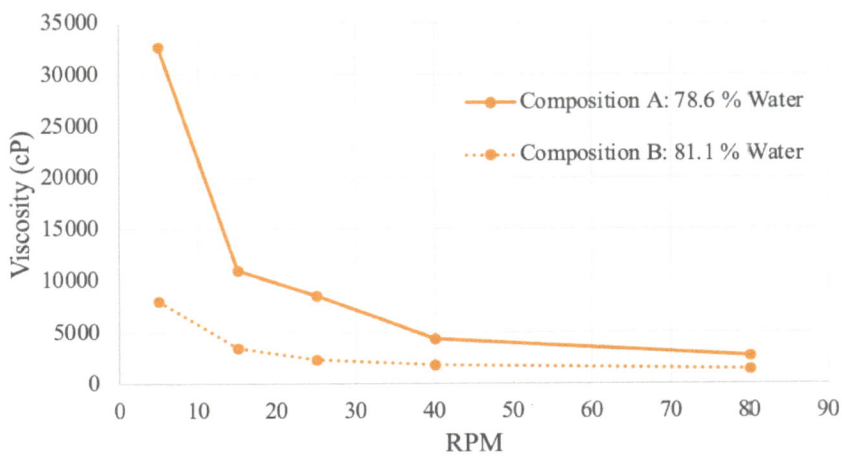

Figure 9. Pea Protein Viscosity Curve.

At an RPM of 5, composition A has a viscosity value of 32,700 cP, while composition B has a viscosity level of 8000 cP. At an RPM of 80, composition A's viscosity decreases to a value of 2700, and composition B's value decreases to 1400. This downward trend may be attributed to the protein particles expanding once they absorb water, resulting in a viscous flow, that then decreases with more rotations and mechanical energy [85]. As illustrated in the graph, when RPM increases and viscosity decreases, the sample is shear-thinning [79]. Further, composition B appears less affected by shear-thinning behavior, as the slope of the curve is overall less steep. This finding implies that shear-thinning behavior is proportionally related to concentration, likely due to molecular bonds and entanglement [82].

A study analyzing the RVA viscosity of pea protein slurry at 15% w/w produces a similar exponential curve shape [85]. The viscosity value begins at 2000 mPa·s, decreasing over time to 500 mPa·s, which is similar to the obtained results [85]. The differences in the initial viscosity value may be attributed to the preparation of the sample. In this study, the protein slurry was preheated to 50 °C and 95 °C [85]. Heating will influence viscosity, as viscosity is directly related to temperature [86]. This literature confirms that increased mechanical energy into the fluid, whether that be with increased time or RPM, is a factor that results in a less viscous fluid [85]. A different study investigating the relationship between shear rate and viscosity for pea protein isolate with 35% moisture content also found the sample to be shear-thinning [87].

3.4.4. Brown Rice Protein Viscosity Results

Rice protein samples were tested in the viscometer using two different concentrations (Figure 10). Composition A had 58.9% water, and composition B had 69.1% water. The rice viscosity results are the only measured samples where a lower powder concentration had a higher measured viscosity. In other words, an increase in water concentration resulted in an increase in viscosity. This observation suggests that the rice protein sample may have different hydrogen bonding patterns, resulting in different velocity distortion compared to the other materials tested [80]. Other compounds within the commercial rice protein, such as carbohydrates, may also interfere with the bonding behavior. A study involving homogenized rice protein and starch samples at various concentrations also produced results where the addition of protein decreased viscosity [88]. Since the rice protein used to make compositions A and B was commercially obtained, it is not unreasonable to assume that possible previous treatments, such as homogenization, may have taken place, which would influence the viscosity.

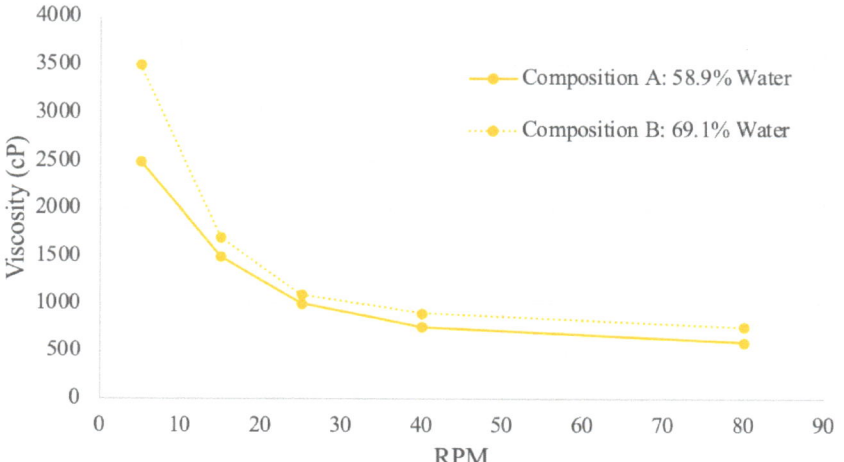

Figure 10. Brown Rice Protein Viscosity Curve.

Both viscosity curves demonstrate that viscosity decreases as RPM increases, which indicates shear-thinning behavior. The increased stress from the raised mechanical energy disorganizes the molecule arrangement, which decreases viscosity [89]. At 5 RPM, composition A has a viscosity of 2500 cP, and composition B has a viscosity of 3500 cP. At 80 RPM, composition A has a viscosity of 600 RPM, and composition B has a viscosity of 760 RPM. Of all the materials sampled, both rice compositions have the least difference in viscosity levels. Shear-thinning viscosity curves for rice protein were also reflected in the literature, even with different sample preparation techniques [88,89]. This shear-thinning flow behavior suggests that the rice protein acts as a pseudoplastic fluid [88].

3.5. Physical Properties and Nutrition

In order to summarize the material properties, Table 5 compares the alternative proteins to each other in terms of their linear processing temperature range and their viscosities. In addition, the Alibaba wholesale retail purchase cost is provided [39,45,50,55]. These ranges are graphed in Figures 11 and 12.

The temperature ranges in Figure 11 were extracted from the linear trend present in the pure protein thermograms produced by the DSC analysis. These temperature ranges are assumed to be relatively more stable due to the lack of thermal activity. Optimum processing temperatures may be found in these regions as the protein samples have more predictable thermal behavior. The soy and spirulina samples have the largest stable temperature ranges. The results suggest that the soy sample can be processed at higher temperatures compared to the other samples due to increased thermal stability over a broader temperature range.

Table 5. Summary of linear temperature and viscosity ranges.

Protein	100% Powder Sample Linear Temperature Range (°C)	Viscosity Range (cP)	Wholesale Retail Purchase Cost (USD/g protein) (Table 1)
Spirulina	70–90	15,100–78,000	0.0142
Soy	87–116	3200–80,000	0.00361
Pea	67–77	1400–32,700	0.00375
Brown Rice	87–97	600–3500	0.00294

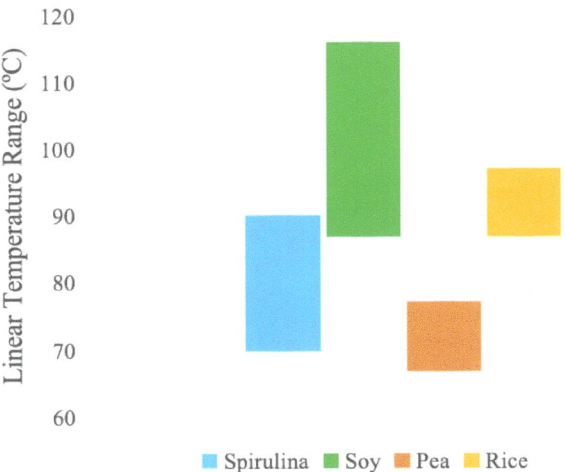

Figure 11. Linear temperature range for each protein sample.

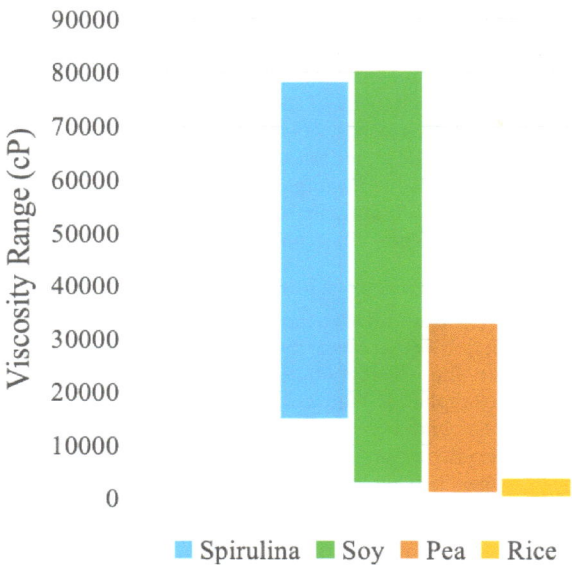

Figure 12. Viscosity range for each protein sample.

Figure 12 displays the viscosity ranges exhibited by both compositions of the protein samples between 5–80 RPM. Soy and spirulina exhibit the largest viscosity values and ranges and may be appropriate materials to use for processing using screw extruders, as these extruders are used for highly viscous non-Newtonian materials [90]. Rice has a much smaller range, which suggests that, of the samples, the viscosity of rice is the least influenced by water concentration. Rice may have different hydrogen bonding patterns than the other samples, which could result in a narrower range [80].

In both the temperature and viscosity data, soy and spirulina have the largest ranges. These protein samples can likely be used across more processing options and parameters compared to pea and rice. Moreover, the pea range is in the region of the spirulina range for both temperature and viscosity, which suggests that pea has similar thermal and rheological responses to spirulina and can possibly be used as a less-expensive substitute. These materials can also be mixed to provide composite protein sources with various beneficial properties (e.g., amino acid profiles, although it should be noted that there are many complexities with amino acids, such as bio-availability, including digestibility, availability, and absorption, which is left for future work).

Various plant protein options contain multiple essential amino acids, and content varies greatly across plant protein sources. Spirulina contains essential amino acids such as leucine and tryptophane; the superfood also has non-essential amino acid content, including glutamic acid and aspartic [91]. Minerals like potassium, calcium, phosphorous, magnesium, zinc, and iron are present in spirulina in significant concentrations [91]. Spirulina also contains vitamin B12, which is notable since animal products tend to be the greatest source of B12 [91]. Despite the richness in vitamins and minerals and the presence of multiple amino acids, microalgae fall short of the recommended essential amino acid percentage at 23% [92]. Therefore, if spirulina (algae protein) was the only protein source consumed, the essential amino acid requirement would not be met.

Legumes, such as soybeans and peas, are generally good sources of protein, carbohydrates, fiber, vitamins, and minerals. Of the essential amino acids, histidine, isoleucine, leucine, lysine, threonine, and valine are generally present in legumes [93]. Certain amino acids—methionine, phenylamine, and tryptophan—are not as abundant across legumes [94,95]. Gorissen et al. demonstrated that soy and pea amino acid percentage levels of 27% and 30%, respectively, meet the WHO/FAO/UN amino acid percentage level, based on protein consumption of 0.66 g/kg body weight/day [92]. These levels, however, are lower than the overall animal-based protein essential amino acid percentage (37 ± 2%) [92].

Cereal grains, such as rice, are widely consumed and are a readily available protein source globally. Essential and non-essential amino acids are present in certain strains of rice, such as rice bran and brown rice [96,97]. Generally, the limiting amino acids in cereals are lysine, methionine, and threonine [98]. Brown rice also meets the WHO/FAO/UNU amino acid requirement, with an essential amino acid percentage in total protein of 28% [92]. This value is also lower than the animal-based protein percentage [92].

4. Discussion

4.1. Plant Protein Processing Options

Heating is a widely used food processing technique, especially in cooking, and it has demonstrated positive impacts on plant protein digestibility and nutritional quality [93,99]. Heating processes utilize thermal treatments for sterilization, flavor and texture enhancement, improvements in functional properties, such as emulsification, and the destruction of undesired compounds [93,100]. The adverse effects of thermal treatment have also been observed, such as the degradation of protein and micronutrients, which can alter amino acid composition [93]. However, protein digestibility can be improved through the denaturation of proteins by heating [93]. The temperature processing parameters for cooking with plant protein usually occur around 100 °C [93,100,101]. Therefore, cooking may be a feasible processing method for compounds comprised of soy and rice protein, as their linear temperature ranges were found to be 87–116 °C and 87–97 °C.

Drying is another common food processing method. The techniques include spray drying, freeze drying, and the use of supercritical fluid technology [102]. In terms of protein functional properties, drying technology is commonly used to improve emulsifying and foaming properties, protein solubility, and water-holding capacity [93]. Spray drying inlet temperatures are typically well over 100 °C; therefore, each protein sample in this study may experience thermal effects if used in a drying process [103,104]. Vacuum drying, however, can also be effective below 100 °C [105].

Autoclaving can be used as a sterilization technique and is a high-pressure cooking method [93]. This process utilizes steam, which is important to note, as thermal stability can be affected by water concentration [70]. The typical autoclaving process parameters for plant-based proteins are 5–15 psi, 112–127 °C, and 10–50 min [93,99]. Therefore, the soy protein sample may be a possible candidate for autoclaving processing.

Extrusion combines mechanical shear, pressure, and heat by using a large rotating screw and high temperatures to result in a high-temperature short time (HTST) food processing technique [106]. Materials extrusion is also a method that can be coupled with additive manufacturing. The denaturation, unfolding, and realignment of plant protein molecules can be present in extrusion, which can improve the functionality of the compound and produce a meat-like texture [106]. The high temperature in extrusion leads to hydrogen and intramolecular disulfide bond breakage, which can promote the formation of new protein aggregates [107]. Overall, extrusion temperatures have a large range but typically fall in the 90–200 °C span, and the screw speeds have RPM values in the hundreds [106]. Zahari et al. produced a meat analogs product by using various soy protein isolate and hemp protein concentrate mixtures at extrusion temperatures ranging from 40–100 °C and screw speeds of 300–800 RPM [108]. Due to the encompassing temperature range, extrusion is a feasible processing method for the spirulina, soy protein, pea protein, and rice protein samples used in this study. The increased industrial RPM values, however, would have mechanical effects by decreasing the viscosity of the protein isolates. Moreover, particle size has also been demonstrated to influence extrusion. Carvalho et al. found that corn meal extrudate expansion increased with particle size, and that the mechanical resistance of the extrudates was significantly larger for the smallest particle size than the largest particle size [109]. If this same result is extended to all plant proteins, the rice protein sample would have the highest level of expansion, and the soy protein sample would have the most mechanical resistance based on the results of the materials used in this study. Moreover, increased density correlates with decreased viscosity, which would influence extrusion behavior [110].

4.2. Viability and Sustainability

Compared to an ovo-lacto vegetarian diet, the average meat-based American diet requires more land, energy, and water resources [5]. Vegetarian diets have been demonstrated to produce less per capita GHG emissions than Mediterranean and pescetarian diets and the average global diet in 2009 [111]. Further, a vegetarian diet is likely to require less additional cropland than meat-including diets [111]. A life cycle analysis performed by Detzal et al. found that a meat-analogous plant-based extrudate had a lower carbon footprint in terms of carbon dioxide equivalents than chicken when using a functional unit of 30 g of protein and 30 g of product [17]. In the same study, when compared to chicken, an optimized plant-based meat alternative had lower impacts in environmental categories, including water processing, land use, climate change, eutrophication, acidification, particulate matter, ozone depletion, and oxidants formation, and was comparable in cumulative nonrenewable energy demand [17]. Additionally, crops, such as soybeans, are used to feed livestock and poultry [112]. Of all soya and grain produced, 75% and 40% is directed towards animal feed, respectively [112]. Countries with a high level of animal products consumed, especially beef, have a larger area of agricultural land usage [113]. Alexander et al. found that the global consumption of an average American omnivorous diet would require 178% more land for agriculture than is currently in use, highlighting the importance of plant-based food options that can be produced on a global scale [113].

The proteins chosen in this study—spirulina, soy, pea, and brown rice—are strong candidates for plant-based food and meat-analogous products because of protein content and geographic availability. Per Table 1, the average protein content as a percentage for spirulina isolate, soy protein isolate, pea protein isolate, and brown rice protein isolate is 67%, 88%, 78%, and 81%, respectively. Those values are higher than the reported plant protein contents for hemp at 51%, lupin at 61%, oat at 64%, and corn at 65% [92]. Further, the protein samples used in this study have a higher protein content than egg at 51%, and a comparable percentage to animal-based proteins of whey at 72–84%, milk around 75%, and casein at 67–78% [92]. Spirulina is found across the globe as it is naturally occurring in salt water, fresh water, brackish water, soil, sand, and marshes [114]. Soya beans are grown primarily in the Americas and Asia, with the top producers being the USA, Brazil, Argentina, China, and India [115]. Soya production is also scattered throughout Europe and Africa [115]. Green peas are mainly produced in Asia, with China and India dominating the production market; Europe, the Americas, and Africa also produce green peas [115]. Asia has the largest share of rice production, with China, India, and Southeast Asian countries as the top producers [115]. The Americas and Africa also have a smaller share of rice production [115].

4.3. Future Work

The literature clearly shows that there would be environmental and health benefits to reducing animal proteins in favor of plant and nonmeat-based proteins. Although nonmeat sources are less expensive per unit protein than animals, as was shown in the introduction, plant-protein based foods are generally more expensive at the retail level than animal proteins. This can be ascribed to a number of reasons, including processing costs. The results presented here are the first step in making nonmeat-based protein affordable at the retail scale, as the base thermal and rheological properties have been elucidated that can allow for greater competition between methods and providers, and lower costs can be assumed. Future work is needed to determine if these are valid assumptions. Further work and process testing, however, is needed to establish these protein isolates in a range of foods.

For example, as the results here and elsewhere have conveyed that temperature and water percentage impact the viscosities of the materials, an in-depth study is needed on both the pure materials and protein mass percent combinations for the available temperature processing ranges and viscosity ranges summarized in Section 3.5. Future work could also assess the impacts of mixing the various volume fractions of the proteins to create ideal properties for a specific material-processing technique. In addition, the geographical availability of each protein source could be evaluated, both for the local optimization of protein source availability during agricultural times as well as considering their use as resilient food [116–119] to help provide adequate nutrition during emergencies [120]. Even excluding emergencies with rapid growth observed in the global population, the demand for meat is increasing and can be offset with plant-based meat alternatives that solve the economic, environmental, and health problems caused by the over-consumption of meat products by humans [121,122]. Far more work is needed to build upon the preliminary results here to meet the complex challenges of replacing meat entirely [122].

The results, however, can be compared to meat. Here, this will be carried out for chicken, as, in 2021, over 132 million tons of poultry meat was consumed globally, which makes it the most common meat-based food [123]. Normally, meat is not converted to a powder and consumed, so only the thermal and rheometric properties will be compared. First, consider that chicken meat in the form of chicken breast patties yields three endothermic transitions, with peak transition temperatures of 53 °C, 70 °C, and 79 °C, respectively [124]. If the alternative proteins analyzed here were to be processed with chicken meat in all cases, the linear temperature range (as shown in Figure 11) would result in endothermic transitions in chicken meat caused by the denaturation of myofibrillar at 53 °C [124]. Only pea protein could be further processed in its linear temperature range before the denaturation of chicken sarcoplasmic (70 and 79 °C) proteins, respectively [124]. Similarly, for the rheometric analysis, chicken meat is normally eaten simply cooked, although there have been some experiments for converting it to space food as a paste. When scientists investigated Chinese yam/chicken semi-liquid paste, they were able to obtain a wide range of viscosities ranging from honey to thick pudding by adding various proportions of chitosan [125]. This same approach could be used with the alternative proteins investigated here so that they could be used for more technically sophisticated processing techniques (e.g., 3D printing [126]). In addition, chitosan can have other beneficial properties, such as being used as a food preservative [127]. This property could also be explored, as it could lead to potentially new products (e.g., alternative protein pastes targeted for the growing 3D printing food market [128,129]). Combining these approaches is already underway, as demonstrated by Wang et al., who combined chicken paste and pea protein to make 3D-printed nuggets [130]. It is clear that the results from this study can also be used as a baseline for food 3D printing studies in the future. In addition, substantial work is underway to increase the aqueous solubility of plant proteins to provide alternative proteins [131]. Future work is needed to determine the impact on the viscosity and DISC of these and other functional properties of plant proteins. Finally, it should be stressed that when comparing alternative proteins to meat, it is the final processed (including cooking) material properties that are the most important for consumption; this area is also needed in the future.

5. Conclusions

This study extended the current state of research into alternative proteins by providing new base material properties for four globally common sources of alternative proteins. These material properties can be used to select a wider range of food processing techniques. Density and particle size were found for (a) spirulina, (b) soy protein, (c) pea protein, and (d) brown rice protein. These properties provide lower limits for resolution for the material-extrusion-based additive manufacturing of the materials. The spirulina sample had a density of 0.49 g/cm^3, a mean area of 28 µm^2, and a mean diameter of 6.0 µm. The soy protein sample had a density of 0.68 g/cm^3, a mean area of 3.8 µm^2, and a mean diameter of 2.2 μm. The pea protein sample had a density of 0.76 g/cm^3, a mean area of 6.6 μm^2, and a mean diameter of 2.9 μm. The brown rice protein sample had a density of 0.54 g/cm^2, a mean area of 340 μm^2, and a mean diameter of 21 μm. The DSC analysis provided thermal processing windows for the materials. The DSC analysis produced dry curves with an amorphous shape and paste curves with a more distinct endothermic peak. Linear temperature ranges were extracted and interpreted as processing parameters for food production. The linear temperature ranges were 70–90 °C for spirulina, 87–116 °C for soy protein, 67–77 °C for pea protein, and 87–97 °C for brown rice protein. Soy had the largest linear temperature range and is likely the most thermally stable due to fewer deviations. The viscosity analysis determined each protein sample experienced shear-thinning and that viscosity increased with decreased water concentration, with rice being an exception to the latter trend. The obtained viscosity range for spirulina was 15,100–78,000 cP, 3200–80,000 cP for soy protein, 1400–32,700 cP for pea protein, and 600–3500 cP for brown rice protein. Additionally, extrusion seems to be a viable method for the further processing of the protein isolates, as this technique has a large temperature operating range and variable screw speed. Extrusion has been used to develop meat-analogous products with plant proteins. Plant-based protein products can be used as an alternative to traditional poultry, livestock, and other meat proteins, as they have a large range of essential and non-essential amino acids and are considered more resource-efficient. The protein isolates analyzed in this study have high protein concentrations and wide geographic availability, making them strong candidates for further research into the materials processing and economic viability of extruded meat-analogous products. Future work will include process optimization and extending this to mixtures of alternative protein materials.

Author Contributions: Conceptualization, J.M.P.; methodology, K.B., T.C., N.A.S. and J.M.P.; validation, K.B., T.C., N.A.S. and J.M.P.; formal analysis, K.B., T.C., N.A.S. and J.M.P.; investigation, K.B., T.C. and N.A.S.; resources, N.A.S. and J.M.P.; data curation, K.B., T.C. and N.A.S.; writing—original draft preparation, K.B., T.C., N.A.S. and J.M.P.; writing—review and editing, K.B., T.C., N.A.S. and J.M.P.; visualization, K.B. and T.C.; supervision, J.M.P.; funding acquisition, J.M.P. All authors have read and agreed to the published version of the manuscript.

Funding: This work was supported by BeeHex, LLC, and the Thompson Endowment.

Institutional Review Board Statement: Not applicable.

Informed Consent Statement: Not applicable.

Data Availability Statement: The original contributions presented in the study are included in the article, further inquiries can be directed to the corresponding author.

Acknowledgments: The authors would like to thank Surface Science Western, Gleb Meirson, and Eric Pilles for technical support.

Conflicts of Interest: The authors declare no conflict of interest. The authors declare that this study received funding from BeeHex, LLC. The funder had the following involvement with the study: collaborative research, although it should be clear that the nature of this foundational research presents no conflict with the funder, and the funder's commercial products were not used for this research.

References

1. United Nations: DESA, Population Division World Population Prospects 2022 Growth Rate. Available online: https://population.un.org/wpp/Graphs/Probabilistic/POP/GrowthRate/900 (accessed on 22 November 2022).
2. Lindgren, E.; Harris, F.; Dangour, A.; Gasparatos, A.; Hiramatsu, M.; Javadi, F.; Loken, B.; Murakami, T.; Scheelbeek, P.; Haines, A. Sustainable Food Systems—A Health Perspective. *Sustain. Sci.* **2018**, *13*, 1505–1517. [CrossRef] [PubMed]
3. Cole, M.; Augustin, M.; Robertson, M.; Manners, J. The Science of Food Security. *NPJ Sci. Food* **2018**, *2*, 14. [CrossRef] [PubMed]
4. Dagevos, H.; Voordouw, J. Sustainability and Meat Consumption: Is Reduction Realistic? *Sustain. Sci. Pract. Policy* **2013**, *9*, 60–69. [CrossRef]
5. Pimentel, D.; Pimentel, M. Sustainability of Meat-Based and Plant-Based Diets and the Environment. *Am. J. Clin. Nutr.* **2003**, *78*, 660S–663S. [CrossRef] [PubMed]

6. Song, M.; Fung, T.T.; Hu, F.B.; Willett, W.C.; Longo, V.D.; Chan, A.T.; Giovannucci, E.L. Association of Animal and Plant Protein Intake With All-Cause and Cause-Specific Mortality. *JAMA Intern. Med.* **2016**, *176*, 1453. [CrossRef] [PubMed]
7. Abete, I.; Romaguera, D.; Vieira, A.R.; Lopez De Munain, A.; Norat, T. Association between Total, Processed, Red and White Meat Consumption and All-Cause, CVD and IHD Mortality: A Meta-Analysis of Cohort Studies. *Br. J. Nutr.* **2014**, *112*, 762–775. [CrossRef] [PubMed]
8. Mayo Clinic. It's Time to Try Meatless Meals. Available online: https://www.mayoclinic.org/healthy-lifestyle/nutrition-and-healthy-eating/in-depth/meatless-meals/art-20048193 (accessed on 6 December 2023).
9. Kim, H.; Caulfield, L.E.; Garcia-Larsen, V.; Steffen, L.M.; Coresh, J.; Rebholz, C.M. Plant-Based Diets Are Associated With a Lower Risk of Incident Cardiovascular Disease, Cardiovascular Disease Mortality, and All-Cause Mortality in a General Population of Middle-Aged Adults. *JAHA* **2019**, *8*, e012865. [CrossRef]
10. Alonso, M.; González-Montaña, J.; Lomillos, J. Consumers' Concerns and Perceptions of Farm Animal Welfare. *Animals* **2020**, *10*, 385. [CrossRef]
11. Beausoleil, N.; Mellor, D.; Baker, L.; Baker, S.; Bellio, M.; Clarke, A.; Dale, A.; Garlick, S.; Jones, B.; Harvey, A.; et al. "Feelings and Fitness" Not "Feelings or Fitness"–The Raison d'être of Conservation Welfare, Which Aligns Conservation and Animal Welfare Objectives. *Front. Vet. Sci.* **2018**, *5*, 296. [CrossRef]
12. Manyi-Loh, C.; Mamphweli, S.; Meyer, E.; Okoh, A. Antibiotic Use in Agriculture and Its Consequential Resistance in Environmental Sources: Potential Public Health Implications. *Molecules* **2018**, *23*, 795. [CrossRef]
13. Serwecińska, L. Antimicrobials and Antibiotic-Resistant Bacteria: A Risk to the Environment and to Public Health. *Water* **2020**, *12*, 3313. [CrossRef]
14. Crippa, M.; Solazzo, E.; Guizzardi, D.; Monforti-Ferrario, F.; Tubiello, F.N.; Leip, A. Food Systems Are Responsible for a Third of Global Anthropogenic GHG Emissions. *Nat. Food* **2021**, *2*, 198–209. [CrossRef]
15. Reisinger, A.; Clark, H. How Much Do Direct Livestock Emissions Actually Contribute to Global Warming? *Glob. Change Biol.* **2018**, *24*, 1749–1761. [CrossRef]
16. Chai, B.; van der Voort, J.; Grofelnik, K.; Eliasdottir, H.; Klöss, I.; Perez-Cueto, F. Which Diet Has the Least Environmental Impact on Our Planet? A Systematic Review of Vegan, Vegetarian and Omnivorous Diets. *Sustainability* **2019**, *11*, 4110. [CrossRef]
17. Detzel, A.; Krüger, M.; Busch, M.; Blanco-Gutiérrez, I.; Varela, C.; Manners, R.; Bez, J.; Zannini, E. Life Cycle Assessment of Animal-based Foods and Plant-based Protein-rich Alternatives: An Environmental Perspective. *J. Sci. Food Agric.* **2022**, *102*, 5111–5120. [CrossRef]
18. Rabès, A.; Seconda, L.; Langevin, B.; Allès, B.; Touvier, M.; Hercberg, S.; Lairon, D.; Baudry, J.; Pointereau, P.; Kesse-Guyot, E. Greenhouse Gas Emissions, Energy Demand and Land Use Associated with Omnivorous, Pesco-Vegetarian, Vegetarian, and Vegan Diets Accounting for Farming Practices. *Sustain. Prod. Consum.* **2020**, *22*, 138–146. [CrossRef]
19. Aschemann-Witzel, J.; Gantriis, R.; Fraga, P.; Perez-Cueto, F. Plant-Based Food and Protein Trend from a Business Perspective: Markets, Consumers, and the Challenges and Opportunities in the Future. *Crit. Rev. Food Sci. Nutr.* **2021**, *51*, 3119–3128. [CrossRef] [PubMed]
20. Plant Based Foods Association 2021 US Retail Sales Data for the Plant-Based Food Industry. Available online: https://www.plantbasedfoods.org/2021-u-s-retail-sales-data-for-the-plant-based-foods-industry/ (accessed on 7 November 2022).
21. Remde, A.; DeTurk, S.; Almardini, A.; Steiner, L.; Wojda, T. Plant-Predominant Eating Patterns–How Effective Are They for Treating Obesity and Related Cardiometabolic Health Outcomes?—A Systematic Review. *Nutr. Rev.* **2022**, *80*, 1094–1104. [CrossRef]
22. Fresán, U.; Sabaté, J. Vegetarian Diets: Planetary Health and Its Alignment with Human Health. *Adv. Nutr.* **2019**, *10*, S380–S388. [CrossRef]
23. Fehér, A.; Gazdecki, M.; Véha, M.; Szakály, M.; Szakály, Z. A Comprehensive Review of the Benefits of and the Barriers to the Switch to a Plant-Based Diet. *Sustainability* **2020**, *12*, 4136. [CrossRef]
24. Von Essen, E. Young Adults' Transition to a Plant-Based Diet as a Psychosomatic Process: A Psychoanalytically Informed Perspective. *Appetite* **2021**, *157*, 105003. [CrossRef] [PubMed]
25. Tziva, M.; Negro, S.; Kalfagianni, A.; Hekkert, M. Understanding the Protein Transition: The Rise of Plant-Based Meat Substitutes. *Environ. Innov. Soc. Transit.* **2021**, *35*, 217–231. [CrossRef]
26. Santo, R.E.; Kim, B.F.; Goldman, S.E.; Dutkiewicz, J.; Biehl, E.M.B.; Bloem, M.W.; Neff, R.A.; Nachman, K.E. Considering Plant-Based Meat Substitutes and Cell-Based Meats: A Public Health and Food Systems Perspective. *Front. Sustain. Food Syst.* **2020**, *4*, 134. [CrossRef]
27. Bakhsh, A.; Lee, S.; Lee, E.; Hwang, Y.; Jo, S. Traditional Plant-Based Meat Alternatives, Current and a Future Perspective: A Review. *J. Agric. Life Sci* **2021**, *28*, 1–11. [CrossRef]
28. Curtain, F.; Grafenauer, S. Plant-Based Meat Substitutes in the Flexitarian Age: An Audit of Products on Supermarket Shelves. *Nutrients* **2019**, *11*, 2603. [CrossRef]
29. Walmart Canada Beyond Meat Plant Based Ground, 340 g. Available online: https://www.walmart.ca/en/ip/beyond-meat-plant-based-ground-340g/6000200853971 (accessed on 26 November 2022).
30. Walmart Canada Search: Ground Beef. Available online: https://www.walmart.ca/search?q=ground%20beef&c=10019 (accessed on 26 November 2022).

31. U.S. Department of Agriculture. *Chicken, Broilers or Fryers, Breast, Meat Only, Cooked, Roasted*; U.S. Department of Agriculture: Washington, DC, USA, 2019.
32. USDA. *Livestock, Poultry, & Grain Market News Broiler/Fryer: USDA Weekly Retail Chicken Feature Activity Report (Fri)*; U.S. Department of Agriculture: Washington, DC, USA, 2022.
33. U.S. Department of Agriculture. *Central Beef, Ground, Patties, Frozen, Cooked, Broiled*; U.S. Department of Agriculture: Washington, DC, USA, 2019.
34. USDA. *Agricultural Marketing Service, Livestock, Poultry, & Grain Market News USDA Weekly Retail Beef Feature Activity*; U.S. Department of Agriculture: Washington, DC, USA, 2022.
35. U.S. Department of Agriculture. *Central Pork, Fresh, Shoulder, Blade, Boston (Roasts), Separable Lean and Fat, Cooked, Roasted*; U.S. Department of Agriculture: Washington, DC, USA, 2019.
36. USDA. *Livestock, Poultry, & Grain Market News National Monthly Pasture Raised Pork Report*; U.S. Department of Agriculture: Washington, DC, USA, 2022.
37. U.S. Department of Agriculture. *Lamb, Ground, Raw*; U.S. Department of Agriculture: Washington, DC, USA, 2019.
38. USDA. *Livestock, Poultry, & Grain Market News National Estimated Lamb Carcass Cutout (PDF) (LM_XL502)*; U.S. Department of Agriculture: Washington, DC, USA, 2022.
39. Alibaba.com. Factory Spirulina Suppliers Organic Green Colors Spirulina Powder Bulk Super Spirulina Powder Tablets. Available online: https://www.alibaba.com/product-detail/Factory-Spirulina-Suppliers-Organic-Green-Colors_1600345281535.html (accessed on 7 November 2022).
40. U.S. Department of Agriculture. *Spirulina Powder*; U.S. Department of Agriculture: Washington, DC, USA, 2021.
41. MRM Nutrition Superfoods—Spirulina Powder. Available online: https://mrmnutrition.com/products/superfoods-raw-spirulina-powder (accessed on 7 November 2022).
42. Healthy Planet Canada Organika Organic Spirulina 500 g. Available online: https://www.healthyplanetcanada.com/organika-organic-spirulina-500g.html (accessed on 7 November 2022).
43. U.S. Department of Agriculture. *Soybeans, Mature Seeds, Dry Roasted*; U.S. Department of Agriculture: Washington, DC, USA, 2019.
44. U.S. Department of Agriculture. *Feed Grains Custom Query Soybean Meal, High Protein*; U.S. Department of Agriculture: Washington, DC, USA, 2022.
45. Alibaba.com. Soy Protein Isolate/ISP Isolated Soy Protein for Beverage Use/Bulk Price Soy Protein Isolates. Available online: https://www.alibaba.com/product-detail/Soy-Protein-Isolate-ISP-Isolated-Soy_1600627163803.html (accessed on 7 November 2022).
46. BulkFoods.com. Soy Isolate Protein 90%. Available online: https://bulkfoods.com/pure-protein-powders/soy-isolate-protein-90.html (accessed on 7 November 2022).
47. Healthy Planet Canada NOW Soy Protein Isolate Unflavoured 544 g. Available online: https://www.healthyplanetcanada.com/now-soy-protein-isolate-unflavoured-544g.html (accessed on 7 November 2022).
48. U.S. Department of Agriculture. *Peas, Green, Raw*; U.S. Department of Agriculture: Washington, DC, USA, 2019.
49. Gittlein, J. *USDA Weekly Bean, Pea, and Lentil Market Review*; U.S. Department of Agriculture: Washington, DC, USA, 2022.
50. Alibaba.com. Hot Sell Non-GMO Organic Pea Protein Powder 85% For Food Supplement. Available online: https://www.alibaba.com/product-detail/Pea-Protein-Hot-Sell-Non-GMO_1600565503077.html (accessed on 7 November 2022).
51. BulkFoods.com. Pea Protein Powder. Available online: https://bulkfoods.com/pure-protein-powders/pea-protein-powder.html (accessed on 7 November 2022).
52. Healthy Planet Canada NOW Pea Protein Unflavoured 907 g. Available online: https://www.healthyplanetcanada.com/now-pea-protein-907g.html (accessed on 7 November 2022).
53. U.S. Department of Agriculture. *Rice Bran, Crude*; U.S. Department of Agriculture: Washington, DC, USA, 2019.
54. U.S. Department of Agriculture. *Feed Grains Custom Query Rice Bran, f. o. b. Mills*; U.S. Department of Agriculture: Washington, DC, USA, 2022.
55. Alibaba.com. Pincredit Supply Plant Protein Supplement Wholesale Brown Rice Protein Powder. Available online: https://www.alibaba.com/product-detail/Protein-Brown-Rice-Pincredit-Supply-Plant_1600558111076.html?spm=a2700.7735675.topad_classic.d_title.2d213cadln1a31 (accessed on 7 November 2022).
56. BulkFoods.com. Brown Rice Protein Powder. Available online: https://bulkfoods.com/pure-protein-powders/brown-rice-protein-powder.html (accessed on 7 November 2022).
57. Healthy Planet Canada North Coast Naturals Brown Rice Protein 340 g. Available online: https://www.healthyplanetcanada.com/north-coast-naturals-brown-rice-protein-340g.html (accessed on 7 November 2022).
58. Zhang, J.; Chen, Q.; Kaplan, D.L.; Wang, Q. High-Moisture Extruded Protein Fiber Formation toward Plant-Based Meat Substitutes Applications: Science, Technology, and Prospect. *Trends Food Sci. Technol.* **2022**, *128*, 202–216. [CrossRef]
59. Rasband, W. ImageJ 2022. Available online: https://imagej.net/ij/ (accessed on 7 November 2022).
60. Shanbhag, A.G. Utilization of Information Measure as a Means of Image Thresholding. *CVGIP: Graph. Models Image Process.* **1994**, *56*, 414–419. [CrossRef]
61. Prewitt, J.M.S.; Mendelsohn, M.L. THE ANALYSIS OF CELL IMAGES*. *Ann. N. Y. Acad. Sci.* **2006**, *128*, 1035–1053. [CrossRef]

62. Cyanotech Corporation. 100% Pure All Natural Hawaiian Spirulina (Fine Powder) Specifications and General Composition. Available online: https://www.cyanotech.com/pdfs/spirulina/specifications.html#:~:text=It%20has%20a%20mild%20seaweed,0.48%20(g/ml).&text=Store%20at%20room%20temperature (accessed on 7 November 2022).
63. John, H.; Mansuri, S.; Giri, S.; Sinha, L. Rheological Properties and Particle Size Distribution of Soy Protein Isolate as Affected by Drying Methods. *NFSIJ* **2018**, *7*, 555721.
64. Overduin, J.; Guérin-Deremaux, L.; Wils, D.; Lambers, T.T. NUTRALYS® Pea Protein: Characterization of in Vitro Gastric Digestion and in Vivo Gastrointestinal Peptide Responses Relevant to Satiety. *Food Nutr. Res.* **2015**, *59*, 25622. [CrossRef] [PubMed]
65. Amagliani, L.; O'Regan, J.; Kelly, A.L.; O'Mahony, J.A. Physical and Flow Properties of Rice Protein Powders. *J. Food Eng.* **2016**, *190*, 1–9. [CrossRef]
66. Schawe, J.; Riesen, R.; Widmann, J.; Schubnell, M.; Jörimann, U. UserCom, Information for Users of Mettler Toledo Thermal Analysis Systems 2000. Available online: https://www.mt.com/dam/mt_ext_files/Editorial/Generic/0/TA_UserCom24_Editorial-Generic_1201690913917_files/usercom24_ta_e_web.pdf (accessed on 7 November 2022).
67. Chronakis, I. Gelation of Edible Blue-Green Algae Protein Isolate (Spirulina Platensis Strain Pacifica): Thermal Transitions, Rheological Properties, and Molecular Forces Involved. *J. Agric. Food Chem.* **2001**, *49*, 888–898. [CrossRef]
68. Poel, G.V.; Istrate, D.; Magon, A.; Mathot, V. Performance and Calibration of the Flash DSC 1, a New, MEMS-Based Fast Scanning Calorimeter. *J. Therm. Anal. Calorim.* **2012**, *110*, 1533–1546. [CrossRef]
69. Sun, X.; Lee, K.O.; Medina, M.A.; Chu, Y.; Li, C. Melting Temperature and Enthalpy Variations of Phase Change Materials (PCMs): A Differential Scanning Calorimetry (DSC) Analysis. *Phase Transit.* **2018**, *91*, 667–680. [CrossRef]
70. Zhong, Z.K.; Sun, X.S. Thermal Behavior and Nonfreezing Water of Soybean Protein Components. *Cereal. Chem.* **2000**, *77*, 495–500. [CrossRef]
71. TA Instruments. *Interpreting Unexpected Events and Transitions in DSC Results*; TA Instruments: New Castle, DE, USA, 2017.
72. Kitabatake, N.; Tahara, M.; Doi, E. Thermal Denaturation of Soybean Protein at Low Water Contents. *Agric. Biol. Chem.* **1990**, *54*, 2205–2212. [CrossRef]
73. Li, S.; Wei, Y.; Fang, Y.; Zhang, W.; Zhang, B. DSC Study on the Thermal Properties of Soybean Protein Isolates/Corn Starch Mixture. *J. Therm. Anal. Calorim.* **2014**, *115*, 1633–1638. [CrossRef]
74. Li, J.-Y.; Yeh, A.-I.; Fan, K.-L. Gelation Characteristics and Morphology of Corn Starch/Soy Protein Concentrate Composites during Heating. *J. Food Eng.* **2007**, *78*, 1240–1247. [CrossRef]
75. Mession, J.-L.; Sok, N.; Assifaoui, A.; Saurel, R. Thermal Denaturation of Pea Globulins (*Pisum Sativum* L.)—Molecular Interactions Leading to Heat-Induced Protein Aggregation. *J. Agric. Food Chem.* **2013**, *61*, 1196–1204. [CrossRef] [PubMed]
76. Shen, S.; Hou, H.; Ding, C.; Bing, D.-J.; Lu, Z.-X. Protein Content Correlates with Starch Morphology, Composition and Physicochemical Properties in Field Peas. *Can. J. Plant Sci.* **2016**, *96*, 404–412. [CrossRef]
77. Oyinloye, T.M.; Yoon, W.B. Stability of 3D Printing Using a Mixture of Pea Protein and Alginate: Precision and Application of Additive Layer Manufacturing Simulation Approach for Stress Distribution. *J. Food Eng.* **2021**, *288*, 110127. [CrossRef]
78. Wang, T.; Xu, P.; Chen, Z.; Zhou, X.; Wang, R. Alteration of the Structure of Rice Proteins by Their Interaction with Soy Protein Isolates to Design Novel Protein Composites. *Food Funct.* **2018**, *9*, 4282–4291. [CrossRef]
79. Bridges, S.; Robinson, L. Chapter 1—Rheology. In *A Practical Handbook for Drilling Fluids Processing*; Gulf Professional Publishing: Oxford, UK, 2022; pp. 3–26.
80. Kar, F.; Arslan, N. Effect of Temperature and Concentration on Viscosity of Orange Peel Pectin Solutions and Intrinsic Viscosity–Molecular Weight Relationship. *Carbohydr. Polym.* **1999**, *40*, 277–284. [CrossRef]
81. Hassanzadeh, H.; Ghanbarzadeh, B.; Galali, Y.; Bagheri, H. The Physicochemical Properties of the Spirulina-wheat Germ-enriched High-protein Functional Beverage Based on Pear-cantaloupe Juice. *Food Sci. Nutr.* **2022**, *10*, 3651–3661. [CrossRef]
82. Liu, P.; Xu, H.; Zhao, Y.; Yang, Y. Rheological Properties of Soy Protein Isolate Solution for Fibers and Films. *Food Hydrocoll.* **2017**, *64*, 149–156. [CrossRef]
83. Zhang, Z.; Liu, Y. Recent Progresses of Understanding the Viscosity of Concentrated Protein Solutions. *Curr. Opin. Chem. Eng.* **2017**, *16*, 48–55. [CrossRef]
84. Qi, G.; Li, N.; Wang, D.; Sun, X.S. Adhesion and Physicochemical Properties of Soy Protein Modified by Sodium Bisulfite. *J. Americ. Oil Chem. Soc.* **2013**, *90*, 1917–1926. [CrossRef]
85. Osen, R.; Toelstede, S.; Wild, F.; Eisner, P.; Schweiggert-Weisz, U. High Moisture Extrusion Cooking of Pea Protein Isolates: Raw Material Characteristics, Extruder Responses, and Texture Properties. *J. Food Eng.* **2014**, *127*, 67–74. [CrossRef]
86. Lee, K.; Lee, J. Chapter 6—Hybrid Thermal Recovery Using Low-Salinity and Smart Waterflood. In *Hybrid Enhanced Oil Recovery Using Smart Waterflooding*; Gulf Professional Publishing: Oxford, UK, 2019; pp. 129–135.
87. Beck, S.M.; Knoerzer, K.; Sellahewa, J.; Emin, M.A.; Arcot, J. Effect of Different Heat-Treatment Times and Applied Shear on Secondary Structure, Molecular Weight Distribution, Solubility and Rheological Properties of Pea Protein Isolate as Investigated by Capillary Rheometry. *J. Food Eng.* **2017**, *208*, 66–76. [CrossRef]
88. Wu, J.; Xu, S.; Yan, X.; Zhang, X.; Li, Q.; Ye, J.; Liu, C. Effect of Homogenization Modified Rice Protein on the Pasting Properties of Rice Starch. *Foods* **2022**, *11*, 1601. [CrossRef] [PubMed]
89. Li, X.; Liu, Y.; Yi, C.; Cheng, Y.; Zhou, S.; Hua, Y. Microstructure and Rheological Properties of Mixtures of Acid-Deamidated Rice Protein and Dextran. *J. Cereal Sci.* **2010**, *51*, 7–12. [CrossRef]

90. Chhabra, R.; Richardson, J. Chapter 3—Flow in Pipes and in Conduits of Non-Circular Cross-Sections. In *Non-Newtonian Flow and Applied Rheology (Second Edition)*; Butterworth-Heinemann: Oxford, UK, 2008; pp. 110–205.
91. Liestianty, D.; Rodianawati, I.; Arfah, R.A.; Assa, A.; Patimah; Sundari; Muliadi. Nutritional Analysis of *Spirulina Sp* to Promote as Superfood Candidate. *IOP Conf. Ser. Mater. Sci. Eng.* **2019**, *509*, 012031. [CrossRef]
92. Gorissen, S.H.M.; Crombag, J.J.R.; Senden, J.M.G.; Waterval, W.A.H.; Bierau, J.; Verdijk, L.B.; Van Loon, L.J.C. Protein Content and Amino Acid Composition of Commercially Available Plant-Based Protein Isolates. *Amino Acids* **2018**, *50*, 1685–1695. [CrossRef] [PubMed]
93. Sá, A.G.A.; Moreno, Y.M.F.; Carciofi, B.A.M. Plant Proteins as High-Quality Nutritional Source for Human Diet. *Trends Food Sci. Technol.* **2020**, *97*, 170–184. [CrossRef]
94. Vasconcelos, I.M.; Campello, C.C.; Oliveira, J.T.A.; Carvalho, A.F.U.; Souza, D.O.B.D.; Maia, F.M.M. Brazilian Soybean *Glycine max* (L.) Merr. Cultivars Adapted to Low Latitude Regions: Seed Composition and Content of Bioactive Proteins. *Rev. Bras. Bot.* **2006**, *29*, 617–625. [CrossRef]
95. Montoya, C.A.; Gomez, A.S.; Lallès, J.-P.; Souffrant, W.B.; Beebe, S.; Leterme, P. In Vitro and in Vivo Protein Hydrolysis of Beans (*Phaseolus vulgaris*) Genetically Modified to Express Different Phaseolin Types. *Food Chem.* **2008**, *106*, 1225–1233. [CrossRef]
96. Amagliani, L.; O'Regan, J.; Kelly, A.L.; O'Mahony, J.A. The Composition, Extraction, Functionality and Applications of Rice Proteins: A Review. *Trends Food Sci. Technol.* **2017**, *64*, 1–12. [CrossRef]
97. Liu, K.; Zheng, J.; Chen, F. Relationships between Degree of Milling and Loss of Vitamin B, Minerals, and Change in Amino Acid Composition of Brown Rice. *LWT-Food Sci. Technol.* **2017**, *82*, 429–436. [CrossRef]
98. Vendemiatti, A.; Rodrigues Ferreira, R.; Humberto Gomes, L.; Oliveira Medici, L.; Antunes Azevedo, R. Nutritional Quality of Sorghum Seeds: Storage Proteins and Amino Acids. *Food Biotechnol.* **2008**, *22*, 377–397. [CrossRef]
99. Boye, J.; Wijesinha-Bettoni, R.; Burlingame, B. Protein Quality Evaluation Twenty Years after the Introduction of the Protein Digestibility Corrected Amino Acid Score Method. *Br. J. Nutr.* **2012**, *108*, S183–S211. [CrossRef] [PubMed]
100. Sarwar Gilani, G.; Wu Xiao, C.; Cockell, K.A. Impact of Antinutritional Factors in Food Proteins on the Digestibility of Protein and the Bioavailability of Amino Acids and on Protein Quality. *Br. J. Nutr.* **2012**, *108*, S315–S332. [CrossRef] [PubMed]
101. Park, S.J.; Kim, T.W.; Baik, B.-K. Relationship between Proportion and Composition of Albumins, and in Vitro Protein Digestibility of Raw and Cooked Pea Seeds (*Pisum Sativum* L.): Composition and in Vitro Digestibility of Pea Protein. *J. Sci. Food Agric.* **2010**, *90*, 1719–1725. [CrossRef] [PubMed]
102. Abdul-Fattah, A.M.; Kalonia, D.S.; Pikal, M.J. The Challenge of Drying Method Selection for Protein Pharmaceuticals: Product Quality Implications. *J. Pharm. Sci.* **2007**, *96*, 1886–1916. [CrossRef]
103. Zhao, Q.; Xiong, H.; Selomulya, C.; Chen, X.D.; Huang, S.; Ruan, X.; Zhou, Q.; Sun, W. Effects of Spray Drying and Freeze Drying on the Properties of Protein Isolate from Rice Dreg Protein. *Food Bioprocess Technol.* **2013**, *6*, 1759–1769. [CrossRef]
104. Burger, T.G.; Singh, I.; Mayfield, C.; Baumert, J.L.; Zhang, Y. The Impact of Spray Drying Conditions on the Physicochemical and Emulsification Properties of Pea Protein Isolate. *LWT* **2022**, *153*, 112495. [CrossRef]
105. Hubbard, B.R.; Putman, L.I.; Techtmann, S.; Pearce, J.M. Open Source Vacuum Oven Design for Low-Temperature Drying: Performance Evaluation for Recycled PET and Biomass. *JMMP* **2021**, *5*, 52. [CrossRef]
106. Nikbakht Nasrabadi, M.; Sedaghat Doost, A.; Mezzenga, R. Modification Approaches of Plant-Based Proteins to Improve Their Techno-Functionality and Use in Food Products. *Food Hydrocoll.* **2021**, *118*, 106789. [CrossRef]
107. Ma, W.; Qi, B.; Sami, R.; Jiang, L.; Li, Y.; Wang, H. Conformational and Functional Properties of Soybean Proteins Produced by Extrusion-Hydrolysis Approach. *Int. J. Anal. Chem.* **2018**, *2018*, 9182508. [CrossRef]
108. Zahari, I.; Ferawati, F.; Helstad, A.; Ahlström, C.; Östbring, K.; Rayner, M.; Purhagen, J.K. Development of High-Moisture Meat Analogues with Hemp and Soy Protein Using Extrusion Cooking. *Foods* **2020**, *9*, 772. [CrossRef]
109. Carvalho, C.W.P.; Takeiti, C.Y.; Onwulata, C.I.; Pordesimo, L.O. Relative Effect of Particle Size on the Physical Properties of Corn Meal Extrudates: Effect of Particle Size on the Extrusion of Corn Meal. *J. Food Eng.* **2010**, *98*, 103–109. [CrossRef]
110. Chen, Z.C.; Ikeda, K.; Murakami, T.; Takeda, T. Effect of Particle Packing on Extrusion Behavior of Pastes. *J. Mater. Sci.* **2000**, *35*, 5301–5307. [CrossRef]
111. Tilman, D.; Clark, M. Global Diets Link Environmental Sustainability and Human Health. *Nature* **2014**, *515*, 518–522. [CrossRef] [PubMed]
112. Kumar, P.; Chatli, M.K.; Mehta, N.; Singh, P.; Malav, O.P.; Verma, A.K. Meat Analogues: Health Promising Sustainable Meat Substitutes. *Crit. Rev. Food Sci. Nutr.* **2017**, *57*, 923–932. [CrossRef] [PubMed]
113. Alexander, P.; Brown, C.; Arneth, A.; Finnigan, J.; Rounsevell, M.D.A. Human Appropriation of Land for Food: The Role of Diet. *Glob. Environ. Chang.* **2016**, *41*, 88–98. [CrossRef]
114. Saranraj, P.; Sivasakthi, S. Spirulina Platensis—Food for Future: A Review. *Asian J. Pharm. Sci. Technol.* **2014**, *4*, 26–33.
115. Food and Agriculture Organization of the United Nations FAOSTAT Crops and Livestock Products from Year 1994 to 2021. 2022. Available online: https://www.fao.org/ (accessed on 7 November 2022).
116. Denkenberger, D.; Pearce, J. *Feeding Everyone No Matter What: Managing Food Security after Global Catastrophe*; Academic Press: Cambridge, MA, USA, 2014.
117. Denkenberger, D.; Pearce, J. Feeding Everyone: Solving the Food Crisis in Event of Global Catastrophes That Kill Crops or Obscure the Sun. *Futures* **2015**, *72*, 57–68. [CrossRef]

118. Baum, S.D.; Denkenberger, D.C.; Pearce, J.M.; Robock, A.; Winkler, R. Resilience to Global Food Supply Catastrophes. *Env. Syst. Decis.* **2015**, *35*, 301–313. [CrossRef]
119. Franc-Dąbrowska, J.; Drejerska, N. Resilience in the Food Sector—Environmental, Social and Economic Perspectives in Crisis Situations. *IFAM* **2022**, *25*, 757–770. [CrossRef]
120. Pham, A.; García Martínez, J.B.; Brynych, V.; Stormbjorne, R.; Pearce, J.M.; Denkenberger, D.C. Nutrition in Abrupt Sunlight Reduction Scenarios: Envisioning Feasible Balanced Diets on Resilient Foods. *Nutrients* **2022**, *14*, 492. [CrossRef]
121. Munialo, C.; Vriesekoop, F. Plant-Based Foods as Meat and Fat Substitutes. *Food Sci. Nutr.* **2023**, *11*, 4898–4911. [CrossRef]
122. Wang, Y.; Lyu, B.; Fu, H.; Li, J.; Ji, L.; Gong, H.; Zhang, R.; Liu, J.; Yu, H. The Development Process of Plant-Based Meat Alternatives: Raw Material Formulations and Processing Strategies. *Food Res. Int.* **2023**, *13*, 112689. [CrossRef] [PubMed]
123. Global Meat Consumption by Type. Available online: https://www.statista.com/statistics/274522/global-per-capita-consumption-of-meat/ (accessed on 23 January 2024).
124. Singh, R.K.; Deshpande, D. Thermally Induced Changes in Quality of Chicken Breast Meat Protein Fractions. *J. Nutr. Food Sci.* **2018**, *8*, 4. [CrossRef]
125. Jiang, J.; Zhang, M.; Bhandari, B.; Cao, P. Development of Chinese Yam/Chicken Semi-Liquid Paste for Space Foods. *LWT* **2020**, *125*, 109251. [CrossRef]
126. Dick, A.; Dong, X.; Bhandari, B.; Prakash, S. The Role of Hydrocolloids on the 3D Printability of Meat Products. *Food Hydrocoll.* **2021**, *119*, 106879. [CrossRef]
127. Cho, H.-R.; Chang, D.-S.; Lee, W.-D.; Jeong, E.-T.; Lee, E.-W. Utilization of Chitosan Hydrolysate as a Natural Food Preservative for Fish Meat Paste Products. *Korean J. Food Sci. Technol.* **1998**, *30*, 817–822.
128. Mantihal, S.; Kobun, R.; Lee, B.-B. 3D Food Printing of as the New Way of Preparing Food: A Review. *Int. J. Gastron. Food Sci.* **2020**, *22*, 100260. [CrossRef]
129. Baiano, A. 3D Printed Foods: A Comprehensive Review on Technologies, Nutritional Value, Safety, Consumer Attitude, Regulatory Framework, and Economic and Sustainability Issues. *Food Rev. Int.* **2022**, *38*, 986–1016. [CrossRef]
130. Wang, T.; Kaur, L.; Furuhata, Y.; Aoyama, H.; Singh, J. 3D Printing of Textured Soft Hybrid Meat Analogues. *Foods* **2022**, *11*, 478. [CrossRef]
131. Grossmann, L.; McClements, D.J. Current Insights into Protein Solubility: A Review of Its Importance for Alternative Proteins. *Food Hydrocoll.* **2023**, *137*, 108416. [CrossRef]

Disclaimer/Publisher's Note: The statements, opinions and data contained in all publications are solely those of the individual author(s) and contributor(s) and not of MDPI and/or the editor(s). MDPI and/or the editor(s) disclaim responsibility for any injury to people or property resulting from any ideas, methods, instructions or products referred to in the content.

Article

Quality Characteristics of Raspberry Fruits from Dormancy Plants and Their Feasibility as Food Ingredients

Sílvia Petronilho [1,2,*], Manuel A. Coimbra [2] and Cláudia P. Passos [2,*]

[1] Chemistry Research Centre-Vila Real, Department of Chemistry, University of Trás-os-Montes and Alto Douro, Quinta de Prados, 5001-801 Vila Real, Portugal
[2] Associated Laboratory for Green Chemistry (LAQV-REQUIMTE), Department of Chemistry, Campus Universitário de Santiago, University of Aveiro, 3810-193 Aveiro, Portugal; mac@ua.pt
* Correspondence: silviapetronilho@ua.pt (S.P.); cpassos@ua.pt (C.P.P.)

Abstract: The raspberry (*Rubus idaeus* L.) is a soft red fruit consumed worldwide due to its bitter-sweet taste and phenolics-associated health benefits. During plant dormancy, raspberry fruits are discarded. However, this work hypothesised that these fruits have the chemical quality to be valorised, which would mitigate their waste if adequately stabilised. This can be achieved by drying. The Pacific Deluxe and Versailles varieties were dried by freeze- and convective-drying (30 °C and 40 °C). The freeze-dried fruits preserved their colour, drupelets structure, and phenolic content. Convective-drying promoted a significant fruit darkening, which was more evident at 30 °C due to the longer drying process, and a loss of drupelets structure. Both temperatures promoted a similar decrease in phenolic content, as determined by HPLC, although the ABTS$^{\bullet+}$ antioxidant activity at 40 °C was lower (IC$_{50}$ = 9 compared to 13 µg AAE/mg dry weight). To incorporate dried raspberries into muffin formulations, while keeping their red colour, it was necessary to change the raising agent from sodium bicarbonate to baker's yeast. Sensory analysis by a non-trained panel revealed good acceptance, showing that fresh or dried raspberry fruits from dormancy had suitable characteristics for use as food ingredients.

Keywords: *Rubus idaeus*; drying; phenolic compounds; antioxidant activity; functional ingredients; bakery products; sensory properties

Citation: Petronilho, S.; Coimbra, M.A.; Passos, C.P. Quality Characteristics of Raspberry Fruits from Dormancy Plants and Their Feasibility as Food Ingredients. *Foods* 2023, *12*, 4443. https://doi.org/10.3390/foods12244443

Academic Editors: Gianluca Nardone, Rosaria Viscecchia and Francesco Bimbo

Received: 22 November 2023
Revised: 5 December 2023
Accepted: 7 December 2023
Published: 11 December 2023

Copyright: © 2023 by the authors. Licensee MDPI, Basel, Switzerland. This article is an open access article distributed under the terms and conditions of the Creative Commons Attribution (CC BY) license (https://creativecommons.org/licenses/by/4.0/).

1. Introduction

The raspberry (*Rubus idaeus* L.) is a soft red fruit, consisting of a cluster of drupelets, consumed worldwide due to its characteristic bittersweet taste and beneficial effects, which are greatly linked to its phenolic-rich profile [1]. This fruit is produced by raspberry plants, which after fruiting undergo dormancy, a state of reduced metabolic activity occurring when the days shorten and the temperature cools down [2]. However, in contrast to wild populations of *R. idaeus* [3], raspberry plants in greenhouses still produce canes with lateral branches able to give flowers and fruits, although in lower yields (ca. 3 times lower) when compared to the harvest season [4]. These fruits are normally not harvested for commercial usage and are wasted during raspberry plant pruning. As a result of the great growth in raspberry fruit production in the last decade (ca. 75% since 2010, with a global production of ca. 896 thousand tonnes of fruits in 2020 [5]), several tonnes of fruits have been neglected from dormancy. Following the circular economy concept, these fruits can be used to obtain added-value products while allowing the minimisation of the output of this waste.

Raspberry fruits are considered to be a source of vitamin C [6], minerals, like iron and potassium [7], phenolic compounds, which are mainly anthocyanins [8], and dietary fibre [9]. As a result of their richness in these biomolecules, raspberry fruits have been related to different bioactive properties, including antioxidant, anti-inflammatory, and anti-cancer activities, among others [10].

Despite their popularity, fresh raspberry fruits are highly perishable due to their high moisture content, which limits the fruits' shelf life to just a few days after harvest. To minimise fruit loss and to ensure its availability all year, raspberry fruits have been channelled to frozen storage or more shelf-stable processed products, such as juice, jam, or puree [11]. Nevertheless, dried fruit consumption is now highly valorised, and dried raspberry fruits are not an exception, being consumed as ready-to-eat snacks or incorporated into formulations of other foodstuffs, including cereal mixtures, and dairy and bakery products [12]. The impacts of different drying technologies on the colour, composition, and antioxidant properties of dried raspberry fruits have been explored, including solar and microwave drying [13], as well as combined drying methods such as osmotic dehydration with vacuum [14] and convective hot air drying [15], both of which are combined with the microwave drying. However, at the industrial level, the most popular procedures are still convective- and freeze-drying [16]. Previous studies have demonstrated that, when compared to convective-dried raspberry fruits, the freeze-dried ones better preserve the colour, nutritive value, and physicochemical composition of the fresh fruits [16,17]. Nevertheless, the influence of these drying technologies on the properties of raspberry fruits from a dormancy state has not yet been elucidated.

In this study, it was hypothesised that dried raspberry fruits harvested in their dormancy state have the same quality attributes as the commercially available fruits. Their drying allows their incorporation into bakery products as functional ingredients rich in phenolic compounds. To prove this hypothesis, two red raspberry varieties (Pacific Deluxe and Versailles) from the dormancy state were dried by freeze- and convective-drying. The influence of the drying techniques on the fruits' characteristics in terms of phenolic composition, colour, and antioxidant activity was assessed. Then, the impact of the incorporation of these fruits in fresh and dried forms into muffin formulations was evaluated according to the chromatic, physical, and sensory properties.

2. Materials and Methods

2.1. Materials

Raspberry fruits from Pacific Deluxe and Versailles red primocane-fruiting varieties, both belonging to *Rubus idaeus* L., were provided by the Portuguese company *Framboesas da Graça* (Ponte de Vagos, Portugal). The red fruits were collected at the company greenhouses in December 2020 (winter season), during the plants' dormancy state. The reagents and standards used were: NaOH (98.6%, Panreac, Barcelona, Spain), formic acid (99%, Chem-Lab, Zedelgem, Belgium), and acetonitrile (99.9%, Carlo Erba Reagents, Cornaredo, MI, Italy). Malvidin-3-glucoside ($\geq$90%, HPLC), phenolphthalein, Folin–Ciocalteu reagent, sodium carbonate ($\geq$99.5%), gallic acid ($\geq$99%, HPLC), and ABTS (>98%) were from Sigma-Aldrich, Madrid, Spain. All the reagents used were of analytical grade.

2.2. Sampling

For each raspberry variety under study, ca. 1000 g of raspberry fruits was picked randomly from the company's greenhouses, following a z-shaped pattern to avoid edge and centre effects. The samples were transported immediately under refrigeration (ca. 4 °C) to the laboratory where the raspberry fruits' physicochemical parameters (fruit weight, pH, total soluble solids (°Brix), titratable acidity, and moisture content) were promptly determined [18]. The remaining fruits were stored at -20 °C for a maximum period of 1 week and then used for the drying processes and incorporation into muffins.

2.3. Raspberry Fruits Drying

For each variety, two drying processes were used, freeze-drying and convective-drying, and five independent analyses were carried out, in which the weight of each sample was recorded before and after drying. The frozen raspberry fruits were placed in a freeze-dryer (Labogene, Scanvac CoolSafe, Allerød, Denmark) for 48 h (P = 0.094 mbar; T = -49 °C). For the convective-drying process, the raspberry fruits were dried in a lab-scale convective oven

(BINDER GmbH, Tuttlingen, Germany) at 30 °C and 40 °C to avoid the thermal degradation of the phenolic compounds [19]. For this, the frozen samples were placed in the oven with a constant airflow (speed at 100% for desired temperature accuracy maintenance), and the weight loss was monitored periodically in an analytical balance with 0.01 g to 0.001 g graduation (Marsden GF24 Balance Weighing Scale, Rotherham, UK) throughout the drying process: every 15 min for the 1st h; every 30 min in the next 2 h; and every 2 h and 4 h until the fruits reached ≤15% humidity, which is the benchmark to avoid microbiological contamination in dried fruits [20].

2.4. Raspberry Fruits Characterisation

2.4.1. Physicochemical Parameters Determination

For each variety, ca. 20 g of fresh fruits was crushed, and the juice was vacuum filtered with 1.2 μm glass microfiber filters to remove any solid materials. Then, its pH, total soluble solids (°Brix), and titratable acidity were measured [18]. The pH of the juice was determined using a digital pH meter (micropH 2002, Crison, Barcelona, Spain). The total soluble solids were quantified using a portable refractometer (FG103/113, Zuzi, Auxilab S.L, Navarra, Spain) with a reading scale of 0–40 ± 0.1 °Brix. The titratable acidity was performed according to the official method AOAC 942.15 [21] by titrimetry, using NaOH 0.1 M and phenolphthalein as the indicator. The moisture content of the fruits was determined by freeze-drying (Labogene, Scanvac CoolSafe, Allerød, Denmark) for 48 h (P = 0.094 mbar; T = −49 °C). Three independent replicates were used for each assay.

2.4.2. Chromatic Properties

For both varieties, the chromatic properties of the fresh and dried fruits were assessed by tristimulus colorimetry (CIELab), using a portable colorimeter (Konica Minolta, CM-2600d/2500d) with a standard D8 light source and a visual angle of 8, calibrated using a white standard ($L^* = 94.61$; $a^* = -0.53$; $b^* = 3.62$), following the supplier's information. Herein, the CIELab coordinates L^* (luminosity), b^* (yellow/blue), and a^* (red/green) were determined, as well as the total colour difference (ΔE) [22–24]. The assay was performed in triplicate, with 5 measurements taken per replicate.

2.4.3. Total Phenolic Content

For each raspberry variety, ca. 20 g of fresh fruits was crushed in a mortar and the final volume (juice + pulp) was measured. For the dried samples, 50 mg was crushed and suspended in 1.5 mL of water to obtain the aqueous extract. Each suspension was centrifuged (5 min, 3000 rpm, room temperature) and filtered in a 0.45 μm nylon membrane to obtain clear juice/aqueous extracts. Then, the total phenolic content was determined, in triplicate, by the Folin–Ciocalteu method [25]. A calibration curve of gallic acid was built (20–250 μg/mL), and the absorbance was measured at 750 nm. The total phenolic content was expressed as μg of gallic acid equivalents (GAE)/mg of sample (dry weight, dw).

2.4.4. Phenolic Profile Analysis

For the phenolic profile analysis, 1 mL of aqueous extracts from the fresh fruits/dried samples, obtained as described in Section 2.4.3, was used. Then, the recovered insoluble pellets obtained after centrifugation were submitted to a hydroalcoholic extraction with the addition of 1 mL of methanol. Both the aqueous and the methanolic extracts were filtered in a C_{18} column (6 mL, Discovery, Supelco, Algés, Portugal) [26]. After evaporation, both extracts were dissolved in distilled water, filtered (0.22 μm Nylon membrane), and analysed by ultra-high-performance liquid chromatography with a diode detector (UHPLC-DAD-ESI/MS^n) (Ultimate 3000, Dionex Co./Thermo Scientific, Waltham, MA, USA). Analysis was conducted using a Hypersil Gold (Thermo Scientific, Waltham, MA, USA) C_{18} column (100 mm length; 2.1 mm i. d.; 1.9 μm particle diameter, end-capped) and an adaptation of the reported chromatographic separation conditions [27]. Briefly, eluent A—0.1% (v/v) formic acid and eluent (B) acetonitrile were used at an elution gradient of 5% to 40% of

(B), at 14.7 min, and 100% (B) at 16.6 min, followed by the return to the initial conditions at 24 min, for a total run time of 34 min. The flow rate was 0.2 mL/min. The phenolic compounds were determined at 520 nm using their retention time and mass spectrum fragments (Table S1). A calibration curve using the anthocyanin malvidin-3-glucoside (0.06–6 mg/mL) was built [26], and the results from three independent aliquots were expressed as µg of malvidin-3-glucose equivalents (Mv 3-gE)/mg of sample (dw).

2.4.5. Antioxidant Activity

The determination of the antioxidant activity, in the fresh and dried raspberry samples, was carried out using the ABTS [2,2′-azino-bis(3-ethylbenzothiazoline-6-sulphonic acid)] method [25]. A calibration curve for ascorbic acid (2–20 µg/mL) was built. The absorbance was read at 734 nm, and the results were expressed as IC_{50} (minimal sample concentration required to inhibit 50% of $ABTS^{\bullet+}$).

2.5. Muffin Preparation and Characterisation
2.5.1. Preparation of the Muffin Formulations

The muffins were prepared according to a homemade recipe [28] (the ingredients can be seen in Table S2). To determine the content of raspberries to be used in the muffin dough, 60, 90, and 120 drupelets of fresh, convective-dried, and freeze-dried raspberries were incorporated into the final formulation. To evaluate the effect of the raising agent on the chromatic and sensory properties of the muffin formulations, sodium bicarbonate (chemical raising agent used in the original recipe) and yeast (natural raising agent) were used.

To prepare the muffins (Figure S1), unsalted butter was mixed with sugar in a mixer (Hoffen Food Expert LH6802) for 1 min at speed 2 using the hard dough tool. Then, the eggs were added, and the mixer speed was gradually increased to the maximum speed of 6 for 5 min. Finally, the flour and sodium bicarbonate were added and mixed at speed 6 for 5 min. Using an ice cream scoop (5.8 cm × 4 cm × 2.5 cm), the dough was placed in a silicone mould (diameter 6.5 cm × height 1.8 cm). After placing the dough in the mould, the raspberries were added, and the dough was uniformised again. This recipe allowed to obtain a yield of 22 muffins per baking. The muffins were placed in a previously heated (180 °C, 5 min) domestic oven (Zanussi Built In, Aveiro, Portugal) for 15 min at 180 °C. Each formulation was carried out in triplicate, and a negative control (without raspberries) was made. After being removed from the oven, the muffins were cooled down to room temperature, coated with aluminium foil, and stored for further characterisation.

2.5.2. Characterisation of the Muffins

To understand the impact of the changes made to the original muffin formulation (incorporation of fresh, freeze-dried, and convective-dried raspberry fruits and raising agent), the dimension, weight, colour, and pH of the muffins were always evaluated 1 day after the muffins' preparation. The physical parameters were recorded using a calliper, as shown in Figure 1. The weight was obtained using a kitchen scale (SilverCrest, Aveiro, Portugal) with a minimum capacity of 1 g and a maximum of 5 kg. Colour measurements were performed as described for the fresh and dried raspberries (Section 2.4.2), with 5 readings per replica (three independent replicates) on the upper and lower sides of the muffins.

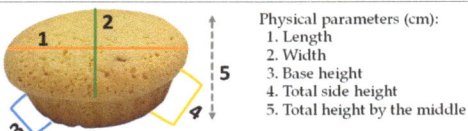

Figure 1. Scheme used to record the physical parameters of the muffins.

The pH was determined based on an adaptation of a previous work [29]. Briefly, the samples removed from the middle of each cooked muffin (ca. 400 mg) were mixed with 20 mL of distilled water. The mixture was vortexed for 3 min and kept standing at room temperature for 1 h to separate the solid from the liquid phases. Then, the mixture was decanted and centrifuged (5 min, 4000 rpm). After centrifugation, the pH of the supernatants was measured using a digital pH meter (micropH 2002, Crison, Barcelona, Spain).

2.5.3. Sensory Analysis of the Muffins

For sensory analysis, the muffin formulations containing yeast as the raising agent and a total of 120 drupelets of Versailles raspberries (fresh and freeze-dried) were used. The formulation without raspberries was used as the reference. The non-trained panel was composed of 5 males and 11 females, between the ages of 20 and 58, from the GlycoFoodChem research group at the University of Aveiro (Portugal). A hedonic scale test was used to measure the sensory quality of the muffins using 5 parameters: 1—dislike a lot, 2—dislike, 3—indifferent, 4—like, and 5—like a lot. The evaluation form given to the panellists listed different parameters (appearance, colour, aroma, sweetness, acidity, saltiness, fruitiness, global taste, softness, hardness, global texture, and global appreciation) and score options with number rankings from 1 to 5. The samples were presented in random order and coded: muffins without raspberries (CN), muffins with 120 fresh raspberry drupelets (MF), and muffins with 120 freeze-dried raspberry drupelets (ML).

2.6. Statistical Analysis

The data were statistically analysed by Student's t-test with a level of significant difference of 95% and $p < 0.05$, using the "test t" tool of Excel 2016. For multiple comparison analysis, the significance of the difference was evaluated with one-way ANOVA at the significance level of $p < 0.05$, followed by Tukey's multiple comparison test using GraphPad Prism 5.01 software (OriginLab Corporation, Northampton, MA, USA, trial version).

3. Results and Discussion

3.1. Fresh Raspberry Fruits Physicochemical Characterisation

The fresh fruits from the Pacific Deluxe dormancy had a weight of 5.73 g per fruit, and their juice had a pH of 3.25, a °Brix of 5.10, and a titratable acidity of 0.90% citric acid equivalents (CAE)/100 mL of juice (Table 1). When compared to the Pacific Deluxe fruits, the Versailles fruits had a higher weight (6.96 g per fruit), and the resulting juice had higher acidity (pH of 2.86 and titratable acidity of 1.25% CAE/100 mL) and sweetness (°Brix of 9.07). The moisture content of the Pacific Deluxe (90.21%) and Versailles (86.98%) fruits was also significantly different. The Versailles fruits seem to be appropriate for conferring acidity and sweetness if used in juices, when compared to the Pacific Deluxe fruits. Despite this variability, the physicochemical parameter values, except for those of the total solids of Pacific Deluxe, were within the range reported for the other *R. idaeus* species collected at commercial maturity: pH of 2.9 to 4.2, total solids of 5.5 to 12.6 °Brix, titratable acidity of 0.8 to 1.9 of CAE/100 mL of juice, and moisture of 82.0 to 90.3% [12,13,17,30–32]. These results allowed the inference that fruits from the dormancy state have similar physicochemical attributes to those of the commercially available fruits.

Table 1. Physicochemical characterisation of fresh raspberry fruits from dormancy state of Pacific Deluxe and Versailles red varieties.

Variety	Weight (g) [1]	pH	°Brix	Titratable Acidity (% CAE/100 mL) [2]	Moisture (%) [3]
Pacific Deluxe	5.73 ± 0.04 [a]	3.25 ± 0.14 [a]	5.10 ± 0.17 [a]	0.90 ± 0.02 [a]	90.21 ± 1.13 [a]
Versailles	6.96 ± 0.26 [b]	2.86 ± 0.06 [b]	9.07 ± 0.12 [b]	1.25 ± 0.03 [b]	86.98 ± 1.58 [b]

[1] Values per berry. [2] CAE—citric acid equivalents. [3] Moisture content determined after freeze-drying. Data presented are expressed as the mean and standard deviation of three independent replicates. In each column, different lowercase letters represent significantly different values ($p < 0.05$).

3.2. Raspberry Fruit Drying

The freeze-drying (−49 °C, 48 h) of the Pacific Deluxe and Versailles fruits resulted in 0.56 and 0.83 g per fruit, respectively, which is consistent with the larger fruit size and lower moisture of the Versailles variety (Table 1). The structured cluster of the drupelets of the fresh fruits was kept when dried, as was the ruby colour (Figure 2).

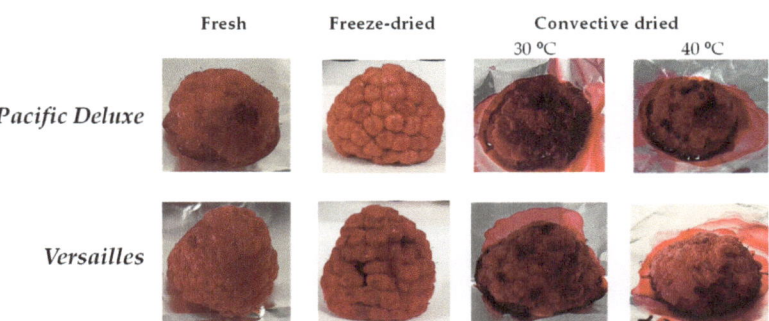

Figure 2. Images of fresh, freeze-dried, and convective-dried (30 °C and 40 °C) raspberry fruits from dormancy state of Pacific Deluxe and Versailles varieties.

The convective-drying of raspberry fruits was performed at 30 °C and 40 °C (Figure 2) since these drying temperatures promote the thermal degradation of the phenolic compounds and the antioxidant activity of red fruits, like raspberries from the Autumn Bliss variety [26] and other non-specified *R. ideaus* varieties [17].

The drying process was followed through the construction of the drying curves as a function of time until reaching a moisture content of ≤10% (Figure 3), which is below the 15% benchmark to avoid microbial contamination [20]. It took around 49 h at 30 °C for both varieties and ca. 10% less (44 h) at 40 °C, which is a decrease that shows the same trend as that of the commercially available raspberries dried at 50 °C, 65 °C, and 130 °C [16]. In contrast to the freeze-dried fruits, the convective-dried fruits acquired a non-uniform dark colour and had volumetric shrinkage and a change in shape, losing their drupelets cluster structure (Figure 2). This was not so evident for the Versailles variety, probably due to their higher weight and total solids content (Table 1).

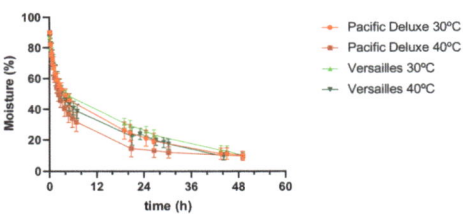

Figure 3. Drying curves of raspberry fruits from dormancy state of Pacific Deluxe and Versailles varieties, obtained at 30 °C and 40 °C.

3.3. Impact of the Drying Processes
3.3.1. Raspberry Fruit Colour

The fresh raspberries from Pacific Deluxe had L^*, a^*, and b^* values of 33.35, 29.16, and 12.78, respectively, reflecting a ruby colour, as represented in Table 2 [33]. When compared to the fresh fruits, the freeze-dried ones had significantly higher a^* (35.74) and lower b^* (10.85) values, which can be translated into a more intense red colour with less yellow. This was explained as being a consequence of the more efficient light diffusion through the raspberry fruit due to the free water replacement by air during freeze-drying [16].

Table 2. Mean values of lightness (L^*), red–green (a^*), yellow–blue (b^*), and total colour difference (ΔE) of fresh, freeze-dried, and convective-dried (30 °C and 40 °C) raspberry fruits from dormancy state of Pacific Deluxe and Versailles varieties.

	L^*	a^*	b^*	ΔE	Colour [1]
Pacific Deluxe					
Fresh	33.35 ± 0.83 [A,a]	29.16 ± 4.26 [A,a]	12.78 ± 1.71 [A,a]	-	
Freeze-dried	32.68 ± 2.53 [a]	35.74 ± 2.64 [b]	10.85 ± 0.94 [b]	6.89	
Dried at 30 °C	20.17 ± 1.81 [b]	22.02 ± 4.77 [c]	7.57 ± 2.12 [c]	15.87	
Dried at 40 °C	25.01 ± 3.38 [c]	29.05 ± 1.66 [a]	11.50 ± 0.97 [b]	8.44	
Versailles					
Fresh	32.86 ± 1.39 [A,a]	31.98 ± 2.32 [A,a]	13.42 ± 1.93 [A,a]	-	
Freeze-dried	25.55 ± 2.08 [b]	34.68 ± 3.05 [a]	11.73 ± 1.20 [a]	7.98	
Dried at 30 °C	25.03 ± 4.45 [b]	29.82 ± 6.33 [a]	11.85 ± 3.39 [a]	8.28	
Dried at 40 °C	28.15 ± 3.72 [b]	30.52 ± 3.50 [a]	11.90 ± 1.73 [a]	5.16	

[1] Fruits colour using an online correspondence CIELab parameters program using RGB (red, green, and blue) values [33]. The same uppercase letter represents no significantly different values between fresh fruits ($p > 0.05$). For each variety, in each column, different lowercase letters represent significantly different values ($p < 0.05$).

The raspberries dried at 30 °C presented a significant decrease in all the chromatic parameters ($L^* = 20.17$, $a^* = 22.02$, $b^* = 7.57$) when compared to the fresh fruits, reflecting the darkness and browning of the ruby colour. However, this darkness was less intense when the fruits were dried at 40 °C, with L^* and b^* values of 25.01 and 11.50, and no significant changes in the red–green (a^*) coordinate. These results were consistent with the literature for the commercial raspberry fruits dried at 50 °C [26]. The colour darkening can be explained by a possible decomposition of the carotenoids and/or non-enzymatic browning related to Maillard reactions, which resulted in the formation of brown compounds [19]. The total colour difference (ΔE) achieved for the freeze-dried ($\Delta E = 6.89$) and convective-dried fruits (30 °C—$\Delta E = 15.9$ and 40 °C—$\Delta E = 8.44$) corroborated the discussed chromatic changes. However, only the ΔE value for the Pacific Deluxe fruits dried at 30 °C was beyond 10, the benchmark used to consider a significant colour degradation in dried fruits [16]. The greater darkening of Pacific Deluxe dried at 30 °C (Figure 2, Table 2) suggested that when the drying time was higher (49 h) the extension of the browning reactions would be higher.

For the Versailles fresh fruits, the obtained L^*, a^*, and b^* values were 32.86, 31.98, and 13.42, respectively. These values are not significantly different from those of Pacific Deluxe. The freeze-dried fruits exhibited a significant decrease in their luminosity ($L^* = 25.55$), and no significant changes were observed for the a^* and b^* coordinates. A similar trend was observed for the fruits dried at 30 °C and 40 °C. The ΔE values varied from 5.16 to 8.28, which were values below 10, thus suggesting no significant degradation of the ruby colour of the dried Versailles fruits [16], as with the Pacific Deluxe samples freeze-dried and dried at 40 °C (Table 2).

3.3.2. Raspberry Fruit Phenolics and Antioxidant Activity

The fresh Pacific Deluxe fruits presented 6.78 µg GAE/mg (dw) of phenolic compounds, as determined by the Folin–Ciocalteu method (Figure 4a), which was in line with the literature concerning other red raspberry varieties [34]. After freeze-drying, no significant changes in the total phenolics content of the raspberry fruits were observed. However, after convective-drying, a significant decrease was determined (3.56 and 2.71 µg GAE/mg (dw) at 30 °C and 40 °C, respectively), although it was not statistically different between the two drying temperatures used. These results revealed a decrease of ca. 48–60% of the phenolics amount when applying convective-drying to the raspberry fruits. A similar trend was observed for the Versailles fruits: no significant changes were determined for the fresh and freeze-dried fruits (4.77 and 5.89 µg GAE/mg (dw), respectively), while a decrease of ca. 45% was registered for the fruits dried at 30 °C and 40 °C (phenolic content of 2.65 and 2.58 µg GAE/mg (dw), respectively) (Figure 4a). These results were in line with literature reports on other *R. ideaus* varieties, where drying temperatures of 50 °C [16,17] and 60 °C [12] led to a decrease in the raspberry fruits' phenolic content between ca. 30 and 63%, while no remarkable changes were observed in the freeze-dried fruits. Beyond temperature, the significant decrease in the total phenolic content during convective-drying can be related to the prolonged drying times required when temperatures are mild [16], which is probably due to the promotion of enzymatic activity able to release phenolics linked to carbohydrates [26], which are more prone to oxidisation [35].

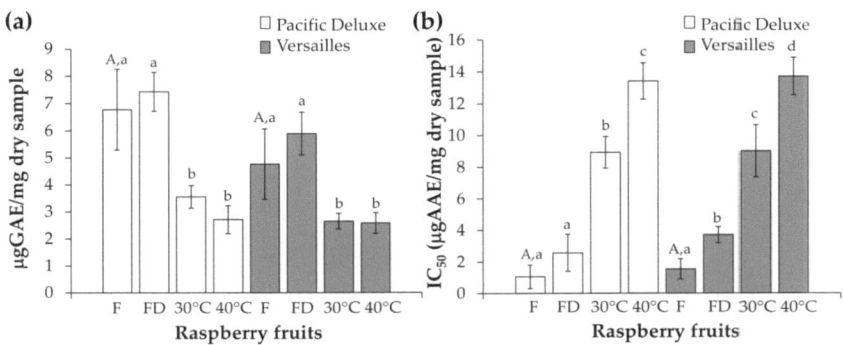

Figure 4. Total phenolic content (**a**) and antioxidant activity (**b**) of fresh (F), freeze-dried (FD), and convective-dried raspberry fruits from dormancy state at 30 °C and 40 °C. The same uppercase letter represents no significantly different values between fresh fruits ($p > 0.05$). For each variety, different lowercase letters represent significantly different values ($p < 0.05$). GAE—gallic acid equivalents and AAE—ascorbic acid equivalents.

The phenolic profile of the fresh, freeze-dried, and convective-dried raspberry fruits from the Pacific Deluxe and Versailles varieties is presented in Table 3 (and there is an example of a chromatogram in Figure S2). A total of four anthocyanins were determined in both varieties, with cyanidin-3-*O*-sophoroside being the most abundant one. This profile was in line with the literature [26] and reflects the anthocyanins' compositional homogeneity among the red varieties studied.

Table 3. Phenolic compounds (μg Mv 3-gE/mg sample, dw) determined in fresh, freeze-dried, and convective-dried (30 °C and 40 °C) raspberry fruits from dormancy state of Pacific Deluxe and Versailles varieties.

	Cy 3-s	Cy 3-gr	Cy 3-(6″-dg)	Cy 3-(6″-s-gl)	Total
Pacific Deluxe					
Fresh	7.54 ± 0.36 [A,a]	2.52 ± 0.08 [A,a]	1.00 ± 0.08 [A,a]	0.46 ± 0.02 [A,a]	11.52 ± 0.38 [A,a]
Freeze-dried	8.01 ± 0.79 [a]	2.87 ± 0.34 [a]	1.10 ± 0.10 [a]	0.53 ± 0.04 [a]	12.52 ± 1.29 [a]
Dried at 30 °C	6.29 ± 0.11 [b]	2.18 ± 0.09 [b]	0.92 ± 0.03 [b]	0.47 ± 0.01 [a]	9.86 ± 0.16 [b]
Dried at 40 °C	4.90 ± 0.58 [c]	1.99 ± 0.27 [b]	0.76 ± 0.05 [c]	0.41 ± 0.04 [a]	8.06 ± 0.49 [c]
Versailles					
Fresh	5.41 ± 0.10 [B,a]	2.82 ± 0.12 [B,a]	1.18 ± 0.05 [B,a]	1.22 ± 0.07 [B,a]	10.63 ± 0.05 [B,a]
Freeze-dried	5.84 ± 0.24 [a]	3.07 ± 0.11 [a]	1.23 ± 0.03 [a]	1.37 ± 0.09 [a]	11.51 ± 0.57 [a]
Dried at 30 °C	3.25 ± 0.11 [b]	1.73 ± 0.24 [b]	0.93 ± 0.21 [b]	0.84 ± 0.11 [b]	6.75 ± 0.23 [b]
Dried at 40 °C	2.95 ± 0.07 [c]	1.46 ± 0.18 [b]	0.81 ± 0.18 [b]	0.70 ± 0.09 [b]	5.92 ± 0.13 [c]

Cy-3-s: Cyanidin-3-*O*-sophoroside; Cy-3-gr: Cyanidin-3-*O*-glucosyl rutinoside; Cy 3-(6″-dg): Cyanidin 3 (6′-dioxalyl-glucoside); Cy 3-(6″-s-gl): Cyanidin-3-(6′-succinyl-glucoside); Mv 3-gE: Malvidin-3-glucose equivalents. Different uppercase letters represent significantly different values between fresh fruits ($p > 0.05$). For each variety, different lowercase letters represent significantly different values ($p < 0.05$).

The fresh fruits from Pacific Deluxe had a total anthocyanin content of 11.52 μg malvidin-3-glucose equivalents (Mv 3-gE)/mg (dw), where ca. 65% corresponded to cyanidin-3-*O*-sophoroside (Cy 3-s = 7.54 µg Mv 3-gE/mg, dw). When compared to the fresh fruits, a similar content and profile was obtained for the freeze-dried Pacific Deluxe samples, while a significant decrease in the total amount of anthocyanins was determined in the convective-dried fruits (a decrease of ca. 14% and 30% in the fruits dried at 30 °C and 40 °C, respectively). This was even more evident for Cy 3-s, where a decrease from 7.54 to 4.90 µg Mv 3-gE/mg (dw) was observed in the fruits dried at 40 °C (Table 3). This followed the trend observed in commercially available *Rubus idaeus* L. raspberry fruits when submitted to drying at 50 °C for 19 h, where Cy-3-s decreased by ca. 78% [17]. This trend was consistent with the results determined by the Folin–Ciocalteu method (Figure 4a), although the decrease percentages obtained by HPLC analysis were lower than those of the colorimetric method, suggesting that, beyond anthocyanins, other phenolics were degraded during the convective-drying.

The fresh Versailles raspberry fruits presented a total anthocyanin content of 10.63 µg Mv 3-gE/mg (dw), which was significantly lower than that of the Pacific Deluxe fresh fruits, except for cyanidin-3-(6′-succinyl-glucoside), which was ca. 2.7 times higher in the Versailles fruits. When compared to the Versailles fresh fruits, no significant changes were determined in the freeze-dried ones (11.51 µg Mv 3-gE/mg, dw), while a decrease of ca. 37% and 44% was reached for the fruits dried at 30 °C and 40 °C, respectively (Table 3), according to the trend observed with the Folin–Ciocalteu method (Figure 4a). These results suggested that for the Versailles fruits the ca. 45% reduction in the total amount of phenolics can be mainly related to the possible degradation of anthocyanins, which is in line with the literature [12,17].

The antioxidant activity of the raspberry fruits from Pacific Deluxe and Versailles was evaluated by determining the samples' ability to scavenge the ABTS$^{\bullet+}$ radical (Figure 4b). Regarding the antiradical activity, the concentration of fresh Pacific Deluxe fruits able to inhibit 50% of ABTS (IC_{50}) was 1.07 µg ascorbic acid equivalents (AAE)/mg (dw). The freeze-dried fruits had an IC_{50} of 2.58 µg AAE/mg (dw), which was not significantly different from that of the fresh fruits. Moreover, IC_{50} values of 8.94 and 13.41 µg AAE/mg (dw) were determined for the convective-dried fruits at 30 °C and 40 °C, respectively; these were significantly higher than those for the fresh fruits. The increase in these values can be translated into a significant decrease in the antiradical activity of the convective-dried fruits. This effect was more evident for the higher temperature tested, where the IC_{50} value was ca. 13 times higher than that of the fresh fruits and thus presented lower antioxidant activity (Figure 4b).

The fresh fruits of the Versailles variety presented an IC_{50} = 1.55 µg AAE/mg (dw), which was not significantly different from that of the Pacific Deluxe fresh fruits. When compared to the fresh fruits, all the Versailles dried fruits exhibited a significant decrease in their antiradical activity. The drying at 40 °C was the procedure that had the higher impact on the decrease in these fruits' antioxidant activities (Figure 4b), which were similar to those of the Pacific Deluxe fruits. These results were consistent with those of the literature, where the antioxidant activity of commercial fresh red raspberry fruits (*R. idaeus* cv. Heritage) was significantly higher than that observed in fruits dried by microwave and solar drying processes [13].

As the antioxidant activity is largely related to the content of phenolic compounds, the higher IC_{50} values observed in the convective-dried samples can be explained by the lower phenolic content registered. As the content of total phenolic compounds determined for the dried raspberries at 30 °C and 40 °C was very similar (Figure 4a), the significant differences in the antioxidant activity suggested that there are other compounds with antioxidant potential that also decrease their activity during drying. This was more evident when the process was carried out at 40 °C (Figure 4b). This may be due to the high content of ascorbic acid in raspberry fruits, which is a compound sensitive to thermal treatments [16].

3.4. Muffin Development

Fresh, freeze-dried, and convective-dried (40 °C) Versailles fruits from dormancy were used as ingredients for the development of muffins. The Versailles variety was selected since lower chromatic changes and a similar phenolic content were observed during drying when compared to the Pacific Deluxe dried fruits (Figure 2, Table 2). Drying at 40 °C was selected instead of 30 °C because no significant changes were observed in the phenolic content of the fruits (Figure 4a), and a 10% reduction in the drying time was achieved (Figure 3); this is a more sustainable option that can allow a decreasing of the processing energy costs.

For muffin production, the content of raspberry drupelets (60, 90, and 120 drupelets) to be included in the muffin dough was optimised. The impact of the fruits on the physical properties, colour, and sensory attributes of the muffins was evaluated, and the selection of the raising agent (chemical—sodium bicarbonate or natural—yeast) was performed. A baking negative control was always carried out, and it corresponded to the formulation made without the incorporation of raspberry fruits.

3.4.1. Influence of Drupelets Incorporation on Physical Properties and Colour of Muffins

Different quantities of fresh, freeze-dried, and convective-dried (40 °C) raspberry fruits (60, 90, and 120 drupelets) were incorporated in a muffin formulation made with sodium bicarbonate as the raising agent (original recipe). The physical properties of the muffins were evaluated to assess the influence of the number of drupelets on the muffins' dimensions (Table 4). When looking at the weight of the muffins, with the incorporation of fresh fruit drupelets in the formulations, a significant increase in weight was observed, ranging from 38.3 g in the muffins without drupelets to a maximum of 51.7 g in the muffins containing 120 fresh drupelets. The muffins' width (parameter 2) also increased, reaching its maximum at 9.1 cm for the muffins with 120 drupelets. According to these results, the muffins became heavier and larger with the increase in the number of fruit drupelets, probably due to the higher water content of the fresh fruits. This may be related to higher water activity and thus should be considered when looking at the microbiological stability of this baked product. No significant changes were observed in the remaining parameters (1—length, 3—base height, 4—total side height, and 5—total height in the middle).

Table 4. Physical parameters of muffins baked with different drupelets amounts (0. 60, 90, 120) of fresh, freeze-dried, and convective-dried (40 °C) Versailles raspberry fruits from dormancy state.

			Parameters (cm)				
	Drupelets	Weight (g)	1	2	3	4	5
Fresh	0	38.3 ± 2.0 [a]	6.8 ± 0.1 [a]	8.3 ± 0.4 [a]	1.5 ± 0.2 [a]	2.6 ± 0.5 [a]	2.7 ± 0.0 [a]
	60	41.3 ± 2.1 [a]	7.4 ± 0.5 [a]	8.2 ± 0.2 [a]	1.5 ± 0.2 [a]	2.7 ± 0.3 [a]	2.7 ± 0.0 [a]
	90	46.3 ± 2.1 [b]	7.4 ± 0.5 [a]	8.8 ± 0.2 [b]	1.5 ± 0.2 [a]	2.6 ± 0.1 [a]	2.7 ± 0.0 [a]
	120	51.7 ± 1.5 [c]	7.3 ± 0.3 [a]	9.1 ± 0.2 [b]	1.5 ± 0.1 [a]	2.5 ± 0.2 [a]	2.8 ± 0.1 [a]
Freeze-dried	0	38.3 ± 2.1 [a]	6.8 ± 0.1 [a]	8.3 ± 0.4 [a]	1.5 ± 0.2 [a]	2.6 ± 0.5 [a]	2.7 ± 0.0 [a]
	60	38.0 ± 1.0 [a]	7.7 ± 0.2 [b]	7.8 ± 0.2 [a]	1.5 ± 0.1 [a]	2.4 ± 0.1 [a]	2.2 ± 0.2 [b]
	90	36.0 ± 1.5 [b]	7.8 ± 0.5 [b]	8.3 ± 0.3 [a]	1.5 ± 0.2 [a]	2.6 ± 0.3 [a]	1.9 ± 0.2 [b]
	120	34.7 ± 1.0 [b]	7.3 ± 0.2 [b]	7.9 ± 0.2 [a]	1.5 ± 0.1 [a]	2.6 ± 0.1 [a]	1.9 ± 0.2 [b]
Dried at 40 °C	0	38.3 ± 3.1 [a]	7.7 ± 0.3 [a]	8.6 ± 0.2 [a]	1.5 ± 0.0 [a]	2.8 ± 0.3 [a]	2.0 ± 0.1 [a]
	60	42.3 ± 2.1 [a]	8.9 ± 1.0 [a]	9.0 ± 0.4 [a]	1.6 ± 0.2 [a]	2.9 ± 0.2 [a]	1.9 ± 0.0 [a]
	90	41.7 ± 2.5 [a]	7.8 ± 0.7 [a]	9.1 ± 0.2 [a]	1.4 ± 0.2 [a]	2.7 ± 0.2 [a]	1.8 ± 0.0 [a]
	120	46.0 ± 1.0 [b]	8.1 ± 0.2 [a]	8.8 ± 0.2 [a]	1.5 ± 0.2 [a]	2.7 ± 0.1 [a]	2.1 ± 0.1 [a]

The physical parameters 1—length, 2—width, 3—base height, 4—total side height, and 5—total height in the middle, given in cm, were determined according to Figure 1. For each type of sample, in each column, different lowercase letters represent significantly different values ($p < 0.05$).

With the incorporation of freeze-dried fruits, in contrast to the use of fresh fruits, a significant decrease in the weight of the muffins was observed, from 38.3 g in the muffins without drupelets, to 36.0 and 34.7 g in the muffins containing 90 and 120 drupelets, respectively. No significant differences were observed among the muffins with these two quantities. Also, the length (parameter 1) and total height in the middle (parameter 5) significantly decreased for all the tested drupelets contents, but no significant changes were observed among the different quantities. These results revealed that the muffins with the freeze-dried fruits became smaller, which was more relevant for the muffins with 90 and 120 drupelets. The lower moisture of these freeze-dried fruits (Table 1) may explain the smaller size of the resulting muffins (Table 4). The opposite was observed in the muffins with the convective-dried fruits, where only the muffins with 120 drupelets became heavier (46.0 g). No significant changes were observed for the other physical parameters (from 1 to 5). In this case, the reduced moisture content of the fruits dried at 40 °C ($\leq$10%, Figure 3) may explain the muffins' higher weight when using a higher number of drupelets, which is like the weight of the muffins containing 90 drupelets of fresh fruits and lower than the weight of the muffins with 120 fresh drupelets (Table 4).

The influence of the fruit incorporation on the colour properties of the muffins on the upper (Table 5) and lower sides (Table S3) showed that the upper side of the muffins, with the incorporation of fresh fruit drupelets in the formulations, significantly decreased in all chromatic values (L^*, a^*, and b^*) when compared to the corresponding control (muffins without drupelets). These differences were not significant among the concentrations of drupelets used. This was manifested in the darkening of the muffins and in the appearance of a greenish colour. The $\Delta E > 10$ for all the assays with the fresh fruits corroborated the significant chromatic changes. With the incorporation of the convective-dried samples, only a significant decrease in the a^* value was observed, which reflected the appearance of a greenish colour in the muffins. Although this difference was not significant among the drupelets amounts used, all the $\Delta E > 10$ revealed a significant colour change. This effect was not observed in the muffins with the freeze-dried samples. In this case, for all the tested drupelets amounts, the L^*, a^*, and b^* values were similar to those of the control (L^* = 66.06, a^* = 6.07, b^* = 38.34), and the ΔE was below 5, which was ca. 2 times lower than the ΔE benchmark for a colour change to be considered significant [16]. When analysing the lower side of the muffins, no significant changes were determined for all the tested conditions (Table S3). The visual aspect of the muffins corroborated these results (Figure S3).

Table 5. Mean values of lightness (L^*), red–green (a^*), yellow–blue (b^*), and total colour difference (ΔE) on the upper side of muffins baked with different drupelets amounts (0, 60, 90, 120) of fresh, freeze-dried, and convective-dried (40 °C) Versailles raspberry fruits from dormancy state.

	Drupelets	L^*	a^*	b^*	ΔE	Colour [1]
Fresh	0	66.06 ± 5.79 [a]	6.07 ± 1.35 [a]	38.34 ± 5.68 [a]	-	
	60	52.53 ± 2.74 [b]	2.35 ± 1.02 [b]	28.61 ± 4.13 [b]	17.08	
	90	53.57 ± 6.19 [b]	2.46 ± 0.61 [b]	26.81 ± 3.89 [b]	17.37	
	120	52.46 ± 4.01 [b]	3.19 ± 1.35 [b]	26.81 ± 2.94 [b]	18.06	
Freeze-dried	0	66.06 ± 5.79 [a]	6.07 ± 1.35 [a]	38.34 ± 5.68 [a]	-	
	60	68.02 ± 2.44 [a]	6.06 ± 1.48 [a]	38.43 ± 5.17 [a]	0.09	
	90	68.31 ± 1.99 [a]	4.64 ± 1.54 [a]	39.62 ± 3.13 [a]	2.96	
	120	69.24 ± 2.40 [a]	5.43 ± 1.35 [a]	41.93 ± 3.07 [a]	4.84	
Dried at 40 °C	0	66.03 ± 5.79 [a]	6.22 ± 1.22 [a]	27.20 ± 5.68 [a]	-	
	60	52.59 ± 5.51 [a]	4.27 ± 0.68 [b]	31.59 ± 2.64 [a]	14.69	
	90	55.98 ± 6.25 [a]	4.78 ± 0.89 [b]	25.03 ± 4.20 [a]	10.38	
	120	46.18 ± 5.25 [a]	4.39 ± 1.07 [b]	35.02 ± 2.85 [a]	24.01	

[1] Muffin colour using an online correspondence CIELab parameters program using RGB (red, green, and blue) values [33]. For each type of sample, in each column, different lowercase letters represent significantly different values ($p < 0.05$).

The major colour changes observed in the muffins made with fresh raspberry drupelets, followed by the muffins containing convective-dried fruits, can be related to the release of moisture contained in the drupelets during cooking, a phenomenon known as syneresis [36,37]. The release of water from the drupelets also led to the release of soluble phenolic compounds, namely anthocyanins, which at the pH of the dough (ca. 10 for all tested formulations, independently of the drupelets amount) gave a greenish colour to the muffins [38]. The syneresis effect was greater as the moisture content of the sample becomes greater, which is the reason why fresh drupelets provoked higher colour changes in the muffins than the ones dried at 40 °C (moisture content ca. 10%), and no changes were observed in the muffins with freeze-dried fruits.

3.4.2. Influence of the Raising Agent on the Colour and Physical Properties of the Muffins

The green colour of the muffins made with fresh and dried raspberry fruits at 40 °C did not make the final greenish colour of the product attractive. For that reason, it was decided to replace the sodium bicarbonate with baker's yeast, a natural raising agent. The impact of this replacement on the pH of the dough and the resulting muffins' colour and physical properties was evaluated. This assay was made only in muffins containing 120 fresh drupelets, where the green colour was strongly evidenced (ΔE = 18.06, Table 5).

When replacing the sodium bicarbonate with yeast, the muffin dough containing fresh raspberry drupelets lost its green colour and exhibited a light-yellow colour (Figure S4a). The ΔE of 4.91 and 2.61 determined on the upper and lower sides of these muffins, respectively, corroborated their colour similarity with the muffins with zero drupelets (Table 6). Due to the raising agent replacement, the pH of the muffin dough changed from ca. 10 (sodium bicarbonate raising agent) to ca. 7 in the muffins made with yeast (Figure S4b). This colour change is related to the reversible structural transformations that anthocyanins undergo when the pH values of the environment change [39,40]. At pH 7, the anthocyanin structure is protonated, presenting a purple quinoid anhydrous base [41,42]. Therefore, it is possible to conclude that the green colour of the muffins containing raspberry drupelets was avoided by replacing the sodium bicarbonate with yeast. Additionally, the substitution of the raising agent resulted in heavier muffins with no significant changes in the other dimensions (Table S4). Based on these results, yeast was used in the recipe for further sensory analysis of the muffins.

Table 6. Mean values of lightness (L^*), red–green (a^*), yellow–blue (b^*), and total colour difference (ΔE) of upper and lower sides of muffins baked with sodium bicarbonate and yeast raising agents with 120 fresh drupelets of Versailles variety from dormancy state.

Raising	Drupelets	L^*	a^*	b^*	ΔE	Colour [1]
Upper side						
Bicarbonate	0	67.01 ± 4.84 [a,A]	6.63 ± 0.70 [a,A]	37.11 ± 2.49 [a,A]	-	
	120	55.98 ± 2.49 [b,B]	4.78 ± 0.96 [b,A]	35.03 ± 2.67 [a,A]	11.37	
Yeast	0	68.56 ± 4.23 [a,A]	7.92 ± 0.33 [a,A]	36.70 ± 1.95 [a,A]	-	
	120	63.91 ± 4.82 [a,A]	6.32 ± 1.64 [a,A]	36.66 ± 2.18 [a,A]	4.91	
Lower side						
Bicarbonate	0	46.17 ± 3.22 [a,A]	15.94 ± 2.04 [a,A]	36.36 ± 1.94 [a,A]	-	
	120	38.38 ± 2.04 [b,B]	14.58 ± 2.38 [a,A]	26.32 ± 1.95 [b,B]	10.95	
Yeast	0	47.77 ± 3.40 [a,A]	18.40 ± 1.10 [a,A]	37.45 ± 2.20 [a,A]	-	
	120	48.79 ± 3.34 [a,A]	17.26 ± 1.17 [a,A]	36.32 ± 1.95 [a,A]	2.61	

[1] Muffin colour using an online correspondence CIELab parameters program using RGB (red, green, and blue) values [33]. In each column, for each type of raising agent, different lowercase letters represent significantly different values ($p < 0.05$), while the uppercase letters represent significantly different values ($p < 0.05$) among the muffins, independently of the raising agent.

3.4.3. Sensory Analysis of Muffins

To evaluate the differences that can be perceived by the incorporation of raspberries in a fresh or dried state, freeze-dried fruits were used to prepare the muffin recipes using the baker's yeast as the raising agent and a total of 120 drupelets of Versailles raspberries (fresh—MF and freeze-dried—ML). The muffins without fruits were also considered (CN). A five-parameter hedonic scale was used by the 16 non-trained panellists. The global appreciation (ranging from 3.9 and 4.1), as well as the global texture (ranging from 3.4 and 3.8), showed that all the products were well appreciated by the panellists (Figure 5).

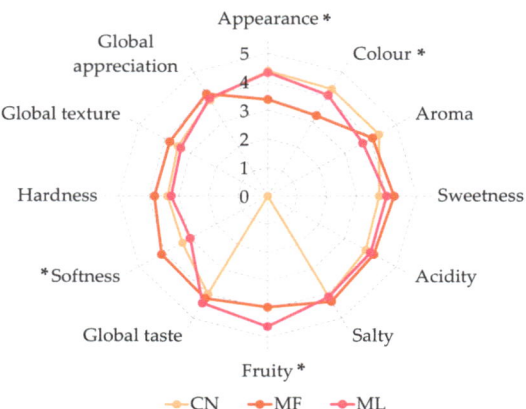

Figure 5. Sensory data, expressed as mean values, of muffins without raspberries (CN), with 120 fresh raspberry drupelets (MF), and with 120 freeze-dried raspberry drupelets (ML), based on the 12 sensory terms used by the non-trained panel (16 panellists). * $p < 0.05$.

The muffins with freeze-dried fruits were significantly better appreciated than the muffins with fresh fruits with regard to their appearance and colour. However, no significant differences were observed for the control, allowing the inference that the incorporation of dried fruits in the formulation is an advantage when compared to the fresh ones. Nevertheless, the muffins prepared with fresh fruits presented significantly higher softness than those prepared with dried fruits and the control.

4. Conclusions

The raspberry fruits harvested during the dormancy state of the plant showed similar characteristics to those of the commercially available fruits, as referred to in the literature. In the studied case regarding the physicochemical differences between the Pacific Deluxe and Versailles fruits, the structure and composition of the fresh fruits were quite similar when the fruits were freeze-dried or when they were submitted to convective-drying at 30 °C or 40 °C. In both cases, the freeze-drying allowed better maintenance of the ruby colour and the cluster of drupelets of the fresh fruits, as well as their total phenolic content and antioxidant properties. Nevertheless, convective-drying at 40 °C, although presenting a darker colour and a lower antioxidant activity than the fresh and freeze-dried fruits, presented properties that allowed them to also be considered as food ingredients. These fruits can be used in muffin formulations. For that, the selection of the raising agent is of major relevance, as raspberry anthocyanins tend to be green due to the alkaline pH of the dough when using sodium bicarbonate. To overcome this drawback, baker's yeast was shown to be an effective substitute. The sensory analysis performed by the non-trained panel attested to the global appreciation and acceptance of the muffins containing dried fruits, showing that raspberry fruits from dormant plants have quality characteristics and that it is feasible to use them as food ingredients.

Supplementary Materials: The following supporting information can be downloaded at: https://www.mdpi.com/article/10.3390/foods12244443/s1, Figure S1: Scheme of the muffin making procedure, Figure S2: Example of the chromatogram obtained for fresh Pacific Deluxe variety from dormancy state: Cy-3-s: Cyanidin-3-O-sophoroside, Cy-3-gr: Cyanidin-3-O-glucosyl rutinoside, Cy 3-(6″-dg): Cyanidin 3 (6′-dioxalyl-glucoside), Cy 3-(6″-s-gl): Cyanidin-3-(6′-succinyl-glucoside), Figure S3: Images of the muffins using 60, 90, and 120 drupelets of fresh, freeze-dried, and convective-dried raspberry fruits at 40 °C, Figure S4: (a) Images of the muffins prepared with different raising agents and with 120 raspberry drupelets: muffins made with sodium bicarbonate (left) and muffins made with yeast (right) and (b) muffin pH and correspondence with the raising agent used, Table S1: Tentatively identified phenolic compounds, at 520 nm, in fresh, freeze-dried, and convective-dried (30 °C and 40 °C) raspberry fruits of Pacific Deluxe and Versailles varieties from dormancy state, Table S2: Ingredients and respective quantity used in muffin formulations, Table S3: Mean values of lightness (L^*), red–green (a^*), yellow–blue (b^*), and total colour difference (ΔE) in the lower side of muffins baked with different drupelets amounts (0, 60, 90, 120) of fresh, freeze-dried, and convective-dried (40 °C) raspberry fruits from Versailles variety from dormancy state, Table S4: Physical parameters of muffins baked with sodium bicarbonate and yeast raising agents with 120 fresh drupelets from Versailles variety from dormancy state.

Author Contributions: Conceptualisation, S.P. and C.P.P.; methodology, S.P. and C.P.P.; validation, S.P. and C.P.P.; formal analysis, S.P. and C.P.P.; investigation, S.P.; resources, M.A.C.; data curation, S.P. and C.P.P.; writing—original draft preparation, S.P.; writing—review and editing, S.P., C.P.P. and M.A.C.; visualisation, S.P. and C.P.P.; supervision, S.P. and C.P.P. All authors have read and agreed to the published version of the manuscript.

Funding: This work received financial support from PT national funds (FCT/MCTES) for the financial support of LAQV-REQUIMTE (UIDB/50006/2020 + UIDP/5006/2020) research unit and CQ-VR at UTAD Vila Real (UIDB/00616/2020 + UIDP/00616/2020) through PT national funds and, when applicable, co-financed by the FEDER, within the PT2020 Partnership Agreement and Compete 2020. FCT is thanked for the postdoc grant SFRH/BPD/117213/2016 (S.P.) and for the Individual Call to Scientific Employment Stimulus (C.P.P., CEECIND/01873/2017).

Institutional Review Board Statement: A permit is not required.

Informed Consent Statement: The non-trained panellists participating in the sensory analysis were informed verbally and a table with their names was signed, although their identities were not disclosed. Before their participation, all non-trained panellists consented verbally to participating in the study. The panellists were volunteers from GlycoFoodChem research group at the University of Aveiro (Portugal).

Data Availability Statement: Data are contained within the article and Supplementary Materials.

Acknowledgments: The authors thank the *Framboesas da Graça* company (Ponte de Vagos, Portugal) for kindly providing the fresh raspberry fruits from Pacific Deluxe and Versailles red varieties, from dormancy state. S.P. thanks the financial support of Norte Portugal Regional Operational Program (NORTE 2020), under the PT 2020 Partnership Agreement, through the European Regional Development Fund (ERDF) and FSE. The authors thank all the volunteers who participated in the study. The authors thank all the volunteers who participated in the sensory analysis.

Conflicts of Interest: The authors declare no conflict of interest. The funders had no role in the design of the study; in the collection, analyses, or interpretation of data; in the writing of the manuscript; or in the decision to publish the results.

References

1. Hidalgo, G.; Almajano, M. Red Fruits: Extraction of Antioxidants, Phenolic Content, and Radical Scavenging Determination: A Review. *Antioxidants* **2017**, *6*, 7. [CrossRef] [PubMed]
2. Sønsteby, A.; Heide, O.M. Flowering and dormancy relations of raspberry and black currant and effects of management and climate warming on production. *Acta Hortic.* **2020**, *1277*, 307–320. [CrossRef]
3. Pritts, M. Soft Fruits. In *Life Sciences*, 2nd ed.; Elsevier Inc.: Amsterdam, The Netherlands, 2016; pp. 268–272.
4. Dale, A.; Sample, A.; King, E. Breaking dormancy in red raspberries for greenhouse production. *Hortscience* **2003**, *38*, 515–519. [CrossRef]
5. Food and Agriculture Organization Corporate Statistical Database (FAOSTAT). *Production of Raspberries in 2020, Pick List by Crops/Regions/Production Quantity*; United Nations: New York, NY, USA, 2022.
6. Ponder, A.; Hallmann, E. The nutritional value and vitamin C content of different raspberry cultivars from organic and conventional production. *J. Food Compos. Anal.* **2020**, *87*, 103429. [CrossRef]
7. Aksic, M.; Nesovic, M.; Ciric, I.; Tesic, Z.; Pezo, L.; Tosti, T.; Gasic, U.; Dojcinovic, B.; Loncar, B.; Meland, M. Chemical Fruit Profiles of Different Raspberry Cultivars Grown in Specific Norwegian Agroclimatic Conditions. *Horticulturae* **2022**, *8*, 765. [CrossRef]

8. Mazur, S.; Nes, A.; Wold, A.; Remberg, S.; Aaby, K. Quality and chemical composition of ten red raspberry (*Rubus idaeus* L.) genotypes during three harvest seasons. *Food Chem.* **2014**, *160*, 233–240. [CrossRef]
9. Baenas, N.; Nunez-Gomez, V.; Navarro-Gonzalez, I.; Sanchez-Martinez, L.; Garcia-Alonso, J.; Periago, M.; Gonzalez-Barrio, R. Raspberry dietary fibre: Chemical properties, functional evaluation and prebiotic in vitro effect. *LWT-Food Sci. Technol.* **2020**, *134*, 110140. [CrossRef]
10. Lopez-Corona, A.; Valencia-Espinosa, I.; Gonzalez-Sanchez, F.; Sanchez-Lopez, A.; Garcia-Amezquita, L.; Garcia-Varela, R. Antioxidant, Anti-Inflammatory and Cytotoxic Activity of Phenolic Compound Family Extracted from Raspberries (*Rubus idaeus*): A General Review. *Antioxidants* **2022**, *11*, 1192. [CrossRef]
11. Padmanabhan, P.; Correa-Betanzo, J.; Paliyath, G. Berries and Related Fruits. In *Encyclopedia of Food and Health*; Caballero, B., Finglas, P.M., Toldrá, F., Eds.; Academic Press: Oxford, UK, 2016.
12. Sette, P.; Franceschinis, L.; Schebor, C.; Salvatori, D. Fruit snacks from raspberries: Influence of drying parameters on colour degradation and bioactive potential. *Int. J. Food Sci. Technol.* **2017**, *52*, 313–328. [CrossRef]
13. Rodriguez, A.; Bruno, E.; Paola, C.; Campanone, L.; Mascheroni, R. Experimental study of dehydration processes of raspberries (*Rubus Idaeus*) with microwave and solar drying. *Food Sci. Technol.* **2019**, *39*, 336–343. [CrossRef]
14. Borquez, R.; Canales, E.; Redon, J. Osmotic dehydration of raspberries with vacuum pretreatment followed by microwave-vacuum drying. *J. Food Eng.* **2010**, *99*, 121–127. [CrossRef]
15. Rodriguez, A.; Rodriguez, M.; Lemoine, M.; Mascheroni, R. Study and comparison of different drying processes for dehydration of raspberries. *Dry. Technol.* **2017**, *35*, 689–698. [CrossRef]
16. Stamenkovic, Z.; Pavkov, I.; Radojcin, M.; Horecki, A.; Keselj, K.; Kovacevic, D.; Putnik, P. Convective Drying of Fresh and Frozen Raspberries and Change of Their Physical and Nutritive Properties. *Foods* **2019**, *8*, 251. [CrossRef] [PubMed]
17. Tekin-Cakmak, Z.; Cakmakoglu, S.; Avci, E.; Sagdic, O.; Karasu, S. Ultrasound-assisted vacuum drying as alternative drying method to increase drying rate and bioactive compounds retention of raspberry. *J. Food Process. Preserv.* **2021**, *45*, e16044. [CrossRef]
18. Petronilho, S.; Rudnitskaya, A.; Coimbra, M.; Rocha, S. Comprehensive Study of Variety Oenological Potential Using Statistic Tools for the Efficient Use of Non-Renewable Resources. *Appl. Sci.* **2021**, *11*, 4003. [CrossRef]
19. Krzykowski, A.; Dziki, D.; Rudy, S.; Gawlik-Dziki, U.; Janiszewska-Turak, E.; Biernacka, B. Wild Strawberry *Fragaria vesca* L.: Kinetics of Fruit Drying and Quality Characteristics of the Dried Fruits. *Processes* **2020**, *8*, 1265. [CrossRef]
20. FAO. Dried Fruit. In *Fruit Processing Toolkit*; Food and Agriculture Organization: Rome, Italy, 2007.
21. AOAC. Fruits and fruit products. In *AOAC Official Method 942.15 Acidity (Titratable) of Fruit Products*; AOAC: Rockville, MD, USA, 2000; p. 11.
22. Radojčin, M.; Babić, M.; Babić, L.; Pavkov, I.; Stojanović, Č. Color parameters change of quince during combined drying. *J. Process. Energy Agric.* **2010**, *14*, 81.
23. Petronilho, S.; Oliveira, A.; Domingues, M.R.; Nunes, F.M.; Coimbra, A.M.; Gonçalves, I. Hydrophobic Starch-Based Films Using Potato Washing Slurries and Spent Frying Oil. *Foods* **2021**, *10*, 2897. [CrossRef]
24. Peixoto, A.M.; Petronilho, S.; Domingues, M.R.; Nunes, F.M.; Lopes, J.; Pettersen, M.K.; Grøvlen, M.S.; Wetterhus, E.M.; Gonçalves, I.; Coimbra, M.A. Potato Chips Byproducts as Feedstocks for Developing Active Starch-Based Films with Potential for Cheese Packaging. *Foods* **2023**, *12*, 1167. [CrossRef]
25. Petronilho, S.; Navega, J.; Pereira, C.; Almeida, A.; Siopa, J.; Nunes, F.M.; Coimbra, M.A.; Passos, C.P. Bioactive Properties of Instant Chicory Melanoidins and Their Relevance as Health Promoting Food Ingredients. *Foods* **2023**, *12*, 134. [CrossRef]
26. Bustos, M.C.; Rocha-Parra, D.; Sampedro, I.; de Pascual-Teresa, S.; León, A.E. The Influence of Different Air-Drying Conditions on Bioactive Compounds and Antioxidant Activity of Berries. *J. Agric. Food Chem.* **2018**, *66*, 2714–2723. [CrossRef] [PubMed]
27. Martín-Gómez, J.; Varo, M.Á.; Mérida, J.; Serratosa, M.P. Influence of drying processes on anthocyanin profiles, total phenolic compounds and antioxidant activities of blueberry (*Vaccinium corymbosum*). *LWT-Food Sci. Technol.* **2020**, *120*, 108931. [CrossRef]
28. Raspberry Chocolate Chip Cookies. Available online: https://www.dessertnowdinnerlater.com/RASPBERRY-CHOCOLATE-CHIP-COOKIES/ (accessed on 10 May 2021).
29. Žilić, S.; Kocadağlı, T.; Vančetović, J.; Gökmen, V. Effects of baking conditions and dough formulations on phenolic compound stability, antioxidant capacity and color of cookies made from anthocyanin-rich corn flour. *LWT-Food Sci. Technol.* **2016**, *65*, 597. [CrossRef]
30. Morodi, V.; Kaseke, T.; Fawole, O. Impact of Gum Arabic Coating Pretreatment on Quality Attributes of Oven-Dried Red Raspberry (*Rubus idaeus* L.) Fruit. *Processes* **2022**, *10*, 1629. [CrossRef]
31. Stevanovic, S.; Petrovic, T.; Markovic, D.; Milovancevic, U.; Stevanovic, S.; Urosevic, T.; Kozarski, M. Changes of quality and free radical scavenging activity of strawberry and raspberry frozen under different conditions. *J. Food Process. Preserv.* **2022**, *46*, e15981. [CrossRef]
32. Purgar, D.; Duralija, B.; Voca, S.; Vokurka, A.; Ercisli, S. A Comparison of Fruit Chemical Characteristics of Two Wild Grown Rubus Species from Different Locations of Croatia. *Molecules* **2012**, *17*, 10390–10398. [CrossRef] [PubMed]
33. Convert L*a*b* Values to the Nearest Standard Colour. Available online: https://www.e-paint.co.uk/convert-lab.asp (accessed on 7 May 2021).
34. Alibabic, V.; Skender, A.; Bajramovic, M.; Sertovic, E.; Bajric, E. Evaluation of morphological, chemical, and sensory characteristics of raspberry cultivars grown in Bosnia and Herzegovina. *Turk. J. Agric. For.* **2018**, *42*, 67–74. [CrossRef]

35. Hossain, M.; Barry-Ryan, C.; Martin-Diana, A.; Brunton, N. Effect of drying method on the antioxidant capacity of six Lamiaceae herbs. *Food Chem.* **2010**, *123*, 85–91. [CrossRef]
36. Mizrahi, S. Syneresis in food gels and its implications for food quality. In *Chemical Deterioration and Physical Instability of Food and Beverages*; Elsevier: Amsterdam, The Netherlands, 2010; pp. 324–348. [CrossRef]
37. Rodriguez-Amaya, D.B. Natural food pigments and colorants. In *Bioactive Molecules in Food*; Mérillon, J.M., Ramawat, K.G., Eds.; Springer: Cham, Switzerland, 2019; pp. 867–901. [CrossRef]
38. Quintanilla, A.; Zhang, H.; Powers, J.; Sablani, S.S. Developing Baking-Stable Red Raspberries with Improved Mechanical Properties and Reduced Syneresis. *Food Bioprocess Technol.* **2021**, *14*, 804. [CrossRef]
39. Zhao, S.; Park, C.H.; Yang, J.; Yeo, H.J.; Kim, T.J.; Kim, J.K.; Park, S.U. Molecular characterization of anthocyanin and betulinic acid biosynthesis in red and white mulberry fruits using high-throughput sequencing. *Food Chem.* **2019**, *279*, 364. [CrossRef]
40. Teixeira, M.; Tao, W.; Fernandes, A.; Faria, A.; Ferreira, I.M.P.L.V.O.; He, J.; Freitas, V.; Mateus, N.; Oliveira, H. Anthocyanin-rich edible flowers, current understanding of a potential new trend in dietary patterns. *Trends Foods Sci. Technol.* **2023**, *138*, 708. [CrossRef]
41. Choi, I.; Lee, J.Y.; Lacroix, M.; Han, J. Intelligent pH indicator film composed of agar/potato starch and anthocyanin extracts from purple sweet potato. *Food Chem.* **2017**, *218*, 122. [CrossRef]
42. Yoshida, C.M.; Maciel, V.B.V.; Mendonça, M.E.D.; Franco, T.T. Chitosan biobased and intelligent films: Monitoring pH variations. *LWT-Food Sci. Technol.* **2014**, *55*, 83. [CrossRef]

Disclaimer/Publisher's Note: The statements, opinions and data contained in all publications are solely those of the individual author(s) and contributor(s) and not of MDPI and/or the editor(s). MDPI and/or the editor(s) disclaim responsibility for any injury to people or property resulting from any ideas, methods, instructions or products referred to in the content.

Article

Effects of Enriched-in-Oleuropein Olive Leaf Extract Dietary Supplementation on Egg Quality and Antioxidant Parameters in Laying Hens

Georgios A. Papadopoulos [1,*], Styliani Lioliopoulou [1], Nikolaos Nenadis [2], Ioannis Panitsidis [3], Ioanna Pyrka [2], Aggeliki G. Kalogeropoulou [2], George K. Symeon [4], Alexios-Leandros Skaltsounis [5], Panagiotis Stathopoulos [5], Ioanna Stylianaki [6], Dimitrios Galamatis [7], Anatoli Petridou [8], Georgios Arsenos [1] and Ilias Giannenas [3]

[1] Laboratory of Animal Husbandry, Faculty of Veterinary Medicine, Aristotle University of Thessaloniki, 54124 Thessaloniki, Greece; slioliopo@vet.auth.gr (S.L.); arsenosg@vet.auth.gr (G.A.)
[2] Laboratory of Food Chemistry and Technology, School of Chemistry, Aristotle University of Thessaloniki, 54124 Thessaloniki, Greece; niknen@chem.auth.gr (N.N.); ioannapyrka@chem.auth.gr (I.P.); aggelkal@chem.auth.gr (A.G.K.)
[3] Laboratory of Nutrition, Faculty of Veterinary Medicine, Aristotle University of Thessaloniki, 54124 Thessaloniki, Greece; panitsid@vet.auth.gr (I.P.); igiannenas@vet.auth.gr (I.G.)
[4] Institute of Animal Science, Hellenic Agricultural Organisation-DEMETER, 58100 Giannitsa, Greece; gsymeon@elgo.gr
[5] Department of Pharmacognosy and Natural Products Chemistry, Faculty of Pharmacy, University of Athens, 15771 Athens, Greece; skaltsounis@pharm.uoa.gr (A.-L.S.); stathopan@pharm.uoa.gr (P.S.)
[6] Laboratory of Pathology, Faculty of Veterinary Medicine, Aristotle University of Thessaloniki, 54124 Thessaloniki, Greece; stylioan@vet.auth.gr
[7] Department of Animal Science, School of Agricultural Sciences, University of Thessaly, 41500 Larissa, Greece; dgalamatis@uth.gr
[8] Laboratory of Evaluation of Human Biological Performance, School of Physical Education and Sport Science at Thessaloniki, Aristotle University of Thessaloniki, 54124 Thessaloniki, Greece; apet@phed.auth.gr
* Correspondence: geopaps@vet.auth.gr

Abstract: The objective of the present study was to evaluate the effects of an olive leaf extract obtained with an up-to-date laboratory method, when supplemented at different levels in laying hens' diets, on egg quality, egg yolk antioxidant parameters, fatty acid content, and liver pathology characteristics. Thus, 96 laying hens of the ISA-Brown breed were allocated to 48 experimental cages with two hens in each cage, resulting in 12 replicates per treatment. Treatments were: T1 (Control: basal diet); T2 (1% olive leaf extract); T3 (2.5% olive leaf extract); T4 (Positive control: 0.1% encapsulated oregano oil). Eggshell weight and thickness were improved in all treatments compared to the control, with T2 being significantly higher till the end of the experiment ($p < 0.001$). Egg yolk MDA content was lower for the T2 and T4 groups, while total phenol content and Haugh units were greater in the T2. The most improved fatty acid profile was the one of T3 yolks. The α-tocopherol yolk content was higher in all groups compared to T1. No effect was observed on cholesterol content at any treatment. Based on the findings, it can be inferred that the inclusion of olive leaf extract at a concentration of 1% in the diet leads to enhancements in specific egg quality attributes, accompanied by an augmentation of the antioxidant capacity.

Keywords: olive; olive byproduct; olive leaf extract; poultry; laying hens; antioxidants; egg quality

1. Introduction

The prohibition imposed by the European Union on the utilization of antibiotic growth promoters, coupled with the increased consciousness among the public regarding the quality standards associated with chicken products, has spurred manufacturers to investigate natural feed additives as potential substitutes for antibiotics [1]. It is generally acknowledged that eggs are the primary animal protein source for humans and are highly nutritious, exhibit a multitude of valuable biochemical properties, and are known worldwide for their substantial antioxidant capabilities [2]. Waste materials and byproducts are used to extract beneficial nutrients to meet customer and societal needs for high-quality, safe, and environmentally friendly processed foods [3–5]. Laying hens may also be exposed to stimuli that increase oxidative stress at different phases of production. Oxidative stress affects laying hens' health and productivity [6,7]. Towards this direction, studies using antioxidant supplements were able to reverse the color of the shell caused by oxidative stress [6,8,9]. Thus, various naturally occurring phytochemicals provided through feed have received attention as poultry antioxidants in recent years [10].

It is known that spices and herbs exhibit the most substantial concentration of polyphenol chemicals when measured by weight [11]. Polyphenols can impede oxidation by preventing free radical production. They may also inhibit oxidation by scavenging free radicals [12]. Furthermore, they improve the antioxidant status of animals by raising the levels of vitamins and antioxidant enzymes in the blood and muscles [13,14]. Several beneficial effects in laying hens have been observed. Specifically, supplementing layer chickens' diets with tea polyphenols enhanced their productivity, the quality of their eggs, and their ability to withstand induced oxidative challenge [9]. Administering tea polyphenols to laying hens enhanced egg production and egg albumen quality [15]. On the other hand, elevated levels of tea polyphenols had a negative impact on the quality of both eggshell and albumen [16]. Elsewhere, green tea polyphenols improved the shape and antioxidant capacity of the uterine in layers subjected to induced oxidative stress, thereby improving eggshell color [9,17]. Others have demonstrated that laying hens' performance, egg quality, and intestine morphometric characteristics were all enhanced by oregano essential oil dietary supplementation [18].

Polyphenol-rich plant leaves have also been studied. The rate and quality of eggs increased with the addition of eucalyptus leaves to the diet. It also boosted the hens' health and blood antioxidant levels. Eucalyptus leaves in the diet protected chickens' liver cells against ethanol-induced oxidative damage [19]. The use of powdered eucalyptus leaves increased the number, mass, feed conversion ratio, and breaking strength of eggs while lowering the laying chickens' heterophil/lymphocyte ratio [20]. More recently, in a separate study, the utilization of Mulberry leaf extract resulted in a reduction of yolk triglyceride and total cholesterol contents while concurrently enhancing both egg yolk color and eggshell strength [21]. Olive leaves possess a substantial antioxidant potential due to their abundance of secoiridoids, simple phenols, phenylethanoids, hydroxycinnamic acid derivatives, and flavonoids [22]. In olive varieties, oleuropein is typically the most abundant phenol. As a component of the phenolic segment of olive leaves, oleuropein is readily isolated [23]. Ahmed et al. (2018) [24] showed that egg yolk color was higher when the hens were fed the greatest level of leaf extract (150 mg/kg) supplementation. Egg yolk cholesterol and saturated fatty acid concentrations were lower for those hens supplemented with the extract. On the other hand, n-3 and n-6 fatty acids were higher while increasing levels of the extract fed [24]. The olive leaf or the use of its extract had no effect on laying hen performance, according to Rezar et al. [25]. In comparison to those fed unsupplemented diets, laying hens fed diets with olive leaf powder had higher body weights and darker egg yolks [26]. Elsewhere, olive leaf powder added to the diet of Japanese quails increased egg production while decreasing metabolic parameters such as serum lipids and cholesterol [27]. Recently, our research group reported that the supplementation of olive leaf extract at the level of 1% in broilers alleviated oxidation in broiler meat [5]. Therefore, it can be deduced that any oxidative challenge experienced by laying hens will have an impact on both their metabolism and the parameters associated with egg quality. This aspect holds significance from a welfare perspective, as contemporary consumers demonstrate concern not only for the quality of obtained products but also for the well-being and health of animals. It should be noted that the results of earlier studies are inconsistent, and only a few have followed a conclusive approach, investigating the analytical composition and polyphenol content of the olive leaf extract and the respective effect in the feed and the eggs produced. Overall, it can be hypothesized that olive leaf extract supplementation could be beneficial for laying hens' health and egg quality characteristics. In view of this, we have implemented in the current study the use of an olive leaf extract that was obtained with an up-to-date laboratory method to ensure an adequate concentration of active phenolic compounds. Moreover, to investigate whether egg quality characteristics could be negatively affected by a pro-oxidant effect of polyphenols, the tested olive leaf extract was evaluated at two levels of supplementation. An additional experimental group, supplemented with oregano essential oil, was used as a positive control group based on the documented effects on laying hens' performance.

2. Materials and Methods

2.1. Ethical Considerations

The experimental procedures were approved by the Research Committee of Aristotle University of Thessaloniki, Greece (approval number 246648/15-10-2021; project number 72623). The experiment's animal phase was designed with careful consideration of all welfare factors specified in the Good Farming Practice Guidelines, according to Directive 2010/63/EC and Commission recommendation 2007/526/EC.

2.2. Raw Materials, Animals, Diets and Experimental Design

The extraction and the HPLC–DAD analyses were carried out at the Department of Pharmacognosy and Natural Products Chemistry, Faculty of Pharmacy, University of Athens, Greece. The extraction of olive leaves was performed according to our previous work [5], and the extract obtained was condensed and dried prior to HPLC-DAD analyses, following COI/T.20/Doc. No 29/Rev.1 2017 elution protocol [28]. The extract contained the active ingredients oleuropein (22.84 g/100 g extract), luteolin-7-O-glucoside, hydroxytyrosol, and verbascoside (<0.8 g/100 g extract), and also triterpenic acids (maslinic and oleanolic acid, 4.97 and 1.08 g/100 g extract, respectively).

The oregano oil used in the present study was prepared by the vis-Naturalis company. According to the latter, it contained 78–85% carvacrol. It was derived after microencapsulation [29] following a spray drying technique, which provided a white powder as the end product.

The investigation was conducted in Galatista, a municipality in Chalkidiki, Greece, in a certified poultry house. The experimental enclosures utilized for accommodating the laying hens were of dimensions 41 cm by 41 cm, resulting in an area of 840.5 cm^2 per hen. This allocation of space exceeded the minimum requirement mandated by the European Union legislation (750 cm^2 available space per hen). In total, 96 laying hens of the ISA-Brown breed, aged 45 weeks, were allocated to 48 experimental cages with two hens in each cage. Each treatment consisted of 12 replicate cages. Diets were formulated to meet nutrient specifications provided by the specified laying hen breeder's guidelines [30]. The average temperature during the experimental period was maintained between 18–22 °C, and relative humidity was registered between 55–70%. A 14 to 16 h of light duration was applied based on the specified laying hen breeder's guidelines [30]. The experimental duration was 6 weeks, preceded by a one-week adaptation period. The dietary treatments included T1-control, no supplement test product; T2-1% olive leaf extract supplementation; T3-2.5% olive leaf extract supplementation; T4-0.1% oregano oil supplementation. All diets were prepared by Vis-Naturalis, Gennimata 17 str-Kalamaria, Thessaloniki, 55132, Greece. The main ingredients and calculated nutrient analysis of the diets are shown in Table 1.

Table 1. Ingredients and nutrients of the diet fed to laying hens during the experimental period.

Ingredients (%)	
Wheat soft	6.0
Corn	51.7
Soybean meal (47% crude protein content)	22.0
Wheat bran	7.5
Soybean oil	1.25
Limestone	9.65
Monocalcium phosphate	0.7
Sodium chloride	0.25
Sodium bicarbonate	0.24
DL-Methionine	0.19
Choline	0.08
Premix of vitamins and minerals *	0.44
Calculated analysis (%)	
Crude protein	17.0
Crude fiber	2.65
Crude fat	3.78
Crude ash	13.05
Metabolizable Energy (kcal/kg of diet)	2764

* Provided per kg diet: Retinyl acetate: 4.2 mg; Cholecalciferol: 0.1 mg; α-tocopherol acetate: 31.25 mg; Menadione: 5.0 mg; Cyanocobalamin: 0.025 mg; folic acid: 1.0 mg; Choline chloride: 450 mg; Pantothenic acid: 12.5 mg; Riboflavin, 6.25 mg; Nicotinic acid: 43.75 mg; Thiamin: 3.0 mg; D-biotin: 0.1 mg; Pyridoxine: 5.0 mg; Manganese: 125 mg; Zinc: 112 mg; Iron: 62 mg; Copper: 10 mg; Iodine: 1.0 mg; Selenium: 0.15 mg.

2.3. Feed Color, a_w, and Proximate Composition Analyses

Feed samples of 400 g were obtained from the main batch (1 tn) and separated into three homogeneous parts. Each sub-sample obtained was analyzed for color (MiniScan XE Plus D/8S Color Analyzer Colorimeter Spectrophotometer, Hunterlab, VA, USA), and ΔE values were evaluated [31]. Regarding a_w, it was measured thrice at 25 °C using an Aqualab 3TE water activity meter (Decagon Devices Inc., Pullman, WA, USA). Measurements of moisture, crude fat, crude protein, and crude ash were based on a previous method [32].

2.4. Feed Fatty Acid Profile Analysis

The fatty-acid content of feed samples was analyzed with gas chromatography, according to Vasilopoulos et al. [33].

2.5. Feed Antioxidant Parameters

Initially, each feed sub-sample was extracted once. The extracts obtained were further processed. The antioxidant parameters measured were the following: Total phenol content (TPC), Total flavonoid content (TFC), DPPH• scavenging, and Cupric ion Reducing Antioxidant Capacity (CUPRAC), as in previously published methodology [5,34].

2.6. Feed Content in Oleuropein

The methanolic extracts obtained were analyzed chromatographically using a Shimadzu Nexera X2 UHPLC System (Shimadzu Corporation, Kyoto, Japan). More details about the UHPLC system are described in our previous published work [5]. For UHPLC data acquisition and analysis, the exact methods are further described in Tsimidou et al. (2019) [35]. Quantification of oleuropein was carried out via the construction of a calibration curve (y = 617.68x + 4947.3, R^2 = 0.9989) using a set of standard solutions (4.55–910 mg/L) analyzed at 280 nm.

2.7. Characterization of Essential Oil Content and Composition

Hydrodistillation was used for oil extraction and disruption of the wall material. The procedures were according to a previous study [36].

2.8. Olive Leaf Extract Characterization

For $α_w$, moisture content, TPC, TFL, and antioxidant capacity of olive leaf extract, the methods followed were the same as described previously for feed extracts. The content of individual constituents was characterized chromatographically (see Section 2.3).

2.9. Egg Quality Parameters

The following egg quality parameters were measured: egg weight (g), yolk weight (g), albumen weight (g), eggshell weight (g), eggshell thickness (mm), longitudinal and transverse axes (mm), shape index, eggshell color, yolk color and Haugh units. Yolk color was measured both with the DSM YolkFanTM scale and by a Chroma meter CR-410 (Konica Minolta, Osaka, Japan), which was used for L*, a*, and b* color values evaluation.

2.10. Yolk Lipid Oxidation

It was assessed by TBARS assay (Thiobarbituric Acid Reactive Substances). Yolk samples weighing 1 g were homogenized with 8 mL of 5% Trichloroacetic acid (TCA) and 5 mL of 0.8% Butylated hydroxytoluene (BHT) solution in hexane. The homogenates were centrifuged (3000 rpm, 3 min), and following centrifugation, 2.5 mL of the bottom layer were collected. The 2.5 mL aliquots collected were transferred in tubes, and 1.5 mL of 0.8% Thiobarbituric Acid aqueous solution was added to each tube. Next, the tubes were incubated in a water bath (70 °C, 30 min) and cooled immediately after the incubation. Absorbance was measured in a spectrophotometer (UV-1700 PharmaSpec, Shimadzu, Japan) at 532 nm. Results were measured as ng of Malondialdehyde (MDA) per gram of yolk (ng MDA/g yolk).

2.11. Yolk Total Phenol Content

The total phenol content of yolks was measured by the Folin–Ciocalteu assay, according to a protocol by Shang et al. [37]. Results were presented as µg of Gallic Acid equivalent (GAE) per g of dried yolk.

2.12. Yolk Total Antioxidant Capacity (TAC) (Phosphomolybdate Method)

Yolk extracts were prepared with the same method described for Total Phenol Content. The phosphomolybdate method was applied to assess TAC, following the protocol by Prieto et al. [38]. Results were expressed as TAC (%).

2.13. Yolk Fatty Acid Profile Analysis

Gas chromatography was used for fatty acid profile analysis of egg yolk. Before extraction, 10 µL of 200 mg/mL pentadecanoic acid (Sigma, St. Louis, MO, USA) in chloroform as internal standard were mixed with 50 mg of yolk. Extraction was performed according to Folch et al. [39].

2.14. Yolk Cholesterol and α-Tocopherol Content

Determination of cholesterol (CHO) and α-Tocopherol (α-T) content in egg yolks was performed in accordance with the protocol of Botsoglou et al. (1998) [40]. Normal phase HPLC-DAD-FLD was performed for the simultaneous determination of CHO and α-T in egg yolk samples. The system used for Chromatographic analysis was a SpectraSystem SCM1000 HPLC system, equipped with a SpectraSYSTEM P4000 pump, a MIDAS autosampler from Spark (Waanderveld, Holland), a communication unit SpectraSYSTEM SN4000, a UV-visible diode array Thermo Separation Products UV6000 detector (Thermo Fischer Scientific, Waltham, MA, USA) coupled to an FL2000/FL3000 FASMA 502 fluorescence detector supplied by Rigas Labs (Thessaloniki, Greece). Data analysis was performed using the ChromQuest ver. 5.0 software (ThermoFischer Scientific, Waltham, MA, USA). The separation was carried out on a normal phase, 250 × 4 mm i.d., 5 µm, LiChrospher-SI column (MZ Analyzentechnik, Mainz, Germany). The elution system consisted of a mixture of hexane:2-propanol (99:1 v/v), with the following elusion conditions: isocratic elution; flow rate of 1.1 mL/min; injection volume: 20 µL. Standard solutions of CHO (100–1000 µg/mL) and α-T (1–10µg/mL) were prepared in hexane. Prior to injection, they were filtrated through PTFE hydrophobic filters (25 mm, 0.45 um) prior to injection. Samples were saponified twice and then analyzed in duplicate (CV% $\leq$ 2, $n = 2 \times 2$). Quantification was carried out using a CHO calibration curve at 210 nm ($y = 316.99x + 2E + 06$, $R2 = 0.9907$) and α-T calibration curve and fluorescence detection (λex 294 nm; λem-1 330 nm) ($y = 1328.8x + 6487.4$, $R2 = 0.9905$).

2.15. Pathology Evaluation

Randomly selected laying hens ($n = 6$ per treatment) were euthanized and subjected to a complete necropsy within 30 min after slaughter. The condition of the carcasses displayed minimal to no signs of autolysis. The liver and genital tract were grossly examined. Livers were examined for enlargement, discoloration, and variations in texture consistency. During the necropsy, liver samples were collected, fixed in 10% neutral buffered formalin, and stained with hematoxylin and eosin (HE). The liver sections were screened for six known microscopic alterations, including hepatocellular vacuolization, hepatocellular necrosis, vascular lesions, inflammation, and biliary reaction. Each histologic modification was scored with a categorical scoring from 0 to 3. Within the scoring system, a score of 0 corresponded to no detection of the specific lesion, while a score from 1–3 denoted greater levels of tissue damage and/or the extent of the observed changes.

2.16. Statistical Analysis

Statistical analysis of data was performed with the use of SPSS software (SPSS 25.0 Version, Chicago, IL, USA). Statistical difference was set at $p < 0.05$, and results were presented as average values ± standard deviation (SD). For analyzing the effects of treatments on the tested variables, one-way ANOVA was employed. Post hoc evaluation was conducted with Tukey's test. The effect of the treatments on hepatic vacuolization scoring in laying hens was evaluated with Chi-square analysis.

3. Results

3.1. Feed Color, a_w, and Proximate Composition

The color values L*, a*, b*, water activity (a_w), and proximate composition of feed samples from the treatments are presented in Table 2.

Table 2. Feed color, a_w, and proximate composition.

Parameters	Treatments				p-Value
	T1	T2	T3	T4	
L*	44.9 ± 0.2 [a]	43.4 ± 0.1 [b]	41.2 ± 0.3 [c]	45.2 ± 0.3 [a]	<0.001
a*	3.2 ± 0.2 [a]	2.3 ± 0.1 [b]	2.4 ± 0.3 [b]	3.0 ± 0.1 [a]	<0.001
b*	14.8 ± 0.1 [a]	15.6 ± 0.2 [b]	16.5 ± 0.5 [c]	14.26 ± 0.1 [a]	<0.001
ΔE	-	1.9 ± 0.2 [a]	4.1 ± 0.3 [b]	0.6 ± 0.2 [c]	<0.001
Water activity (a_w)	0.583 ± 0.012 [a]	0.594 ± 0.004 [a,b]	0.584 ± 0.002 [a]	0.601 ± 0.004 [b]	0.032
% Moisture	10.23 ± 0.18	10.24 ± 0.04	10.30 ± 0.51	10.25 ± 0.30	0.852
% Crude fat	2.7 ± 0.2	2.9 ± 0.2	2.7 ± 0.2	2.8 ± 0.1	0.384
% Crude protein	17.3 ± 0.1 [a]	16.5 ± 0.6 [a,b]	15.9 ± 0.3 [b]	16.2 ± 0.4 [b]	0.018
% Crude ash	13.0 ± 0.1 [a]	14.0 ± 0.2 [b]	13.5 ± 0.2 [c]	11.6 ± 0.5 [d]	<0.001

T1: Control; T2: basal diet with 1% olive leaf extract; T3: basal diet with 2.5% olive leaf extract; T4: basal diet with 0.1% encapsulated oregano oil. Values are means ± SD ($n = 5$ for L* to ΔE, $n = 3$ for all proximate composition values). Values in the same row with different superscripts differ significantly ($p < 0.05$).

Calculation of the ΔE values for the three different feeds compared to T1 indicated that the addition of the olive leaf extract affected the appearance of the feeds due to its yellow–green coloration and no influence was found for the addition of the encapsulated oregano oil. In agreement with previous findings for preparing broilers' feeds [5], the corresponding ΔE values indicated that mixing was adequate and that T3 contained a higher dose of the extract than T2. Although higher than those obtained in our previous work, water activity and moisture content values were still relatively low and in agreement with published values [41]. Fat, protein, and ash content were rather similar in all treatments.

3.2. Feed Fatty Acid Profile

Results obtained from fatty acid profile analysis of feed samples from the dietary treatments are presented in Table 3.

Table 3. Fatty acid composition (%) of crude fat contained in feed samples.

Fatty Acid	Treatments				p-Value
	T1	T2	T3	T4	
16:0	10.9 ± 0.2	10.5 ± 0.2	10.7 ± 0.1	10.9 ± 0.2	0.075
18:0	1.3 ± 0.2	1.3 ± 0.2	1.5 ± 0.1	1.6 ± 0.2	0.182
18:1 (n-9)	32.6 ± 0.7	33.9 ± 0.8	32.3 ± 0.7	32.8 ± 0.0	0.065
18:2	54.8 ± 0.4 [a,b]	53.9 ± 0.3 [b]	55.0 ± 0.6 [a]	54.3 ± 0.3 [a]	0.046
18:3 (n-6)	0.4 ± 0.0	0.6 ± 0.0	0.5 ± 0.1	0.5 ± 0.1	0.052

T1: Control; T2: basal diet with 1% olive leaf extract; T3: basal diet with 2.5% olive leaf extract; T4: basal diet with 0.1% encapsulated oregano oil. Values are means ± SD (n = 3). Values in the same row with different superscripts differ significantly ($p < 0.05$).

The fat content of the feed was rich in linolenic acid (~55%), followed by oleic acid (~33%), and a content in saturates approx. 12%, as expected, considering the ingredients in Table 1.

3.3. Feed Antioxidant Parameters

Oleuropein content (OLE), total phenol content (TPC), total flavonoid content (TFL), and in vitro antioxidant activity (DPPH• scavenging and cupric ion reducing antioxidant capacity, CUPRAC) of the feeds are given in Table 4.

Table 4. Oleuropein content (OLE), total phenol content (TPC), total flavonoid content (TFL), DPPH scavenging (DPPH), and cupric ion reducing antioxidant capacity (CUPRAC) of feed samples.

Parameters	Treatments				p-Value
	T1	T2	T3	T4	
OLE (mg/kg)	-	1852 ± 15 [a]	4283 ± 27 [b]	-	<0.001
TPC (mg GAE/g)	1.07 ± 0.04 [a]	2.80 ± 0.05 [b]	5.10 ± 0.06 [c]	1.16 ± 0.04 [d]	<0.001
TFL (μg QUE/g)	26.9 ± 1.5 [a]	192.7 ± 2.2 [b]	293.4 ± 1.9 [c]	33.8 ± 1.3 [d]	<0.001
DPPH• (μmol TE/g)	4.7 ± 0.1 [a]	14.1 ± 0.2 [b]	27.8 ± 0.4 [c]	5.2 ± 0.1 [d]	<0.001
CUPRAC (μmol TE/g)	11.1 ± 1.2 [a]	27.8 ± 2.2 [b]	48.4 ± 1.0 [c]	12.3 ± 1.3 [a]	<0.001

T1: Control; T2: basal diet with 1% olive leaf extract; T3: basal diet with 2.5% olive leaf extract; T4: basal diet with 0.1% encapsulated oregano oil. Values are means ± SD (n = 3). Values in the same row with different superscripts differ significantly ($p < 0.05$).

Oleuropein was present in the diets that contained olive leaf extract (T2, T3) in a dose-dependent manner (in T3, ~2.3-fold higher OLE level was found compared to T2), as verified by HPLC. Due to the presence of the extract, TPC, TFL, and the antioxidant potential of the feeds were elevated compared to T1 and the feed containing the encapsulated oregano oil (T4). In the latter, all values were comparable to those of the T1. T3, as expected, was the richest in bioactives and with a higher antioxidant potential.

3.4. Characterization of Essential Oil Content and Composition

The GC–MS analysis revealed that thymol and carvacrol were the main components of the essential oil. The major compounds identified are presented in Table 5. Antioxidant parameters evaluated in essential oil are also presented in Table 5.

Table 5. Essential oil content, composition, total phenol content (TPC), DPPH scavenging (DPPH), and cupric ion reducing antioxidant capacity (CUPRAC).

Essential Oil Major Compounds (%)	
Carvacrol	63.91
Thymol	24.9
Caryophyllene	2.75
Caryophyllene oxide	2.68
Bisabolene	2.68
Borneol	1.68
Terpineol	0.79
Trans dihydrocarvone	0.26
Essential Oil Antioxidant Parameters	
TPC (mg GAE/g oil)	60.2 ± 8.8
DPPH$^\bullet$ (mmol TE/g oil)	0.42 ± 0.03
CUPRAC (mmol TE/g oil)	0.68 ± 0.05

Values for antioxidant parameters are means $\pm$ SD ($n = 3$).

3.5. Olive Leaf Extract Characterization

The parameters measured in olive leaf extract are presented in Table 6.

Table 6. Olive leaf extract water activity (a_w), total phenol content (TPC), total flavonoid content (TFL), DPPH scavenging (DPPH), and cupric ion reducing antioxidant capacity (CUPRAC).

Olive Leaf Extract Characterization	
Water activity (a_w)	0.238
TPC (mg GAE/g)	161.89 ± 7.0
TFL (µg QUE/g)	16.71 ± 0.07
DPPH$^\bullet$ (µmol TE/g)	1.05 ± 0.09
CUPRAC (µmol TE/g)	1.53 ± 0.13

Values for antioxidant parameters are means $\pm$ SD ($n = 3$).

3.6. Egg Quality Parameters

For the overall experimental period, all treated eggs (T2, T3, T4) had improved eggshell weight and eggshell thickness than T1 ($p < 0.001$ for both parameters). The transverse axis was shorter in the T4 group ($p = 0.022$). Shape index was increased in T2 eggs ($p = 0.029$). T4 treatment affected eggshell color, which became lighter compared to the other treatments ($p < 0.001$). Haugh units were increased in T2 treatment ($p < 0.001$). Regarding egg yolk color, it was lighter in T3 yolks ($p = 0.011$), an effect which was also apparent by the lower a* values observed for the T3 treatment ($p = 0.03$). Yolk lightness (L*) was higher in the T2 group compared to the T4 group ($p = 0.014$). The results are presented in Table 7 for the 1st, 4th, and 6th week and the overall period. Overall period values were calculated, taking into account the measurements from all experimental weeks. As shown in Table 7, most egg quality parameters were not affected at the end of the 1st week, except for yolk color, which was lighter in the T2 and T3 groups in comparison with T1 ($p = 0.031$). At the end of the 4th week, the lightness (L*) of egg yolk was higher in T3 and T4 groups compared to T2 ($p < 0.001$). Moreover, the changes in eggshell weight and thickness became apparent in the 4th week's measurements, with all treatments positively affecting eggshell weight and thickness compared to T1 ($p = 0.004$ and $p = 0.012$, respectively). At the end of the 6th week, all treatments continued to result in numerically higher eggshell weight and thickness compared to T1, but significant effects were noticed only for T2 treatment ($p < 0.001$ for both parameters), as shown in Table 7.

Table 7. Egg quality parameters in the 1st, 4th, and 6th week of the experiment and for the overall experimental period.

Parameters	T1	T2	T3	T4	p-Value
		1st Week			
Egg weight (g)	65.9 ± 4.92	64.7 ± 6.46	63.3 ± 5.59	62.6 ± 6.31	0.550
Yolk weight (g)	16.4 ± 0.60	16.3 ± 1.37	15.9 ± 1.20	16.4 ± 1.48	0.774
Albumen weight (g)	43.3 ± 4.29	42.2 ± 5.77	41.1 ± 4.09	39.8 ± 5.58	0.400
Eggshell weight (g)	6.19 ± 0.644	6.27 ± 0.859	6.25 ± 1.098	6.40 ± 0.714	0.949
Eggshell thickness (mm)	0.41 ± 0.040	0.40 ± 0.052	0.40 ± 0.053	0.41 ± 0.029	0.957
Longitudinal axis (mm)	59.5 ± 2.99	57.8 ± 1.89	58.6 ± 1.90	57.7 ± 1.92	0.183
Transverse axis (mm)	45.3 ± 1.25	45.4 ± 1.81	44.9 ± 1.09	44.8 ± 1.72	0.734
Shape index	76.3 ± 4.37	78.6 ± 2.76	76.6 ± 2.91	77.7 ± 2.65	0.315
Eggshell color	17.1 ± 2.40	18.6 ± 3.47	17.4 ± 3.45	19.4 ± 2.44	0.274
Yolk color fan score	12.6 ± 0.51 [a]	11.8 ± 1.03 [b]	11.8 ± 0.94 [b]	12.1 ± 0.69 [ab]	0.031
Haugh units	83.2 ± 8.68	88.1 ± 9.17	82.8 ± 7.12	85.2 ± 3.97	0.320
L*	74.7 ± 1.99	76.3 ± 2.18	74.7 ± 2.59	76.0 ± 1.11	0.133
a*	20.9 ± 1.77	19.3 ± 2.72	18.1 ± 3.03	18.8 ± 2.22	0.069
b*	55.1 ± 4.64	57.1 ± 4.54	53.4 ± 6.84	55.3 ± 5.30	0.438
		4th Week			
Egg weight (g)	64.1 ± 5.52	66.0 ± 5.59	65.4 ± 3.70	63.3 ± 7.12	0.318
Yolk weight (g)	17.6 ± 1.75	17.4 ± 1.99	16.7 ± 1.51	17.4 ± 1.04	0.496
Albumen weight (g)	42.6 ± 5.12	41.7 ± 5.08	41.9 ± 3.25	41.4 ± 7.93	0.953
Eggshell weight (g)	5.52 ± 0.819 [b]	6.65 ± 1.026 [a]	6.43 ± 0.391 [a]	6.30 ± 0.808 [a]	0.004
Eggshell thickness (mm)	0.36 ± 0.043 [b]	0.40 ± 0.048 [a]	0.40 ± 0.018 [a]	0.40 ± 0.036 [a]	0.012
Longitudinal axis (mm)	58.9 ± 2.14	58.5 ± 2.20	58.9 ± 1.51	59.0 ± 3.00	0.954
Transverse axis (mm)	44.9 ± 1.46	45.5 ± 1.72	45.3 ± 1.18	44.8 ± 2.16	0.626
Shape index	76.2 ± 2.96	77.8 ± 2.80	76.9 ± 1.78	76.0 ± 2.54	0.238
Eggshell color	19.0 ± 3.56	17.9 ± 4.19	16.1 ± 1.91	19.0 ± 2.83	0.077
Yolk color fan score	11.4 ± 0.77	11.9 ± 0.73	11.5 ± 0.94	12.0 ± 0.82	0.147
Haugh units	75.6 ± 9.13	79.4 ± 9.96	75.3 ± 7.22	74.6 ± 6.13	0.425
L*	76.1 ± 3.12 [ab]	74.0 ± 2.61 [b]	78.5 ± 1.70 [a]	77.6 ± 1.60 [a]	<0.001
a*	17.1 ± 2.78	17.3 ± 2.29	16.3 ± 2.63	19.0 ± 2.62	0.075
b*	58.0 ± 6.21	56.8 ± 5.48	60.9 ± 4.07	60.7 ± 2.45	0.079
		6th Week			
Egg weight (g)	64.7 ± 6.50	67.8 ± 5.26	63.2 ± 3.96	62.1 ± 8.14	0.116
Yolk weight (g)	17.0 ± 2.66	17.3 ± 1.77	15.9 ± 1.08	16.6 ± 1.86	0.271
Albumen weight (g)	42.2 ± 3.83	43.6 ± 4.73	41.2 ± 3.03	39.6 ± 6.91	0.211
Eggshell weight (g)	5.56 ± 0.935 [b]	6.95 ± 0.933 [a]	6.14 ± 0.538 [b]	5.95 ± 0.657 [b]	<0.001
Eggshell thickness (mm)	0.36 ± 0.047 [c]	0.43 ± 0.052 [a]	0.40 ± 0.019 [b]	0.39 ± 0.027 [b]	<0.001
Longitudinal axis (mm)	59.9 ± 2.71	59.7 ± 2.24	58.6 ± 1.61	58.0 ± 2.55	0.116
Transverse axis (mm)	44.9 ± 1.39	45.8 ± 1.14	44.8 ± 1.03	44.7 ± 2.12	0.224
Shape index	75.1 ± 2.89	76.8 ± 2.57	76.6 ± 2.28	77.1 ± 1.80	0.145
Eggshell color	17.9 ± 4.81	18.8 ± 3.20	16.9 ± 3.46	19.2 ± 2.90	0.398
Yolk color fan score	12.6 ± 0.65	12.7 ± 0.75	12.2 ± 0.90	12.3 ± 0.95	0.303
Haugh units	80.9 ± 11.71	83.3 ± 7.53	76.7 ± 8.34	78.2 ± 9.92	0.305
L*	75.1 ± 1.68	75.8 ± 2.65	75.5 ± 3.02	76.3 ± 1.83	0.603
a*	19.3 ± 3.14	19.5 ± 3.28	18.1 ± 2.94	18.3 ± 2.58	0.577
b*	54.8 ± 4.92	56.7 ± 6.31	55.3 ± 5.75	52.5 ± 6.09	0.327

Table 7. Cont.

Parameters	Treatments				p-Value
	T1	T2	T3	T4	
	Overall Period				
Egg weight (g)	65.2 ± 5.82	66.1 ± 5.82	65.4 ± 4.34	63.8 ± 7.89	0.066
Yolk weight (g)	17.2 ± 2.05	17.1 ± 1.74	16.6 ± 1.45	16.7 ± 1.44	0.089
Albumen weight (g)	42.7 ± 4.30	42.3 ± 4.95	42.3 ± 3.27	40.8 ± 7.00	0.124
Eggshell weight (g)	5.86 ± 0.859 [b]	6.60 ± 0.952 [a]	6.34 ± 0.761 [a]	6.24 ± 0.816 [a]	<0.001
Eggshell thickness (mm)	0.37 ± 0.042 [b]	0.41 ± 0.048 [a]	0.40 ± 0.037 [a]	0.40 ± 0.036 [a]	<0.001
Longitudinal axis (mm)	59.1 ± 3.01	58.6 ± 2.28	59.1 ± 1.84	58.6 ± 2.54	0.337
Transverse axis (mm)	45.1 ± 1.39 [a]	45.6 ± 1.48 [a]	45.2 ± 1.05 [a]	44.8 ± 1.99 [b]	0.022
Shape index	76.5 ± 4.56 [b]	77.8 ± 2.68 [a]	76.5 ± 2.38 [b]	76.5 ± 2.38 [b]	0.029
Eggshell color	18.1 ± 4.25 [b]	17.9 ± 3.54 [b]	16.9 ± 3.16 [b]	20.4 ± 6.40 [a]	<0.001
Yolk color fan score	12.1 ± 0.88 [a]	12.1 ± 0.90 [a]	11.8 ± 0.83 [b]	12.1 ± 0.82 [a]	0.011
Haugh units	79.6 ± 10.15 [b]	83.9 ± 7.71 [a]	77.7 ± 8.18 [b]	78.3 ± 9.46 [b]	<0.001
L*	74.9 ± 3.11 [ab]	73.5 ± 4.09 [a]	74.7 ± 3.79 [ab]	75.4 ± 3.32 [b]	0.014
a*	18.6 ± 3.00 [a]	18.2 ± 2.85 [a]	17.0 ± 2.65 [b]	18.3 ± 2.72 [a]	0.003
b*	55.0 ± 5.46	55.1 ± 7.13	55.8 ± 6.19	54.8 ± 6.23	0.779

T1: Control; T2: basal diet with 1% olive leaf extract; T3: basal diet with 2.5% olive leaf extract; T4: basal diet with 0.1% encapsulated oregano oil. Values are means ± SD (n = 12 for the 1st, 4th, and 6th week and n = 36 for the overall period). Values in the same row with different superscripts differ significantly ($p < 0.05$)

3.7. Egg Yolk Lipid Oxidation, Total Phenol Content (TPC) and Total Antioxidant Capacity (TAC)

In Day 1 of analysis, MDA levels in yolk were significantly lower in T2 and T4 groups compared to T1 and T3 ($p < 0.001$). On Day 5 of the analysis, there were no differences in yolk MDA levels among the treatments. Results are presented in Table 8. Egg yolk TPC expressed as μg of Gallic Acid equivalents per gram of dry yolk (μg GAE/g) was higher in T2 treatment compared to T3 and T4 ($p = 0.026$). Results are presented in Table 8. As shown in Table 8, Total Antioxidant capacity (TAC) was increased in T1 and T2 treatments in comparison with T4 ($p = 0.003$).

Table 8. Egg yolk MDA content evaluated on the 1st day after collection (Day 1) and after maintaining the samples in a refrigerator for 5 days (Day 5), total phenol content (TPC), and total antioxidant capacity (TAC).

Parameters	Treatments				p-Value
	T1	T2	T3	T4	
MDA (ng MDA/g) Day 1	42.0 ± 17.42 [a]	11.8 ± 10.41 [b]	33.6 ± 12.70 [a]	17.1 ± 8.26 [b]	<0.001
MDA (ng MDA/g) Day 5	45.8 ± 28.38	25.3 ± 11.90	33.5 ± 13.95	33.7 ± 18.99	0.088
TPC (μg GAE/g dry yolk)	203.9 ± 74.64 [ab]	230.0 ± 76.40 [a]	171.2 ± 62.08 [b]	189.5 ± 68.36 [b]	0.026
TAC (%)	7.4 ± 2.15 [a]	7.8 ± 3.52 [a]	4.8 ± 6.06 [ab]	1.8 ± 1.05 [b]	0.003

T1: Control; T2: basal diet with 1% olive leaf extract; T3: basal diet with 2.5% olive leaf extract; T4: basal diet with 0.1% encapsulated oregano oil. Values are means ± SD (n = 12). Values in the same row with different superscripts differ significantly ($p < 0.05$).

3.8. Egg Yolk Fatty Acid Profile

Egg yolk fatty acid analysis showed that the majority of the individual fatty acids were not affected by the treatments. However, T3 treatment increased the proportions of Margaric (17:0), Linoleic (C18:2n6c), α-Linolenic (C18:3n3) and cis,cis-11,14-Eicosadienoic (C20:2n6) fatty acids in egg yolk compared to T1 ($p = 0.016$; $p = 0.010$; $p = 0.001$; $p = 0.020$, respectively). The results are presented in Table 9.

Table 9. Egg yolk fatty acid profile (%).

Fatty Acid	T1	T2	T3	T4	p-Value
C14:0	0.31 ± 0.043	0.32 ± 0.027	0.31 ± 0.027	0.34 ± 0.038	0.229
C14:1n5	0.10 ± 0.034	0.10 ± 0.026	0.08 ± 0.017	0.10 ± 0.032	0.618
C16:0	24.65 ± 1.320	24.43 ± 0.965	24.39 ± 1.060	24.99 ± 0.618	0.638
C16:1n9	0.70 ± 0.150	0.70 ± 0.096	0.69 ± 0.120	0.71 ± 0.104	0.972
C16:1n7	3.26 ± 0.723	3.23 ± 0.557	3.05 ± 0.569	3.41 ± 0.458	0.676
C17:0	0.17 ± 0.020 [b]	0.18 ± 0.014 [ab]	0.20 ± 0.019 [a]	0.17 ± 0.018[b]	0.016
C17:1n7	0.11 ± 0.014	0.11 ± 0.018	0.12 ± 0.009	0.11 ± 0.020	0.382
C18:0	8.07 ± 0.616	8.17 ± 0.386	8.49 ± 0.508	8.01 ± 0.447	0.242
C18:1n9t	0.20 ± 0.453	0.18 ± 0.022	0.16 ± 0.010	0.19 ± 0.028	0.054
C18:1n9c	43.53 ± 2.206	42.98 ± 1.429	41.32 ± 1.521	42.74 ± 1.782	0.101
C18:1n7c	2.51 ± 0.133	2.57 ± 0.177	2.34 ± 0.166	2.55 ± 0.232	0.072
C18:2n6t	0.04 ± 0.017	0.04 ± 0.007	0.03 ± 0.005	0.04 ± 0.010	0.813
C18:2n6c	11.95 ± 1.438 [b]	12.49 ± 0.866 [ab]	13.96 ± 1.017 [a]	11.93 ± 1.582 [b]	0.010
C20:0	0.04 ± 0.012	0.04 ± 0.015	0.05 ± 0.026	0.04 ± 0.008	0.330
C18:3n6	0.09 ± 0.021	0.09 ± 0.018	0.07 ± 0.025	0.09 ± 0.013	0.202
C18:3n3	0.33 ± 0.048 [b]	0.38 ± 0.044 [ab]	0.47 ± 0.097 [a]	0.34 ± 0.054 [b]	0.001
C18:4n3	0.03 ± 0.016	0.04 ± 0.013	0.04 ± 0.005	0.04 ± 0.009	0.409
C20:1n9	0.28 ± 0.020	0.28 ± 0.025	0.27 ± 0.025	0.26 ± 0.016	0.662
C21:0	0.02 ± 0.005	0.02 ± 0.004	0.02 ± 0.003	0.02 ± 0.004	0.418
C20:2n6	0.13 ± 0.034 [b]	0.15 ± 0.019 [ab]	0.17 ± 0.023 [a]	0.14 ± 0.030 [ab]	0.020
C22:0	0.08 ± 0.021	0.08 ± 0.016	0.07 ± 0.010	0.07 ± 0.010	0.384
C20:3n6	0.13 ± 0.022	0.13 ± 0.011	0.14 ± 0.015	0.13 ± 0.021	0.812
C20:4n6	1.76 ± 0.125	1.70 ± 0.189	1.80 ± 0.154	1.72 ± 0.139	0.550
C23:0	0.05 ± 0.031	0.04 ± 0.011	0.03 ± 0.004	0.04 ± 0.015	0.260
C20:5n3	0.02 ± 0.006	0.02 ± 0.005	0.02 ± 0.003	0.02 ± 0.004	0.746
C22:4n6	0.18 ± 0.027	0.18 ± 0.030	0.20 ± 0.062	0.19 ± 0.038	0.706
C22:5n6	0.58 ± 0.155	0.61 ± 0.113	0.65 ± 0.168	0.69 ± 0.216	0.594
C22:5n3	0.10 ± 0.018	0.010 ± 0.023	0.11 ± 0.036	0.11 ± 0.025	0.464
C22:6n3	0.60 ± 0.048	0.63 ± 0.099	0.73 ± 0.194	0.79 ± 0.271	0.125

T1: Control; T2: basal diet with 1% olive leaf extract; T3: basal diet with 2.5% olive leaf extract; T4: basal diet with 0.1% encapsulated oregano oil. Values are means ± SD (n = 8). Values in the same row with different superscripts differ significantly ($p < 0.05$).

Following the determination of major fatty acid categories, the statistical analysis results are presented in Figures 1 and 2. No differences were detected in saturated and unsaturated fatty acid percentages among treatments. The monounsaturated fatty acid (MUFA) percentage was lower ($p < 0.01$) in the T3 group compared to T1. It was also lower compared to T2 and T4 ($p < 0.05$). The opposite effect was observed for T3 treatment on total polyunsaturated fatty acids (PUFA), which were found to be significantly higher compared to T1 ($p < 0.01$). Moreover, PUFA levels were lower in T4 treatment compared to T3 ($p < 0.05$).

In comparison with T1, T3 treatment resulted in higher n-3 and n-6 fatty acid content in egg yolks ($p < 0.05$). Moreover, according to statistical analysis, n-6 fatty acid levels were lower in T4 treatment compared to T3 ($p < 0.05$). The n-6:n-3 ratio was not affected by the treatments, as shown in Figure 2C.

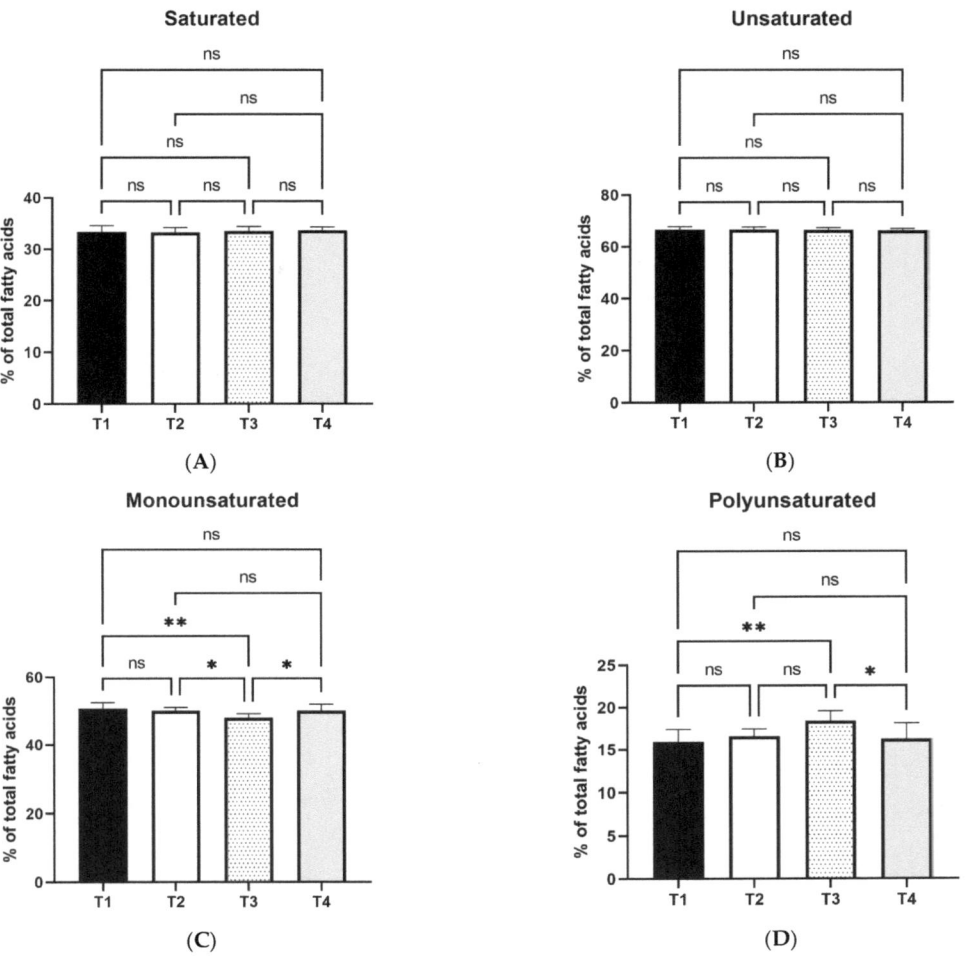

Figure 1. Effects of two levels of dietary olive leaf extract or oregano oil on egg yolk saturated (**A**), unsaturated (**B**), monounsaturated (**C**), and polyunsaturated (**D**) fatty acid percentage. T1: Control; T2: basal diet with 1% olive leaf extract; T3: basal diet with 2.5% olive leaf extract; T4: basal diet with 0.1% encapsulated oregano oil. *: mean values differ significantly between them ($p < 0.05$); **: mean values differ significantly between them ($p < 0.01$); ns: not significant. ($n = 8$).

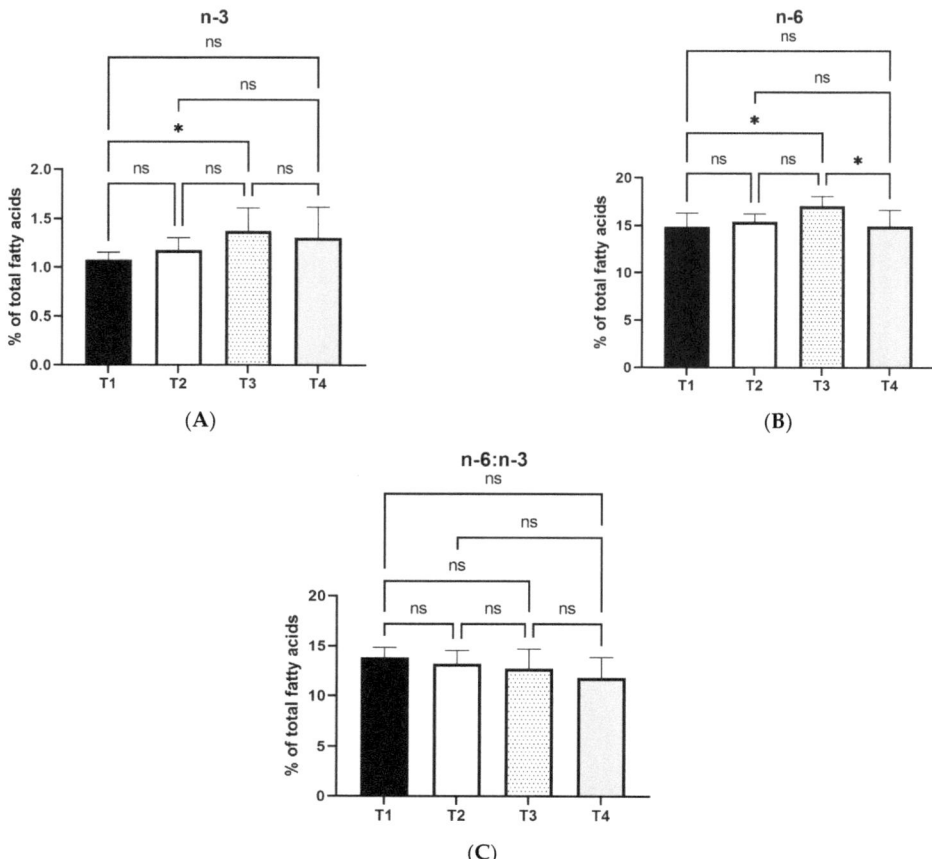

Figure 2. Effects of two levels of dietary olive leaf extract or oregano oil on n-3 (**A**), n-6 (**B**), and n-6:n-3 ratio (**C**) of egg yolks. T1: Control; T2: basal diet with 1% olive leaf extract; T3: basal diet with 2.5% olive leaf extract; T4: basal diet with 0.1% encapsulated oregano oil. *: mean values differ significantly between them ($p < 0.05$); ns: not significant. ($n = 8$).

3.9. Egg Yolk Cholesterol and α-Tocopherol Content

Yolk cholesterol content did not differ significantly among the groups (Table 10). Regarding α-tocopherol, all treatments had increased levels compared to T1 (Table 10). More precisely, according to statistical analysis, T3 yolks had significantly higher α-tocopherol content than T2 and T4, which in turn had significantly higher values than T1 ($p < 0.001$).

Table 10. Egg yolk cholesterol and α-tocopherol content.

	Treatments				
Parameters	T1	T2	T3	T4	*p*-Value
Cholesterol (mg/100 g)	620.1 ± 186.5	613.6 ± 173.6	617.5 ± 156.6	547.1 ± 167.4	0.805
α-tocopherol (mg/100 g)	2.43 ± 0.5 [c]	3.59 ± 0.9 [b]	4.81 ± 0.9 [a]	3.51 ± 0.9 [b]	<0.001

T1: Control; T2: basal diet with 1% olive leaf extract; T3: basal diet with 2.5% olive leaf extract; T4: basal diet with 0.1% encapsulated oregano oil. Values are means ± SD ($n = 8$). Values in the same row with different superscripts differ significantly ($p < 0.05$).

3.10. Pathology Evaluation

Only liver discoloration was grossly observed, while histologically, multifocal areas of hepatic vacuolization were noted. The severity and the extent of the lesions ranged from mild to moderate, representing scores 1 and 2 on the four-tier system employed. However, among the groups, the lesions did not exhibit any distinct pattern in their distribution and did not show statistically significant differences (Figure 3, Table 11).

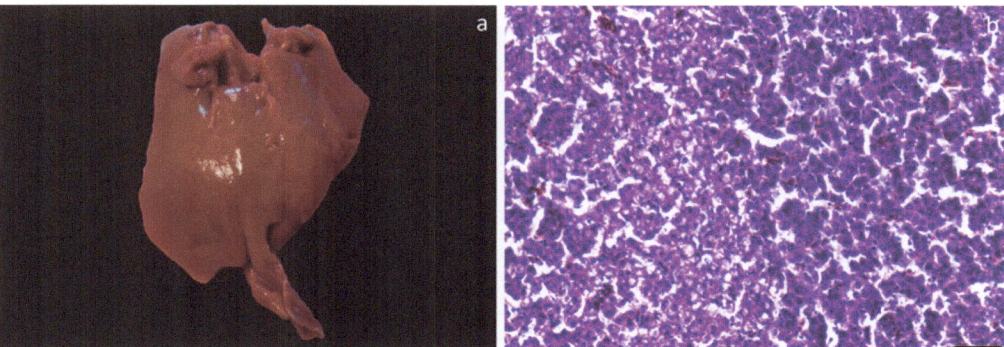

Figure 3. (**a**) Gross appearance of a mildly discolorated liver; (**b**) Focally extensive hepatocellular vacuolization (score 1). H&E, Bar = 50 μm.

Table 11. The impact of the treatments on hepatic vacuolization scoring in laying hens is depicted as the percentage (%) distribution of cases within each treatment group across scoring categories. Score 1 indicates a mild presence of the lesion, while score 2 corresponds to a moderate presence of the lesion.

	Treatments				
Score	T1	T2	T3	T4	Asymptotic Significance (X^2 Value)
1	100%	50%	83.3%	66.6%	0.217
2	0%	50%	16.7%	33.4%	

T1: Control; T2: basal diet with 1% olive leaf extract; T3: basal diet with 2.5% olive leaf extract; T4: basal diet with 0.1% encapsulated oregano oil (n = 6).

4. Discussion

As already mentioned, the olive supplement used in this study was a resin-purified aqueous isopropanol olive leaf extract, which was supplemented in laying hens in two levels. There was also a group supplemented with encapsulated oregano oil as a positive control. Oregano essential oil composition was generally in line with the literature for *Origanum vulgare* ssp. *hirtum* [42,43]. Given that the pure original oregano oil was unavailable for analysis, the values presented in Section 3.4 may be carefully evaluated. For instance, the absence of p-cymene and certain variations from the company's declaration might be connected to the procedure of isolating the material from the encapsulated form [44].

Starting with the results of the treatments on egg quality, eggshell weight was increased in all treatments compared to T1, while some eggshell shape parameters (transverse axis, shape index) were also affected by the treatments. These changes in eggshell parameters may be attributed to the mineral content of olive leaves and oregano, especially calcium (Ca) and Phosphorus (P). It is widely known that Ca is an essential mineral for eggshell formation and quality, as eggshells consist mainly of $CaCO_3$, with a percentage of about 94% [45]. It was found that Ca and P are among the major minerals found in olive leaves [46]. The latter authors also proposed olive leaf extract derived from three varieties as an affordable source of minerals such as Ca. In a previous study [47], two levels of olive oil were supplemented in laying hens, and the results showed increased eggshell-breaking strength and shell thickness. It was proposed by the latter authors that vitamin D, contained in olive oil, may be responsible for these effects; however, we cannot derive the same conclusion for our study, as the raw material used was different. In another study, olive pulp, a byproduct rich in minerals, resulted in a lower percentage of broken eggshells when supplemented in laying hens [4]. Possibly, a similar mechanism is involved in our study.

Oregano oil (T4) treatment resulted in lighter-colored eggshells, an effect that has not been previously reported, to our knowledge. Eggshell coloration is defined mostly genetically, but factors such as hen age, diseases, nutrition, stress, and environment can also affect it [48–50]. The main pigment responsible for the brown eggshell coloration is protoporphyrin IX, whereas biliverdin and its zinc chelates have little effect [51]. The last step in the protoporphyrin IX biosynthesis pathway is the auto-oxidation of protoporphyrinogen, a colorless molecule, to protoporphyrin IX [52]. In our case, possibly the antioxidants in oregano oil reduced this procedure, resulting in lighter-colored eggshells. On the other hand, no similar effects on eggshell coloration were noticed in the groups supplemented with olive leaf extract (T2, T3). The eggshells from these groups were slightly darker than those of T1. Even though olive leaf extract also contains high amounts of antioxidants, its different composition may be responsible for these inconsistencies. It has been reported that trace minerals such as iron (Fe) and magnesium (Mg) have positive effects with respect to brown eggshell coloration [50,53,54]. These minerals were found in high concentrations in olive leaves in the study of de Oliveira et al. (2023) [46], which provides a possible interpretation for our findings. Taking into account consumers' preference for darker eggshells, maintaining the eggshell coloration in eggs from hens supplemented with 1% and 2.5% olive leaf extract is an important finding.

Haugh units are the most widely used measure of egg internal quality [55]. It is known that they deteriorate during storage time, and they also depend on hen age, but nutrition does not appear to have any great effect on albumen quality [55]. In some cases, diet modification can improve Haugh units, as shown for rosehip and flaxseed meal supplementation in laying hens [56]. In the present study, T2 treatment resulted in increased Haugh units, which is contrary to previous studies regarding olive byproduct supplementation in laying hens, where Haugh units were unaffected [4,57–60]. The different types and inclusion levels of byproducts used in these studies may be responsible for these inconsistencies. Moreover, in a previous study [61], 9% olive pulp supplementation in laying hens resulted in eggs with lower Haugh units compared to the control and lower supplementation level group (4.5%). In even higher inclusion levels of olive pulp, Haugh units were lower compared to control in the studies of Mohebbifar et al. [62] and Afsari et al. [63]. In the present study, the positive effect on Haugh units was noticed in the lower inclusion level, which is in accordance with the aforementioned studies. This finding might be due to the antioxidant compounds of olive leaf extract, which probably reduced albumen quality deterioration by reducing the lipid and protein oxidation procedures in the lower inclusion level.

Another important parameter that affects consumers' preference is yolk coloration. Most European consumers prefer yolks with darker hues [64]. It is also known that yolk coloration depends on the accumulation of carotenoids in the diet, as carotenoids are synthesized de novo by some plant species, bacteria, algae, and fungi [65]. Laying hens cannot synthesize xanthophylls, which are the main carotenoids with pigmenting properties [65]. As olive leaves are considered an excellent source for the recovery of carotenoids [66], it was hypothesized that yolk coloration would be more intense in the groups supplemented with olive leaf extract. However, the results showed that the T2 treatment did not change yolk coloration compared to T1, while the higher inclusion level (T3) led to a lower yolk color fan score and redness value (a*). In a previous study [26], it was found that 2% and 3% olive leaf powder supplementation in laying hens improved egg yolk coloration. Literature about how dietary supplementation of olive leaf extract in laying hens affects egg quality is very limited. The carotenoid content of different olive leaf extracts may vary depending on the extraction method [67], which could be partially responsible for the unexpected finding of our study. Moreover, the lighter-colored yolks of the T3 group may indicate lower carotenoid content, which could be attributed to oxidation phenomena. Indeed, some bioactive compounds found in olive leaves, for example, phenolic compounds such as oleuropein, may exhibit pro-oxidant effects when supplemented at higher doses [68,69]. More plausible explanations may be either the reduction of the activity of the digestive enzymes by phenolic compounds when present at high levels in the diet [70,71], having as a consequence the limitation of the liberation of carotenoids from the feed matrix, or the reduction of micellization and the competition with carotenoids for introduction into the micelles [71]. The latter is a prerequisite before the absorption and transfer to target tissues. Regarding T4 treatment, oregano oil maintained yolk coloration. This finding is in line with a previous study [72], where oregano essential oil supplemented at two levels (50 or 100 mg/kg) in laying hens did not alter yolk coloration.

Lipid oxidation negatively affects animal product quality, organoleptic properties, nutritional value, and shelf-life [73]. MDA is a secondary lipid oxidation product, which is considered the main product for the evaluation of lipid peroxidation [74,75]. In the present study, both T2 and T4 treatments reduced yolk lipid oxidation in fresh eggs. In a previous study, dietary treatment with 10 g/kg olive leaves in laying hens resulted in lower yolk MDA values after 40 days of storage, but olive leaf extract supplementation led to values similar to control [25]. In another study, 10g/kg olive leaves supplementation in laying hens feeding with linseed oil-enriched diets did not alter MDA levels of yolks but reduced the concentration of lipid hyperoxides, which are primary lipid oxidation products [75]. It can be assumed that the olive leaf extract used in this study, which contained phenols and flavonoids responsible for its antioxidant properties, reduced lipid oxidation in the yolk. The same antioxidant molecules, however, could act as pro-oxidants in higher inclusion levels, which could explain the non-dose-dependent effect in the T3 group [68,69]. A similar interpretation was proposed in a previous study by our research group in broilers, where the same olive leaf extract was supplemented in two levels (1% and 2.5%), and the lower dose was more beneficial for the lipid oxidation of meat [5]. The effectiveness of dietary oregano essential oil in delaying lipid oxidation has been previously reported in egg yolks [76,77] and poultry meat [5,72,78–80].

Olive leaves contain a large variety of phenolic compounds with antioxidant properties, including simple phenols, flavonoids, and secoiridoids [81]. Literature about the deposition of total phenolic compounds in egg yolk is scarce. It has been shown that phenolic compounds found in poultry feed can be transferred and deposited in egg yolk [82,83], but the circumstances and the factors that are involved in phenolics' bioavailability and deposition in egg yolk are still under investigation. In the present study, T4 feed samples had similar TPC to T1, but TPC was more than 2.5-fold higher in T2 diet and almost 5-fold higher in T3 diet in comparison with T1. Regarding egg yolk TPC, it did not differ significantly in any treatment in comparison with T1. This result may be attributed to the hydrophilic nature of the main phenolic compounds of olive leaves, i.e., oleuropein and hydroxytyrosol, which may limit their deposition in egg yolk [81,84]. However, the lower olive leaf extract inclusion level (T2 treatment) resulted in significantly higher TPC in yolk compared to the higher (T3). This finding indicates lower oxidation of phenolic compounds and may be related to the assumption that some molecules may act as antioxidants in low doses and pro-oxidants in higher doses. This mechanism has already been discussed previously, as this finding is in line with our previous results of T2 treatment in other antioxidant parameters (carotenoid content, MDA).

As emerging egg technology produces functional eggs by modifying the diets of laying hens, there is growing interest in preventing the oxidative deterioration of eggs [85]. In the present study, the yolk fatty acid profile was improved in the T3 treatment, which exhibited higher PUFA percentage and n-3 and n-6 content, with a corresponding reduction in MUFA percentage. The increased percentage of some individual PUFA (linoleic, α-linolenic, cis-cis-11,14 eicosadienoic) observed in the T3 treatment agrees with these findings. It is known that highly unsaturated eggs may be more prone to oxidation than conventional eggs [85]. Estimation of total antioxidant capacity (TAC) in yolk revealed that yolks from

T2 and T3 groups, treated with olive leaf extract, maintained their antioxidant capacity, as it did not differ significantly from the one of T1. Despite the T3 group having the most unsaturated profile, the TAC values of yolk were reduced only numerically compared to T1 and T2. The lowest TAC values were those of T4 yolks, which were significantly lower compared to T1 and T2, and maybe this effect is due to differences in the content of antioxidants, pro-oxidants, and substrates prone to oxidation. In previous studies, 2% or 2.5% olive oil supplementation in laying hens increased unsaturated fatty acids (UFA) content in yolk [47] or PUFA content and the proportions of oleic and linolenic acids in yolk, respectively [60]. Dietary olive pulp also improved egg fatty acid profile in laying hens in the study of Dedousi et al. [4], with increased PUFA percentage and reduced SFA being the main findings. Regarding olive leaf extract's effect on yolk fatty acid profile, the literature is very limited. Olive leaf extracts have been used successfully for stabilization purposes in refined olive oil, food lipids, and table olives [86–88]. Due to the presence of compounds with antioxidant properties in high concentrations, the olive leaf extract used in our study may have exhibited protective effects on yolk lipids, as shown by the higher PUFA percentage, n-3 and n-6 content, when supplemented in the higher dose. This observation warrants further investigation.

Yolk cholesterol levels were not affected by the treatments even though the limited data in the literature regarding the supplementation of olive oil, leaves (extract or powder), olive pomace, or even oleuropein indicated a beneficial effect. More specifically, in the study of Ahmed et al. [24], who added ~160–500-fold lower levels of the extract (50, 100, and 150 mg/kg) than in our study to the diet of Bandarah chicken, found a positive effect. The effect was more pronounced by increasing the levels of the extract. In another study [60], where two varieties of olive oil were supplemented in laying hens, one with high TPC and the other with low TPC, the results showed that only the high-TPC oil reduced plasma and yolk cholesterol levels compared to a control diet. Olive leaf powder up to 3% in laying hens' diet tended to reduce yolk cholesterol [26]. Based on the existing literature, it can be assumed that the different olive products and byproducts, the different doses used, and the presence of oxidants, prooxidants, and nutrition/anti-nutrition factors in the diet may be responsible for the inconsistencies of the findings. More specifically, polyphenols, catechins, and flavonols have been associated with lowering cholesterol effects in egg yolk [89], while the type of dietary fat can also affect egg yolk cholesterol content [90]. Few studies have approached more precisely the possible underlying mechanisms, like the one of Iannaccone et al. (2019) [91], who found that dietary supplementation of dried olive pomace in laying hens modulated several biological pathways related to cholesterol biosynthesis. However, it should be mentioned that yolk cholesterol content is very resistant to changes, as embryo development normally requires a minimum necessary cholesterol concentration in yolk [92], so manipulating yolk cholesterol by hen nutrition is usually unsuccessful. As olive leaf extract is rich in phenolic compounds, further research is proposed to explore if there is potential in reducing yolk cholesterol levels and under which circumstances.

Olive leaves are considered an alternative α-tocopherol source, and appropriate extraction methods can lead to extracts with high α-tocopherol concentration [93]. Generally, yolk α-tocopherol content reflects the tocopherol concentration of the diet [94]. The findings of the present study were as expected, with α-tocopherol yolk enrichment being dose-dependent and, thus, higher in T3 treatment. They also agree with previous studies [67,85], where eggs of hens supplemented with olive leaves had higher α-tocopherol yolk content compared to the control group. Besides the apparent dose-dependent effect, it has been reported by many authors that the presence of other antioxidants in the diet could spare α-tocopherol and further increase its bioavailability by protecting it from oxidative damage during digestion in the intestine [75,94–96]. In a previous study, sage supplementation resulted in astonishingly higher yolk α-tocopherol content than expected [97], and the latter authors suggested that more tocopherol is available and absorbed by the intestine when other antioxidants are present in the diet of laying hens. The effect seen for oregano oil treatment (T4), which increased yolk α-tocopherol content compared to T1, could be related to the fact that the oil, though used at low levels, is encapsulated; thus, its phenols (mainly thymol and carvacrol), lipophilic in nature, are available in the intestine [98] favoring tocopherol protection. Taken together, it can be suggested that olive leaf extract supplementation in laying hens can increase α-tocopherol content in yolk not only due to its α-tocopherol content but also because it seems to protect and enhance the absorption of the α-tocopherol found in other feed ingredients. Enriching eggs with α-tocopherol could be important for consumer health, as α-tocopherol is the most bioactive form of vitamin E and has been associated with lowering the risk of chronic disease and protection against negative effects of aging and cognitive decline [99].

The liver was subjected to gross and histological analysis, which demonstrated that the incorporation of olive leaf extract had a beneficial effect. This inclusion did not have any negative impact on the lipid liver metabolism of the laying hens. Additionally, it resulted in the preservation of a heightened antioxidant status, as well as the promotion of enhanced fatty acid quality in egg yolks. Furthermore, the cholesterol levels remained stable throughout this process.

5. Conclusions

Overall, it can be concluded that a 1% level of dietary inclusion of a resin-purified aqueous isopropanol olive leaf extract in laying hens can improve albumen quality, reduce yolk lipid oxidation procedures, and increase yolk TPC. The greater supplementation level (2.5%) improved the yolk fatty acid profile but resulted in brighter yolk coloration. Oregano oil treatment, used as a positive control, reduced yolk MDA values of fresh eggs. All treatments increased eggshell weight and yolk α-tocopherol content, which was greater in the treatment with the higher dose of olive leaf extract (2.5%).

Author Contributions: Conceptualization, G.A. and A.-L.S.; methodology, G.A.P., S.L., I.P. (Ioannis Panitsidis) and G.K.S.; software, G.A.P., I.P. (Ioanna Pyrka) and A.G.K.; validation, N.N., A.-L.S. and P.S.; formal analysis, N.N., I.P. (Ioannis Panitsidis), I.P. (Ioanna Pyrka), A.G.K., P.S., I.S., A.P. and I.G.; investigation, G.A.P., S.L., I.P. (Ioannis Panitsidis), D.G., G.A. and I.G.; resources, G.A.P., A.-L.S. and G.A.; data curation, G.A.P., S.L., N.N., I.P. (Ioannis Panitsidis), I.P. (Ioanna Pyrka), A.G.K., P.S., I.S. and A.P.; writing—original draft preparation, G.A.P., S.L., N.N., D.G. and I.G.; writing—review and editing, G.A.P., S.L., N.N., D.G., P.S. and I.G.; visualization, I.S.; supervision, G.A.P., N.N., A.-L.S. and G.A.; project administration, G.A.P., N.N., A.-L.S. and G.A.; funding acquisition, A.-L.S. and G.A. All authors have read and agreed to the published version of the manuscript.

Funding: Exploitation of olives processing by-products and wastes to produce innovative biofunctional feeds and quality animal products" was a research project co-financed by the European Regional Development Fund of the European Union and Greek national funds through the Operational Program Competitiveness, Entrepreneurship and Innovation, under the call RESEARCH–CREATE–INNOVATE (project code: T2EDK-03891).

Data Availability Statement: Data are available upon request.

Conflicts of Interest: The authors declare no conflict of interest. The funders had no role in the design of the study, in the collection, analyses, or interpretation of data, in the writing of the manuscript, or in the decision to publish the results.

References

1. Abd El-Hack, M.E.; El-Saadony, M.T.; Salem, H.M.; El-Tahan, A.M.; Soliman, M.M.; Youssef, G.B.; Taha, A.E.; Soliman, S.M.; Ahmed, A.E.; El-Kott, A.F.; et al. Alternatives to antibiotics for organic poultry production: Types, modes of action and impacts on bird's health and production. *Poult. Sci.* **2022**, *101*, 101696. [CrossRef]
2. Wang, J.; Jia, R.; Celi, P.; Ding, X.; Bai, S.; Zeng, Q.; Mao, X.; Xu, S.; Zhang, K. Green tea polyphenol epigallocatechin-3-gallate improves the antioxidant capacity of eggs. *Food Funct.* **2020**, *11*, 534–543. [CrossRef] [PubMed]
3. Berbel, J.; Posadillo, A. Review and analysis of alternatives for the valorisation of agro-industrial olive oil by-products. *Sustainability* **2018**, *10*, 237. [CrossRef]
4. Dedousi, A.; Kritsa, M.Z.; Đukić Stojčić, M.; Sfetsas, T.; Sentas, A.; Sossidou, E. Production performance, egg quality characteristics, fatty acid profile and health lipid indices of produced eggs, blood biochemical parameters and welfare indicators of laying hens fed dried olive pulp. *Sustainability* **2022**, *14*, 3157. [CrossRef]

5. Vasilopoulou, K.; Papadopoulos, G.A.; Lioliopoulou, S.; Pyrka, I.; Nenadis, N.; Savvidou, S.; Symeon, G.; Dotas, V.; Panitsidis, I.; Arsenos, G.; et al. Effects of Dietary Supplementation of a Resin-Purified Aqueous-Isopropanol Olive Leaf Extract on Meat and Liver Antioxidant Parameters in Broilers. *Antioxidants* **2023**, *12*, 1723. [CrossRef]
6. Abbas, A.O.; Alaqil, A.A.; El-Beltagi, H.S.; Abd El-Atty, H.K.; Kamel, N.N. Modulating laying hens productivity and immune performance in response to oxidative stress induced by E. coli challenge using dietary propolis supplementation. *Antioxidants* **2020**, *9*, 893. [CrossRef] [PubMed]
7. Ding, X.; Cai, C.; Jia, R.; Bai, S.; Zeng, Q.; Mao, X.; Xu, S.; Zhang, K.; Wang, J. Dietary resveratrol improved production performance, egg quality, and intestinal health of laying hens under oxidative stress. *Poult. Sci.* **2022**, *101*, 101886. [CrossRef]
8. Wang, J.P.; He, K.R.; Ding, X.M.; Luo, Y.H.; Bai, S.P.; Zeng, Q.F.; Su, Z.W.; Xuan, Y.; Zhang, K.Y. Effect of dietary vanadium and vitamin C on egg quality and antioxidant status in laying hens. *J. Anim. Physiol. Anim. Nutr.* **2016**, *100*, 440–447. [CrossRef] [PubMed]
9. Yuan, Z.H.; Zhang, K.Y.; Ding, X.M.; Luo, Y.H.; Bai, S.P.; Zeng, Q.F.; Wang, J.P. Effect of tea polyphenols on production performance, egg quality, and hepatic antioxidant status of laying hens in vanadium-containing diets. *Poult. Sci.* **2016**, *95*, 1709–1717. [CrossRef]
10. Akbarian, A.; Michiels, J.; Degroote, J.; Majdeddin, M.; Golian, A.; De Smet, S. Association between heat stress and oxidative stress in poultry; mitochondrial dysfunction and dietary interventions with phytochemicals. *J. Anim. Sci. Biotechnol.* **2016**, *7*, 37. [CrossRef]
11. Pérez-Jiménez, J.; Neveu, V.; Vos, F.; Scalbert, A. Identification of the 100 richest dietary sources of polyphenols: An application of the Phenol-Explorer database. *Eur. J. Clin. Nutr.* **2010**, *64*, S112–S120. [CrossRef] [PubMed]
12. Tsao, R. Chemistry and biochemistry of dietary polyphenols. *Nutrients* **2010**, *2*, 1231–1246. [CrossRef] [PubMed]
13. Lipiński, K.; Mazur, M.; Antoszkiewicz, Z.; Purwin, C. Polyphenols in monogastric nutrition–A review. *Ann. Anim. Sci.* **2017**, *17*, 41–58. [CrossRef]
14. Abdel-Moneim, A.M.E.; Shehata, A.M.; Alzahrani, S.O.; Shafi, M.E.; Mesalam, N.M.; Taha, A.E.; Swelum, A.A.; Arif, M.; Fayyaz, M.; Abd El-Hack, M.E. The role of polyphenols in poultry nutrition. *J. Anim. Physiol. Anim. Nutr.* **2020**, *104*, 1851–1866. [CrossRef] [PubMed]
15. Wang, X.; Wu, S.; Cui, Y.; Qi, G.; Wang, J.; Zhang, H. Effects of dietary tea polyphenols on performance, egg quality and antioxidant ability of laying hens. *Chin. J. Anim. Nutr.* **2017**, *29*, 193–201.
16. Zhu, Y.F.; Wang, J.P.; Ding, X.M.; Bai, S.P.; Qi, S.R.N.; Zeng, Q.F.; Zhang, K.Y. Effect of different tea polyphenol products on egg production performance, egg quality and antioxidant status of laying hens. *Anim. Feed Sci. Technol.* **2020**, *267*, 114544. [CrossRef]
17. Wang, J.; Yuan, Z.; Zhang, K.; Ding, X.; Bai, S.; Zeng, Q.; Peng, H.; Celi, P. Epigallocatechin-3-gallate protected vanadium-induced eggshell depigmentation via P38MAPK-Nrf2/HO-1 signaling pathway in laying hens. *Poult. Sci.* **2018**, *97*, 3109–3118. [CrossRef]
18. Ramirez, S.Y.; Peñuela-Sierra, L.M.; Ospina, M.A. Effects of oregano (*Lippia origanoides*) essential oil supplementation on the performance, egg quality, and intestinal morphometry of Isa Brown laying hens. *Vet. World* **2021**, *14*, 595. [CrossRef]
19. Chen, Y.; Chen, H.; Li, W.; Miao, J.; Chen, N.; Shao, X.; Cao, Y. Polyphenols in Eucalyptus leaves improved the egg and meat qualities and protected against ethanol-induced oxidative damage in laying hens. *J. Anim. Physiol. Anim. Nutr.* **2018**, *102*, 214–223. [CrossRef]
20. Abd El-Motaal, A.M.; Ahmed, A.M.H.; Bahakaim, A.S.A.; Fathi, M.M. Productive performance and immunocompetence of commercial laying hens given diets supplemented with eucalyptus. *Int. J. Poult. Sci.* **2008**, *7*, 445–449. [CrossRef]
21. Zhang, B.; Wang, Z.; Huang, C.; Wang, D.; Chang, D.; Shi, X.; Chen, Y.; Chen, H. Positive effects of Mulberry leaf extract on egg quality, lipid metabolism, serum biochemistry, and antioxidant indices of laying hens. *Front. Vet. Sci.* **2022**, *9*, 1005643. [CrossRef]
22. Martínez-Navarro, M.E.; Kaparakou, E.H.; Kanakis, C.D.; Cebrián-Tarancón, C.; Alonso, G.L.; Salinas, M.R.; Tarantilis, P.A. Quantitative Determination of the Main Phenolic Compounds, Antioxidant Activity, and Toxicity of Aqueous Extracts of Olive Leaves of Greek and Spanish Genotypes. *Horticulturae* **2023**, *9*, 55. [CrossRef]
23. Ryan, D.; Antolovich, M.; Prenzler, P.; Robards, K.; Lavee, S. Biotransformations of phenolic compounds in *Olea europaea* L. *Sci. Hortic.* **2002**, *92*, 147–176. [CrossRef]
24. Ahmed, M.M.; Elsaadany, A.S.; Shreif, E.Y.; El-Barbary, A.M. Effect of dietary olive leaves extract (oleuropein) supplementation on productive, physiological and immunological parameters in bandarah chickens 2-during production period. *Egypt. Poult. Sci. J.* **2018**, *37*, 277–292.
25. Rezar, V.; Levar, A.; Salobir, J. The effect of olive by products and their extracts on antioxidant status of laying hens and oxidative stability of eggs enriched with n-3 fatty acids. *Poljoprivreda* **2015**, *21* (Suppl. S1), 216–219. [CrossRef]
26. Cayan, H.; Erener, G. Effect of olive leaf (*Olea europaea*) powder on laying hens performance, egg quality and egg yolk cholesterol levels. *Asian-australas. J. Anim. Sci.* **2015**, *28*, 538. [CrossRef] [PubMed]
27. Christaki, E.; Bonos, E.; Florou-Paneri, P. Effect of dietary supplementation of olive leaves and/or α-tocopheryl acetate on performance and egg quality of laying Japanese quail (*Coturnix japonica*). *Asian J. Anim. Vet. Adv.* **2011**, *6*, 1241–1248. [CrossRef]
28. International Olive Council. Determination of Biophenols in Olive Oils by HPLC. COI/T.20/Doc. No 29/Rev.1. 2017. Available online: https://www.internationaloliveoil.org/wp-content/uploads/2019/11/COI-T.20-Doc.-No-29-Rev-1-2017.pdf (accessed on 10 July 2023).
29. Partheniadis, I.; Zarafidou, E.; Litinas, K.E.; Nikolakakis, I. Enteric release essential oil prepared by co-spray drying methacrylate/polysaccharides—Influence of starch type. *Pharmaceutics* **2020**, *12*, 571. [CrossRef] [PubMed]

30. Commercial Management Guide. Cage Housing: Version L0260-6. Available online: https://layinghens.hendrixgenetics.com/documents/980/Management_guide_commercial_cage_English_vs_L0260-6_pdf (accessed on 25 May 2023).
31. Delta E101. Available online: http://zschuessler.github.io/DeltaE/learn/ (accessed on 10 July 2023).
32. European Commission. Commission Regulation (EC) No 152/2009 of 27 January 2009 (and amendments) laying down the methods of sampling and analysis for the official control of feed. *Off. J. Eur. Union* **2009**, *54*, 1–173.
33. Vasilopoulos, S.; Dokou, S.; Papadopoulos, G.A.; Savvidou, S.; Christaki, S.; Kyriakoudi, A.; Dotas, V.; Tsiouris, V.; Bonos, E.; Skoufos, I.; et al. Dietary supplementation with pomegranate and onion aqueous and cyclodextrin encapsulated extracts affects broiler performance parameters, welfare and meat characteristics. *Poultry* **2022**, *1*, 74–93. [CrossRef]
34. Pyrka, I.; Mantzouridou, F.T.; Nenadis, N. Optimization of olive leaves' thin layer, intermittent near-infrared-drying. *Innov. Food Sci. Emerg. Technol.* **2023**, *84*, 103264. [CrossRef]
35. Tsimidou, M.Z.; Sotiroglou, M.; Mastralexi, A.; Nenadis, N.; García-González, D.L.; Gallina Toschi, T. In house validated UHPLC protocol for the determination of the total hydroxytyrosol and tyrosol content in virgin olive oil fit for the purpose of the health claim introduced by the EC Regulation 432/2012 for "Olive Oil Polyphenols". *Molecules* **2019**, *24*, 1044. [CrossRef]
36. Plati, F.; Papi, R.; Paraskevopoulou, A. Characterization of oregano essential oil (*Origanum vulgare* L. subsp. hirtum) particles produced by the novel nano spray drying technique. *Foods* **2021**, *10*, 2923. [CrossRef]
37. Shang, H.; Zhang, H.; Guo, Y.; Wu, H.; Zhang, N. Effects of inulin supplementation in laying hens diet on the antioxidant capacity of refrigerated stored eggs. *Int. J. Biol. Macromol.* **2020**, *153*, 1047–1057. [CrossRef]
38. Prieto, P.; Pineda, M.; Aguilar, M. Spectrophotometric quantitation of antioxidant capacity through the formation of a phosphomolybdenum complex: Specific application to the determination of vitamin E. *Anal. Biochem.* **1999**, *269*, 337–341. [CrossRef]
39. Folch, J.; Lees, M.; Sloane-Stanley, G.H. A simple method for isolation and purification of total lipids from animal tissues. *J. Biol. Chem.* **1957**, *226*, 497–509. [CrossRef] [PubMed]
40. Botsoglou, N.; Fletouris, D.; Psomas, I.; Mantis, A. Rapid gas chromatographic method for simultaneous determination of cholesterol and α-tocopherol in eggs. *J. AOAC Int.* **1998**, *81*, 1177–1184. [CrossRef]
41. Eisenberg, S. Relative stability of selenites and selenates in feed premixes as a function of water activity. *J. AOAC Int.* **2007**, *90*, 349–353. [CrossRef]
42. Kokkini, S.; Karousou, R.; Hanlidou, E.; Lanaras, T. Essential oil composition of Greek (*Origanum vulgare* ssp. hirtum) and Turkish (*O. onites*) oregano: A tool for their distinction. *J. Essent. Oil Res.* **2004**, *16*, 334–338. [CrossRef]
43. Nakas, A.; Giannarelli, G.; Fotopoulos, I.; Chainoglou, E.; Peperidou, A.; Kontogiannopoulos, K.N.; Tsiaprazi-Stamou, A.; Varsamis, V.; Gika, H.; Hadjipavlou-Litina, D.; et al. Optimizing the Distillation of Greek Oregano—Do Process Parameters Affect Bioactive Aroma Constituents and In Vitro Antioxidant Activity? *Molecules* **2023**, *28*, 971. [CrossRef]
44. Storniolo, C.E.; Sacanella, I.; Lamuela-Raventos, R.M.; Moreno, J.J. Bioactive compounds of mediterranean cooked tomato sauce (Sofrito) modulate intestinal epithelial cancer cell growth through oxidative stress/arachidonic acid cascade regulation. *ACS Omega* **2020**, *5*, 17071–17077. [CrossRef] [PubMed]
45. Murakami, F.S.; Rodrigues, P.O.; Campos, C.M.T.D.; Silva, M.A.S. Physicochemical study of $CaCO_3$ from egg shells. *Food Sci. Technol.* **2007**, *27*, 658–662. [CrossRef]
46. de Oliveira, N.M.; Lopes, L.; Chéu, M.H.; Soares, E.; Meireles, D.; Machado, J. Updated Mineral Composition and Potential Therapeutic Properties of Different Varieties of Olive Leaves from *Olea europaea*. *Plants* **2023**, *12*, 916. [CrossRef]
47. Zhang, Z.F.; Kim, I.H. Effects of dietary olive oil on egg quality, serum cholesterol characteristics, and yolk fatty acid concentrations in laying hens. *J. Appl. Anim. Res.* **2014**, *42*, 233–237. [CrossRef]
48. Lang, M.R.; Wells, J.W. A review of eggshell pigmentation. *Worlds Poult. Sci. J.* **1987**, *43*, 238–246. [CrossRef]
49. Drabik, K.; Karwowska, M.; Wengerska, K.; Próchniak, T.; Adamczuk, A.; Batkowska, J. The variability of quality traits of table eggs and eggshell mineral composition depending on hens' breed and eggshell color. *Animals* **2021**, *11*, 1204. [CrossRef]
50. Lu, M.Y.; Xu, L.; Qi, G.H.; Zhang, H.J.; Qiu, K.; Wang, J.; Wu, S.G. Mechanisms associated with the depigmentation of brown eggshells: A review. *Poult. Sci.* **2021**, *100*, 101273. [CrossRef]
51. Samiullah, S.; Roberts, J.R.; Chousalkar, K. Eggshell color in brown-egg laying hens—A review. *Poult. Sci.* **2015**, *94*, 2566–2575. [CrossRef]
52. Sparks, N.H. Eggshell pigments–from formation to deposition. *Avian Biol. Res.* **2011**, *4*, 162–167. [CrossRef]
53. Inkee, P.; Hankyu, L.; Sewon, P. Effects of organic iron supplementation on the performance and iron content in the egg yolk of laying hens. *J. Poult. Sci.* **2009**, *46*, 198–202.
54. Seo, Y.; Shin, K.; Rhee, A.; Chi, Y.; Han, J.; Paik, I. Effects of dietary Fe-soy proteinate and MgO on egg production and quality of eggshell in laying hens. *Asian Australas. J. Anim. Sci.* **2010**, *23*, 1043–1048. [CrossRef]
55. Williams, K.C. Some factors affecting albumen quality with particular reference to Haugh unit score. *World's Poult. Sci. J.* **1992**, *48*, 5–16. [CrossRef]
56. Vlaicu, P.A.; Untea, A.E.; Turcu, R.P.; Panaite, T.D.; Saracila, M. Rosehip (*Rosa canina* L.) Meal as a Natural Antioxidant on Lipid and Protein Quality and Shelf-Life of Polyunsaturated Fatty Acids Enriched Eggs. *Antioxidants* **2022**, *11*, 1948. [CrossRef]
57. Zarei, M.; Ehsani, M.; Torki, M. Productive Performance of Laying Hens Fed Wheat-Based Diets Included Olive Pulp with or without a Commercial Enzyme Product. *Afr. J. Biotechnol.* **2011**, *10*, 4303–4312.
58. Ghasemi, R.; Torki, M.; Ghasemi, H.A.; Zarei, M. Single or Combined Effects of Date Pits and Olive Pulps on Productive Traits, Egg Quality, Serum Lipids and Leucocytes Profiles of Laying Hens. *J. Appl. Anim. Res.* **2014**, *42*, 103–109. [CrossRef]

59. Al-Harthi, M.A. The Effect of Different Dietary Contents of Olive Cake with or without Saccharomyces Cerevisiae on Egg Production and Quality, Inner Organs and Blood Constituents of Commercial Layers. *Eur. Poult. Sci.* **2015**, *79*, 83. [CrossRef]
60. Laudadio, V.; Ceci, E.; Lastella, N.; Tufarelli, V. Dietary high-polyphenols extra-virgin olive oil is effective in reducing cholesterol content in eggs. *Lipids Health Dis.* **2015**, *14*, 1–7. [CrossRef]
61. Zangeneh, S.; Torki, M. Effects of B-Mannanase Supplementing of Olive Pulp-Included Diet on Performance of Laying Hens, Egg Quality Characteristics, Humoral and Cellular Immune Response and Blood Parameters. *Glob. Vet.* **2011**, *7*, 391–398.
62. Mohebbifar, A.; Afsari, M.; Torki, M. Egg Quality Characteristics and Productive Performance of Laying Hens Fed Olive Pulp Included Diets Supplemented with Enzyme. *Glob. Vet.* **2011**, *6*, 409–416.
63. Afsari, M.; Mohebbifar, A.; Torki, M. Effects of Dietary Inclusion of Olive Pulp Supplemented with Probiotics on Productive Performance, Egg Quality and Blood Parameters of Laying Hens. *Annu. Res. Rev. Biol.* **2014**, *4*, 198–211. [CrossRef]
64. Hernandez, J.M.; Beardsworth, P.; Weber, G. Egg quality-meeting consumer expectations. *Int. Poult. Prod.* **2005**, *13*, 3.
65. Nys, Y. Dietary carotenoids and egg yolk coloration. *Archiv. Geflugelkd.* **2000**, *64*, 45–54.
66. Lorini, A.; Aranha, B.C.; da Fonseca, A.B.; Otero, D.M.; Jacques, A.C.; Zambiazi, R.C. Metabolic profile of olive leaves of different cultivars and collection times. *Food Chem.* **2021**, *345*, 128758. [CrossRef] [PubMed]
67. Žugčić, T.; Abdelkebir, R.; Alcantara, C.; Collado, M.C.; Garcia-Perez, J.V.; Meléndez-Martínez, A.J.; Režek Jambrak, A.; Lorenzo, J.M.; Barba, F.J. From extraction of valuable compounds to health promoting benefits of olive leaves through bioaccessibility, bioavailability and impact on gut microbiota. *Trends Food Sci. Technol.* **2019**, *83*, 63–77. [CrossRef]
68. Odiatou, E.M.; Skaltsounis, A.L.; Constantinou, A.I. Identification of the factors responsible for the in vitro pro-oxidant and cytotoxic activities of the olive polyphenols oleuropein and hydroxytyrosol. *Cancer Lett.* **2013**, *330*, 113–121. [CrossRef]
69. Scicchitano, S.; Vecchio, E.; Battaglia, A.M.; Oliverio, M.; Nardi, M.; Procopio, A.; Costanzo, F.; Biamonte, F.; Faniello, M.C. The double-edged sword of oleuropein in ovarian cancer cells: From antioxidant functions to cytotoxic effects. *Int. J. Mol. Sci.* **2023**, *24*, 842. [CrossRef]
70. Griffiths, D.W. The Inhibition of Digestive Enzymes by Polyphenolic Compounds. In *Nutritional and Toxicological Significance of Enzyme Inhibitors in Foods. Advances in Experimental Medicine and Biology*; Friedman, M., Ed.; Springer: Boston, MA, USA, 1986; Volume 199.
71. Marques, M.C.; Hacke, A.; Neto, C.A.C.; Mariutti, L.R. Impact of phenolic compounds in the digestion and absorption of carotenoids. *Curr. Opin. Food Sci.* **2021**, *39*, 190–196. [CrossRef]
72. Florou-Paneri, P.; Palatos, G.; Govaris, A.; Botsoglou, D.; Giannenas, I.; Ambrosiadis, I. Oregano herb versus oregano essential oil as feed supplements to increase the oxidative stability of turkey meat. *Int. J. Poult. Sci.* **2005**, *4*, 866–871.
73. Mohajer, A.; Sadighara, P.; Mohajer, M.; Farkhondeh, T.; Samarghandian, S. A comparison of antioxidant effects of some selected fruits with butylated hydroxytoluene on egg yolk. *Curr. Res. Nutr. Food Sci.* **2019**, *15*, 525–527. [CrossRef]
74. Grotto, D.; Maria, L.S.; Valentini, J.; Paniz, C.; Schmitt, G.; Garcia, S.C.; Pomblum, V.J.; Rocha, J.B.T.; Farina, M. Importance of the lipid peroxidation biomarkers and methodological aspects for malondialdehyde quantification. *Quim. Nova* **2009**, *32*, 169–174. [CrossRef]
75. Botsoglou, E.; Govaris, A.; Fletouris, D.; Botsoglou, N. Effect of supplementation of the laying hen diet with olive leaves (*Olea europea* L.) on lipid oxidation and fatty acid profile of α-linolenic acid enriched eggs during storage. *Br. Poult. Sci.* **2012**, *53*, 508–519. [CrossRef]
76. Florou-Paneri, P.; Nikolakakis, I.; Giannenas, I.; Koidis, A.; Botsoglou, E.; Dotas, V.; Mitsopoulos, I. Hen performance and egg quality as affected by dietary oregano essential oil and tocopheryl acetate supplementation. *Int. J. Poult. Sci.* **2005**, *4*, 449–454.
77. Migliorini, M.J.; Boiago, M.M.; Stefani, L.M.; Zampar, A.; Roza, L.F.; Barreta, M.; Arno, A.; Robazza, W.S.; Giuriatti, J.; Galvao, A.C.; et al. Oregano essential oil in the diet of laying hens in winter reduces lipid peroxidation in yolks and increases shelf life in eggs. *J. Therm. Biol.* **2019**, *85*, 102409. [CrossRef] [PubMed]
78. Botsoglou, N.A.; Florou-Paneri, P.; Christaki, E.; Fletouris, D.J.; Spais, A.B. Effect of dietary oregano essential oil on performance of chickens and on iron-induced lipid oxidation of breast, thigh and abdominal fat tissues. *Br. Poult. Sci.* **2002**, *43*, 223–230. [CrossRef] [PubMed]
79. Giannenas, I.A.; Florou-Paneri, P.; Botsoglou, N.A.; Christaki, E.; Spais, A. Effect of supplementing feed with oregano and/or alpha-tocopheryl acetate on growth of broiler chickens and oxidative stability of meat. *J. Anim. Feed Sci.* **2005**, *14*, 521. [CrossRef]
80. Marcincak, S.; Cabadaj, R.; Popelka, P.; Soltysova, L. Antioxidative effect of oregano supplemented to broilers on oxidative stability of poultry meat. *Slov. Vet. Res.* **2008**, *45*, 61–66.
81. Talhaoui, N.; Taamalli, A.; Gómez-Caravaca, A.M.; Fernández-Gutiérrez, A.; Segura-Carretero, A. Phenolic compounds in olive leaves: Analytical determination, biotic and abiotic influence, and health benefits. *Food Res. Int.* **2015**, *77*, 92–108. [CrossRef]
82. Benakmoum, A.; Larid, R.; Zidani, S. Enriching egg yolk with carotenoids and phenols. *Int. J. Food Sci. Nutr. Eng.* **2013**, *7*, 489–493.
83. Lioliopoulou, S.; Papadopoulos, G.A.; Giannenas, I.; Vasilopoulou, K.; Squires, C.; Fortomaris, P.; Mantzouridou, F.T. Effects of dietary supplementation of pomegranate peel with xylanase on egg quality and antioxidant parameters in laying hens. *Antioxidants* **2023**, *12*, 208. [CrossRef]
84. Fernández, E.; Vidal, L.; Canals, A. Rapid determination of hydrophilic phenols in olive oil by vortex-assisted reversed-phase dispersive liquid-liquid microextraction and screen-printed carbon electrodes. *Talanta* **2018**, *181*, 44–51. [CrossRef] [PubMed]
85. Botsoglou, N.A.; Yannakopoulos, A.L.; Fletouris, D.J.; Tserveni-Goussi, A.S.; Fortomaris, P.D. Effect of dietary thyme on the oxidative stability of egg yolk. *J. Agric. Food Chem.* **1997**, *45*, 3711–3716. [CrossRef]

86. Paiva-Martins, F.; Correia, R.; Félix, S.; Ferreira, P.; Gordon, M.H. Effects of enrichment of refined olive oil with phenolic compounds from olive leaves. *J. Agric. Food Chem.* **2007**, *55*, 4139–4143. [CrossRef] [PubMed]
87. De Leonardis, A.; Aretini, A.; Alfano, G.; Macciola, V.; Ranalli, G. Isolation of a hydroxytyrosol-rich extract from olive leaves (*Olea europaea* L.) and evaluation of its antioxidant properties and bioactivity. *Eur. Food Res. Technol.* **2008**, *226*, 653–659. [CrossRef]
88. Lalas, S.; Athanasiadis, V.; Gortzi, O.; Bounitsi, M.; Giovanoudis, I.; Tsaknis, J.; Bogiatzis, F. Enrichment of table olives with polyphenols extracted from olive leaves. *Food Chem.* **2011**, *127*, 1521–1525. [CrossRef]
89. Azeke, M.A.; Ekpo, K.E. Egg yolk cholesterol lowering effects of garlic and tea. *J. Med. Plant Res.* **2009**, *3*, 1113–1117. [CrossRef]
90. Elkin, R.G. Reducing shell egg cholesterol content. I. Overview, genetic approaches, and nutritional strategies. *Worlds Poult. Sci. J.* **2006**, *62*, 665–687.
91. Iannaccone, M.; Ianni, A.; Ramazzotti, S.; Grotta, L.; Marone, E.; Cichelli, A.; Martino, G. Whole blood transcriptome analysis reveals positive effects of dried olive pomace-supplemented diet on inflammation and cholesterol in laying hens. *Animals* **2019**, *9*, 427. [CrossRef]
92. Faitarone, A.B.G.; Garcia, E.A.; Roça, R.D.O.; Ricardo, H.D.A.; De Andrade, E.N.; Pelícia, K.; Vercese, F. Cholesterol levels and nutritional composition of commercial layers eggs fed diets with different vegetable oils. *Braz. J. Poult. Sci.* **2013**, *15*, 31–37. [CrossRef]
93. Şahin, S.; Bilgin, M. Olive tree (*Olea europaea* L.) leaf as a waste by-product of table olive and olive oil industry: A review. *J. Sci. Food Agric.* **2018**, *98*, 1271–1279. [CrossRef]
94. Jiang, Y.H.; McGeachin, R.B.; Bailey, C.A. a-Tocopherol, b-carotene, and retinol enrichment of chicken eggs. *Poult. Sci.* **1994**, *73*, 1137–1143. [CrossRef]
95. Botsoglou, E.; Govaris, A.; Fletouris, D.; Botsoglou, N. Lipid oxidation of stored eggs enriched with very long chain n−3 fatty acids, as affected by dietary olive leaves (*Olea europea* L.) or α-tocopheryl acetate supplementation. *Food Chem.* **2012**, *134*, 1059–1068. [CrossRef]
96. Goñí, I.; Brenes, A.; Centeno, C.; Viveros, A.; Saura-Calixto, F.; Rebolé, A.; Arija, I.; Estevez, R. Effect of dietary grape pomace and vitamin E on growth performance, nutrient digestibility, and susceptibility to meat lipid oxidation in chickens. *Poult. Sci.* **2007**, *86*, 508–516. [CrossRef]
97. Loetscher, Y.; Kreuzer, M.; Messikommer, R.E. Late laying hens deposit dietary antioxidants preferentially in the egg and not in the body. *J. Appl. Poult. Res.* **2014**, *23*, 647–660. [CrossRef]
98. Grgić, J.; Šelo, G.; Planinić, M.; Tišma, M.; Bucić-Kojić, A. Role of the encapsulation in bioavailability of phenolic compounds. *Antioxidants* **2020**, *9*, 923. [CrossRef]
99. Maras, J.E.; Bermudez, O.I.; Qiao, N.; Bakun, P.J.; Boody-Alter, E.L.; Tucker, K.L. Intake of α-tocopherol is limited among US adults. *J. Am. Diet. Assoc.* **2004**, *104*, 567–575. [CrossRef]

Disclaimer/Publisher's Note: The statements, opinions and data contained in all publications are solely those of the individual author(s) and contributor(s) and not of MDPI and/or the editor(s). MDPI and/or the editor(s) disclaim responsibility for any injury to people or property resulting from any ideas, methods, instructions or products referred to in the content.

Article

Strategies to Formulate Value-Added Pastry Products from Composite Flours Based on Spelt Flour and Grape Pomace Powder

Mariana-Atena Poiana [1], Ersilia Alexa [1,*], Isidora Radulov [2], Diana-Nicoleta Raba [3], Ileana Cocan [1], Monica Negrea [1], Corina Dana Misca [1], Christine Dragomir [1], Sylvestre Dossa [1] and Gabriel Suster [3]

[1] Faculty of Food Engineering, University of Life Sciences "King Michael I" from Timisoara, Aradului Street No 119, 300645 Timisoara, Romania; marianapoiana@usvt.ro (M.-A.P.); ileanacocan@usvt.ro (I.C.); monicanegrea@usvt.ro (M.N.); corinamisca@usvt.ro (C.D.M.); christine.dragomir98@gmail.com (C.D.); sylvestredossa04@gmail.com (S.D.)

[2] Faculty of Agriculture, University of Life Sciences "King Michael I" from Timisoara, Aradului Street No 119, 300645 Timisoara, Romania; isidora_radulov@usvt.ro

[3] Faculty of Tourism and Rural Management, University of Life Sciences "King Michael I" from Timisoara, Aradului Street No 119, 300645 Timisoara, Romania; diana.raba@usvt.ro (D.-N.R.); gabrielsuster@usvt.ro (G.S.)

* Correspondence: ersiliaalexa@usvt.ro; Tel.: +40-722-696-357

Abstract: In recent years, sustainability has promoted new research to develop reformulation strategies for value-added food products by exploiting grape pomace. Grape pomace powder (GP) was used to substitute spelt flour (SF) at 0, 5, 10, 15, 20 and 25% to obtain three types of fortified pastry products: biscuits and cakes involving a chemical leavening agent, and rolls leavened by yeast. Proximate composition, total phenolic content (TPC), total flavonoids content (TFC), 1,1-diphenyl-2-picrylhydrazyl (DPPH) radical scavenging activity and ferric-reducing antioxidant power (FRAP) along with physical characteristics and sensory analysis of the enriched products were considered. The retention rate of the functional attributes of formulations in response to baking was also evaluated. Significant improvements in TPC, TFC and both antioxidant tests were achieved in the fortified products by the incremental incorporation of GP. With a substitution of 25% SF by GP, the following increases were recorded in biscuits, cakes and rolls over the control samples: 7.198-, 7.733- and 8.117-fold for TPC; 8.414-, 7.000- and 8.661-fold for TFC; 16.334-, 17.915- and 18.659-fold for FRAP and 16.384-, 17.908- and 18.775-fold for DPPH. The retention rates of TPC, TFC, FRAP and DPPH relative to the corresponding dough were 41–63%, 37–65%, 48–70% and 45–70%. The formulas leavened by yeast revealed higher functionality than those produced with a chemical raising agent. With the increase in GP, the elasticity and porosity gradually decreased for cakes and rolls, while the spread ratio of biscuits increased. Regarding sensory evaluation, all formulations with incorporated GP up to 10% were rated at an extremely pleasant acceptability level. The solutions derived from this study have great practical applicability for the development of new pastry formulations with improved functionality from GP valorisation.

Keywords: spelt flour; grape pomace; composite flour; pastry products; phytochemical compounds; antioxidant properties

1. Introduction

Fortification with functional ingredients derived from agro-food by-products is a valuable strategy for enhancing the nutritional properties of foods, well aligned with the growing interest in obtaining foods with health-promoting properties [1,2]. In line with circular economy concepts, there is now a strong focus on zero-waste technologies, with by-products being effectively reintegrated into the food chain as a source of biologically active species to produce functional products [3–5]. Wine making leads to the generation of huge quantities of grape pomace—around 7–9 million tonnes/year worldwide [6]. Grape pomace represents a phenolic-rich matrix with multiple benefits to human health, since after the winemaking, about 70% of the phenolic fraction from grapes, consisting mainly of tannins, phenolic acids, anthocyanins and resveratrol, remains in the waste [7–9]. Significant efforts have been made in recent years to prove the functional potential of grape pomace and to explore its use as a value-added food ingredient for the development of new foods for the prevention of nutrition-related diseases [10–12]. Grape pomace valorisation is important both ecologically and as a sustainable source of high-quality bioactive compounds with significant antioxidant activity [13,14]. Its incorporation into cosmetic, food or pharmaceutical products represents a market opportunity for wine producers and a strategy for increasing the dietary intake of phenolic compounds [15,16].

A number of food products have been successfully enriched with phenolic compounds by incorporating grape pomace, either as natural extracts or as powders from seeds, skins, or whole grape marc. Bakery products represent the category with the most applications of grape pomace, but there are currently no commercial products on the market that contain grape pomace as a substitute for conventional wheat flour [17,18].

Since flour products are consumed all over the world, they can be an excellent basis for supplementation with bioactive-rich materials [19,20]. Improving the functionality of these products remains of great interest, even though it has been reported that additions of more than 15% may negatively affect the taste of the developed products [18].

Grape pomace has been included in various bakery products as a partial replacement for wheat flour to decrease the output gluten and increase elasticity and dietary fibre level [21], as follows: grape seed flour was incorporated in cereal bars, pancakes, noodles [22], butter biscuits [23] and bread [24], and grape pomace was incorporated into bread [20,21,25], biscuits [26], crackers [27], pasta [19,28,29], cakes [30] and cookies [31,32], while grape skin powder has been included in pasta recipes [33]. The incorporation of grape pomace could be used as a feasible way to develop new pastry formulations for the following reasons: it is an alternative source of high-value dietary fibre and phenolic compounds; it is a substitute for modified food starch; and it results in products with reduced gluten content and an increase in bioactive compounds [34].

Although there are numerous studies on the changes induced in the chemical, technological and sensory properties of bakery and pastry products through the addition of grape pomace, the type of flour used has not always been mentioned, nor the grape variety from which the marc originated, or whether it was included in whole or fractionated form or whether the basic flour was composed solely of wheat flour [19,35]. As such, the lack of standardisation of studies is a major factor that affects the comparison of the properties of the products formulated by including grape pomace in the recipe [19]. Today, the milling and bakery industry is increasingly focused on developing nutritionally and functionally high-value products based on sustainable and locally accessible resources that meet consumer demands for a healthy diet. Composite flours, defined as mixtures of powdered ingredients consisting of cereal flours or milling products combined with powders from fruit, vegetables and food by-products of plant origin, have become increasingly popular in the pastry industry due to their nutritional and functional value, with the advantage of being used immediately, without prior preparation, facilitating and speeding up the manufacturing process [36].

The interest in spelt wheat (*Triticum spelta* L.) has increased in the last few years, especially for organically grown wheat [37,38]. This is mainly due to the superior nutritional composition of spelt wheat compared to common wheat flour, reflected by a higher concentration of mineral nutrients, fibre, protein, lipids, antioxidant properties and phenolic compounds [39,40]. There are no reported studies on the impact of fortifying pastry products obtained from spelt flour with phenolic compounds provided by grape pomace on their nutritional, bioactive and sensory attributes.

This study investigated the partial replacement of spelt flour with grape pomace powder in the recipe of three types of pastry products, such as biscuits and cakes where a chemical raising agent was involved, and rolls leavened by yeast. In this respect, some important operational issues are addressed: What impact was achieved by increasing the level of GP on the nutritional, bioactive and sensory properties of the pastry formulas? How does the baking process influence the rate of retention of functional properties of products in relation to the corresponding dough? Are there differences in the bioactive compound content and antioxidant properties of pastry products depending on the production process, and which of the three types of pastry promotes the highest level of functional properties? What is the recommended percentage of GP to be incorporated as the partial replacement of spelt flour without affecting the sensory attributes of pastry formulas? To answer these questions, an integrative study was carried out starting from the proximate composition of spelt flour, grape pomace powder and pastry products, followed by investigations into the content of total phenolic compounds, total flavonoids, ferric-reducing antioxidant power and DPPH radical scavenging activity in the products and corresponding dough, as well as the sensory evaluation of pastry formulas.

2. Materials and Methods

2.1. Chemicals and Reagents

Folin–Ciocalteu reagent, ferric chloride hexahydrate and sodium carbonate anhydrous were purchased from Merck (Darmstadt, Germany); standards of gallic acid and quercetin, 6-hydroxy-2,5,7,8-tetramethylchroman-2-carboxylic acid (Trolox), 1,1-diphenyl-2-picrylhydrazyl (DPPH), 2,4,6-tris(2-pyridyl)-s-triazine (TPTZ), ferrous sulphate heptahydrate, sodium acetate anhydrous, glacial acetic acid, hydrochloric acid 0.1 M, aluminium nitrate nonahydrate and sodium nitrite were purchased from Sigma-Aldrich (Taufkirchen, Germany); ethanol 96% was acquired from Chimreactiv (Bucharest, Romania). All reagents were of analytical grade.

2.2. Ingredients and Manufacture of Pastry Products

The whole spelt wheat (*Triticum aestivum* ssp. *Spelta*) flour (SF) was purchased from SC PRONAT SRL, Sanandrei, Timis County, Romania.

Red grapes (*Vitis vinifera* L., cultivar Merlot) were provided from Recas vineyard, (Timis County, Romania, vintage year 2021). Raw grape pomace was obtained following a laboratory-scale vinification process, after a maceration phase of 10 days and pressing. The raw material was dried at 60 °C for a total period of 24 h, 8 h per day for 3 days in a row, in a Binder convective oven (Binder GmbH, Tuttlingen, Germany) until the moisture content was less than 5% to ensure microbial safety [41,42]. After cooling at 20 °C, the dried grape pomace was milled with a Grindomix GM 200 cutting mill (Retsch GmbH, Haan, Germany), and then the material was sieved with a 60-mesh sieve and the grape pomace powder (GP) was vacuum-packed in polypropylene bags and stored at room temperature in the dark until analysis or until the composite flours were obtained, as described by Tolve et al. [25]. The microbiological analysis of GP performed by standard methods for counting total aerobic mesophilic germ count (ISO 4833:2003) [43], *Enterobacteriaceae* count (ISO 21528-2:2004) [44], yeast and mould count (ISO 21527:2008) [45] and the presumptive number of *Bacillus cereus* germs (SR EN ISO 21871:2006) [46] revealed the following results: total aerobic mesophilic germ count: 2×10^2 colony-forming units (CFU)/g, *Enterobacteriaceae* count: 0 CFU/g, yeasts and mould count: 4×10 CFU/g and *Bacillus* cereus count: 1×10^2 CFU/g. Considering the low moisture of the sample, GP is protected against microbial development. Legislation does not present microbiological limits for GP and this is the reason for referring to the legislative values for flours used in baking, especially as GP will be incorporated into the recipe of pastry products. GP showed low microbial counts, below the maximum level allowed by Regulation (EC) No. 2073/2005 [47], and is compatible with its further use when also considering that the manufacture of these products involves high-temperature treatment.

Five composite flours were prepared by replacing SF with GP in the proportions of 5, 10, 15, 20 and 25% (w/w) and labelled as SF95GP5 (95% SF + 5% GP), SF90GP10 (90% SF + 10% GP),

SF85GP15 (85% SF + 15% GP), SF80GP20 (80% SF + 20% GP) and SF75GP25 (75% SF + 25% GP). The composite flour was used in the production of three types of enriched pastries that differ in the type of dough, such as biscuits, cakes and rolls, in the pilot bakery unit of the University of Life Sciences "King Michael I", Timisoara. Five fortified formulations and one control (100% SF) were developed for each type of pastry according to the recipes shown in Table 1 and were labelled as follows: BSF, BSF95GP5, BSF90GP10, BSF85GP15, BSF80GP20 and BSF75GP25 for biscuits, CSF, CSF95GP5, CSF90GP10, CSF85GP15, CSF80GP20 and CSF75GP25 for cakes, and RSF, RSF95GP5, RSF90GP10, RSF85GP15, RSF80GP20 and RSF75GP25 for rolls.

Table 1. Recipe for manufacturing biscuits, cakes and rolls (control samples and fortified formulas).

Ingredients	BSF	BSF95GP5	BSF90GP10	BSF85GP15	BSF80GP20	BSF75GP25
SF (g)	250	-	-	-	-	-
Composite flour (g)	-	250	250	250	250	250
Sugar (g)	85	85	85	85	85	85
Butter (g)	125	125	125	125	125	125
Eggs (g)	60	60	60	60	60	60
Baking powder (g)	3.500	3.500	3.500	3.500	3.500	3.500
Salt (g)	1	1	1	1	1	1
Total materials (g)	524.500	524.500	524.5	524.5	524.5	524.5
Biscuits (g)	455.967	454.724	453.718	452.821	451.833	450.668
	CSF	CSF95GP5	CSF90GP10	CSF85GP15	CSF80GP20	CSF75GP25
SF (g)	105	-	-	-	-	-
Composite flour (g)	-	105	105	105	105	105
Sugar (g)	105	105	105	105	105	105
Butter (g)	105	105	105	105	105	105
Eggs (g)	105	105	105	105	105	105
Baking powder	1.500	1.500	1.500	1.500	1.500	1.500
Salt	1	1	1	1	1	1
Total materials (g)	422.500	422.500	422.500	422.500	422.500	422.500
Cakes (g)	392.460	391.592	390.770	389.689	388.870	387.932
	RSF	RSF95GP5	RSF90GP10	RSF85GP15	RSF80GP20	RSF75GP25
SF (g)	250	-	-	-	-	-
Composite flour (g)	-	250	250	250	250	250
Sugar (g)	50	50	50	50	50	50
Milk (g)	100	100	100	100	100	100
Butter (g)	37.500	37.500	37.500	37.500	37.500	37.500
Eggs (g)	60	60	60	60	60	60
Yeast (g)	20	20	20	20	20	20
Salt (g)	4.400	4.400	4.400	4.400	4.400	4.400
Total materials (g)	521.900	523.900	521.900	521.900	521.900	521.900
Rolls (g)	478.080	476.101	474.328	472.589	470.836	468.860

SF: spelt flour; BSF, CSF, RSF (control: biscuits, cakes and rolls); BSF95GP5, CSF95GP5, RSF95GP5, (biscuits, cakes and rolls: 95% spelt flour + 5% grape pomace); BSF90GP10, CSF90GP10, RSF90GP10 (biscuits, cakes and rolls: 90% spelt flour + 10% grape pomace); BSF85GP15, CSF85GP15, RSF85GP15 (biscuits, cakes and rolls: 85% spelt flour + 15% grape pomace); BSF80GP20, CSF80GP20, RSF80GP20 (biscuits, cakes and rolls: 80% spelt flour + 20% grape pomace); BSF75GP25, CSF75GP25, RSF75GP25 (biscuits, cakes and rolls: 75% spelt flour + 25% grape pomace).

The ingredients involved in the production recipes such as sugar (Margaritar, Agrana Romania S.R.L., Romania), milk with 3.5% fat (ProdLacta, SC PRODLACTA SA, Brasov, Romania), butter with 80% fat (ProdLacta, SC PRODLACTA SA, Brasov, Romania), eggs (Agricola, Bacau, Romania), baking powder (Dr. Oetker Original Backin, Dr. Oetker SRL, Arges County, Romania), salt (Salrom, Bucuresti, Romania) and fresh yeast (Pakmaya, ROMPAK SRL, Pascani, Romania) were purchased from a local supermarket. Baking powder was used as a chemical raising agent for the biscuits and cakes, while yeast (*Saccharomyces cerevisiae*) was used as a natural leavening agent to obtain the rolls.

Control samples of each type of product were prepared according to traditional methods, applied on a small-scale level. The recipes were previously tested in the bakery and pastry processing unit of the Didactic and Experimental Station of the University of Life Sciences "King Michael I", Timisoara. The same methods were used for the pastry formulas with GP incorporation, except that the SF was replaced by composite flours. The production process of the pastry products is shown in Figure 1.

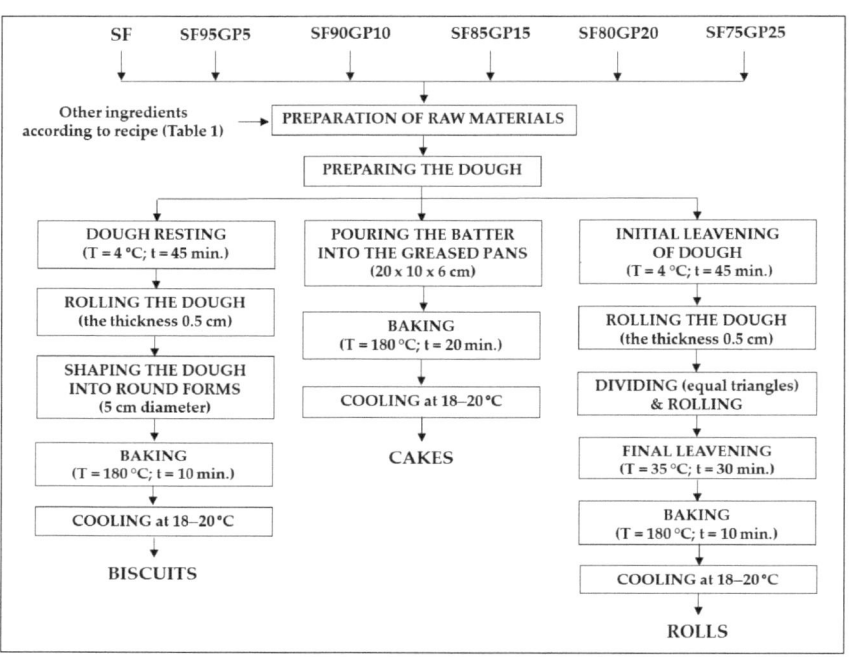

Figure 1. Technological flowchart for production of pastry products. SF: spelt flour; GP: grape pomace; SF95GP5: 95% spelt flour + 5% grape pomace; SF90GP10: 90% spelt flour + 10% grape pomace; SF85GP15: 85% spelt flour + 15% grape pomace; SF80GP20: 80% spelt flour + 20% grape pomace; SF75GP25: 75% spelt flour + 25% grape pomace.

The dough for the biscuits and rolls was prepared by mixing the ingredients for 5 min with a planetary mixer (Moulinex, 800 W, Moulinex SA, Paris, France) a in a single-phase procedure. Before use, the fresh yeast was crumbled into a bowl, and then a tablespoon of sugar and a little lukewarm milk was added and stirred to dissolve before being left to stand for 15 min, during which time the mixture increased in volume and bubbles formed on the surface. The batter for the cakes was prepared using the following procedure: the egg whites were whipped for 5 min with salt using a hand mixer with five speeds (Bosch MFQ49300, 850 W, Robert Bosch GmbH, Stuttgart, Germany) at top speed. Then, the sugar was added gradually and mixed for 1 min at full speed. Separately, the egg yolks were mixed for 1 min at high speed with butter previously kept at room temperature for 1 h and then this mixture was incorporated into the foam from egg whites. SF, or composite flours, depending on the formulation, was slowly added to the resulting mixture along with the baking powder. The dough obtained was labelled according to each pastry product: DBSF, DBSF95GP5, DBSF90GP10, DBSF85GP15, DBSF80GP20 and DBSF75GP25 for biscuits; DCSF, DCSF95GP5, DCSF90GP10, DCSF85GP15, DCSF80GP20 and DCSF75GP25 for cakes; and DRSF, DRSF95GP5, DRSF90GP10, DRSF85GP15, DRSF80GP20 and DRSF75GP25 for rolls. Baking of the pastries was carried out in an electric convection oven (Esmach, 1200 W, 50 Hz, Esmach Ali Group SRL, Grisignano Di Zocco, Italy) set at 180 °C. All pastry products showed a similar baking behaviour. Each pastry formula was produced in three batches on the same day. Samples were taken from each batch and dough formula, packed in polypropylene food storage bags, sealed and stored at −20 °C until chemical analysis. After baking and cooling the products at room temperature for 12 h, a representative sample for each batch and pastry formulation was made up from randomly selected subsamples, packed in polypropylene bags, sealed and stored at −20 °C until chemical analysis. The remaining pastry products were packaged in plastic food storage containers and distributed for sensory analysis. Figure 2 illustrates the baked pastries, control samples and the formulas with GP as an SF substitute.

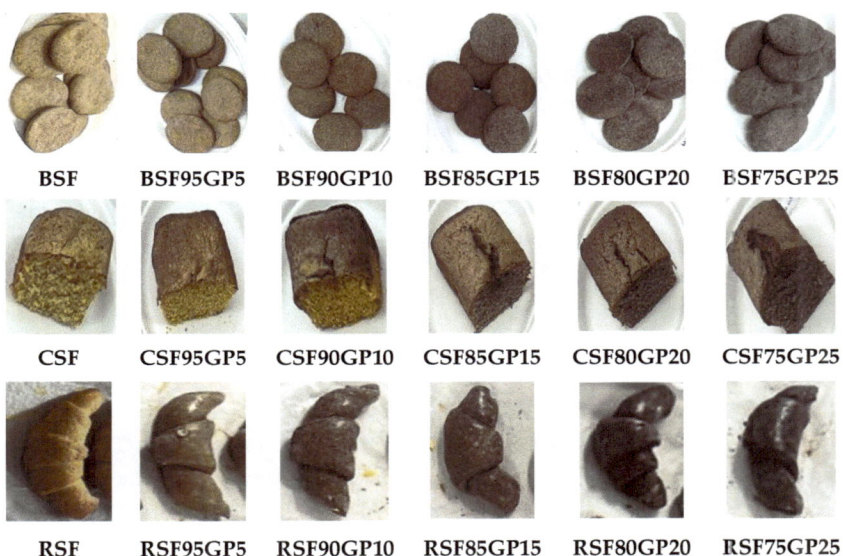

Figure 2. Pastry products after baking. BSF, CSF, RSF (control: biscuits, cakes and rolls); BSF95GP5, CSF95GP5, RSF95GP5, (biscuits, cakes and rolls: 95% spelt flour + 5% grape pomace); BSF90GP10, CSF90GP10, RSF90GP10 (biscuits, cakes and rolls: 90% spelt flour + 10% grape pomace); BSF85GP15, CSF85GP15, RSF85GP15 (biscuits, cakes and rolls: 85% spelt flour + 15% grape pomace); BSF80GP20, CSF80GP20, RSF80GP20 (biscuits, cakes and rolls: 80% spelt flour + 20% grape pomace); BSF75GP25, CSF75GP25, RSF75GP25 (biscuits, cakes and rolls: 75% spelt flour + 25% grape pomace).

2.3. Proximate Composition and Energy Value Evaluation of SF, GP and Pastry Products

The proximate composition of SF, GP and pastry products was evaluated in accordance with the standard method described by the Association of Official Analytical Chemists [48]. The carbohydrate content (%) was calculated by subtracting the protein, ash, lipid and moisture from 100. The energy value was calculated in accordance with Das et al. [49], taking into account that 1 g of carbohydrates provides 4 calories, 1 g of protein provides 4 calories and 1 g fat provides 9 calories, as shown in Equation (1).

$$\text{Energy value (kcal/100 g)} = \text{carbohydrates (\%)} \times 4 + \text{lipids (\%)} \times 9 + \text{proteins (\%)} \times 4 \tag{1}$$

2.4. Phytochemical Content and Antioxidant Activity of SF, GP and Pastry Products

2.4.1. Alcoholic Extract Preparation

The total phenolic content, total flavonoid content and antioxidant activity were assessed in ethanol extracts following the procedure described by Litwinek et al. [50] with minor modifications. From each sample of SF, GP, pastry formulas and corresponding dough, 1 g was weighed into lidded containers and mixed with 10 mL of 70% (v/v) ethanol.

The extraction was carried out for 120 min under continuous stirring at room temperature using an IDL magnetic stirrer (IDL GmbH & Co KG, Nidderau, Germany), and then the mixtures were centrifuged at 10,000 rpm for 10 min (Hettich EBA 21, Andreas Hettich GmbH & Co. KG., Tuttlingen, Germany). The supernatant was carefully collected and the residue was further washed with 70% (v/v) ethanol and subjected to extraction under continuous stirring for a further 60 min at room temperature, and then centrifuged at 10,000 rpm for 10 min. Afterwards, the supernatants were mixed and stored in darkness at $-20\ °C$ until analysis. For each sample, the extraction was performed in triplicate and each replicate was subsequently used in the analysis.

2.4.2. Evaluation of Total Phenolic Content

The method applied to quantify the total phenolic content (TPC) was based on the oxidation of hydroxyl groups of phenolic compounds in alkaline media using Folin–Ciocalteu reagent, as described by Tolve et al. [25] and Blanch at al. [51]. Prior to analysis, the alcoholic extracts were diluted with distilled water, as follows: 1:2.5 (v/v) for SF, pastry products and dough, and 1:50 (v/v) for GP. Further, 0.5 mL of diluted extracts were mixed with 2.5 mL of Folin–Ciocalteu reagent previously diluted 1:10 (v/v) with distilled water. Next, 2 mL of 7.5% Na_2CO_3 solution was dosed and the obtained mixture was allowed to incubate for 30 min at 50 °C in the INB500 thermostat (Memmert GmbH, Schwabach, Germany), after which the absorbance was recorded at a wavelength of 750 nm using the Specord 205 UV–Vis spectrophotometer from Analytik Jena Inc. (Jena, Germany) versus a blank sample prepared under the same experimental conditions. TPC was calculated based on a calibration curve plotted by using standard gallic acid solutions with concentrations ranging from 0.1 to 1.0 μM gallic acid equivalents (GAE)/mL. TPC was reported as mg GAE/100 g dry weight (DW) of sample.

2.4.3. Evaluation of Total Flavonoids Content (TFC)

The total flavonoids content (TFC) of the samples was assessed according to the procedure described by Al-Farsi et al. [52] with minor modifications. Briefly, a 3.0 mL aliquot of ethanol extract was mixed with 4.5 mL of distilled water and 1 mL of 0.3% $NaNO_2$ solution. The mixture was allowed to incubate for 6 min at 20 °C, and then 1 mL of 10% $Al(NO_3)_3$ was dosed and after another 6 min, 10 mL of 4% (w/w) NaOH solution was added. The volume of the mixture was made up to 25 mL with 70% (v/v) ethanol and the absorbance was read at 510 nm after 15 min, against a blank sample of 70% (v/v) ethanol. The calibration curve was plotted using standard quercetin (QE) solutions in the concentration range 0.5–50 μg/mL. TFC was reported as mg QE/100 g DW of sample.

2.4.4. Assessment of Antioxidant Activity by 1,1-Diphenyl-2-picrylhydrazyl (DPPH) Assay

The free radical scavenging activity of the samples was evaluated by DPPH assay [3], using a 0.1 mM DPPH solution prepared in 70% (v/v) ethanol. This method is frequently used to evaluate the antioxidant activity of the samples under investigation based on their ability to scavenge free radicals. The ethanol extracts previously obtained were further diluted with 70% (v/v) ethanol at a ratio of 1:5 (v/v) for SF, pastry products and dough, respectively, and 1:100 (v/v) for GP. Next, a 1.0 mL aliquot of the diluted extracts was mixed with 2.5 mL of 0.1 mM DPPH solution in 70% (v/v) ethanol. The mixtures were homogenised using a hot plate stirrer (IDL, IDL GmbH & Co KG, Nidderau, Germany), and incubated for 30 min in the dark, at a temperature of 20 °C. The absorbance of the mixture was read at a wavelength of 517 nm versus 70% (v/v) ethanol. Under the same working conditions, a control sample was also prepared, consisting of a mixture of 1 mL of 70% (v/v) ethanol and 2.5 mL of 0.1 mM DPPH solution in 70% (v/v) ethanol. Equation (2) was used to compute the DPPH radical scavenging activity:

$$\text{DPPH Scavenging Activity (\%)} = \frac{A_c - A_s}{A_c} \times 100 \qquad (2)$$

where A_c represents the absorbance of the control sample and A_s represents the absorbance read in the presence of the test sample. The calibration curve DPPH scavenging activity (%) versus Trolox concentration (μg/mL) was plotted using standard solutions of Trolox with concentrations in the range 1.0–25 μg Trolox/mL [53]. Based on the calibration curve, in the first step, the Trolox equivalent (TE) concentration was calculated. Then, taking into account the molar mass of Trolox and the concentration of the sample solution in g/mL, the antioxidant activity was determined and expressed in μM Trolox equivalent (TE)/g DW of sample.

2.4.5. Assessment of Antioxidant Activity by Ferric-Reducing Antioxidant Power (FRAP) Assay

The total antioxidant potential of the samples was assessed based on the ferric reducing antioxidant power (FRAP) assay based on the ability of antioxidant compounds from ethanol extracts to reduce Fe^{3+} from colourless ferric complex (Fe^{3+}–tripyridyltriazine) to Fe^{2+} due to the action of electron-donating antioxidant species at a low pH [54]. The deep blue-coloured Fe^{2+}–tripyridyltriazine complex shows a maximum absorbance at 593 nm [54]. Basically, the working solution was prepared from 100 mL acetate buffer (pH = 3.6), 10 mL of 10 mM TPTZ solution in 40 mM HCl and 10 mL of 20 mM $FeCl_3 \cdot 6H_2O$ solution. Prior to analysis, the extracts were diluted with distilled water, as follows: 1:2.5 (v/v) for SF, 1:50 (v/v) for GP and 1:2.5 (v/v) for pastries and dough. Then, 0.5 mL of diluted extracts were left to react with 2.5 mL of working solution at a temperature of 37 °C for 30 min, and then the absorbance was read at a wavelength of 593 nm versus a blank sample obtained under the same operational conditions. The results were calculated as μM Fe^{2+} equivalent/g DW of sample based on a calibration curve plotted using $FeSO_4 \cdot 7H_2O$ solutions with concentrations ranging from 0.05 to 0.5 μM Fe^{2+} equivalents/mL.

2.5. Assessment of Physical Characteristics

The cake and roll formulas were subjected to porosity and core elasticity assessment according to SR 91:2007 [55]. For biscuits, the spread ratio (SR) was determined as the ratio of biscuit diameter to biscuit height [56].

2.6. Sensory Evaluation

The sensory analysis was carried out by a panel consisting of 27 assessors (12 males and 15 females) recruited from the staff and students at University of Life Sciences "King Michael I", Timisoara (Romania), non-smokers, aged between 20 and 50, regular consumers of bakery products, without known cases of food allergies or food intolerances. Participation in this study was voluntary. The panellists were trained prior to the analysis to identify the attributes to be evaluated. The sensory analysis of the products followed laboratory ethical guidelines and written informed consent was obtained from each evaluator in conformity with the European Union guidelines on Ethics and Food-Related Research [57]. The samples were presented in cardboard plates with three-digit characters, one at a time to each evaluator. The panellists were asked to rinse their mouths with still water between sample evaluations. The sensory characteristics (appearance, flavour, texture, taste and overall acceptability) of the coded samples were assessed based on their liking degree using a five-point hedonic scale from 1—dislike extremely to 5—like extremely. The score ranges and level of acceptability were grouped as follows: 1.00–1.49 (extremely unpleasant); 1.5–2.49 (slightly unpleasant); 2.50–3.49 (neither pleasant nor unpleasant); 3.5–4.49 (slightly pleasant); 4.5–5.00 (extremely pleasant) [58].

2.7. Statistical Analysis

The data were obtained from three independently performed experiments, each of which were analysed in three replicates. The results represent an average of three independent experiments and were expressed as mean ± standard deviation (SD). One-way analysis of variance (ANOVA) was conducted, followed by multiple comparisons of means using the post hoc Tukey test and Levene's test for equal variances to evaluate the statistical significance of differences among formulations. Assumptions regarding homogeneity of variance, normality of residuals or residuals have the same distribution and independence of residuals were met. The differences were considered statistically significant at a probability less than 0.05 ($p < 0.05$).

3. Results and Discussion

3.1. Proximate Composition of GP, SF and Pastry Products

The analytical results for the proximate composition of Merlot grape pomace powder (GP), spelt flour and pastry products are presented in Table 2.

Table 2. Proximate composition of spelt flour, grape pomace powder and pastry products (control samples and fortified formulas).

Sample	Moisture (g/100 g)	Ash (g/100 g)	Protein (g/100 g)	Lipids (g/100 g)	Carbohydrates (g/100 g)	Sugar (g/100 g)	Energy Value (kcal/100 g)
SF	12.291 ± 0.026 [a]	1.608 ± 0.008 [b]	14.594 ± 0.066 [a]	2.506 ± 0.012 [b]	69.001	1.619 ± 0.014 [a]	356.934
GP	4.872 ± 0.011 [b]	6.634 ± 0.023 [a]	13.421 ± 0.054 [b]	8.861 ± 0.043 [a]	66.212	1.517 ± 0.011 [b]	398.281
Pastry products							
BSF	6.024 ± 0.030 [a]	1.841 ± 0.006 [f]	8.803 ± 0.027 [a]	21.788 ± 0.071 [e]	61.544	18.708 ± 0.075 [a]	477.480
BSF95GP5	5.767 ± 0.034 [b]	1.927 ± 0.010 [e]	8.678 ± 0.039 [b]	21.916 ± 0.066 [e]	61.712	18.664 ± 0.066 [a]	478.804
BSF90GP10	5.558 ± 0.037 [c]	2.019 ± 0.014 [d]	8.567 ± 0.033 [c]	22.037 ± 0.068 [d]	61.819	18.625 ± 0.053 [a]	479.877
BSF85GP15	5.371 ± 0.028 [d]	2.119 ± 0.011 [c]	8.456 ± 0.040 [d]	22.169 ± 0.075 [c]	61.885	18.582 ± 0.070 [a]	480.885
BSF80GP20	5.164 ± 0.022 [e]	2.208 ± 0.012 [b]	8.348 ± 0.032 [e]	22.297 ± 0.056 [b]	61.983	18.541 ± 0.061 [a]	481.997
BSF75GP25	4.919 ± 0.025 [f]	2.290 ± 0.015 [a]	8.242 ± 0.038 [f]	22.431 ± 0.080 [a]	62.118	18.519 ± 0.073 [a]	483.319
CSF	15.622 ± 0.039 [a]	0.987 ± 0.003 [f]	6.831 ± 0.034 [a]	27.669 ± 0.057 [c]	48.891	23.651 ± 0.089 [a]	471.909
CSF95GP5	15.435 ± 0.054 [b]	1.011 ± 0.005 [e]	6.767 ± 0.033 [a]	27.726 ± 0.071 [c]	49.061	23.624 ± 0.069 [a]	472.846
CSF90GP10	15.257 ± 0.037 [c]	1.052 ± 0.007 [d]	6.708 ± 0.025 [b]	27.795 ± 0.054 [c]	49.188	23.578 ± 0.075 [a]	473.739
CSF85GP15	15.022 ± 0.046 [d]	1.099 ± 0.011 [c]	6.657 ± 0.037 [c]	27.867 ± 0.079 [c]	49.355	23.531 ± 0.065 [a]	474.851
CSF80GP20	14.843 ± 0.050 [e]	1.149 ± 0.009 [b]	6.602 ± 0.027 [d]	27.928 ± 0.084 [b]	49.478	23.473 ± 0.073 [a]	475.672
CSF75GP25	14.637 ± 0.036 [f]	1.201 ± 0.012 [a]	6.537 ± 0.034 [e]	28.001 ± 0.098 [a]	49.624	23.452 ± 0.083 [a]	476.653
RSF	26.163 ± 0.046 [a]	2.212 ± 0.006 [f]	10.668 ± 0.040 [a]	9.857 ± 0.025 [f]	51.100	10.961 ± 0.076 [a]	335.785
RSF95GP5	25.856 ± 0.050 [b]	2.324 ± 0.005 [e]	10.536 ± 0.046 [b]	10.006 ± 0.030 [e]	51.278	10.928 ± 0.063 [a]	337.310
RSF90GP10	25.579 ± 0.059 [c]	2.435 ± 0.009 [d]	10.409 ± 0.034 [b]	10.155 ± 0.026 [d]	51.422	10.901 ± 0.058 [a]	338.719
RSF85GP15	25.305 ± 0.064 [d]	2.540 ± 0.010 [c]	10.237 ± 0.058 [c]	10.309 ± 0.041 [c]	51.609	10.867 ± 0.055 [a]	340.165
RSF80GP20	25.027 ± 0.046 [e]	2.653 ± 0.008 [b]	10.138 ± 0.051 [c]	10.456 ± 0.038 [b]	51.726	10.821 ± 0.076 [a]	341.560
RSF75GP25	24.711 ± 0.053 [f]	2.761 ± 0.011 [a]	10.009 ± 0.047 [d]	10.612 ± 0.027 [a]	51.907	10.779 ± 0.081 [a]	343.172

SF: spelt flour; GP: grape pomace; BSF, CSF, RSF (control: biscuits, cakes and rolls); BSF95GP5, CSF95GP5, RSF95GP5, (biscuits, cakes and rolls: 95% spelt flour + 5% grape pomace); BSF90GP10, CSF90GP10, RSF90GP10 (biscuits, cakes and rolls: 90% spelt flour + 10% grape pomace); BSF85GP15, CSF85GP15, RSF85GP15 (biscuits, cakes and rolls: 85% spelt flour + 15% grape pomace); BSF80GP20, CSF80GP20, RSF80GP20 (biscuits, cakes and rolls: 80% spelt flour + 20% grape pomace); BSF75GP25, CSF75GP25, RSF75GP25 (biscuits, cakes and rolls: 75% spelt flour + 25% grape pomace). The values represent the mean of three independent experiments ± standard deviation (SD). The values with different superscripts in a column are statistically different (one-way ANOVA, $p < 0.05$).

It can be noted that the drying of raw grape pomace at 60 °C resulted in a final moisture content of 4.872%. Drying of plant origin by-products for flour production usually takes place below 65 °C to reduce losses of phenolic compounds, anthocyanins and proteins [35,42].

The SF moisture was far higher than that of GP. Hence, a significant ($p < 0.05$) moisture reduction with the increasing level of GP incorporation in the pastry products was observed, reducing from 6.024 g/100 g (control) to 4.919 g/100 g (biscuits with 25% GP), 15.622 g/100 g (control) to 14.637 g/100 g (cakes with 25% GP) and from 26.163 g/100 g to 24.711 g/100 g (rolls with 25% GP). Moisture reduction in pastry products with different amounts of GP was also reported by other authors [30,31].

The chemical composition of GP was similar to that found by other authors who reported values of 6.4% for ash, 13.87% for protein [59] and 7.30% for lipids [60]. The data in Table 2 also show the high protein content of SF, in line with Escarnot et al. [39], who reported an average content of 14.04% at 90% dry matter content. With regard to pastry products, the changes induced in their proximate composition can be observed with a gradual increase in the level of incorporated GP from 5 to 25%.

GP-enriched formulations have higher ash and lipid content than the control, while the protein content slightly decreases. This finding is closely related to the contribution made by GP addition and has been confirmed by other authors, as follows: Troilo et al. [16] reported increases in lipid and ash content and decreases in protein when using GP in the production of functional muffins, and the same trend was revealed by Karnop et al. [31] on the physico-chemical and functional properties of cookies enriched with Bordeaux GP; Nakov et al. [30] found improvements in lipid, ash and protein content by including GP powder in cake formulations, while Acun and Gül [61] reported increased amounts of these substances in cakes when the level of GP included was 5, 10 and 15%.

The lipid content in SF was around 3.5 times lower compared to GP, where the lipids come from the grape seeds. The increasing lipid content in GP-enriched products is a result of the abundant amount of lipids provided by the incorporated GP. Consequently, the lipid content increased from 21.788 g/100 g (control) to 22.431 g/100 g for biscuits with 25% GP, from 27.669 g/100 g (control) to 28.001 g/100 g for cakes with 25% GP and from 9.857 g/100 g (control) to 10.612 g/100 g for rolls with 25% GP incorporation.

The content of minerals (ash) in GP was more than four times higher than in SF (6.634 vs. 1.608 g/100 g). The mineral substances provided by grape pomace led to increases in the ash content of the products: from 1.841 g/100 g (control) to a maximum of 2.290 g/100 g (biscuits with 25% GP), from 0.987 g/100 g (control) to a maximum of 1.201 g/100 g (cakes with 25% GP) and from 2.212 g/100 g (control) to a maximum of 2.761 g/100 g (rolls with 25% GP).

A slight decrease in the protein content with increasing levels of GP incorporation was observed, as GP has a lower protein content compared to SF. The improvements in protein content reported by other authors in cakes [30] are due to the fact that common wheat flour with a lower protein content than SF was used. In this situation, the incorporation of GP improved the protein content of fortified products.

The carbohydrate content of pastry formulas with different levels of GP was slightly higher compared to the control samples, with increases of 0.6–0.8 g/100 g. Small decreases in sugar content of about 0.2 g/100 g were obtained with an increasing percentage of GP, but without statistical significance ($p > 0.05$).

The energy value was not strongly influenced by increasing the level of GP incorporation. Increases of 5.839 kcal/100 g, 4.744 kcal/100 g and 7.387 kcal/100 g were recorded in biscuits, cakes and rolls, compared to corresponding control samples, for a 25% GP inclusion level.

The results showed the improvements in the nutritional profile of pastry products, particularly in terms of lipid and ash content, with increasing levels of grape pomace incorporation.

3.2. Phytochemical Content and Antioxidant Activity of GP and SF

In the present study, the total phenolic content (TPC), total flavonoid content (TFC) and antioxidant activity evaluated by FRAP and DPPH assays were tested for SF GP, and the results are reported in Table 3.

Phenolic compounds are the main class of bioactive substances found in SF, forming part of the plant defence system with a variety of functions [40,62]. The data in Table 3 indicate a TPC for SF of 130.211 mg GAE/100 g DW, consistent with the results of Wang et al. [40], ranging from 120.72 to 190.42 mg GAE/100 g DW, with this variability being associated with the agronomic practices, spelt wheat varieties and milling protocols.

Table 3. Total phenolic content, total flavonoids content and antioxidant activity of spelt flour and grape pomace powder.

Sample	TPC (mg GAE/100 g DW)	TFC (mg QE/100 g DW)	FRAP (μM Fe^{2+}/g DW)	DPPH (μM TE/g DW)
SF	130.211 ± 0.584 [b]	81.156 ± 0.437 [b]	3.689 ± 0.028 [b]	4.937 ± 0.041 [b]
GP	4708.683 ± 4.053 [a]	3975.457 ± 3.291 [a]	305.925 ± 1.817 [a]	409.378 ± 2.019 [a]

SF: spelt flour; GP: grape pomace powder. The values represent the mean of three independent experiments ± standard deviation (SD). The values with different superscripts in a column are statistically different (one-way ANOVA, $p < 0.05$).

Our results showed that GP achieved a TPC value of 4708.683 mg GAE/100 g DW, in line with data reported by Rockenbach et al. [63], varying between 3300 and 7500 mg GAE/100 g DW. Other authors found TPC values ranging from 120 to 7480 mg GAE/100 g DW [64], 2300 to 3800 mg GAE/100 g DW [12] and 2700 to 5300 mg GAE/100 g DW [65] depending on the grape variety. Research evidence shows that the prevalent group of bioactive molecules with strong antioxidant activity in GP are phenolic compounds, mostly derived from grape skins and seeds and belonging to the class of flavonoids and non-flavonoid compounds, with a wide range of biological effects [9,10].

The flavonoid content of GP (3975.457 mg QE/100 g DW) was significantly higher than that of SF (81.156 mg QE/100 g DW). Similar TFC values in GP were reported by Cui et al. [66] (2833–3719 mg QE/100 g DW) and Putnik et al. [67] (3628 mg QE/100 g DW). Regarding SF, Ivanišová et al. [68] reported a TFC ranging from 33 to 155 mg QE/100 g DW, while Sumczynski et al. [69] showed a value of 37 mg QE/100 g DW.

For the antioxidant activity of GP, a value of 305.925 μM Fe^{2+}/g DW by FRAP assay and 409.378 μM TE/g DW by DPPH assay was obtained. Our results for GP are well aligned with the data of Rockenbach et al. [63], ranging from 118 to 250 μmol of Fe^{2+}/g DW (FRAP), respectively in the range 188–506 μM TE/g DW (DPPH). With regard to the SF, our data for FRAP (3.689 μM Fe^{2+}/g DW) and DPPH (4.937 μM TE/g DW) closely match the results reported by Wang et al. [40], ranging from 2.2 to 7.4 μmol of Fe^{2+}/g DW (FRAP) and from 4.5 to 13.8 μM TE/g DW (DPPH). Similar DPPH values of SF were also reported by Abdel-Aal and Rabalski [70].

The data in Table 3 reflect significantly ($p < 0.05$) higher values of TPC, TFC, FRAP and DPPH for GP compared to SF. This finding fully justifies the use of GP as a high-value substitute for SF to obtain composite flour with enhanced functionality. The production of grape pomace powder offers an attractive destination for industrial waste, as this flour has shown great potential to improve the functional value of other conventional flours or foodstuffs. As composite flours are of particular interest in the development of innovative high-quality food products, the functional properties of the component materials are a key parameter in assessing their suitability for this purpose.

3.3. Phytochemical Content and Antioxidant Activity of Dough and Pastry Products

In our study, the functional potential of GP was exploited in three types of pastry products, which differ both in terms of manufacturing recipe and technological process. In addition, the impact of the leavening agents (baking powder as a chemical raising agent for biscuits and cakes, and yeast as biological leavening agent for rolls) on the investigated properties is also considered. This approach integrates the raw material requirements into the assessment of the functional properties of the pastry products.

Changes in phytochemical content and antioxidant activity following progressive supplementation of GP were tested in both dough and pastry. The phytochemical content and antioxidant activity of the dough obtained during the production process of the pastry products are shown in Table 4.

Table 4. Total phenol content, total flavonoid content and antioxidant activity of dough (control samples and fortified formulas).

Sample	TPC (mg GAE/100 g DW)	TFC (mg QE/100 g DW)	FRAP (μM Fe^{2+}/g DW)	DPPH (μM TE/g DW)
DBSF	66.323 ± 0.362 [f, B]	48.137 ± 0.294 [f, B]	1.879 ± 0.024 [f, B]	2.515 ± 0.029 [f, B]
DBSF95GP5	180.267 ± 0.767 [e, B]	132.465 ± 0.604 [e, B]	8.917 ± 0.108 [e, B]	11.978 ± 0.143 [e, B]
DBSF90GP10	291.244 ± 1.486 [d, B]	216.605 ± 1.211 [d, B]	16.138 ± 0.173 [d, B]	22.518 ± 0.181 [d, B]
DBSF85GP15	397.101 ± 1.877 [c, B]	302.448 ± 1.547 [c, B]	24.109 ± 0.208 [c, B]	32.772 ± 0.193 [c, B]
DBSF80GP20	500.551 ± 1.935 [b, B]	384.141 ± 1.628 [b, B]	32.259 ± 0.246 [b, B]	42.841 ± 0.260 [b, B]
DBSF75GP25	599.799 ± 2.104 [a, B]	479.135 ± 1.714 [a, B]	40.200 ± 0.361 [a, B]	54.256 ± 0.412 [a, B]
DCSF	37.871 ± 0.275 [f, C]	27.501 ± 0.224 [f, C]	1.017 ± 0.021 [f, C]	1.361 ± 0.025 [f, C]
DCSF95GP5	98.084 ± 0.439 [e, C]	73.841 ± 0.331 [e, C]	5.401 ± 0.059 [e, C]	6.481 ± 0.073 [e, C]
DCSF90GP10	158.134 ± 0.501 [d, C]	127.487 ± 0.593 [d, C]	9.887 ± 0.115 [d, C]	12.185 ± 0.136 [d, C]
DCSF85GP15	222.448 ± 0.547 [c, C]	181.512 ± 0.602 [c, C]	13.864 ± 0.137 [c, C]	17.733 ± 0.159 [c, C]
DCSF80GP20	282.755 ± 1.219 [b, C]	238.541 ± 0.903 [b, C]	17.586 ± 0.154 [b, C]	23.181 ± 0.178 [b, C]
DCSF75GP25	348.479 ± 1.425 [a, C]	290.651 ± 1.071 [a, C]	21.265 ± 0.188 [a, C]	29.358 ± 0.185 [a, C]
DRSF	73.128 ± 0.463 [f, A]	53.128 ± 0.342 [f, A]	2.279 ± 0.031 [f, A]	3.049 ± 0.037 [f, A]
DRSF95GP5	201.818 ± 0.854 [e, A]	157.487 ± 0.611 [e, A]	13.223 ± 0.147 [e, A]	14.526 ± 0.147 [e, A]
DRSF90GP10	314.397 ± 1.207 [d, A]	251.012 ± 1.004 [d, A]	22.157 ± 0.164 [d, A]	27.307 ± 0.191 [d, A]
DRSF85GP15	426.532 ± 1.446 [c, A]	358.009 ± 1.317 [c, A]	31.070 ± 0.183 [c, A]	39.742 ± 0.239 [c, A]
DRS0GP20	535.685 ± 1.913 [b, A]	466.108 ± 1.629 [b, A]	40.961 ± 0.209 [b, A]	51.952 ± 0.308 [b, A]
DRSF75GP25	630.041 ± 2.307 [a, A]	562.083 ± 1.853 [a, A]	49.6674 ± 0.252 [a, A]	63.795 ± 0.363 [a, A]

DBSF, DCSF, DRSF (control dough: biscuits, cakes and rolls); DBSF95GP5, DCSF95GP5, DRSF95GP5, (dough for biscuits, cakes and rolls: 95% spelt flour + 5% grape pomace); DBSF90GP10, DCSF90GP10, RRSF90GP10 (dough for biscuits, cakes and rolls: 90% spelt flour + 10% grape pomace); DBSF85GP15, DCSF85GP15, DRSF85GP15 (dough for biscuits, cakes and rolls: 85% spelt flour + 15% grape pomace); DBSF80GP20, DCSF80GP20, DRSF80GP20 (dough for biscuits, cakes and rolls: 80% spelt flour + 20% grape pomace); DBSF75GP25, DCSF75GP25, DRSF75GP25 (dough for biscuits, cakes and rolls: 75% spelt flour + 25% grape pomace). The values represent the mean of three independent experiments ± standard deviation (SD). Values with different superscripts in a column are statistically different (one-way ANOVA, $p < 0.05$). Lowercase letters (a–f) differentiate the formulas within each dough type, while uppercase letters (A–C) differentiate the three types of dough obtained from the same composite flour.

The complexity of dough composition is often overlooked when a functional material is incorporated into a bakery product, making it difficult to understand the differences that occur in the bioactive compound content of raw materials and final products. Also, the amount of functional material required to obtain a certain amount of product is often neglected when assessing the bioactive properties of finished products. The evaluation of pastry formulas in close relation to the amount of composite flour required to produce each type of product is one of the aspects that supports the added value of this approach. In our case, the amount of composite flour required for a particular pastry product has a major impact on the differences registered in the investigated properties. The data in Table 4 showed that the highest values were recorded for rolls, followed by biscuits and cakes. The values were directly correlated to the amount and composition of flour mixtures based on SF and GP required for a particular pastry formulation.

Based on the recipe of the pastry products (Table 1), the amount of composite flour required to prepare 100 g of dough is 47.664 g for biscuits, 24.852 g for cakes and 47.902 g for rolls. The quantities of composite flours needed to prepare 100 g of products were also calculated and varied in the range of 54.828–55.473 g for biscuits, 26.754–27.067 g for cakes and 52.292–53.321 g for rolls. This justifies the lower values recorded in cake dough, considering that obtaining 100 g of product requires about 50% of the amount needed to obtain biscuits or rolls. In the case of biscuit and roll dough, if we take into account the amount of composite flour involved in the recipe, it would have been expected that close values would be obtained for the two types of dough. The higher values were basically obtained in the dough for the rolls, and this is explained by the presence of a liquid ingredient (milk) in their recipe, which led to an increase in moisture content, making the reported values per 100 g DW higher for the rolls. Within each product type, the TPC, TFC and both antioxidant tests increased significantly ($p < 0.05$) in fortified dough compared to the control with increasing levels of GP incorporation. A closer look at the three types of dough made from the same composite flour revealed significant differences ($p < 0.05$) between the functional properties.

The TPC of the GP-enriched pastry formulas versus control samples are shown in Figure 3.

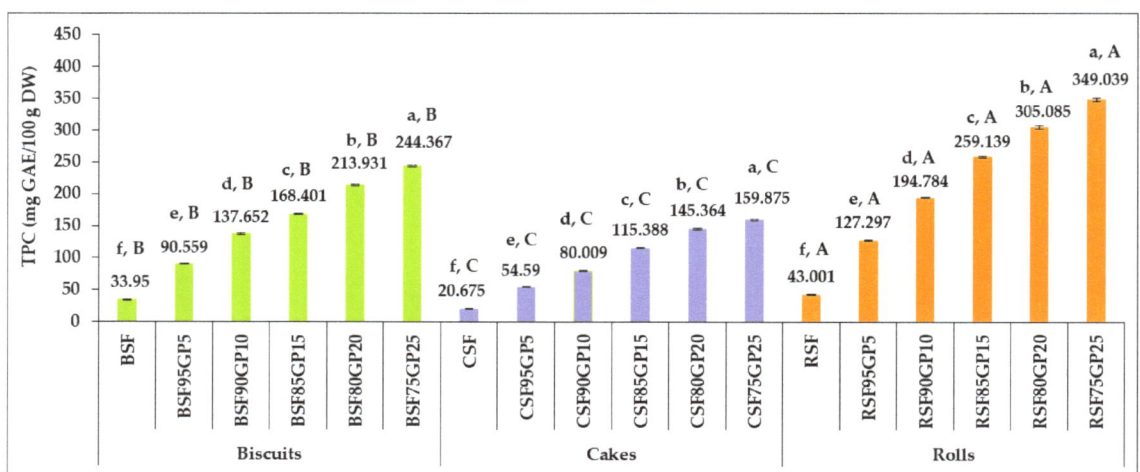

Figure 3. Changes in the total phenolic content of pastry formulas, in response to increasing the percentage of incorporated grape pomace. BSF, CSF, RSF (control: biscuits, cakes and rolls); BSF95GP5, CSF95GP5, RSF95GP5, (biscuits, cakes and rolls: 95% spelt flour + 5% grape pomace); BSF90GP10, CSF90GP10, RSF90GP10 (biscuits, cakes and rolls: 90% spelt flour + 10% grape pomace); BSF85GP15, CSF85GP15, RSF85GP15 (biscuits, cakes and rolls: 85% spelt flour + 15% grape pomace); BSF80GP20, CSF80GP20, RSF80GP20 (biscuits, cakes and rolls: 80% spelt flour + 20% grape pomace); BSF75GP25, CSF75GP25, RSF75GP25 (biscuits, cakes and rolls: 75% spelt flour + 25% grape pomace). The values represent the mean of three independent experiments ± standard deviation (SD). The values for bars with different letters are statistically different (one-way ANOVA, $p < 0.05$). Lowercase letters (a–f) differentiate the formulas within each pastry type, while uppercase letters (A–C) differentiate the three pastry types obtained from the same composite flour.

It can be seen that the increasing GP concentrations in the enriched pastry formulations led to a progressive augmentation of TPC. The highest values were noted in rolls, followed by biscuits and cakes. The TPC increased 2.667-fold, 2.640-fold and 2.960-fold as GP replacement in biscuits, cakes and rolls increased from 0% to 5%. As the level of GP incorporation increased from 0% to 25%, TPC increased 7.198-fold, 7.733-fold and 8.117-fold over the control samples. Our findings are in agreement with those of other studies that have shown increases in the TPC of pastry products when the GP level included in the recipe was increased: Nakov et al. [30] reported significant improvements in cakes as the percentage of incorporated GP powder increased from 4% to 10%, Maner et al. [71] found an improvement in the functional properties of cookies by increasing the level of added GP flour in cookies from 5% to 20%, and Karnop et al. [31] also found considerable increases in TPC by replacing the wheat flour in a cookie recipe with 20, 25 and 30% GP. Mildner-Szkudlarz et al. [26] reported increases in TPC from 211 mg GAE/100 g DM for an addition of 10% GP to 445 mg GAE/100 g DM in the case of 30% GP incorporation.

The results presented in Figure 4 show that the fortification of pastry products with GP had a positive effect on the total flavonoids content, with the highest improvements achieved at a GP incorporation level of 25%.

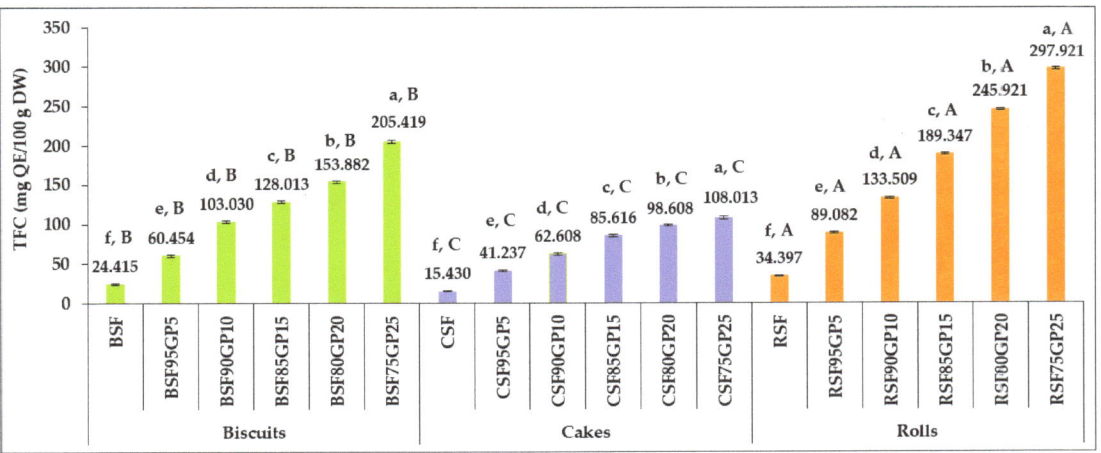

Figure 4. Changes in the total flavonoids content of pastry formulas, in response to increasing the percentage of incorporated grape pomace. BSF, CSF, RSF (control: biscuits, cakes and rolls); BSF95GP5, CSF95GP5, RSF95GP5, (biscuits, cakes and rolls: 95% spelt flour + 5% grape pomace); BSF90GP10, CSF90GP10, RSF90GP10 (biscuits, cakes and rolls: 90% spelt flour + 10% grape pomace); BSF85GP15, CSF85GP15, RSF85GP15 (biscuits, cakes and rolls: 85% spelt flour + 15% grape pomace); BSF80GP20, CSF80GP20, RSF80GP20 (biscuits, cakes and rolls: 80% spelt flour + 20% grape pomace); BSF75GP25, CSF75GP25, RSF75GP25 (biscuits, cakes and rolls: 75% spelt flour + 25% grape pomace). The values represent the mean of three independent experiments ± standard deviation (SD). The values for bars with different letters are statistically different (one-way ANOVA, $p < 0.05$). Lowercase letters (a–f) differentiate the formulas within each pastry type, while uppercase letters (A–C) differentiate the three pastry types obtained from the same composite flour.

The TFC increased 2.476-, 2.672- and 2.590-fold as GP replacement in biscuits, cakes and rolls increased from 0% to 5%. When the GP incorporation reached 25%, the TFC increased 8.414-, 7.000- and 8.661-fold in biscuits, cakes and rolls over the control samples.

Our results are consistent with those reported by Maner et al. [71], who observed significant improvements in the TFC of cookies, in the range of 0.32–1.347 mg catechin equivalent/g, by increasing the level of grape seed powder incorporated. A similar trend was recorded in the study carried out by Nakov et al. [30], when a gradual increase in TFC was recorded by replacing wheat flour with GP at the rates of 4%, 6%, 8% and 10%.

Significant differences ($p < 0.05$) were obtained between TPC and TFC of pastry samples with different levels of incorporation as well as between the formulations of three types of products prepared from the same composite flour.

The data in Figures 2 and 3 show that an incremental increase in the level of GP incorporation in the recipe did not result in the same growth rates for TPC and TFC. Increases in the dose of GP of 2-, 3-, and 5-fold in the products did not lead to proportional increases in TPC and TFC. This finding was also reported in the studies carried out by Gaita et al. [33] and by Sęczyk et al. [72] when the changes induced in the antioxidant properties of pasta by the addition of plant origin functional materials were investigated. Several factors, including the binding of phenolic compounds to components of the food matrix, contribute to limiting the increases in their content. The incorporation of materials rich in phenolic compounds resulted in significant modifications in the relationships between the chemical constituents of farinaceous products, affecting their bioavailability. The fortification efficiency is restricted by the interactions that arise between phenolic species and the proteins in the matrix [72]. The powerful affinity of phenolic species to form different types of bonds with proteins present in the product composition can in fact result in a decline in their free forms, negatively affecting the bioavailability. The TPC is dependent on their particular ability to combine with the proteins from wheat flour or with other constituents of the matrix. On this basis, the interaction of bioactive compounds from GP with the constituents of the SF matrix should be carefully considered for the design of new pastries.

Figure 5 illustrates the changes in antioxidant properties assessed by FRAP and DPPH assays of the three types of GP-enriched pastry products compared to the control.

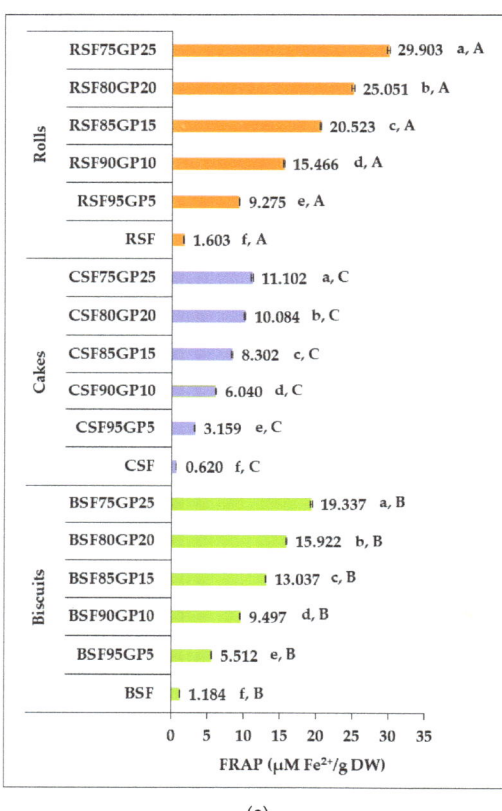
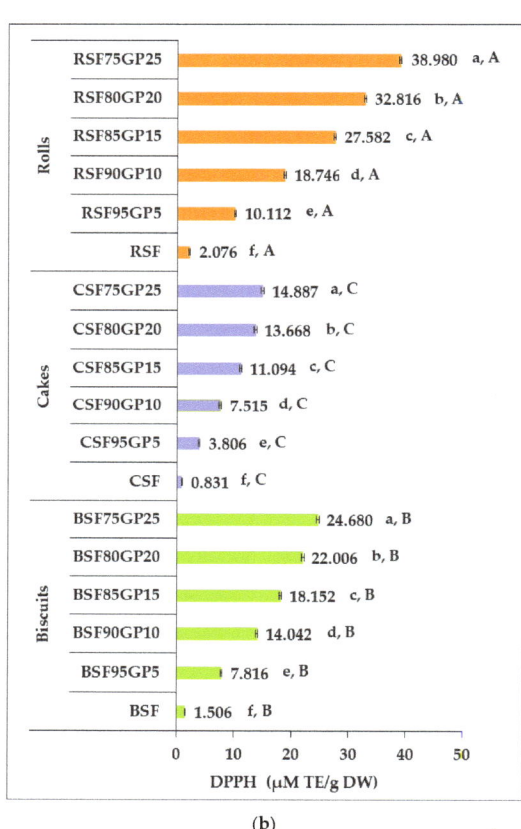

(a) (b)

Figure 5. Changes in the antioxidant activity of pastry formulas in response to increasing the percentage of incorporated grape pomace: (**a**) FRAP value; (**b**) DPPH value. BSF, CSF, RSF (control: biscuits, cakes and rolls); BSF95GP5, CSF95GP5, RSF95GP5, (biscuits, cakes and rolls: 95% spelt flour + 5% grape pomace); BSF90GP10, CSF90GP10, RSF90GP10 (biscuits, cakes and rolls: 90% spelt flour + 10% grape pomace); BSF85GP15, CSF85GP15, RSF85GP15 (biscuits, cakes and rolls: 85% spelt flour + 15% grape pomace); BSF80GP20, CSF80GP20, RSF80GP20 (biscuits, cakes and rolls: 80% spelt flour + 20% grape pomace); BSF75GP25, CSF75GP25, RSF75GP25 (biscuits, cakes and rolls: 75% spelt flour + 25% grape pomace). The values represent the mean of three independent experiments ± standard deviation (SD). The values for bars with different letters are statistically different (one-way ANOVA, $p < 0.05$). Lowercase letters (a–f) differentiate the formulas within each pastry type, while uppercase letters (A–C) differentiate the three pastry types obtained from the same composite flour.

The antioxidant activity increased markedly by the inclusion of GP into the pastry formulation. Biscuits, cakes and rolls with different incorporation levels were found to be significantly higher ($p < 0.05$) in both FRAP and DPPH values over the control. This finding can be attributed to the increase in TPC because phenolic substances contained in winery grape waste have well-demonstrated antioxidant activities [4,5].

The FRAP value increases achieved in biscuits, cakes and rolls for the incorporation of 25% GP were 16.334-, 17.915- and 18.659-fold over the control samples, while the DPPH increases in BSF75GP25, CSF75GP25 and RSF75GP25 were 16.384-, 17.908- and 18.775-fold over the control samples. The samples with the highest level of GP inclusion showed the highest antioxidant activity. Similarly, Maner et al. [71] noted an increase in FRAP value with the addition of 5, 10, 15 and 20% GP in a cake recipe by 2.5, 6.4, 11 and 16 times compared to the control. Other authors [16,20,26,61] reported significant improvements in the DPPH value of bakery products with increasing GP dose. Hence, fortification with GP should allow the manufacture of pastry formulations with improved nutritional and functional properties.

With regard to the three types of products, the highest values of FRAP and DPPH were recorded in rolls, followed by biscuits and cakes, for all levels of GP incorporation. The differences recorded between the antioxidant activity of pastry formulas fortified with the same level of GP were statistically significant ($p < 0.05$).

The methods used to assess antioxidant activity differ in principle, as DPPH is used to scavenge the free radicals by phenolic compounds in GP-enriched pastries, while FRAP is used to measure the ability of antioxidant compounds in samples to reduce the ferric ions in solution to the ferrous ions, to prevent the oxidation reaction [73]. Gómez-Brandón et al. [7] stated that the phytochemicals responsible for the antioxidant activity in GP are most likely phenolic compounds, especially phenolic acids, anthocyanins and other flavonoid compounds. This finding supports the idea that increased levels of TPC and TFC in pastries, as a result of GP incorporation, strongly contribute to the enhancement of their FRAP value. The results reported by Ky et al. [74] revealed a high contribution of TPC to DPPH radical scavenging activity of grapes and grape pomaces, which indicates that the level of TPC could be a good indication of their functionality.

All of this reinforces the idea that the ability of the formulated products to scavenge free radicals was strongly dependent on the level of TPC and TFC provided by the incorporated GP. Along with increasing TPC and TFC, a significant increase in the antioxidant activity of pastry products can be achieved.

3.4. Retention Rate of Phytochemical Content and Antioxidant Activity in Pastry Products in Response to Baking

In addition to the fact that the functional attributes of pastry formulas are closely related to the amount of composite flour required for their production, of particular importance, that also supports the added value of this study, is the calculation of the retention rate of the phytochemical content and antioxidant attributes of the products, in relation to the corresponding dough, in order to assess the losses caused by the baking process. Table 5 summarises the retention rates of TPC, TFC and antioxidant activity in pastry products relative to dough as a result of baking.

TPC retention rates reported to the corresponding dough were observed in the range 41–51% for biscuits, 46–56% for cakes and 55–63% for rolls. Flavonoid retention rates were 40–51% for biscuits, 37–56% for cakes and 53–68% for rolls. The FRAP value recorded retention rates in the biscuits, cakes and rolls of 48–63%, 52–61% and 60–70%, while the DPPH value showed retention rates of 45–65%, 50–63 and 61–70%. Our results revealed significant losses of functional properties in response to heat treatment during baking, which closely match the data reported by Nakov et al. [30], revealing losses ranging from 31.19% to 49.15%. Despite the fact that baking the pastries at 180 °C induced significant losses in their functional properties, they still retained a high level of bioactive compounds.

Table 5. Retention rate of phytochemical content and antioxidant activity of pastry products in response to baking.

Sample	Retention Rate of TPC (%)	Retention Rate of TFC (%)	Retention Rate of FRAP (%)	Retention Rate of DPPH (%)
BSF	51.189 ± 0.125 [a, C]	50.721 ± 0.114 [a, C]	63.006 ± 0.154 [a, B]	59.905 ± 0.147 [c, C]
BSF95GP5	50.236 ± 0.153 [b, C]	45.638 ± 0.138 [c, C]	61.811 ± 0.189 [b, B]	65.254 ± 0.161 [a, B]
BSF90GP10	47.263 ± 0.102 [c, C]	47.566 ± 0.103 [b, C]	58.849 ± 0.127 [c, C]	62.360 ± 0.134 [b, B]
BSF85GP15	42.408 ± 0.129 [d, C]	42.326 ± 0.129 [d, C]	54.075 ± 0.165 [d, C]	55.389 ± 0.169 [d, C]
BSF80GP20	42.739 ± 0.121 [d, C]	40.059 ± 0.113 [e, C]	49.356 ± 0.140 [e, C]	51.367 ± 0.146 [e, C]
BSF75GP25	40.741 ± 0.097 [e, C]	42.873 ± 0.102 [f, B]	48.102 ± 0.111 [f, C]	45.488 ± 0.107 [f, C]
CSF	54.593 ± 0.132 [b, B]	56.108 ± 0.137 [a, B]	60.954 ± 0.149 [a, C]	61.099 ± 0.151 [c, B]
CSF95GP5	55.656 ± 0.170 [a, B]	55.846 ± 0.158 [a, B]	58.489 ± 0.178 [c, C]	58.714 ± 0.179 [d, C]
CSF90GP10	50.596 ± 0.117 [e, B]	49.110 ± 0.108 [b, B]	61.090 ± 0.132 [a, B]	61.677 ± 0.133 [b, C]
CSF85GP15	51.872 ± 0.158 [c, B]	47.168 ± 0.144 [c, B]	59.882 ± 0.183 [b, B]	62.558 ± 0.172 [a, B]
CSF80GP20	51.410 ± 0.146 [d, B]	41.338 ± 0.118 [d, B]	57.334 ± 0.162 [d, B]	58.960 ± 0.167 [d, B]
CSF75GP25	45.878 ± 0.109 [f, B]	37.163 ± 0.107 [e, C]	52.209 ± 0.143 [e, B]	50.709 ± 0.124 [e, B]
RSF	58.802 ± 0.144 [d, A]	64.744 ± 0.159 [a, A]	70.332 ± 0.172 [a, A]	68.087 ± 0.167 [c, A]
RSF95GP5	63.075 ± 0.172 [a, A]	56.564 ± 0.173 [b, A]	70.144 ± 0.184 [a, A]	69.617 ± 0.149 [a, A]
RSF90GP10	61.955 ± 0.134 [b, A]	53.188 ± 0.137 [c, A]	69.801 ± 0.150 [b, A]	68.647 ± 0.152 [b, A]
RSF85GP15	60.755 ± 0.179 [c, A]	52.889 ± 0.143 [c, A]	66.052 ± 0.167 [c, A]	69.403 ± 0.132 [a, A]
RS0GP20	56.952 ± 0.161 [e, A]	52.761 ± 0.160 [c, A]	61.157 ± 0.173 [d, A]	63.166 ± 0.154 [d, A]
RSF75GP25	55.399 ± 0.133 [f, A]	53.003 ± 0.164 [c, A]	60.198 ± 0.143 [e, A]	61.102 ± 0.140 [e, A]

BSF, CSF, RSF (control: biscuits, cakes and rolls); BSF95GP5, CSF95GP5, RSF95GP5, (biscuits, cakes and rolls: 95% spelt flour + 5% grape pomace powder); BSF90GP10, CSF90GP10, RSF90GP10 (biscuits, cakes and rolls: 90% spelt flour + 10% grape pomace); BSF85GP15, CSF85GP15, RSF85GP15 (biscuits, cakes and rolls: 85% spelt flour + 15% grape pomace); BSF80GP20, CSF80GP20, RSF80GP20 (biscuits, cakes and rolls: 80% spelt flour + 20% grape pomace); BSF75GP25, CSF75GP25, RSF75GP25 (biscuits, cakes and rolls: 75% spelt flour + 25% grape pomace). The values represent the mean of three independent experiments ± standard deviation (SD). Data with different superscripts in a column are statistically different (one-way ANOVA, $p < 0.05$). Lowercase letters (a–f) differentiate the formulas within each pastry type, while uppercase letters (A–C) differentiate the three pastry types obtained from the same composite flour.

A significant discrepancy was observed in the retention rates of the phytochemical content and antioxidant activity of pastries made from the same composite flour in response to baking. Considering that the baking temperature for all products was 180 °C, the differences could be associated with the processes taking place in the dough.

Dough formulations and baking conditions significantly affect the antioxidant properties and phenolic compound stability [75]. A closer look at the results in Table 5 showed a significantly higher preservation of the investigated properties in the rolls compared to the biscuits and cakes. The roll formulas leavened by yeast showed enhanced phytochemical amounts and superior antioxidant activity than the biscuit and cake formulas obtained with chemical raising agents for all levels of GP incorporation. The higher retention rates in the rolls could be assigned to the dough fermentation involved in their production. Changes in the leavening system have been found to reduce losses of phenolic compounds [76]. The leavening process is essential to develop the quality properties of pasty products. Baking powder provides a complete leavening system to produce gases via a reaction of a base like sodium bicarbonate and a weak organic acid [75]. Throughout the fermentation process developed in rolls, the gas produced by yeast activity diffuses into the dough and increases the number of air bubbles. During the first baking phase, gas bubbles are formed due to rising temperature, air expansion and carbon dioxide formation by the leavening agent [77]. Gas retention capacity is essential for dough development, significantly impacting the overall quality of the product [77]. The addition of plant-based materials rich in phenolic compounds has been shown to improve both dough functionality and gas retention by interactions between the matrix proteins and phenolic compounds [78]. According to Santetti et al. [77] and Chi and Cho [79], the addition of functional material together with the fermentation process by yeast enhances the bioavailability of phenolic compounds and the antioxidant activity of the dough. In addition, it indicates that the most effective fermentation times for the release of bioactive compounds is between 30 min and 60 min. Incorporating GP into the recipe enables an increase in dough functionality due to the biologically active compounds provided. However, more research is needed to explore the formation/degradation/hydrolysis reactions of phenolic compounds during baking.

In the case of cakes, even though the baking time was twice as long as for biscuits and rolls, this was not reflected in the impairment of their functional properties. On the one hand, this can be explained by the fact that phenolic compounds are retained by the cell matrix and can establish new types of interactions during heat treatment with other organic compounds, such as polysaccharides, which have a significant contribution to their increased stability [35,41]. On the other hand, heat treatment at elevated temperatures can potentially release the bound phenols, enhancing their availability [35]. In addition, it is known that high-temperature Maillard reaction products generated during the baking exhibit antioxidant properties [50].

The retention rate of TPC and TFC is an important criterion when considering the design of new pastry formulations with improved functional properties. It is therefore important to understand more deeply the impact of processing on these compounds in order to maximise their retention in final products. To this end, further analysis of individual polyphenolic compounds in pastry formulations with different levels of GP incorporation is needed.

3.5. Physical Characteristics of Pastry Products

Table 6 shows the physical characteristics of the bakery products, namely, porosity and elasticity for cakes and leavened rolls, and spread ratio for biscuits.

Table 6. Physical characteristics of pastry products (control samples and fortified formulas).

Physical Characteristics	Pastry Products					
	BSF	BSF95GP5	BSF90GP10	BSF85GP15	BSF80GP20	BSF75GP25
SR	5.001 ± 0.012 [e]	5.109 ± 0.011 [d]	5.158 ± 0.013 [c]	5.206 ± 0.010 [b]	5.229 ± 0.013 [a]	5.257 ± 0.011 [a]
	CSF	CSF95GP5	CSF90GP10	CSF85GP15	CSF80GP20	CSF75GP25
Porosity (%)	84.697 ± 0.282 [a]	84.431 ± 0.273 [a]	83.224 ± 0.258 [b]	82.877 ± 0.249 [b]	80.149 ± 0.227 [c]	78.113 ± 0.221 [d]
Elasticity (%)	97.146 ± 0.301 [a]	97.148 ± 0.294 [a]	95.653 ± 0.285 [b]	94.129 ± 0.279 [c]	92.181 ± 0.232 [d]	91.435 ± 0.229 [e]
	RSF	RSF95GP5	RSF90GP10	RSF85GP15	RS0GP20	RSF75GP25
Porosity (%)	73.395 ± 0.223 [a]	68.403 ± 0.207 [b]	68.226 ± 0.198 [b]	67.724 ± 0.186 [c]	67.431 ± 0.173 [c]	66.762 ± 0.169 [d]
Elasticity (%)	75.008 ± 0.249 [a]	72.229 ± 0.233 [b]	68.903 ± 0.219 [c]	62.511 ± 0.188 [d]	61.507 ± 0.164 [e]	60.104 ± 0.151 [f]

SR: spread ratio; BSF, CSF, RSF (control: biscuits, cakes and rolls); BSF95GP5, CSF95GP5, RSF95GP5, (biscuits, cakes and rolls: 95% spelt flour + 5% grape pomace powder); BSF90GP10, CSF90GP10, RSF90GP10 (biscuits, cakes and rolls: 90% spelt flour + 10% grape pomace); BSF85GP15, CSF85GP15, RSF85GP15 (biscuits, cakes and rolls: 85% spelt flour + 15% grape pomace); BSF80GP20, CSF80GP20, RSF80GP20 (biscuits, cakes and rolls: 80% spelt flour + 20% grape pomace); BSF75GP25, CSF75GP25, RSF75GP25 (biscuits, cakes and rolls: 75% spelt flour + 25% grape pomace). The values represent the mean of three independent experiments ± standard deviation (SD). Data with different superscripts in a row are statistically different (one-way ANOVA, $p < 0.05$).

The results reflect decreases in elasticity and porosity of cakes and rolls with the inclusion of GP in the recipe. The maximum value for elasticity was recorded for the control cake sample (97.146%) and decreased by 5.879% for a 25% GP incorporation. In the case of the roll formulations, the decrease in elasticity was higher compared to that of the cake, being 19.870% for a 25% level of GP incorporation. A similar behaviour was recorded for porosity, which decreased by the progressive incorporation of GP. The maximum porosity value was obtained for the control sample (84.697% for cakes and 73.395% for rolls), and the values decreased compared to the control samples by 7.774% for cakes and by 9.037% in the rolls, for a 25% GP incorporation. Similar results have been reported in the literature revealing significant decreases in bread elasticity and porosity by adding non-cereal matrices at a level of 10–30%, as reported by Plustea et al. [80], who incorporated lupin flour in the bread recipe and by Dossa et al. [81], who used baobab pulp flour as a replacement for wheat flour. The structural properties of bakery products are influenced by technological parameters and the constituent phases. The elasticity and porosity are determined by dough preparation steps and, in particular, by processes involving gluten development and starch gelling. The gluten content and its ability to retain fermentation gases, as well as the amount of hydrolysable starch and its structure, influence textural parameters [80]. Considering that, by incorporating GP, the two chemical components responsible for the texture of the products are diminished, the decrease in the elasticity and porosity of enriched formulas by adding a non-cereal material is well argued.

For biscuits, the incremental incorporation of GP led to an increase in the spread ratio (SR); similar results have been reported by other authors when other non-cereal flours were added to the dough composition [1,56], probably due to the reduction in the gluten network and the increase in fibre intake [1].

3.6. Sensory Evaluation of Pastry Products

The sensory analysis of the pastry products was performed to assess consumer acceptability using a five-point hedonic scale, and the average scores for the attributes (appearance, aroma, texture, taste and overall acceptability) of the samples are summarised in Table 7.

Table 7. Global values of the sensory attributes of enriched pastry formulas versus control using a five-point hedonic scale.

Sample	Scores (5-Point Hedonic Scale)				
	Appearance	Flavour	Texture	Taste	Overall Acceptability
BSF	4.444 ± 0.506 [a]	4.259 ± 0.447 [a]	4.444 ± 0.506 [a]	4.444 ± 0.506 [a]	4.370 ± 0.565 [a]
BSF95GP5	4.481 ± 0.509 [a]	4.370 ± 0.492 [a]	4.481 ± 0.509 [a]	4.481 ± 0.509 [a]	4.519 ± 0.509 [a]
BSF90GP10	4.593 ± 0.501 [a]	4.481 ± 0.509 [a]	4.519 ± 0.509 [a]	4.593 ± 0.501 [a]	4.630 ± 0.492 [a]
BSF85GP15	4.185 ± 0.396 [b]	4.296 ± 0.465 [a]	4.296 ± 0.465 [a]	4.185 ± 0.483 [b]	4.370 ± 0.492 [a]
BSF80GP20	3.889 ± 0.320 [c]	4.148 ± 0.362 [a]	4.259 ± 0.447 [a]	3.963 ± 0.338 [c]	4.111 ± 0.320 [b]
BSF75GP25	3.852 ± 0.362 [c]	4.074 ± 0.267 [b]	4.148 ± 0.362 [b]	3.815 ± 0.396 [d]	3.926 ± 0.267 [c]
CSF	4.370 ± 0.492 [a]	4.259 ± 0.447 [a]	4.407 ± 0.501 [a]	4.481 ± 0.509 [a]	4.481 ± 0.509 [a]
CSF95GP5	4.593 ± 0.501 [a]	4.444 ± 0.506 [a]	4.593 ± 0.501 [a]	4.519 ± 0.509 [a]	4.556 ± 0.506 [a]
CSF90GP10	4.667 ± 0.480 [a]	4.593 ± 0.501 [a]	4.630 ± 0.492 [a]	4.593 ± 0.501 [a]	4.630 ± 0.492 [a]
CSF85GP15	4.296 ± 0.465 [b]	4.370 ± 0.492 [a]	4.222 ± 0.424 [b]	4.370 ± 0.492 [a]	4.481 ± 0.509 [a]
CSF80GP20	4.185 ± 0.483 [c]	4.259 ± 0.526 [a]	4.148 ± 0.362 [b]	4.222 ± 0.424 [b]	4.259 ± 4.259 [a]
CSF75GP25	4.074 ± 0.267 [c]	4.185 ± 0.396 [b]	4.074 ± 0.385 [b]	4.148 ± 0.362 [c]	4.185 ± 0.396 [b]
RSF	4.259 ± 0.447 [a]	4.111 ± 0.320 [a]	4.222 ± 0.424 [a]	4.222 ± 0.424 [a]	4.296 ± 0.465 [a]
RSF95GP5	4.370 ± 0.492 [a]	4.333 ± 0.480 [a]	4.370 ± 0.492 [a]	4.259 ± 0.447 [a]	4.370 ± 0.492 [a]
RSF90GP10	4.485 ± 0.509 [a]	4.407 ± 0.501 [a]	4.481 ± 0.509 [a]	4.370 ± 0.492 [a]	4.519 ± 0.509 [a]
RSF85GP15	4.148 ± 0.362 [b]	4.222 ± 0.424 [a]	4.185 ± 0.396 [a]	4.111 ± 0.320 [a]	4.185 ± 0.396 [b]
RS0GP20	4.037 ± 0.192 [b]	4.148 ± 0.362 [a]	4.074 ± 0.267 [b]	3.926 ± 0.267 [b]	4.111 ± 0.320 [b]
RSF75GP25	3.815 ± 0.396 [c]	4.037 ± 0.192 [b]	3.963 ± 0.338 [c]	3.778 ± 0.424 [c]	3.889 ± 0.320 [c]

BSF, CSF, RSF (control: biscuits, cakes and rolls); BSF95GP5, CSF95GP5, RSF95GP5, (biscuits, cakes and rolls: 95% spelt flour + 5% grape pomace); BSF90GP10, CSF90GP10, RSF90GP10 (biscuits, cakes and rolls: 90% spelt flour + 10% grape pomace); BSF85GP15, CSF85GP15, RSF85GP15 (biscuits, cakes and rolls: 85% spelt flour + 15% grape pomace); BSF80GP20, CSF80GP20, RSF80GP20 (biscuits, cakes and rolls: 80% spelt flour + 20% grape pomace); BSF75GP25, CSF75GP25, RSF75GP25 (biscuits, cakes and rolls: 75% spelt flour + 25% grape pomace). The values represent the mean of three independent experiments ± standard deviation (SD). The different letters (a–d) shown in the same column for each sensory attribute represent statistically significant differences among pastry formulas (one-way ANOVA, $p < 0.05$).

The graphical representation of the sensory profile of GP-enriched pastry formulas versus control samples is illustrated in Figure 6.

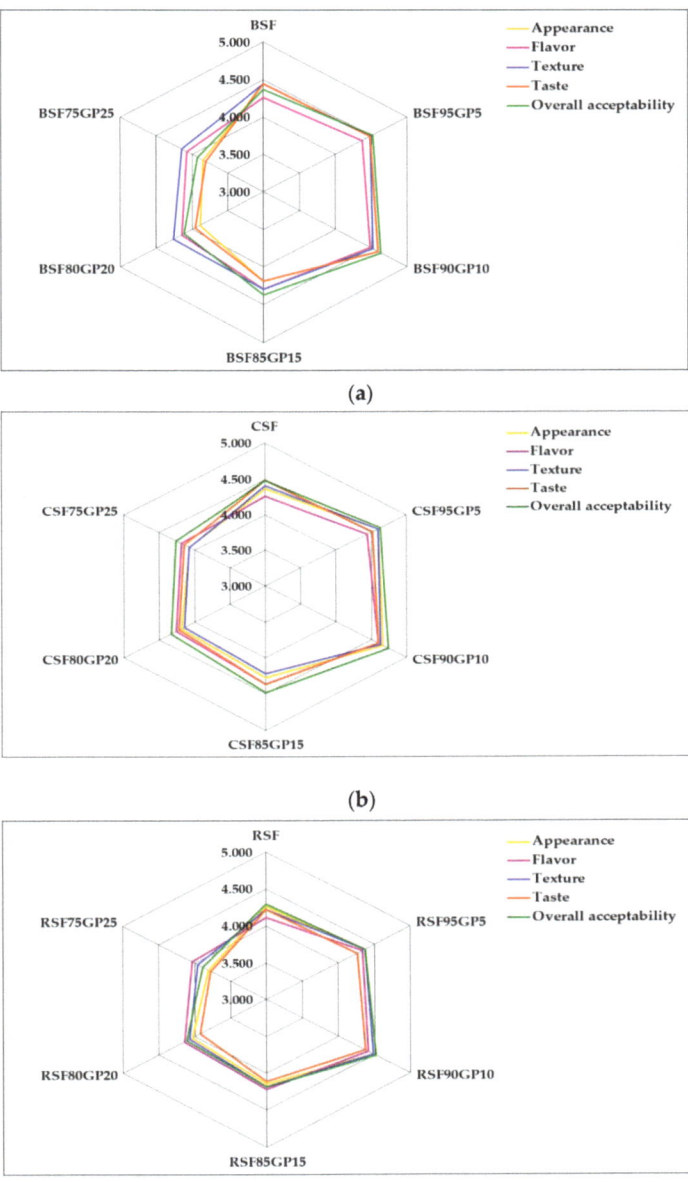

Figure 6. Sensory profile of quality attributes of enriched pastry formulas versus control, using a five-point hedonic scale (n = 27): (**a**): biscuits; (**b**) cakes; (**c**): rolls. BSF, CSF, RSF (control: biscuits, cakes and rolls); BSF95GP5, CSF95GP5, RSF95GP5 (biscuits, cakes and rolls: 95% spelt flour + 5% grape pomace); BSF90GP10, CSF90GP10, RSF90GP10 (biscuits, cakes and rolls: 90% spelt flour + 10% grape pomace); BSF85GP15, CSF85GP15, RSF85GP15 (biscuits, cakes and rolls: 85% spelt flour + 15% grape pomace); BSF80GP20, CSF80GP20, RSF80GP20 (biscuits, cakes and rolls: 80% spelt flour + 20% grape pomace); BSF75GP25, CSF75GP25, RSF75GP25 (biscuits, cakes and rolls: 75% spelt flour + 25% grape pomace).

In terms of the taste attribute of the biscuits, the scores increased in the following order: BSF90GP10 > BSF95GP5 > BSF > BSF85GP15 > BSF80GP20 > BSF75GP25, and in terms of flavour, the samples were ranked as follows: BSF90GP10 > BSF95GP5 > BSF85GP15 > BSF > BSF80GP20 > BSF80GP25. Regarding the overall acceptability, the score increased in the following order: BSF90GP10 > BSF95GP5 > BSF + BSF85GP15 > BSF80GP20 > BSF75GP25 (Figure 6a). The most appreciated GP-enriched biscuit formula was BSF90GP10, followed by BSF95GP5, which can be explained by the fact that the incorporation of up to 10% GP in the recipe conferred sufficient grape flavour and was appreciated by consumers.

The results were similar to previously reported data showing that, in terms of flavour, the addition of 10% GP scored highest in the sensory analysis conducted by Lou et al. [82], also showing increased values with decreasing proportion of GP. Sharma et al. [83] highlighted that cookies with up to 15% GP recorded high sensory values, i.e., texture, mouth sensation, aroma, taste and overall acceptability. It was also pointed out that biscuits fortified with 10% red GP showed higher overall acceptability [84].

In terms of cake formulas, the most appreciated sample was CSF90GP10, followed by CSF95GP5, as for biscuits. With regard to all sensory attributes, the scores increased in the following order: CSF90GP10 > CSF95GP5 > CSF; CSF85GP15 > CSF80GP20 > CSF75GP25 (Figure 6b). The results are consistent with those reported by Nakov et al. [30], who noted that the best scores were recorded by cakes with a low level of GP addition (4%) while the samples with 6% GP showed the best texture. Incorporating GP at a level of 4–6% in cake recipes improves the nutritional value of the products and provides better sensory characteristics compared to the control sample.

In the case of GP-enriched roll formulas, the best scores for sensory attributes were also obtained for the sample with 10% GP. In terms of taste and overall acceptability, the scores of the roll samples increased in the following order: RSF90GP10 > RSF95GP5 > RSF > RSF85GP15 > RSF80GP20 > RSF75GP25 (Figure 6c). These results are in agreement with studies carried out by Boff et al. [19], who pointed out that bakery products with a maximum of 10% addition of GP flour seem to be accepted by consumers.

Among the three types of GP-enriched pastry products, the highest scores were registered for sample CSF90GP10 with 4.667 (appearance), 4.630 (texture) and 4.630 (overall acceptability), followed by sample BSF90GP10 in terms of taste and appearance (4.593) and overall acceptability (4.630), and then roll sample RSF90GP10 with a score of 4.519 for overall acceptability, 4.485 for appearance and 4.481 for texture.

The least appreciated samples of each product type were RSF75GP25 with a score of 3.778 for taste and 3.889 for overall acceptability, followed by BSF75GP25 with 3.815 for taste and 3.926 for overall acceptability, and CSF75GP25 with a score of 4.074 for appearance and texture and 4.185 for overall acceptability (Figure 6b).

All pastries with GP incorporation up to 10% were rated at an extremely pleasant acceptability level (4.5–5.0). The appearance scores of all samples showed a similar trend to those reported by other authors [85], with lower scores being recorded as the level of GP incorporated increased, which could probably be explained by the darker colour of the samples. In connection with this finding, the use of GP as a natural colouring agent in bakery products has been suggested, as it has a great influence on obtaining the characteristic browned appearance of the products [7,19]. All pastry products were well accepted up to a 10% level of GP incorporation, with no significant differences between the control samples and those obtained from composite flour with 5 and 10% GP.

4. Conclusions

The results of this study provide strong evidence for the use of GP as a partial replacement of spelt wheat flour to obtain fortified pastry formulas, as it is a source of phenolic compounds and has high antioxidant activity. The biscuits, cakes and rolls enriched with grape pomace powder revealed an improved nutritional profile than the control, particularly in terms of lipid and ash content. The incremental incorporation of GP up to 25% in the recipe led to significantly higher amounts of total phenolic content and total flavonoids and allowed increasing antioxidant activity compared to the control. The three types of pastry products significantly differ in their functional properties, with the highest values for rolls, followed by biscuits and cakes. This finding could be assigned to the amount of composite flour based on SF and GP included in their recipe, but also to the leavening agent involved in their production. Roll formulations prepared with yeast as a natural leavening agent revealed higher levels of TPC, TFC, FRAP and DPPH than the biscuits and cakes formulated with baking powder as a chemical raising agent. The dough fermentation process seemed to improve the functionality of pastry products and can be a determining factor for the design of new value-added formulations. The retention rate of phytochemical content and antioxidant activity of pastry products in response to baking revealed a preservation in the range of 41–63% for TPC, 37–65% for TFC, 48–70% for FRAP value and 45–70% for DPPH value relative to the corresponding dough. The porosity and elasticity of the cakes and rolls decreased via augmenting the GP level, while the spread ratio of the biscuits increased. Although GP incorporation influenced the sensory attributes of the fortified products, all pastries with a GP incorporation of up to 10% were rated at an extremely pleasant acceptability level. These data are useful as inputs in the formulation of new food products with improved functionality. The proposed use of GP, in addition to providing food with improved nutritional and functional properties, will also mitigate the environmental problems associated with their disposal.

Author Contributions: Conceptualisation, M.-A.P. and E.A.; methodology, M.-A.P., E.A., D.-N.R. and I.R.; software, I.C., M.N., D.-N.R., C.D., S.D. and G.S.; formal analysis, M.-A.P, E.A., D.-N.R., I.C., M.N., S.D. and C.D.M.; investigation, M.-A.P., E.A., I.R., D.-N.R., I.C., M.N., S.D. and C.D.M.; resources, I.R. and G.S.; data curation, D.-N.R., C.D., M.N., I.C. and G.S.; writing—original draft preparation, M.-A.P., E.A., I.R., D.-N.R., I.C., M.N., C.D.M., C.D., S.D. and G.S.; writing—review and editing, M.-A.P. and E.A.; supervision, M.-A.P. and I.R. All authors have read and agreed to the published version of the manuscript.

Funding: This research is supported by the West Regional Development Agency, Romania, through the Project "Achieving technology transfer to obtain innovative functional foods enriched in bioactive compounds—CTTU 2020", SMIS Code: 140030, submitted in the competition Regional Operational Program 2014–2020, Priority Axis 1: Promoting technology transfer and by the Ministry of Research, Innovation and Digitization from Romania through the project entitled: "Increasing the impact of excellence research on the capacity for innovation and technology transfer within USAMVB Timisoara", code 6PFE, submitted in the competition Program 1—Development of the national system of research—development, Subprogram 1.2—Institutional performance, Institutional development projects—Development projects of excellence in R.D.I.

Institutional Review Board Statement: The study was conducted in accordance with the Declaration of Helsinki and approved by the Bioethics Committee of the University of Life Sciences "King Michael I", Timisoara, Aradului Street No 119, 300645 Timisoara, Romania; Project Code: POR/2020/1/1.1.A./2/140030; (No 206/04 April 2023).

Informed Consent Statement: Informed consent was obtained for from all subjects involved in the study.

Data Availability Statement: The report of the analyses performed for the samples in the paper can be found at the Interdisciplinary Research Platform (PCI) at the University of Life Sciences "King Michael I", Timisoara.

Acknowledgments: The authors of this paper acknowledge the technical support provided by the Interdisciplinary Research Platform at the University of Life Sciences "King Michael I", Timisoara, where the analyses were performed.

Conflicts of Interest: The authors declare no conflict of interest. The funders had no role in the design of the study; in the collection, analyses, or interpretation of data; in the writing of the manuscript; or in the decision to publish the results.

References

1. Nakov, G.; Brandolini, A.; Estivi, L.; Bertuglia, K.; Ivanova, N.; Jukić, M.; Komlenić, D.K.; Lukinac, J.; Hidalgo, A. Effect of Tomato Pomace Addition on Chemical, Technological, Nutritional, and Sensorial Properties of Cream Crackers. *Antioxidants* **2022**, *11*, 2087. [CrossRef]
2. Trigo, J.P.; Alexandre, E.M.; Saraiva, J.A.; Pintado, M.E. High value-added compounds from fruit and vegetable by-products–Characterization, bioactivities, and application in the development of novel food products. *Crit. Rev. Food Sci. Nutr.* **2022**, *60*, 1388–1416. [CrossRef]
3. Antonic, B.; Dordevic, D.; Jancikova, S.; Holeckova, D.; Tremlova, B.; Kulawik, P. Effect of Grape Seed Flour on the Antioxidant Profile, Textural and Sensory Properties of Waffles. *Processes* **2021**, *9*, 131. [CrossRef]
4. Tournour, H.H.; Segundo, M.A.; Magalhaes, L.M.; Barreiros, L.; Queiroz, J.; Cunha, L.M. Valorization of grape pomace: Extraction of bioactive phenolics with antioxidant properties. *Ind. Crops Prod.* **2015**, *74*, 397–406. [CrossRef]
5. Lau, K.Q.; Sabran, M.R.; Shafie, S.R. Utilization of vegetable and fruit by-products as functional ingredient and food. *Front. Nutr.* **2021**, *8*, 661693. [CrossRef] [PubMed]
6. Beres, C.; Costa, G.N.; Cabezudo, I.; Da Silva-James, N.K.; Teles, A.S.; Cruz, A.P.; Mellinger-Silva, C.; Tonon, R.V.; Cabral, L.M.; Freitas, S.P. Towards integral utilization of grape pomace from winemaking process: A review. *Waste Manag.* **2017**, *68*, 581–594. [CrossRef] [PubMed]
7. Gómez-Brandón, M.; Lores, M.; Insam, H.; Domínguez, J. Strategies for recycling and valorization of grape marc. *Crit. Rev. Biotechnol.* **2019**, *39*, 437–450. [CrossRef]
8. Iuga, M.; Mironeasa, S. Potential of grape byproducts as functional ingredients in baked goods and pasta. *Compr. Rev. Food Sci. Food Saf.* **2020**, *19*, 2473–2505. [CrossRef] [PubMed]
9. Yu, J.; Ahmedna, M. Functional components of grape pomace: Their composition, biological properties and potential applications. *Int. J. Food Sci. Technol.* **2013**, *48*, 221–237. [CrossRef]
10. García-Lomillo, J.; González-SanJosé, M.L. Applications of wine pomace in the food industry: Approaches and functions. *Compr. Rev. Food Sci. Food Saf.* **2017**, *16*, 3–22. [CrossRef]
11. Pinelo, M.; Rubilar, M.; Jerez, M.; Sineiro, J.; Núñez, M.J. Effect of solvent, temperature, and solvent-to-solid ratio on the total phenolic content and antiradical activity of extracts from different components of grape pomace. *J. Agric. Food Chem.* **2005**, *53*, 2111–2117. [CrossRef]
12. Monteiro, G.C.; Minatel, I.O.; Junior, A.P.; Gomez-Gomez, H.A.; de Camargo, J.P.C.; Diamante, M.S.; Pereira Basilio, L.S.; Tecchio, M.A.; Lima, G.P.P. Bioactive compounds and antioxidant capacity of grape pomace flours. *LWT-Food Sci. Technol.* **2021**, *135*, 110053. [CrossRef]
13. Ferrer-Gallego, R.; Silva, P. The Wine Industry By-Products: Applications for Food Industry and Health Benefits. *Antioxidants* **2022**, *11*, 2025. [CrossRef] [PubMed]
14. Bender, A.B.; Speroni, C.S.; Salvador, P.R.; Loureiro, B.B.; Lovatto, N.M.; Goulart, F.R.; Lovattoc, M.T.; Mirandad, M.Z.; Silvab, L.P.; Penna, N.G. Grape pomace skins and the effects of its inclusion in the technological properties of muffins. *J. Culin. Sci. Technol.* **2017**, *15*, 143–157. [CrossRef]
15. Chowdhary, P.; Gupta, A.; Gnansounou, E.; Pandey, A.; Chaturvedi, P. Current trends and possibilities for exploitation of Grape pomace as a potential source for value addition. *Environ. Pollut.* **2021**, *278*, 116796. [CrossRef]
16. Troilo, M.; Difonzo, G.; Paradiso, V.M.; Pasqualone, A.; Caponio, F. Grape Pomace as Innovative Flour for the Formulation of Functional Muffins: How Particle Size Affects the Nutritional, Textural and Sensory Properties. *Foods* **2022**, *11*, 1799. [CrossRef]
17. Muhlack, R.A.; Potumarthi, R.; Jeffery, D.W. Sustainable wineries through waste valorisation: A review of grape marc utilisation for value-added products. *Waste Manag.* **2018**, *72*, 99–118. [CrossRef]
18. Torbica, A.; Škrobot, D.; Hajnal, E.J.; Belović, M.; Zhang, N. Sensory and physico-chemical properties of wholegrain wheat bread prepared with selected food by-products. *LWT-Food Sci. Technol.* **2019**, *114*, 108414. [CrossRef]
19. Boff, J.M.; Strasburg, V.J.; Ferrari, G.T.; de Oliveira Schmidt, H.; Manfroi, V.; de Oliveira, V.R. Chemical, Technological, and Sensory Quality of Pasta and Bakery Products Made with the Addition of Grape Pomace Flour. *Foods* **2022**, *11*, 3812. [CrossRef] [PubMed]
20. Hayta, M.; Özuğur, G.; Etgü, H.; Şeker, İ.T. Effect of Grape (*Vitis vinifera* L.) Pomace on the Quality, Total Phenolic Content and Anti-Radical Activity of Bread. *J. Food Process. Preserv.* **2014**, *38*, 980–986. [CrossRef]
21. Šporin, M.; Avbelj, M.; Kovač, B.; Možina, S.S. Quality characteristics of wheat flour dough and bread containing grape pomace flour. *Food Sci. Technol. Int.* **2018**, *24*, 251–263. [CrossRef]
22. Rosales Soto, M.U.; Brown, K.; Ross, C.F. Antioxidant activity and consumer acceptance of grape seed flour-containing food products. *Int. J. Food Sci. Technol.* **2012**, *47*, 592–602. [CrossRef]
23. Samohvalova, O.; Grevtseva, N.; Brykova, T.; Grigorenko, A. The effect of grape seed powder on the quality of butter biscuits. *East. Eur. J. Enterp. Technol.* **2016**, *3*, 61–66. [CrossRef]
24. Meral, R.; Doğan, İ.S. Grape seed as a functional food ingredient in bread-making. *Int. J. Food Sci. Nutr.* **2013**, *64*, 372–379. [CrossRef] [PubMed]
25. Tolve, R.; Simonato, B.; Rainero, G.; Bianchi, F.; Rizzi, C.; Cervini, M.; Giuberti, G. Wheat Bread Fortification by Grape Pomace Powder: Nutritional, Technological, Antioxidant, and Sensory Properties. *Foods* **2021**, *10*, 75. [CrossRef] [PubMed]

26. Mildner-Szkudlarz, S.; Bajerska, J.; Zawirska-Wojtasiak, R.; Górecka, D. White grape pomace as a source of dietary fibre and polyphenols and its effect on physical and nutraceutical characteristics of wheat biscuits. *J. Sci. Food Agric.* **2013**, *93*, 389–395. [CrossRef]
27. Palma, M.L.; Nunes, M.C.; Gameiro, R.; Rodrigues, M.; Gothe, S.; Tavares, N.; Pego, C.; Nicolai, M.; Pereira, P. Preliminary sensory evaluation of salty crackers with grape pomace flour. *Biomed. Biopharm. Res.* **2020**, *17*, 33–43. [CrossRef]
28. Sant'Anna, V.; Christiano, F.D.P.; Marczak, L.D.F.; Tessaro, I.C.; Thys, R.C.S. The effect of the incorporation of grape marc powder in fettuccini pasta properties. *LWT-Food Sci. Technol.* **2014**, *58*, 497–501. [CrossRef]
29. Oliveira, B.E.; Contini, L.; Garcia, V.A.D.S.; Cilli, L.P.D.L.; Chagas, E.G.L.; Andreo, M.A.; Vanin, F.M.; Carvalho, R.A.; Sinnecker, P.; Venturini, A.C.; et al. Valorization of grape by-products as functional and nutritional ingredients for healthy pasta development. *J. Food Process. Preserv.* **2022**, *46*, e17245. [CrossRef]
30. Nakov, G.; Brandolini, A.; Hidalgo, A.; Ivanova, N.; Stamatovska, V.; Dimov, I. Effect of grape pomace powder addition on chemical, nutritional and technological properties of cakes. *LWT-Food Sci. Technol.* **2020**, *134*, 109950. [CrossRef]
31. Karnopp, A.R.; Figueroa, A.M.; Los, P.R.; Teles, J.C.; Simões, D.R.S.; Barana, A.C.; Kubiaki, F.T.; Oliveira, J.G.B.d.; Granato, D. Effects of whole-wheat flour and bordeaux grape pomace (*Vitis labrusca* L.) on the sensory, physicochemical and functional properties of cookies. *Food Sci. Technol.* **2015**, *35*, 750–756. [CrossRef]
32. Fontana, M.; Murowaniecki Otero, D.; Pereira, A.M.; Santos, R.B.; Gularte, M.A. Grape Pomace Flour for Incorporation into Cookies: Evaluation of Nutritional, Sensory and Technological Characteristics. *J. Culin. Sci. Technol.* **2022**, 1–20. [CrossRef]
33. Gaita, C.; Alexa, E.; Moigradean, D.; Conforti, F.; Poiana, M.A. Designing of high value-added pasta formulas by incorporation of grape pomace skins. *Rom. Biotechnol. Lett.* **2020**, *25*, 1607–1614. [CrossRef]
34. Rainero, G.; Bianchi, F.; Rizzi, C.; Cervini, M.; Giuberti, G.; Simonato, B. Breadstick fortification with red grape pomace: Effect on nutritional, technological and sensory properties. *J. Sci. Food Agric.* **2022**, *102*, 2545–2552. [CrossRef]
35. Larrosa, A.P.Q.; Otero, D.M. Flour made from fruit by-products: Characteristics, processing conditions, and applications. *J. Food Process. Preserv.* **2021**, *45*, e15398. [CrossRef]
36. Hasmadi, M.; Noorfarahzilah, M.; Noraidah, H.; Zainol, M.K.; Jahurul, M.H.A. Functional properties of composite flour: A review. *Food Res.* **2020**, *4*, 1820–1831. [CrossRef]
37. Kohajdová, Z.; Karovicova, J. Nutritional value and baking application of spelt wheat. *Acta Sci. Pol. Technol. Aliment.* **2008**, *7*, 5–14.
38. Biel, W.; Stankowski, S.; Jaroszewska, A.; Pużyński, S.; Bośko, P. The influence of selected agronomic factors on the chemical composition of spelt wheat (*Triticum aestivum* ssp. *spelta* L.) grain. *J. Integr. Agric.* **2016**, *15*, 1763–1769. [CrossRef]
39. Escarnot, E.; Jacquemin, J.M.; Agneessens, R.; Paquot, M. Comparative study of the content and profiles of macronutrients in spelt and wheat, a review. *Biotechnol. Agron. Soc. Environ.* **2012**, *16*, 243–256.
40. Wang, J.; Chatzidimitriou, E.; Wood, L.; Hasanalieva, G.; Markellou, E.; Iversen, P.O.; Seala, C.; Baranskib, M.; Vigarj, V.; Ernstj, L.; et al. Effect of wheat species (*Triticum aestivum* vs. *T. spelta*), farming system (organic vs. conventional) and flour type (wholegrain vs white) on composition of wheat flour–Results of a retail survey in the UK and Germany–2. Antioxidant activity, and phenolic and mineral content. *Food Chem.* **2020**, *6*, 100091. [CrossRef]
41. Jung, J.; Cavender, G.; Zhao, Y. Impingement drying for preparing dried apple pomace flour and its fortification in bakery and meat products. *J. Food Sci. Technol.* **2015**, *52*, 5568–5578. [CrossRef]
42. Santos, D.; da Silva, J.A.L.; Pintado, M. Fruit and vegetable by-products' flours as ingredients: A review on production process, health benefits and technological functionalities. *LWT-Food Sci. Technol.* **2022**, *154*, 112707. [CrossRef]
43. ISO 4833:2003; Microbiology of Food and Animal Feeding Stuffs—Horizontal Method for the Enumeration of Microorganisms—Colony-Count Technique at 30 Degrees C. ISO: Geneva, Switzerland, 2003.
44. ISO 21528-2:2004; Microbiology of Food and Animal Feeding Stuffs—Horizontal Methods for the Detection and Enumeration of Enterobacteriaceae—Part 2: Colony-Count Method. ISO: Geneva, Switzerland, 2004.
45. ISO 21527:2008; Microbiology of Food and Animal Feeding Stuffs—Horizontal Method for the Enumeration of Yeasts and Moulds—Part 2: Colony Count Technique in Products with Water Activity Less than or Equal to 0.95. ISO: Geneva, Switzerland, 2008.
46. ISO 21871:2006; Microbiology of Food and Animal Feeding Stuffs—Horizontal Method for the Determination of Low Numbers of Presumptive Bacillus cereus—Most Probable Number Technique and Detection Method. ISO: Geneva, Switzerland, 2006.
47. Commission Regulation (EC) No 2073/2005 of 15 November 2005 on Microbiological Criteria for Foodstuffs. Official Journal of the European Union L 338, 22.12.2005. pp. 1–26. Available online: https://www.eumonitor.eu/9353000/1/j4nvk5yhcbpeywk_j9vvik7m1c3gyxp/vi8rm2zgvzuf (accessed on 15 March 2023).
48. Association of Official Analytical Chemists (AOAC). *Official Methods of Analysis*, 17th ed.; AOAC: Washington, DC, USA, 2000.
49. Das, P.C.; Khan, M.J.; Rahman, M.S.; Majumder, S.; Islam, M.N. Comparison of the physico-chemical and functional properties of mango kernel flour with wheat flour and development of mango kernel flour based composite cakes. *NFS J.* **2019**, *17*, 1–7. [CrossRef]
50. Litwinek, D.; Gumul, D.; Łukasiewicz, M.; Zięba, T.; Kowalski, S. The Effect of Red Potato Pulp Preparation and Stage of Its Incorporation into Sourdough or Dough on the Quality and Health-Promoting Value of Bread. *Appl. Sci.* **2023**, *13*, 7670. [CrossRef]
51. Blanch, G.P.; Ruiz del Castillo, M.L. Effect of Baking Temperature on the Phenolic Content and Antioxidant Activity of Black Corn (*Zea mays* L.) Bread. *Foods* **2021**, *10*, 1202. [CrossRef] [PubMed]

52. Al-Farsi, M.; Al-Amri, A.; Al-Hadhrami, A.; Al-Belushi, S. Color, flavonoids, phenolics and antioxidants of Omani honey. *Heliyon* **2018**, *4*, e00874. [CrossRef] [PubMed]
53. Mekky, H.; El Sohafy, S.; Abu El-Khair, R.A.; El Hawiet, A.E. Total polyphenolic content and antioxidant activity of Silybum marianum cultures grown on different growth regulators. *Int. J. Pharm. Pharm. Sci.* **2017**, *9*, 44–47. [CrossRef]
54. Metzner Ungureanu, C.-R.; Poiana, M.-A.; Cocan, I.; Lupitu, A.I.; Alexa, E.; Moigradean, D. Strategies to Improve the Thermo-Oxidative Stability of Sunflower Oil by Exploiting the Antioxidant Potential of Blueberries Processing Byproducts. *Molecules* **2020**, *25*, 5688. [CrossRef] [PubMed]
55. SR 91:2007; Romanian Standard for Bread, Confectionery and Bakery Specialties—Methods of Analysis. ASRO—Romanian Standards Association: Bucharest, Romania, 2007.
56. Raymundo, A.; Fradinho, P.; Nunes, M.C. Effect of Psyllium fibre content on the textural and rheological characteristics of biscuit and biscuit dough. *Bioact. Carbohydr. Diet. Fibre* **2014**, *3*, 96–105. [CrossRef]
57. Alfonsi, A.; Coles, D.; Hasle, C.; Koppel, J.; Ladikas, M.; Schmucker von Koch, J.; Schroeder, D.; Sprumont, D.; Verbeke, W.; Zaruk, D. *Guidance Note: Ethics and Food-Related Research*; European Commission Ethics Review Sector: Brussels, Belgium, 2012.
58. Pestorić, M.; Škrobot, D.; Žigon, U.; Šimurina, O.; Filipčev, B.; Belović, M.; Mišan, A. Sensory profile and preference mapping of cookies enriched with medicinal herbs. *Int. J. Food Prop.* **2017**, *20*, 350–361. [CrossRef]
59. Beres, C.; Freitas, S.P.; de Oliveira Godoy, R.L.; de Oliveira, D.C.R.; Deliza, R.; Iacomini, M.; Mellinger-Silva, C.; Cabral, L.M.C. Antioxidant dietary fibre from grape pomace flour or extract: Does it make any difference on the nutritional and functional value? *J. Funct. Foods* **2019**, *56*, 276–285. [CrossRef]
60. Yi, C.; Shi, J.; Kramer, J.; Xue, S.; Jiang, Y.; Zhang, M.; Ma, I.; Pohorly, J. Fatty acid composition and phenolic antioxidants of winemaking pomace powder. *Food Chem.* **2009**, *114*, 570–576. [CrossRef]
61. Acun, S.; Gül, H. Effects of grape pomace and grape seed flours on cookie quality. *Qual. Assur. Saf. Crop. Foods* **2014**, *6*, 81–88. [CrossRef]
62. Keriene, I.; Mankeviciene, A.; Bliznikas, S.; Jablonskyte-Rasce, D.; Maikštėnienė, S.; Česnulevičienė, R. Biologically active phenolic compounds in buckwheat, oats and winter spelt wheat. *Zemdirb. Agric.* **2015**, *102*, 289–296. [CrossRef]
63. Rockenbach, I.; Rodrigues, E.; Gonzaga, L.V.; Genovese, M.I.; Gonçalves, A.E.; Fett, R. Phenolic compounds content and antioxidant activity in pomace from selected red grapes (*Vitis vinifera* L. and *Vitis labrusca* L.) widely produced in Brazil. *Food Chem.* **2011**, *127*, 174–179. [CrossRef]
64. Iora, S.R.; Maciel, G.M.; Zielinski, A.A.; da Silva, M.V.; Pontes, P.V.D.A.; Haminiuk, C.W.; Granato, D. Evaluation of the bioactive compounds and the antioxidant capacity of grape pomace. *Int. J. Food Sci. Technol.* **2015**, *50*, 62–69. [CrossRef]
65. Negro, C.; Aprile, A.; Luvisi, A.; De Bellis, L.; Miceli, A. Antioxidant Activity and Polyphenols Characterization of Four Monovarietal Grape Pomaces from Salento (Apulia, Italy). *Antioxidants* **2021**, *10*, 1406. [CrossRef]
66. Cui, W.; Wang, Y.; Sun, Z.; Cui, C.; Li, H.; Luo, K.; Cheng, A. Effects of steam explosion on phenolic compounds and dietary fiber of grape pomace. *LWT-Food Sci. Technol.* **2023**, *173*, 114350. [CrossRef]
67. Putnik, P.; Bursać Kovačević, D.; Radojčin, M.; Dragović-Uzelac, V. Influence of acidity and extraction time on the recovery of flavonoids from grape skin pomace optimized by response surface methodology. *Chem. Biochem. Eng. Q.* **2016**, *30*, 455–464. [CrossRef]
68. Ivanišová, E.; Ondrejovič, M.; Šilhár, S. Antioxidant activity of milling fractions of selected cereals. *Nova Biotechnol. Chim.* **2012**, *11*, 45–56. [CrossRef]
69. Sumczynski, D.; Bubelova, Z.; Sneyd, J.; Erb-Weber, S.; Mlcek, J. Total phenolics, flavonoids, antioxidant activity, crude fibre and digestibility in non-traditional wheat flakes and muesli. *Food Chem.* **2015**, *174*, 319–325. [CrossRef]
70. Abdel-Aal, E.S.M.; Rabalski, I. Bioactive Compounds and their Antioxidant Capacity in Selected Primitive and Modern Wheat Species. *Open Agric. J.* **2008**, *2*, 7–14. [CrossRef]
71. Maner, S.; Sharma, A.K.; Banerjee, K. Wheat flour replacement by wine grape pomace powder positively affects physical, functional and sensory properties of cookies. *Proc. Natl. Acad. Sci. India Sect. B Biol. Sci.* **2017**, *87*, 109–113. [CrossRef]
72. Sęczyk, Ł.; Świeca, M.; Gawlik-Dziki, U. Changes of antioxidant potential of pasta fortified with parsley (*Petroselinum Crispum* mill.) leaves in the light of protein-phenolics interactions. *Acta Sci. Pol. Technol. Aliment.* **2015**, *14*, 29–36. [CrossRef] [PubMed]
73. Kruczek, M.; Gumul, D.; Korus, A.; Buksa, K.; Ziobro, R. Phenolic Compounds and Antioxidant Status of Cookies Supplemented with Apple Pomace. *Antioxidants* **2023**, *12*, 324. [CrossRef]
74. Ky, I.; Lorrain, B.; Kolbas, N.; Crozier, A.; Teissedre, P.-L. Wine by-Products: Phenolic Characterization and Antioxidant Activity Evaluation of Grapes and Grape Pomaces from Six Different French Grape Varieties. *Molecules* **2014**, *19*, 482–506. [CrossRef]
75. Žilić, S.; Kocadağlı, T.; Vančetović, J.; Gökmen, V. Effects of baking conditions and dough formulations on phenolic compound stability, antioxidant capacity and color of cookies made from anthocyanin-rich corn flour. *LWT* **2016**, *65*, 597–603. [CrossRef]
76. Francavilla, A.; Joye, I.J. Anthocyanin Content of Crackers and Bread Made with Purple and Blue Wheat Varieties. *Molecules* **2022**, *27*, 7180. [CrossRef]
77. Santetti, G.S.; Dacoreggio, M.V.; Silva, A.C.M.; Biduski, B.; Bressiani, J.; Oro, T.; de Francisco, A.; Gutkoski, L.C.; Amboni, R.D.M.C. Effect of yerba mate (Ilex paraguariensis) leaves on dough properties, antioxidant activity, and bread quality using whole wheat flour. *J. Food Sci.* **2021**, *86*, 4354–4364. [CrossRef]

78. Ozdal, T.; Capanoglu, E.; Altay, F. A review on protein-phenolic interactions and associated changes. *Food Res. Int.* **2013**, *51*, 954–970. [CrossRef]
79. Chi, C.H.; Cho, S.J. Improvement of bioactivity of soybean meal by solid-state fermentation with Bacillus amyloliquefaciens versus Lactobacillus spp. and Saccharomyces cerevisiae. *LWT-Food Sci. Technol.* **2016**, *68*, 619–625. [CrossRef]
80. Plustea, L.; Negrea, M.; Cocan, I.; Radulov, I.; Tulcan, C.; Berbecea, A.; Popescu, I.; Obistioiu, D.; Hotea, I.; Suster, G.; et al. Lupin (*Lupinus* spp.)-Fortified Bread: A Sustainable, Nutritionally, Functionally, and Technologically Valuable Solution for Bakery. *Foods* **2022**, *11*, 2067. [CrossRef] [PubMed]
81. Dossa, S.; Negrea, M.; Cocan, I.; Berbecea, A.; Obistioiu, D.; Dragomir, C.; Alexa, E.; Rivis, A. Nutritional, Physico-Chemical, Phytochemical, and Rheological Characteristics of Composite Flour Substituted by Baobab Pulp Flour (*Adansonia digitata* L.) for Bread Making. *Foods* **2023**, *12*, 2697. [CrossRef]
82. Lou, W.; Zhou, K.; Li, B.; Nataliya, G. Rheological, pasting and sensory properties of biscuits supplemented with grape pomace powder. *Food Sci. Technol.* **2022**, *42*, e78421. [CrossRef]
83. Sharma, A.K.; Dagadkhair, R.A.; Somkuwar, R.G. Evaluation of grape pomace and quality of enriched cookies after standardizing baking conditions: Evaluation of grape pomace and quality of enriched cookies. *J. AgriSearch* **2018**, *5*, 50–55. [CrossRef]
84. Azami, S.; Roufegari-Nejad, L. The Effect of Red Grape Pomace Powder Replacement on Physical Characteristics and Acrylamide Content of Biscuit. *Iranian J. Nutr. Sci. Food Technol.* **2019**, *14*, 109–117.
85. Ajila, C.M.; Leelavathi, K.; Prasada Rao, U.J.S. Improvement of dietary fiber content and antioxidant properties in soft dough biscuits with the incorporation of mango peel powder. *J. Cereal Sci.* **2008**, *48*, 319–326. [CrossRef]

Disclaimer/Publisher's Note: The statements, opinions and data contained in all publications are solely those of the individual author(s) and contributor(s) and not of MDPI and/or the editor(s). MDPI and/or the editor(s) disclaim responsibility for any injury to people or property resulting from any ideas, methods, instructions or products referred to in the content.

Article

Determination of Coenzyme Q10 Content in Food By-Products and Waste by High-Performance Liquid Chromatography Coupled with Diode Array Detection

Cristina Anamaria Semeniuc [1], Floricuța Ranga [1], Andersina Simina Podar [2,*], Simona Raluca Ionescu [1], Maria-Ioana Socaciu [1], Melinda Fogarasi [1], Anca Corina Fărcaș [1], Dan Cristian Vodnar [1] and Sonia Ancuța Socaci [1,*]

[1] Faculty of Food Science and Technology, University of Agricultural Sciences and Veterinary Medicine of Cluj-Napoca, 3-5 Mănăștur St., 400372 Cluj-Napoca, Romania; cristina.semeniuc@usamvcluj.ro (C.A.S.); floricutza_ro@yahoo.com (F.R.); rallucab@yahoo.com (S.R.I.); maria-ioana.socaciu@usamvcluj.ro (M.-I.S.); melinda.fogarasi@usamvcluj.ro (M.F.); anca.farcas@usamvcluj.ro (A.C.F.); dan.vodnar@usamvcluj.ro (D.C.V.)

[2] Centre for Technology Transfer-BioTech, 64 Calea Florești, 400509 Cluj-Napoca, Romania

* Correspondence: andersina-simina.podar@usamvcluj.ro (A.S.P.); sonia.socaci@usamvcluj.ro (S.A.S.); Tel.: +40-264-596-384 (A.S.P. & S.A.S.)

Abstract: Coenzyme Q10 (CoQ10) is a vitamin-like compound found naturally in plant- and animal-derived materials. This study aimed to determine the level of CoQ10 in some food by-products (oil press cakes) and waste (fish meat and chicken hearts) to recover this compound for further use as a dietary supplement. The analytical method involved ultrasonic extraction using 2-propanol, followed by high-performance liquid chromatography with diode array detection (HPLC-DAD). The HPLC-DAD method was validated in terms of linearity and measuring range, limits of detection (LOD) and quantification (LOQ), trueness, and precision. As a result, the calibration curve of CoQ10 was linear over the concentration range of 1–200 µg/mL, with an LOD of 22 µg/mL and an LOQ of 0.65 µg/mL. The CoQ10 content varied from not detected in the hempseed press cake and the fish meat to 84.80 µg/g in the pumpkin press cake and 383.25 µg/g in the lyophilized chicken hearts; very good recovery rates and relative standard deviations (RSDs) were obtained for the pumpkin press cake (100.9–116.0% with RSDs between 0.05–0.2%) and the chicken hearts (99.3–106.9% CH with RSDs between 0.5–0.7%), showing the analytical method's trueness and precision and thus its accuracy. In conclusion, a simple and reliable method for determining CoQ10 levels has been developed here.

Keywords: coenzyme Q10; food by-products; food waste; ultrasonic extraction; 2-propanol extracts; high-performance liquid chromatography; diode-array detection

Citation: Semeniuc, C.A.; Ranga, F.; Podar, A.S.; Ionescu, S.R.; Socaciu, M.-I.; Fogarasi, M.; Fărcaș, A.C.; Vodnar, D.C.; Socaci, S.A. Determination of Coenzyme Q10 Content in Food By-Products and Waste by High-Performance Liquid Chromatography Coupled with Diode Array Detection. *Foods* 2023, 12, 2296. https://doi.org/10.3390/foods12122296

Academic Editors: Rosaria Viscecchia, Francesco Bimbo and Gianluca Nardone

Received: 6 May 2023
Revised: 5 June 2023
Accepted: 6 June 2023
Published: 7 June 2023

Copyright: © 2023 by the authors. Licensee MDPI, Basel, Switzerland. This article is an open access article distributed under the terms and conditions of the Creative Commons Attribution (CC BY) license (https:// creativecommons.org/licenses/by/ 4.0/).

1. Introduction

Coenzyme Q10 (CoQ10) is a lipid-soluble molecule found in the mitochondria of each cell of eukaryotic and prokaryotic organisms [1,2]. It is involved in mitochondrial processes, such as respiration, cellular ATP (adenosine triphosphate) synthesis, maintaining heart muscle strength, neutralizing free radicals in the fight against aging, and stimulating the immune system [3]. With a chemical structure of 2,3-dimethoxy-5-methyl-6-decaprenyl-1,4-benzoquinone, CoQ10 is also known as ubiquinone-10 due to the "ubiquitous" occurrence of this compound in cells [4–6]. CoQ10 is also present in food products but is denatured by thermal processing because it is thermosensitive [7,8].

A previous paper [9], a literature review, highlighted vegetable oils, fish oil, organs, and meat as the richest sources of CoQ10. Based on this information, press cakes could be considered potential sources of this compound, given that they still contain oils, and cold-press oil extraction creates high amounts of these by-products [10]. In addition, the meat and fish industries generate large quantities of solid waste [11,12] since their primary products have a short shelf-life when stored in refrigerated conditions. Therefore, it would be appropriate to investigate whether fish meat and chicken hearts from supermarket shelves still contain CoQ10 after expiration when they are considered waste.

Several methods are available in the literature for CoQ10 determination in food matrices by high-performance liquid chromatography with diode array detection (HPLC-DAD) [8,13–18], which first consists of analyte extraction with different solvents, followed by the instrumental analysis of the obtained extracts [9]. These include direct extraction with an ethanol/hexane mixture [13–15,18], supercritical carbon dioxide extraction [16], accelerated solvent extraction [17], and extraction by saponification [8]. Choosing the appropriate analytical method for a given problem must consider the analyst's needs which must be weighed against the advantages and disadvantages of the available techniques [19]. In addition, revalidation or verification is necessary whenever a method is changed or applied to a new circumstance (such as a different sample matrix), depending on the change's degree and the unique circumstance's nature [20]. Therefore, typical characteristics such as accuracy, precision, specificity, detection limit, quantification limit, linearity, and range should be considered when validating an analytical method [21]. The extraction methods mentioned above are suitable for isolating CoQ10 from food matrices. However, some of these employ significant amounts of hazardous reagents, negatively impacting the environment and human health. As a result, they generate large quantities of toxic chemical waste. Therefore, it is preferable to obtain extracts by avoiding or reducing toxic solvents to minimize the side effects of analytical methods [17].

Isopropanol (2-propanol or propan-2-ol), a polar solvent, can also be used as a CoQ10 extractant from food matrices [22,23]. It is at the top of the list of green chemicals and is considered environmentally safe in industrial solvent selection guidelines [24]. Moreover, CoQ10 is more soluble in 2-propanol than in ethanol, while triglycerides, possible interference compounds of the food matrix, are not very soluble in this solvent [15]. To increase the yield of CoQ10, before extraction, sample pretreatment with ultrasound can also be applied [25]. According to Directive 2009/32/EC of the European Parliament and the Council [26], 2-propanol is already approved as an extraction solvent for processing raw materials, foodstuffs, food components, or ingredients, with a maximum residue limit of 10 mg/kg food in the extracted foodstuff or food ingredient. Assuming that it is used in soft drinks at a concentration of 600 mg/l, in 2005, the European Food Safety Authority's Scientific Panel on food additives, flavorings, processing aids, and materials in contact with food estimated a mean potential consumption of 1.3 mg/kg bw/day, which is below the acceptable daily intake [27].

This study aims to assess the feasibility of using some food by-products (oil press cakes) and waste (fish meat and chicken hearts), presumed to be rich in CoQ10, as natural sources of this bioactive compound. It, therefore, proposes an analytical procedure for recovering and quantifying CoQ10 from such matrices through ultrasonic extraction with 2-propanol, preceded by HPLC coupled to DA detection. For method validation, the experimental design also includes linearity and spike-and-recovery experiments. No studies have so far investigated the CoQ10 content in oil press cakes using an HPLC-DAD technique. Altogether, the findings of this study provide valuable information for companies in the extractive industries interested in the commercial-scale extraction of CoQ10 from vegetable and animal materials.

2. Materials and Methods

2.1. Sample Collection

The chicken hearts were purchased from S.C. Puiul Regal S.R.L. (Gilău, Romania) in a quantity of 2.1 kg on 30 May 2022 and kept in a refrigerator (at 4 °C) until the expiration date written on the packaging, 4 June 2022, when they became waste. On the day following expiry (5 June 2022), they were minced with a meat grinder (N12; Lancom Distribution S.R.L., Bucharest, Romania), homogenized using a silicone spatula, divided into 50 and 100 g portions, and then stored in sealed polyethylene bags. Six 100 g shares were lyophilized in a laboratory freeze-dryer (LyoQuest-55; Azbil Telstar Technologies S.L.,

Barcelona, Spain) under the following operating conditions [28], and the rest of them were kept at $-18\ °C$ until extraction:
- Freezing conditions: freezing temperature $-80\ °C$ and freezing time 24 h;
- Sublimation conditions: vacuum pressure 0.01 mbar, sublimation temperature $-55\ °C$, and sublimation time ~3 days (until a constant weight (± 0.005 g)). The lyophilization yield was 22.1%.

The fish, chilled whole rainbow trout (*Oncorhynchus mykiss*), were purchased from Bistromar La Timona S.R.L. (Bucharest, Romania) in a quantity of 1.573 kg on 3 June 2022 and kept in a refrigerator (at 4 °C) until the expiration date written on the packaging, 4 June 2022, when they became waste. The next day (5 June 20220), the fish individuals were cut into pieces using a stainless-steel knife, then minced with a meat grinder (N12; Lancom Distribution S.R.L., Bucharest, Romania), homogenized using a silicone spatula, divided into 50 and 100 g portions, and stored in sealed polyethylene bags. Six parts of 100 g were lyophilized in the freeze-dryer under the abovementioned operating conditions, and the others were kept at $-18\ °C$ until extraction. The lyophilization yield for fish meat was 30.8%.

Next, the lyophilizates of the chicken hearts and fish meat were placed in amber glass jars, hermetically sealed using the lids, and kept in a refrigerator (at 4 °C) until use when they were milled to fine powders using an electric grinder (Titan Mil 300 DuoClean; Grupo Cecotec Innovaciones S.L., Valencia, Spain).

The oil press cakes of rapeseed, sunflower, pumpkin, and walnut were received by a donation from Taf Presoil S.R.L. (Luncani, Romania) as pellets, 1 kg of each type, and kept in a refrigerator (at 4 °C) up to extraction when they were milled with an electric grinder (Titan Mil 300 DuoClean; Grupo Cecotec Innovaciones S.L., Valencia, Spain) to fine powders.

2.2. Reagents and Standards

The HPLC-grade methanol (34860), HPLC-grade 2-propanol (34863), HPLC-grade ethanol absolute (34852-M), and CoQ10 standard (C9538) were acquired from Sigma-Aldrich Co. (St. Louis, MO, USA) while the 2-propanol pure p.a. (117515002) from was acquired Chempur (Piekary Śląskie, Poland).

2.3. Preparation of 2-Propanol Extracts

This was carried out using the slightly modified method of Stiff et al. [23], which consisted of direct extraction with 2-propanol, introducing an additional sonication step as recommended by Zu et al. [25] (see Figure 1). First, into a 15 mL conical centrifuge tube, 0.1 g of minced/finely ground sample was weighted to the nearest 0.1 mg (analytical balance ABJ-220-4NM; Kern & Sohn GmbH, Balingen, Germany). Next, 2 mL of 2-propanol was added and vortexed for 30 s (vortex mixer 6776; Corning Life Sciences, Monterrey, Mexico).

The mixture was sonicated (ultrasonic bath USC 300 THD; VWR International, Singapore, Malaysia) at 200 W intensity and 45 kHz frequency for 15 min, then orbitally shaken (orbital shaker 3005; GFL Gesellschaft für Labortechnik mbH, Burgwedel, Germany) at 150 rpm for 30 min, vortexed for 30 s, and centrifuged (benchtop centrifuge Universal 320 R; Andreas Hettich GmbH & Co. KG, Tuttlingen, Germany) at $8981\times g$ (9000 rpm) for 15 min at 20 °C, and the supernatant was collected in a weighted test tube.

The solvent was evaporated under a stream of nitrogen at 40 °C (thermoblock TA 120 P2; FALC Instruments S.R.L., Treviglio, Italy) until a constant weight was obtained (mass change not exceeding 1 mg).

The dried extract was resuspended in 1 mL 2-propanol, vortexed for 30 s, filtered through a polyamide syringe filter (0.45 μm pore size, 25 mm diameter), and kept at $-18\ °C$ until chromatographic analysis.

The extraction was performed in triplicate for each sample.

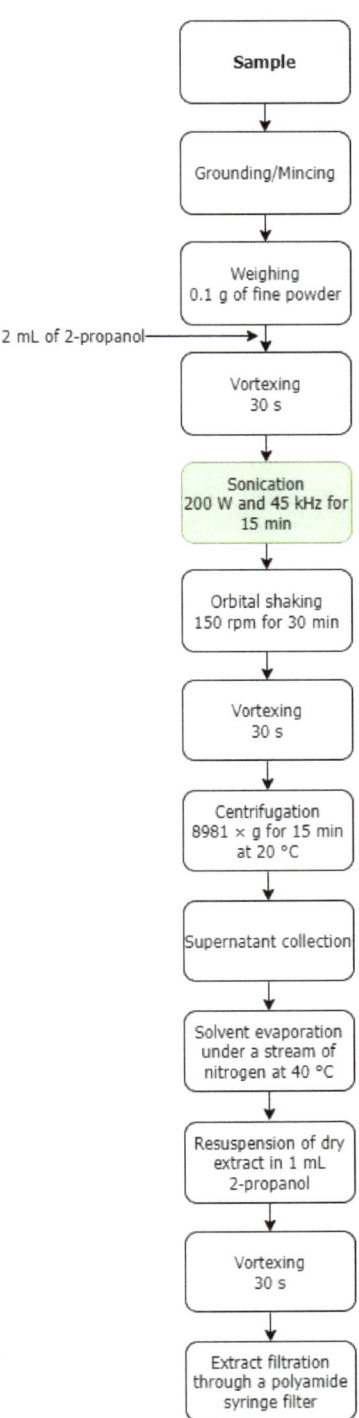

Figure 1. Workflow of the proposed extraction procedure: ultrasonic extraction with 2-propanol.

2.4. HPLC-DAD Analysis of the Extracts

Separation of CoQ10 was performed on a Kinetex XB-C18 column (150 mm L × 4.6 mm ID × 5 µm particle size; Phenomenex, Torrance, CA, USA) using a mixture of methanol/2-propanol/ethanol (70:15:15, $v/v/v$) as mobile phase in isocratic elution and a liquid chromatography system (1200 HPLC; Agilent Technologies Inc., Palo Alto, CA, USA) equipped with a diode array detector (DAD) according to a previous method described by Mattila and Kumpulainen [14], Souchet and Laplante [15], Ercan and El [29], Tobin et al. [30], and Román-Pizarro et al. [18], with minor modifications. The system also included a quaternary pump, a degasser, an autosampler, and a thermostatted column compartment.

Twenty microliters of extract were injected into the HPLC system to perform the instrumental analysis. The mobile phase flow rate was programmed to 1.2 mL/min and the column oven temperature was programmed to 25 °C. The DAD was set to 275 nm, and data acquisition was performed for 15 min using ChemStation software (Rev B.04.02 SP1; Agilent Technologies Inc., Palo Alto, CA, USA).

For calibration curve construction, a stock solution of CoQ10 was prepared at 1000 µg/mL in 2-propanol and subsequently diluted with the same solvent to prepare six working solutions at 0,25, 1.0, 50, 100, 150, and 200 µg/mL. The stock solution and working standards were stored at −18 °C in the dark for further use.

Before injection into HPLC, each sample extract or working standard was filtered through polyamide syringe filters (0.45 µm pore size and 25 mm diameter). CoQ10 was identified by comparing its retention time to the standard's one (studied under the same conditions). The results were expressed in µg CoQ10/g sample. All samples and working standards were analyzed in triplicate.

Figure 2 shows the overlap of some chromatograms, such as that of the RPC (with the highest content of CoQ10 from all vegetable matrices) and CH (the animal matrix where CoQ10 was detected), with that of the CoQ10 standard.

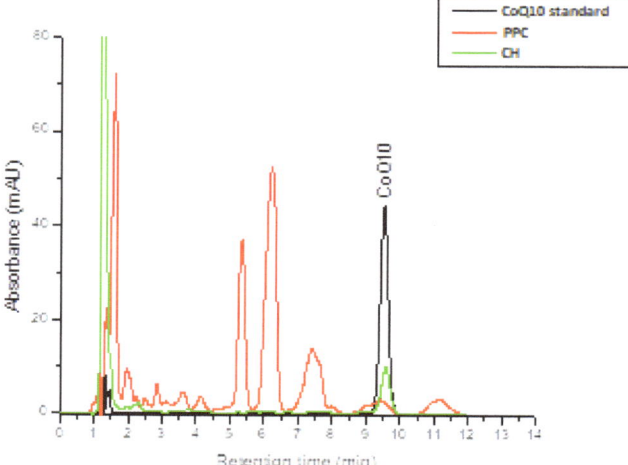

Figure 2. Overlay of chromatograms obtained by chromatographic separation of pumpkin press cake (PPC) and chicken heart (CH) extracts with the CoQ10 standard (detection wavelength: 275 nm).

2.5. Statistical Analysis

The statistically significant difference between the mean CoQ10 levels of oil press cakes, fish meat, and chicken hearts was determined by performing one-way analysis of variance (ANOVA) with a post hoc Tukey's test at a 95% confidence level ($p < 0.05$) using Minitab statistical software (version 19.1.1; LEAD Technologies, Inc., Charlotte, NC, USA).

3. Results and Discussion

HPLC with DA detection at 275 nm is the most common technique to estimate CoQ10 in foods, generally using calibration with an external standard [9]. For example, in 2000, it was used by Mattila et al. [13] to measure CoQ10 levels in pork heart, beef meat, and Baltic herring flesh by applying direct extraction with n-hexane/ethanol (5 1, v/v). One year later, the same authors [14] employed direct ethanol/n-hexane (2:5, v/v) extraction for samples such as cauliflower, potato, tomato, carrot, pea, bean, orange, clementine, apple, blackcurrant, lingonberry, strawberry, reindeer meat, pork heart, beef meat, beef heart, chicken meat, pollack flesh, hen's egg, skimmed milk (1.5% fat), and yogurt while for samples such as rapeseed oil, pork liver, beef liver, Emmental cheese, and Edam cheese, saponification with an aqueous potassium hydroxide solution took place prior to n-hexane extraction. The preparation protocol proposed by Souchet and Laplante in 2007 [15] to determine the level of CoQ10 in mackerel flesh, herring flesh, mackerel heart, and herring heart included sample homogenization with sodium dodecyl sulfate and a sodium chloride solution followed by extraction with ethanol/hexane (2:5, v/v). After two years, Laplante et al. [16] applied the same method to quantify CoQ10 in whole mackerel and whole herring; the oils of mackerel and herring, which were also assessed, were extracted by enzymatic hydrolysis from lyophilized fish samples using Protamex^{TM} and supercritical CO_2 and were then dissolved in 2-propanol before HPLC analysis. In 2012, Xue et al. [17] developed a method for dosing the level of CoQ10 in rape, apricot, tea, and mixed bee pollen by applying an online clean-up of accelerated extraction using absolute ethanol. The method of Mattila and Kumpulainen [14] was also used by Román-Pizarro et al. [18] in 2017 for samples such as parsley, spinach, avocado, peanut, pistachio, pork liver, and beef liver, utilizing a mixture of ethanol/n-hexane (2:5, v/v) for extraction. In 2018, Mandrioli et al. [19] investigated the level of CoQ10 in whole cow milk using the internal standard method (CoQ9); the sample was first saponified with an ethanolic potassium hydroxide solution and a pyrogallol solution, then extracted with petroleum ether/diethyl ether (9:1, v/v). Although efficient, these extraction protocols either use toxic solvents, too many or in too large quantities or require too many steps for extraction, thus making them challenging to implement in the large-scale extraction of CoQ10. Therefore, this study proposes a simple extraction procedure that uses a single solvent, 2-propanol, which is non-toxic and environmentally friendly; the recovered solvent can be used for repeated extractions within the same matrix, thus reducing the production costs.

The results of linearity and spike-and-recovery experiments carried out in this study are shown and discussed in Sections 3.1 and 3.3 of this paper, respectively; those regarding the determination of CoQ10 content in rapeseed press cake (RPC), sunflower press cake (SPC), pumpkin press cake (PPC), linseed press cake (LPC), walnut press cake (WPC), hempseed press cake (HPC), whole fish (WF), lyophilized whole fish (LWF), chicken hearts (CH), and lyophilized chicken hearts (LCH) are in Section 3.2.

3.1. Linearity and Measuring Range, Limits of Detection (LOD), and Quantification (LOQ)

Linearity and Measuring Range. The linearity of an analytical method is the interval in the measurement range in which the output signal of the analyte linearly correlates with its determined concentration. A measuring range is a set of values (analyte concentrations) where a measuring instrument's error is less than the value assumed [31]. The linearity check should confirm that the analytical method produces a linear response; as an acceptance criterion of testing, the regression coefficient must be more than 0.99 over the measuring range [20].

Six standard solutions of CoQ10, with concentrations of 0.25, 1.0, 50. 100, 150, and 200 µg/mL, were analyzed using the HPLC-DAD method described in Section 2.4, with three successive injections performed for each. The calibration curve was constructed by plotting the average value of each standard's peak area against its concentration (µg/mL). However, the regression equation ($y = 16.759x - 40.466$) was linear only in the 1–200 µg/mL range, with it having a good coefficient of regression ($R^2 = 0.9974$). An almost similar linear range (0.25–200 µg/mL) was reported by Xue et al. [17] when the HPLC-DAD method mentioned above was used to quantify CoQ10 in bee pollen.

Limit of Detection (LOD) and Limit of Quantification (LOQ). The lowest concentration (smallest amount) of an analyte that can be detected with statistically significant certainty is known as the LOD. The amount, or the smallest concentration, of an analyte that can be determined using a specific analytical procedure with an assumed accuracy, precision, and uncertainty is the LOQ [31].

The measurement results obtained for standard solutions of CoQ10 with the three lowest concentrations (0.25, 1.0, and 50 µg/mL) were used to determine the detection and quantification limits, as described by Konieczka and Namiesnik [31] and Michiu et al [32]. For this purpose, another calibration curve was outlined just based on these data; then, the LOD and LOQ were calculated with the formulas below (1), (2):

$$LOD(\mu g/mL) = 3.3 \times \frac{\sigma_a}{b} \qquad (1)$$

$$LOQ(\mu g/mL) = 10 \times \frac{\sigma_a}{b} \qquad (2)$$

where σ_a is the intercept's standard deviation and b is the slope.

Values of 0.22 µg/mL for LOD and 0.65 µg/mL for LOQ were thus obtained, comparable with those reported by Mandrioli et al. [8] (0.35 µg/mL for LOD and 1.18 µg/mL for LOQ) with the HPLC-DAD method that they used. The following conditions must be met to verify the LOD estimated: $LOD < C_{min}$ (concentration of the lowest standard used for the LOD determination) and $10 \times LOD > C_{min}$ [31]. These were achieved in the present study.

3.2. Levels of CoQ10 in Food By-Products and Waste

The CoQ10 content was determined for all three sample replicates on two consecutive days (see Table 1). The intra- and inter-day repeatability of measurements, in terms of relative standard deviation (RSD), was calculated for each sample by the following Formula (3):

$$RSD(\%) = \frac{\sigma \times 100}{X_{mean}} \qquad (3)$$

where σ is the standard deviation calculated using the STDEV function in Excel, and X_{mean} is the mean CoQ10 level of the three replicates.

To estimate the results precisely, the mean value of all six measurements (three on the 1st day of analysis and three on the 2nd day) and the pooled standard deviation (σ_{pooled}) was calculated (4) for each sample.

$$\sigma_{pooled} = \sqrt{\frac{(2 \times \sigma_1^2) + (2 \times \sigma_2^2)}{3+3}} \qquad (4)$$

where σ_1 is the standard deviation of the three values obtained on the 1st day of measurement, 2 means the degrees of freedom (3-1), and 3 is the number of measurements performed daily.

Next, the pooled relative standard deviation (RSD_{pooled}) was computed for each sample with the below Formula (5):

$$RSD_{pooled}(\%) = \frac{\sigma_{pooled} \times 100}{X_{mean}} \qquad (5)$$

where σ_{pooled} is the pooled standard deviation calculated above, and X_{mean} is the mean CoQ10 level of all six replicates (from both measurement days).

Table 1. CoQ10 levels in the oil press cakes, fish meat, and chicken hearts.

Sample/MC (%)	1st Day of Measurement		2nd Day of Measurement		Mean ± σ_{pooled} (μg CoQ10/g Sample)	RSD$_{pooled}$ (%)
	Mean ± σ (μg CoQ10/g Sample)	RSD (%)	Mean ± σ (μg CoQ10/g Sample)	RSD (%)		
RPC/10.4%	57.06 ± 2.316	4.1	56.47 ± 1.717	3.0	56.77 ± 1.664 [d]	2.9
SPC/9.8%	48.80 ± 1.386	2.8	49.78 ± 1.480	3.0	49.29 ± 1.171 [e]	2.4
PPC/9.9%	84.30 ± 0.978	1.2	85.29 ± 2.151	2.5	84.80 ± 1.364 [c]	1.6
LPC/10.7%	53.50 ± 0.913	1.7	53.89 ± 0.906	1.7	53.70 ± 0.743 [de]	1.4
WPC/9.8%	36.56 ± 0.845	2.3	36.56 ± 0.352	1.0	36.56 ± 0.528 [f]	1.4
HPC/11.2%	n.d.	-	n.d.	-	n.d.	-
WF/69.0%	n.d.	-	n.d.	-	n.d.	-
LWF/3.4%	n.d.	-	n.d.	-	n.d.	-
CH/77.0%	119.95 ± 11.141	10.0	116.83 ± 11.419	9.8	114.39 ± 9.211 [b]	8.1
LCH/3.2%	384.52 ± 0.680	0.2	381.97 ± 0.872	0.2	383.25 ± 0.639 [a]	0.2

MC—moisture content (%) determined using oven drying at 103 °C to a constant weight [33]; RPC—rapeseed press cake; SPC—sunflower press cake; PPC—pumpkin press cake; LPC—linseed press cake; WPC—walnut press cake; HPC—hempseed press cake; WF—whole fish; LWF—lyophilized whole fish; CH—chicken hearts; LCH—lyophilized chicken hearts; n.d.—not detected; σ—standard deviation; RSD—relative standard deviation; σ_{pooled}—pooled standard deviation; RSD$_{pooled}$—pooled relative standard deviation. Results are expressed as mean ± standard deviation of triplicate data (n = 3). Different letters in the row indicate a statistically significant difference at $p < 0.05$ (Tukey's test).

Oil press cakes. The richest source of CoQ10 is PPC (84.80 μg CoQ10/g sample), followed by RPC (56.77 μg CoQ10/g sample). On the other hand, there is no significant difference in the CoQ10 level of RPC and LPC (53.70 μg CoQ10/g sample) or between that of LPC and SPC (49.29 μg CoQ10/g sample). The lowest concentration of CoQ10 was found in WPC (36.56 μg CoQ10/g sample), while in HPC, it was not detected. Most studies have only focused on determining CoQ10 in raw materials and oils [9,34], and, to our knowledge, this is the first one conducted on oil press cakes. In the study of Rodríguez-Acuña et al. [34], CoQ10 was isolated by solid-phase extraction (SPE) from crude oil of rapeseed and refined oil of sunflower, showing levels of 46.4 and 8.7 μg/g, respectively.

Fish meat. CoQ10 was not detected in the WF and LWF samples analyzed in this study. In contrast to our findings, Souchet and Laplante [15] found concentrations of 18.6 μg CoQ10/g raw meat and 88.4 μg CoQ10/g lyophilized meat in whole mackerel, and 9.9 μg CoQ10/g raw meat and 50.9 μg CoQ10/g lyophilized meat in whole herring. Given that mackerel and herring are fatty fishes (above 8% fat content) and the rainbow trout investigated here is a mid-fat fish (2–8% fat content) [35], it can be concluded that the CoQ10 content in fish is proportional to its fat content.

In the study of Laplante et al. [16], the performance and production costs of supercritical CO_2 extraction versus enzymatic hydrolysis followed by centrifugation were compared to recover the oils and CoQ10 from mackerel and herring. The enzymatic hydrolysis process gave higher oil and CoQ10 yields with both fish species and generated lower production costs. A 65 μg/g concentration was found in mackerel oil and 275 μg/g was found in herring oil.

Chicken hearts. The highest content of CoQ10 was found in the chicken hearts, with a level of 114.39 μg CoQ10/g for the CH sample and 383.25 μg CoQ10/g for the LCH sample. Through lyophilization, the CoQ10 amount recovered from the chicken hearts was approximately 3.3 times higher. Although expensive, freeze-drying reduces the weight and volume of waste, providing better storage stability and higher extraction yield for CoQ10. The result obtained for CH is consistent with those of Kubo et al. [22], who reported a similar concentration, 107 μg CoQ10/g, in raw chicken hearts. However, CoQ10 is quickly oxidized when exposed to light or in contact with air [36], with its stability being affected by storage time and conditions. Therefore, extracting CoQ10 as close as possible to the moment the food matrix has become waste is essential.

In a more recent study, Villanueva-Bermejo and Temelli (2011) [37] investigated the level of CoQ10 in oil recovered from chicken hearts by supercritical CO2 extraction. They found a concentration of 670 µg/g in the oil obtained during the constant extraction rate and 2310 µg/g in the one obtained during the falling extraction rate. Hence, more CoQ10 can be recovered from a food matrix by extracting its oil. However, supercritical fluid technology is expensive due to the high investment costs [38]. Oil extraction using a green solvent followed by solvent evaporation could be, thus, a less expensive alternative.

3.3. Accuracy (Trueness and Precision)

The ISO 5725-4:2020 standard [39] uses two terms, "trueness" (the closeness of agreement between the expected value of a measurement result and the actual value) and "precision" (the closeness of agreement between independent measurement results obtained using predetermined conditions), to describe the accuracy of a measurement method.

The analytical method trueness, reported as percentage recovery (see Table 2), was assessed on samples of PPC and CH spiked with the same volume of CoQ10 standard solution at three different concentrations so that in final extracts, it reached 0.5, 1.5, and 5.0 µg CoQ10/mL, which corresponded to 5, 10, and 50 µg CoQ10/g sample, respectively. The recovery rate was calculated using the following Formula (6) [40]:

$$Recovery\ rate(\%) = \frac{C_{spiked}}{C_{expected}} \times 100 \qquad (6)$$

where C_{spiked} is the concentration of CoQ10 found in the spiked sample (µg CoQ10/mL) and $C_{expected}$ is the expected (theoretical or calculated) CoQ10 concentration in the spiked sample (µg CoQ10/mL).

$$C_{expected}(\mu g\ CoQ10/mL) = C_{unspiked} + C_{added} \qquad (7)$$

where $C_{unspiked}$ is the CoQ10 concentration found in the unspiked sample (µg CoQ10/mL) and C_{added} is the concentration of CoQ10 added to the spiked sample (µg CoQ10/mL).

Table 2. Recovery rates (trueness values) of CoQ10 from samples of PPC and CH with RSDs (precision values).

Sample	Spiking Concentration (µg CoQ10/mL)	Found Concentration (µg CoQ10/mL)	RSD (%)	Recovery Rate (%)	RSD (%)
PPC	-	12.05 ± 0.042	0.4	-	-
	0.5	14.56 ± 0.042	0.3	116.0 ± 0.054	0.05
	1.5	15.63 ± 0.042	0.3	115.4 ± 0.053	0.05
	5.0	17.21 ± 0.084	0.5	100.9 ± 0.242	0.2
CH	-	14.29 ± 0.0	0.0	-	-
	0.5	15.66 ± 0.084	0.5	105.9 ± 0.569	0.5
	1.5	16.88 ± 0.127	0.7	106.9 ± 0.802	0.7
	5.0	19.15 ± 0.127	0.7	99.3 ± 0.656	0.7

PPC—pumpkin press cake; CH—chicken hearts; RSD—relative standard deviation. The results are expressed as the mean ± standard deviation of the triplicate data ($n = 3$).

For all analytes within the scope of a method, mean recoveries from initial validation should fall between 70 and 120%, with the associated repeatability (RSD) of less than or equal to 20% [41]. The recovery of an analyte depends on the matrix complexity, the sample preparation procedure, and the analyte concentration [42]. A too-low recovery underestimates the calculated LOD [31].

The average recovery rate of CoQ10 from the PPC matrix ranged between 100.9% and 116.0% (see Table 2), while from the CH sample, it ranged from 99.3 to 106.9%, within the European Commission SANTE 11312/2021 guideline requirements (mean 70–120%; RSD ≤ 20%) [41]. Similar recovery rates (100.4–113.0% for mackerel red flesh; 98.0–100.4% for herring's heart) were reported by Souchet and Laplante [15] in fish samples spiked with 1–15 µg CoQ10 by using an HPLC-DAD method. The percentage RSD between the recovery values (see Table 2), which indicates the precision of our method, was in the range of 0.05 and 0.2% for the PPC sample and, for the CH one, it was between 0.5 and 0.7%. All these results meet the acceptance criteria mentioned in the previous paragraph.

4. Conclusions

A valuable analytical method has been developed here to determine CoQ10 in oil press cakes, fish meat, and chicken hearts by modifying a previous extraction procedure-introducing an additional sonication step. Briefly, this consisted of sample homogenization with 2-propanol, followed by sonication, short maceration, centrifugation, supernatant collection, solvent evaporation, dry-extract resuspension in 2-propanol, and filtration before HPLC analysis. As a result, a very good recovery rate for CoQ10, with excellent repeatability, was achieved. In addition, the solvent used, 2-propanol, is environmentally friendly, economically viable, and feasible for scaling up the production of CoQ10. The highest level of CoQ10 was found in CH and then in PPC. Therefore, these matrices can be used as raw materials to extract CoQ10 for further applications in different industries (foods, supplements, pharmaceutics, and cosmetics).

This study's findings provide information regarding CoQ10-rich food by-products and waste, a green extraction procedure, and an HPLC method for CoQ10 analysis, which might be helpful to natural extract producers. Future work will focus on lipid fraction extraction from these matrices using 2-propanol to characterize them towards obtaining natural dietary supplements based on CoQ10.

5. Patents

Patent application A/00138 from 24 March 2023: "Process for the preparation of some natural dietary supplements based on coenzyme Q10". Inventors: Cristina-Anamaria Semeniuc, Andersina-Simona Podar, Sonia-Ancuța Socaci, Floricuța Ranga, Simona-Raluca Ionescu, Maria-Ioana Socaciu, Melinda Fogarasi, Dan-Cristian Vodnar, and Anca-Corina Fărcaș

Author Contributions: Conceptualization, C.A.S. and S.A.S.; methodology, C.A.S.; formal analysis, F.R., A.S.P., S.R.I., M.-I.S., M.F. and A.C.F.; resources, C.A.S., S.A.S. and D.C.V.; data curation, C.A.S.; writing—original draft preparation, C.A.S. and A.S.P.; writing—review and editing, S A.S.; visualization, D.C.V.; supervision, S.A.S.; project administration, C.A.S.; funding acquisition, S.A.S. and D.C.V. All authors have read and agreed to the published version of the manuscript.

Funding: This work was supported by a grant from the Romanian Ministry of Education and Research, CNCS-UEFISCDI, project number PN-III-P4-ID-PCE-2020-1847, within PNCDI III.

Data Availability Statement: The data presented in this study are available on request from the corresponding author.

Acknowledgments: We are grateful for the administrative and financial support received from the University of Agricultural Sciences and Veterinary Medicine of Cluj-Napoca, Romania.

Conflicts of Interest: The authors declare no conflict of interest.

References

1. Rodriguez-Estrada, M.T.; Poerio, A.; Mandrioli, M.; Lercker, G.; Trinchero, A.; Tosi, M.R.; Tugnoli, V. Determination of coenzyme Q10 in functional and neoplastic human renal tissues. *Anal. Biochem.* **2006**, *357*, 150–152. [CrossRef] [PubMed]
2. Turkowicz, M.J.; Karpińska, J. Analytical problems with the determination of coenzyme Q_{10} in biological samples. *BioFactors* **2013**, *39*, 176–185. [CrossRef]
3. Atla, S.R.; Raja, B.; Dontamsetti, B.R. A new method of synthesis of coenzyme Q10 from isolated solanesol from tobacco waste. *Int. J. Pharm. Pharm. Sci.* **2014**, *6*, 499–502.
4. Niklowitz, P.; Döring, F.; Paulussen, M.; Menke, T. Determination of coenzyme Q10 tissue status via high-performance liquid chromatography with electrochemical detection in swine tissues (*Sus scrofa domestica*). *Anal. Biochem.* **2013**, *437*, 88–94. [CrossRef]
5. Rao, G.; Shen, G.; Xu, G. Ultrasonic assisted extraction of coenzyme Q10 from litchi (*Litchi chinensis* Sonn.) pericarp using response surface methodology. *J. Food Process Eng.* **2011**, *34*, 671–681. [CrossRef]
6. Bao, K.; Zhang, C.; Xie, S.; Feng, G.; Liao, S.; Cai, L.; He, J.; Guo, Y.; Jiang, C. A simple and accurate method for the determination of related substances in coenzyme Q10 soft capsules. *Molecules* **2019**, *24*, 1767. [CrossRef] [PubMed]

7. Cao, X.L.; Xu, Y.T.; Zhang, G.M.; Xie, S.M.; Dong, Y.M.; Ito, Y. Purification of coenzyme Q_{10} from fermentation extract: High-speed counter-current chromatography versus silica gel column chromatography. *J. Chromatogr. A* **2006**, *1127*, 92–96. [CrossRef]
8. Mandrioli, M.; Semeniuc, C.A.; Boselli, E.; Rodriguez-Estrada, M.T. Ubiquinone in Italian high-quality raw cow milk. *Ital. J. Food Sci.* **2018**, *30*, 144–155.
9. Podar, A.S.; Semeniuc, C.A.; Ionescu, S.R.; Socaciu, M.-I.; Fogarasi, M.; Fărcaș, A.C.; Vodnar, D.C.; Socaci, S.A. An overview of analytical methods for quantitative determination of coenzyme Q10 in foods. *Metabolites* **2023**, *13*, 272. [CrossRef]
10. Smeu, I.; Dobre, A.A.; Cucu, E.M.; Mustățea, G.; Belc, N.; Ungureanu, E.L. Byproducts from the vegetable oil industry: The challenges of safety and sustainability. *Sustainability* **2022**, *14*, 2039. [CrossRef]
11. Jayathilakan, K.; Sultana, K.; Radhakrishna, K.; Bawa, A.S. Utilization of byproducts and waste materials from meat, poultry and fish processing industries: A review. *J. Food Sci. Technol.* **2012**, *49*, 278–293. [CrossRef] [PubMed]
12. Socaciu, M.-I.; Fogarasi, M.; Simon, E.L.; Semeniuc, C.A.; Socaci, S.A.; Podar, A.S.; Vodnar, D.C. Effects of whey protein isolate-based film incorporated with tarragon essential oil on the quality and shelf-life of refrigerated brook trout. *Foods* **2021**, *10*, 401. [CrossRef] [PubMed]
13. Mattila, P.; Lehtonen, M.; Kumpulainen, J. Comparison of in-line connected diode array and electrochemical detectors in the high-performance liquid chromatographic analysis of coenzymes Q(9) and Q(10) in food materials. *JAFC* **2000**, *48*, 1229–1233. [CrossRef]
14. Mattila, P.; Kumpulainen, J. Coenzymes Q_9 and Q_{10}: Contents in foods and dietary intake. *J. Food Compos. Anal.* **2001**, *14*, 409–417. [CrossRef]
15. Souchet, N.; Laplante, S. Seasonal variation of Co-enzyme Q10 content in pelagic fish tissues from Eastern Quebec. *J. Food Compos. Anal.* **2007**, *20*, 403–410. [CrossRef]
16. Laplante, S.; Souchet, N.; Bryl, P. Comparison of low-temperature processes for oil and coenzyme Q10 extraction from mackerel and herring. *Eur. J. Lipid Sci. Technol.* **2009**, *111*, 135–141. [CrossRef]
17. Xue, X.; Zhao, J.; Chen, L.; Zhou, J.; Yue, B.; Li, Y.; Wu, L.; Liu, F. Analysis of coenzyme Q10 in bee pollen using online cleanup by accelerated solvent extraction and high performance liquid chromatography. *Food Chem.* **2012**, *133*, 573–578. [CrossRef]
18. Román-Pizarro, V.; Fernández-Romero, J.M.; Gómez-Hens, A. Automatic determination of coenzyme Q10 in food using cresyl violet encapsulated into magnetoliposomes. *Food Chem.* **2017**, *221*, 864–870. [CrossRef]
19. Chemistry LibreTexts. Available online: https://chem.libretexts.org/Bookshelves/Analytical_Chemistry/Instrumental_Analysis_(LibreTexts)/01%3A_Introduction/1.04%3A_Selecting_an_Analytical_Method (accessed on 1 May 2023).
20. Australian Pesticides and Veterinary Medicines Authority. *Guidelines for the Validation of Analytical Methods for Active Constituent*, United Nations Office on Drugs and Crime. *Guidance for the Validation of Analytical Methodology and Calibration of Equipment used for Testing of Illicit Drugs in Seized Materials and Biological Specimens*; United Nations: New York, NY, USA, 2009. Available online: https://www.unodc.org/documents/scientific/validation_E.pdf (accessed on 26 January 2023).
21. European Medicines Agency. *ICH Topic Q 2 (R1). Validation of Analytical Procedures: Text and Methodology-Step 5*; EMEA: London, UK, 1995. Available online: https://www.ema.europa.eu/documents/scientific-guideline/ich-q-2-r1-validation-analytical-procedures-text-methodology-step-5_en.pdf (accessed on 24 January 2023).
22. Kubo, H.; Fujii, K.; Kawabe, T.; Matsumoto, S.; Kishida, H.; Hosoe, K. Food content of ubiquinol-10 and ubiquinone-10 in the Japanese diet. *J. Food Compos. Anal.* **2008**, *21*, 199–210. [CrossRef]
23. Stiff, M.R.; Weissinger, A.K.; Danehower, D.A. Analysis of CoQ_{10} in cultivated tobacco by a high-performance liquid chromatography–ultraviolet method. *J. Agric. Food Chem.* **2011**, *59*, 9054–9058. [CrossRef]
24. Yilmaz, E.; Soylak, M. Type of Green Solvents Used in Separation and Preconcentration Methods. In *New Generation Green Solvents for Separation and Preconcentration of Organic and Inorganic Species*, 1st ed.; Soylak, M., Yilmaz, E., Eds.; Elsevier: Amsterdam, The Netherlands, 2020; pp. 207–266.
25. Zu, Y.; Zhao, C.; Li, C.; Zhang, L. A rapid and sensitive LC-MS/MS method for determination of coenzyme Q_{10} in tobacco (*Nicotiana tabacum* L.) leaves. *J. Sep. Sci.* **2006**, *29*, 1607–1612. [CrossRef] [PubMed]
26. European Union. Directive 2009/32/EC of the European Parliament and of the Council of 23 April 2009 on the approximation of the laws of the Member States on extraction solvents used in the production of foodstuffs and food ingredients. *Off. J. Eur. Commun.* **2009**, *L141*, 3–11.
27. Anton, R.; Barlow, S.; Boskou, D.; Castle, L.; Crebelli, R.; Dekant, W.; Engel, K.-H.; Forsythe, S.; Grunow, W.; Larsen, J.C.; et al. Opinion of the Scientific Panel on Food Additives, Flavourings, Processing Aids and Materials in Contact with Food on a request from the Commission related to Propan-2-ol as a carrier solvent for Flavourings. *EFSA J.* **2005**, *202*, 1–10.
28. González, C.M.; Llorca, M.; Quiles, A.; Hernando, I.; Moraga, G. Water sorption and glass transition in freeze-dried persimmon slices. Effect on physical properties and bioactive compounds. *LWT Food Sci. Technol.* **2020**, *130*, 109633. [CrossRef]
29. Ercan, P.; El, S.N. Changes in content of coenzyme Q10 in beef muscle, beef liver and beef heart with cooking and in vitro digestion. *J. Food Compos. Anal.* **2011**, *24*, 1136–1140. [CrossRef]
30. Tobin, B.D.; O'Sullivan, M.G.; Hamill, R.; Kerry, J.P. Effect of cooking and in vitro digestion on the stability of co-enzyme Q10 in processed meat products. *Food Chem.* **2014**, *150*, 187–192. [CrossRef]
31. Konieczka, P.; Namiesnik, J. *Quality Assurance and Quality Control in the Analytical Chemical Laboratory: A Practical Approach*, 1st ed.; CRC Press: Boca Raton, FL, USA, 2009; pp. 131–213.
32. Michiu, D.; Socaciu, M.-I.; Fogarasi, M.; Jimborean, A.M.; Ranga, F.; Mureșan, V.; Semeniuc, C.A. Implementation of an analytical method for spectrophotometric evaluation of total phenolic content in essential oils. *Molecules* **2022**, *27*, 1345. [CrossRef]

33. Semeniuc, C.A.; Mandrioli, M.; Tura, M.; Socaci, B.S.; Socaciu, M.-I.; Fogarasi, M.; Michiu, D.; Jimborean, A.M.; Mureşan, V.; Ionescu, S.R.; et al. Impact of lavender flower powder as a flavoring ingredient on volatile composition and quality characteristics of Gouda-type cheese during ripening. *Foods* **2023**, *12*, 1703. [CrossRef]
34. Rodríguez-Acuña, R.; Brenne, E.; Lacoste, F. Determination of coenzyme Q10 and Q9 in vegetable oils. *J. Agric. Food Chem.* **2008**, *56*, 6241–6245. [CrossRef]
35. Taşbozan, O.; Gökçe, M.A. Fatty Acids in Fish. In *Fatty Acids*; Catala, A., Ed.; IntechOpen Limited: London, UK, 2017; pp. 143–159.
36. Temova Rakuša, Ž.; Kristl, A.; Roškar, R. Stability of reduced and oxidized coenzyme Q10 in finished products. *Antioxidants* **2021**, *10*, 360. [CrossRef]
37. Villanueva-Bermejo, D.; Temelli, F. Extraction of oil rich in coenzyme Q10 from chicken by-products using supercritical CO_2. *J. Supercrit. Fluids* **2021**, *174*, 105242. [CrossRef]
38. Cassanelli, M.; Prosapio, V.; Norton, I.; Mills, T. Design of a cost-reduced flexible plant for supercritical fluid-assisted applications. *Chem. Eng. Technol.* **2018**, *41*, 1368–1377. [CrossRef]
39. ISO 5725-4:2020; Accuracy (Trueness and Precision) of Measurement Methods and Results Part 4: Basic Methods for the Determination of the Trueness of a Standard Measurement Method. ISO: Geneva, Switzerland, 2020. Available online: www.iso.org/standard/69421.html (accessed on 22 January 2023).
40. Manzi, P.; Durazzo, A. Rapid determination of coenzyme Q10 in cheese using high-performance liquid chromatography. *Dairy Sci. Technol.* **2015**, *95*, 533–539. [CrossRef]
41. European Commission. *Analytical Quality Control and Method Validation Procedures for Pesticide Residues Analysis in Food and Feed SANTE 11312/2021*; SANTE: Brussels, Belgium, 2021. Available online: https://food.ec.europa.eu/system/files/2022-02/pesticides_mrl_guidelines_wrkdoc_2021-11312.pdf (accessed on 22 January 2023).
42. *Agricultural and Veterinary Chemical Products*; APVMA: Kingston, Australia, 2004. Available online: https://apvma.gov.au/sites/default/files/docs/guideline-69-analytical-methods.pdf (accessed on 22 January 2023).

Disclaimer/Publisher's Note: The statements, opinions and data contained in all publications are solely those of the individual author(s) and contributor(s) and not of MDPI and/or the editor(s). MDPI and/or the editor(s) disclaim responsibility for any injury to people or property resulting from any ideas, methods, instructions or products referred to in the content.

Article

An Empirical Model for Predicting the Fresh Food Quality Changes during Storage

Reham Abdullah Sanad Alsbu [1], Prasad Yarlagadda [2] and Azharul Karim [1,*]

[1] School of Mechanical, Medical and Process Engineering, Queensland University of Technology, Brisbane, QLD 4001, Australia; rehamabdullahsanad.alsbua@hdr.qut.edu.au
[2] School of Engineering, University of Southern Queensland, Springfield Central, QLD 4300, Australia; y.prasad@usq.edu.au
* Correspondence: azharul.karim@qut.edu.au

Abstract: It is widely recognized that the quality of fruits and vegetables can be altered during transportation and storage. Firmness and loss of weight are the crucial attributes used to evaluate the quality of various fruits, as many other quality attributes are related to these two attributes. These properties are influenced by the surrounding environment and preservation conditions. Limited research has been conducted to accurately predict the quality attributes during transport and storage as a function of storage conditions. In this research, extensive experimental investigations have been conducted on the changes in quality attributes of four fresh apple cultivars (Granny Smith, Royal Gala, Pink Lady, and Red Delicious) during transportation and storage. The study evaluated the weight loss and change in firmness of these apples varieties at different cooling temperatures ranging from 2 °C to 8 °C to assess the impact of storing at these temperatures on the quality attributes. The results indicate that the firmness of each cultivar continuously decreased over time, with the R^2 values ranging from 0.9489–0.8691 for red delicious, 0.9871–0.9129 for royal gala, 0.9972–0.9647 for pink lady, and 0.9964–0.9484 for granny smith. The rate of weight loss followed an increasing trend with time, and the high R^2 values indicate a strong correlation. The degradation of quality was evident in all four cultivars, with temperature having a significant impact on firmness. The decline in firmness was found to be minimal at 2 °C, but increased as the storage temperature increased. The loss of firmness also varied among the four cultivars. For instance, when stored at 2 °C, the firmness of pink lady decreased from an initial value of 8.69 kg·cm² to 7.89 kg·cm² in 48 h, while the firmness of the same cultivar decreased from 7.86 kg·cm² to 6.81 kg·cm² after the same duration of storage. Based on the experimental results, a multiple regression quality prediction model was developed as a function of temperature and time. The proposed models were validated using a new set of experimental data. The correlation between the predicted and experimental values was found to be excellent. The linear regression equation yielded an R^2 value of 0.9544, indicating a high degree of accuracy. The model can assist stakeholders in the fruit and fresh produce industry in anticipating quality changes at different storage stages based on the storage conditions.

Keywords: fresh food quality; apple; Australia; firmness; weight loss; experimental investigations

Citation: Sanad Alsbu, R.A.; Yarlagadda, P.; Karim, A. An Empirical Model for Predicting the Fresh Food Quality Changes during Storage. *Foods* **2023**, *12*, 2113. https://doi.org/10.3390/foods12112113

Academic Editors: Gianluca Nardone, Rosaria Viscecchia and Francesco Bimbo

Received: 10 April 2023
Revised: 20 May 2023
Accepted: 21 May 2023
Published: 24 May 2023

Copyright: © 2023 by the authors. Licensee MDPI, Basel, Switzerland. This article is an open access article distributed under the terms and conditions of the Creative Commons Attribution (CC BY) license (https://creativecommons.org/licenses/by/4.0/).

1. Introduction

Fresh food supply chain demands special attention, as plant-based food materials (PBFM) are considered important sources of bioactive compounds and minerals essential for mankind [1,2]. Losses of fruits and vegetables in the food chain are estimated to be about a third of the total global production [3]. During transportation and storage, PBFMs undergo physical and biochemical modifications and a significant outcome of this modification is shrinkage and firmness: a decrease in volume, shape, porosity, and changes in hardness [4]. Of the many adverse effects of shrinkage to be found in the available literature, a particularly significant one is the reduction of rehydration capability, which modifies texture [5].

Due to the consumer emphasis on high quality products, producers and distributors are currently prioritizing the preservation of fresh food quality during the transportation and storage through the utilization of advanced cooling and preservation technologies [6–9]. As a consequence of this transformation, the supply chains for fruits and vegetables are moving away from the traditional model that involved intermediaries such as wholesalers and progressing toward a more direct and collaborative relationship between farmers and supermarkets. This shift allows supermarkets to establish a reliable supply of fresh produce directly from farms, ensuring consistent quality [10].

This study focuses on the impact of storage and transportation conditions on the quality of fresh produce, using apple as a sample. Although changes in many quality attributes, including biochemical properties, take place during storage, the study investigates only the physical qualities and more specifically the firmness and weight loss of apples under various environmental conditions during the storage and transportation stages. These two parameters are considered crucial and have a strong correlation with other quality indicators [11]. The primary aim of this study is to develop and validate an empirical quality degradation model for predicting changes in quality of fresh produce throughout the supply chain.

The quality of fresh fruits and vegetables can be defined as the extent to which a set of inherent characteristics satisfies customer expectations [12]. This definition is also supported by [13], who describe quality as a dynamic composite of physiochemical properties and consumer perception. Harvest time, picking and handling techniques, storage temperature, and other factors significantly impact the quality. The timing of harvest is particularly critical, as early harvesting leads to changes in flavor, color, size and, storability, while late harvesting results in softening and reduces shelf life of apples [14]. Fruit maturity at harvest time and storage conditions significantly affect fruit firmness, which can be measured using various destructive and non-destructive methods [15]. Despite differences in cultivars, firmness change during storage is a common occurrence [14].

Apple (Malus Domestica) is among the top four most widely produced and consumed fruits worldwide, with an annual global production exceeding 87 million tons [16]. Cultivated in subtropical and tropical environments, the European Apple Inventory lists more than ten thousand apple cultivars, resulting in an extensive range of quality traits [17,18]. Firmness is a key attribute for assessing apple quality, as it is closely associated with water content and is the most noticeable change during shelf life [19].

Flesh firmness has been widely used to assess the quality of apples at different temperatures, as demonstrated in experiments by [13] who reported changes in the texture of apples with temperature. Similarly, ref. [20] studied the effect of storage temperature and ripening stage on the firmness of fresh cut tomatoes. Several mathematical models have been utilized to predict the quality indicators and determine the quality status of a product at a given stage. Ref. [21] reviewed the kinetics of thermal softening in foods and found that first-order kinetic expressions were suitable for expressing the degree of softening at a constant temperature. The Arrhenius relationship has been utilized to determine the effect of temperature on the quality of fruits, where the relationship between firmness and temperature is approximately linear [22,23]. The degradation of food quality during storage or transport is determined by storage time (t), storage temperature T, and other constants such as the gas constant and activation energy.

Quality is changed over time according to the following Equation (1):

$$\frac{dq}{dt} = kq^n, \qquad (1)$$

where q represents the quality indicator of the product, k is the rate of quality degradation which depends on environment conditions such as temperature, n is the order of the reaction which can be either 0 or 1 (zero order and first order reactions). Value of n determines whether reaction rate is dependent on the amount of quality (q) left. When n equals 1 (first-order), the quality of the food decays exponentially because of microbial growth (e.g., fresh meat and fish). Fresh fruits and vegetables follow the zero-order reaction when the quality decays linearly. The quality degradation for both orders is illustrated in Figure 1 [24].

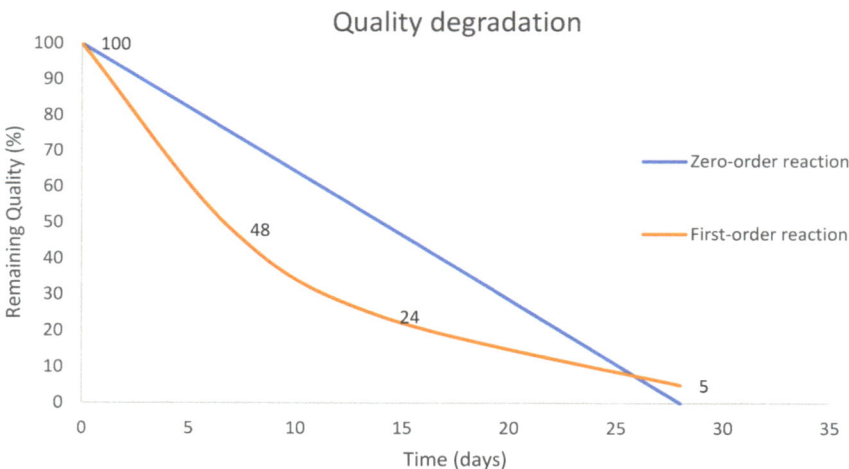

Figure 1. Quality degradation model.

Temperature of transport or storage is a critical factor affecting and controlling food quality in the supply chain, and the quality changes can be expressed using the following Arrhenius equation:

$$\ln K = \ln A - E_a/RT \qquad \text{leading to: } K = A \times e^{[-E_a/RT]} \qquad (2)$$

where K is the quality degradation rate, E_a is the activation energy (an empirical parameter characterizing the exponential temperature dependence), R is the universal gas constant, A is the rate constant, and T is the absolute temperature (in Kelvin degrees). Equation (1) can be used to describe the change in firmness of the apple fruit given that firmness depends on the temperature of the fruit of the ambient environment [23].

Based on Equations (1) and (2), quality level of fruit can be predicted at certain time in the supply chain using the initial quality (q_0), the time t_i and the degradation rate k_i (depending on the temperature T_i), leading to:

$$q = q_0 - \sum k_i t_i \qquad (3)$$

for zero-order reaction. Substituting the equations for the rate of degradation we get the following equations to express the degradation of quality in specific time at certain temperature:

$$q(t) = q_0 - k_0 \, t \exp(-a/T), \, t \leq t_1 \qquad (4)$$

where t_1 is the delay time before the product being stored or transported. And a is a constant of E_a/R.

Ref. [25] modelled the quality of tomato based on the change of color under constant and variable temperature. From that model we can observe the following equation:

$$f(q) = f(q_0) + \rho t \qquad (5)$$

where q is the quality attribute, and q_0 is the initial quality attribute, t is time of storage or transportation, ρ is the reaction rate or slope depending on temperature, and f is a function of q which is here the firmness.

The preservation of fruit quality throughout the food supply chain is crucial for minimizing quality loss and maintaining high nutritional value. Given the importance of fresh produce quality in customer-focused competitive strategies, both suppliers and supermarkets make considerable efforts to maintain the quality of fresh produce to increase profits and reduce wastage. However, there is a lack of studies on experimentally determining the quality changes in the supply chain in Australia. This gap in knowledge presents a significant challenge for effectively selecting appropriate storage technology and transportation conditions. The present study aims to fill this research gap by investigating the impact of storage temperature and duration on apple quality, with firmness being the key quality indicator. Although changes in many quality attributes, including biochemical properties, take place during storage, the study investigates only the physical qualities and more specifically the firmness and weight loss of apples under various environmental conditions during the storage and transportation stages. These two parameters are considered crucial and have a strong correlation with other quality indicators.

Prediction quality changes in the supply chain is an important tool to minimize the food loss. Traditional models such as the Arrhenius model have limited practical applicability in evaluating fruit quality due to their reliance on difficult-to-determine parameters such as initial quality attributes and degradation rates. The study proposes a multi-regression model that considers four main apple cultivars, which have been developed through extensive experimentation and statistical analysis. This empirical model will accurately predict changes in firmness over time as a function of temperature, allowing managers to estimate quality changes during storage or transportation under different conditions. The developed models provide a more realistic representation of product quality. By gaining a more accurate understanding of the relationship between quality and environmental conditions, this study will help to improve product quality and reduce waste in the fruit supply chain.

2. Materials and Methods

An experimental investigation was conducted to examine the changes in quality of locally harvested apples during transportation and storage. The study involved four cultivars of apple namely 'Golden Delicious', 'Granny Smith', 'Royal Gala', and 'Pink lady' obtained from a commercial farm in Stanthorpe, located in southern Queensland. Apples were grown using integrated fruit production methods and were harvested at their optimum maturity.

Upon harvesting, the apples were stored in large bins and transported into a cold storage facility, where they were kept until the next harvesting season to meet market demand. Harvesting occurred between early February and late May, with each cultivar being harvested at a different time. The fruits were sorted and only the fault-free apples were placed in the cold storage.

Random sampling was employed to collect 100 apples from each cultivar at the farm. The samples were transported to the laboratories of the Faculty of Engineering (O block) and Faculty of Health (Q block) at the Queensland University of Technology, where they were divided into five groups of 20 samples each and stored for two days at four different temperatures: 2 °C, 4 °C, 6 °C, and 8 °C. The firmness and weight loss were analyzed four times at 12-h intervals during the two-day storage period.

2.1. Measuring the Weight Loss

Weight loss was determined by periodical weighing of samples. The weight was measured using a digital balance from A&D company ltd (model FZ-300i, San Jose, CA, USA). Twenty apples from each group were taken for the weight loss test. The weight loss (%) was calculated using the formula:

Weight loss (%) = ((Fruit initial weight − Fruit weight after interval)/Fruit initial weight) × 100

2.2. Testing the Firmness

To measure the firmness, a fruit pressure tester (Penetrometer) with a plunger of diameter 8 mm was used. Firmness results were expressed in kg·cm^{-2}. Penetrometer is one of the destructive methods used to measure the firmness and maturity of a fruit based on the force placed on a known diameter into a growing medium. The hand operated penetrometer used in this study is made by QA supplies LLC (Norfolk, VA, USA).

To test the firmness of the apples, several steps were followed. Initially, a random sample of apples was taken from boxes of each size, representing four different cultivars. The apples were then prepared by peeling the skin from opposite sides, ensuring a uniform thickness of the tested area. The peeling process was carried out on the cheek of the apple, between the stem and calyx. Each apple was tested twice, at opposite sides of its largest diameter.

The measurement of firmness was conducted by holding the apple against a bench top and placing the plunger tip of a penetrometer on the fruit's surface. With a consistent and uniform force, the plunger was pressed into the apple. The firmness reading was recorded, and the same procedure was repeated for the opposite side of the apple. This test was repeated on two additional samples.

Once all the firmness readings were obtained, the next step involved interpreting the results. This included calculating the mean firmness by considering all the readings obtained from the multiple apple samples tested.

Additionally, it was recommended that the temperature of the fruit be maintained at 10 °C or higher during the testing process. This temperature was suggested to ensure practical and accurate results since colder fruits tend to yield higher firmness readings.

Figure 2 shows the experimental facility and the procedure followed to measure the firmness and weight loss.

Figure 2. Experimental procedure.

2.3. Validation of the Model

In order to verify the accuracy of the empirical models formulated, the extrapolated model outputs were evaluated against experimental results. To achieve this, a further set of experiments was conducted using a random selection of four apple cultivars, and initial firmness and weight measurements were obtained. The fruits were subjected to measurement of firmness and weight loss at intervals of 12, 24, 36, and 48 h. To ensure greater precision, the experiment was performed in duplicate.

2.4. Statistical Analysis

In this study, four cultivars of apples were selected, and the experiment was designed using a completely random scheme with a total of 16 samples subjected to four different storage conditions. The data obtained from the experiments were analyzed using SPSS (Statistical Package for the Social Science). An ANOVA test was performed, and a significance level of $p < 0.05$ was achieved for each cultivar.

3. Results and Discussion

In this comprehensive study, one of the objectives was to assess the influence of postharvest storage and transportation temperature on the quality of four distinct fresh apple cultivars. To achieve this, firmness and weight loss measurements were used as the indicators of physical quality. The obtained results are presented in Figures 3 and 4, which depict the variations in firmness and weight loss for each cultivar investigated across a wide range of storage temperatures, specifically 2 °C, 4 °C, 6 °C, and 8 °C.

Figure 3 effectively captures the changes in firmness for each apple cultivar at the four different temperatures. An ANOVA test of the results demonstrated the statistical significance of the data ($p < 0.05$) for each cultivar. The analysis of the data revealed interesting insights into the firmness patterns exhibited by the cultivars under varying storage conditions. Notably, the R^2 values, which indicate the strength of the relationship between temperature and firmness, ranged from 0.8691 to 0.9489 for red delicious, 0.9129 to 0.9871 for royal gala, 0.9647 to 0.9972 for pink lady, and 0.9484 to 0.9964 for granny smith. These high R^2 values signify a robust correlation between storage temperature and firmness, underscoring the critical role of temperature control in preserving the quality of apples during postharvest handling. These results are supported by the findings of [26]. All the test data, together with the standard deviation, has been presented in Appendix A.

Furthermore, Figure 4 presents the weight loss profiles of the apple cultivars over time at different storage temperatures. The data clearly reveal an upward trend in the percentage of weight loss as time progresses. Again, the results were significant ($p < 0.05$) as demonstrated by ANOVA tests. The researchers conducted linear regression analyses to derive equations that quantify the relationship between weight loss and time. The resulting coefficients of determination (R^2) ranged from 0.8407 to 0.9944 for pink lady, 0.7889 to 0.9971 for red delicious, 0.7531 to 0.9733 for granny smith, and 0.7286 to 0.9942 for royal gala. These coefficients provide a quantitative measure of the influence of temperature on weight loss for each cultivar, highlighting the significance of temperature management in mitigating weight loss and prolonging the shelf life of apples. Ref. [27] also reported similar weight loss during storage of pomegranate.

Analyzing the findings in greater detail, it is evident that both firmness and weight loss exhibit a clear declining trend over time. Before storage, the initial firmness of random samples was measured as a baseline for comparison. Notably, at a storage temperature of 2 °C, all cultivars displayed the highest firmness compared to the other temperatures investigated. The rate of firmness decline varied among the four cultivars, reflecting the inherent differences in characteristics between them. Pink lady consistently exhibited the highest firmness, measuring 8.69 kg·cm^{-2} at the start of storage and maintaining a firmness of 7.89 kg·cm^{-2} even after 48 h. Following closely behind was Granny Smith, which recorded the second highest level of firmness.

The comprehensive evaluation of firmness and weight loss in the four apple cultivars after storage at different temperatures provides valuable insights into the quality degradation process. The observed decline in firmness across all temperatures, with the highest rate at 8 °C and the lowest at 2 °C, emphasizes the adverse effects of higher storage temperatures on apple quality. The initial firmness measurements, coupled with subsequent assessments over time, effectively demonstrated the degradation of quality across all four cultivars. It is noteworthy that storage temperature exerted a substantial influence on firmness, serving as an indicator of quality deterioration.

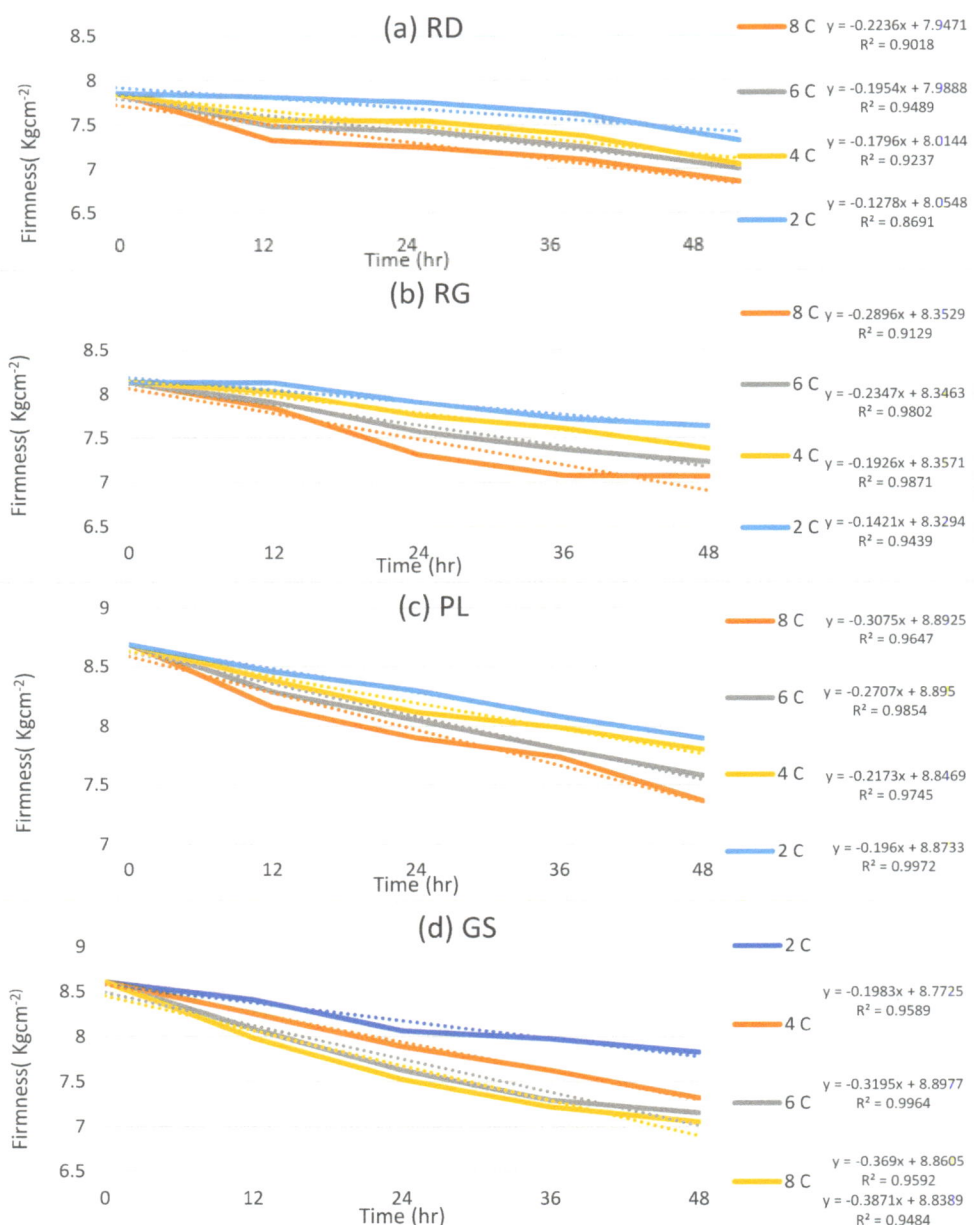

Figure 3. Change in firmness with time at different temperatures (**a**): Red Delicious, (**b**): Royal Gala, (**c**): Pink Lady, (**d**): Granny smith.

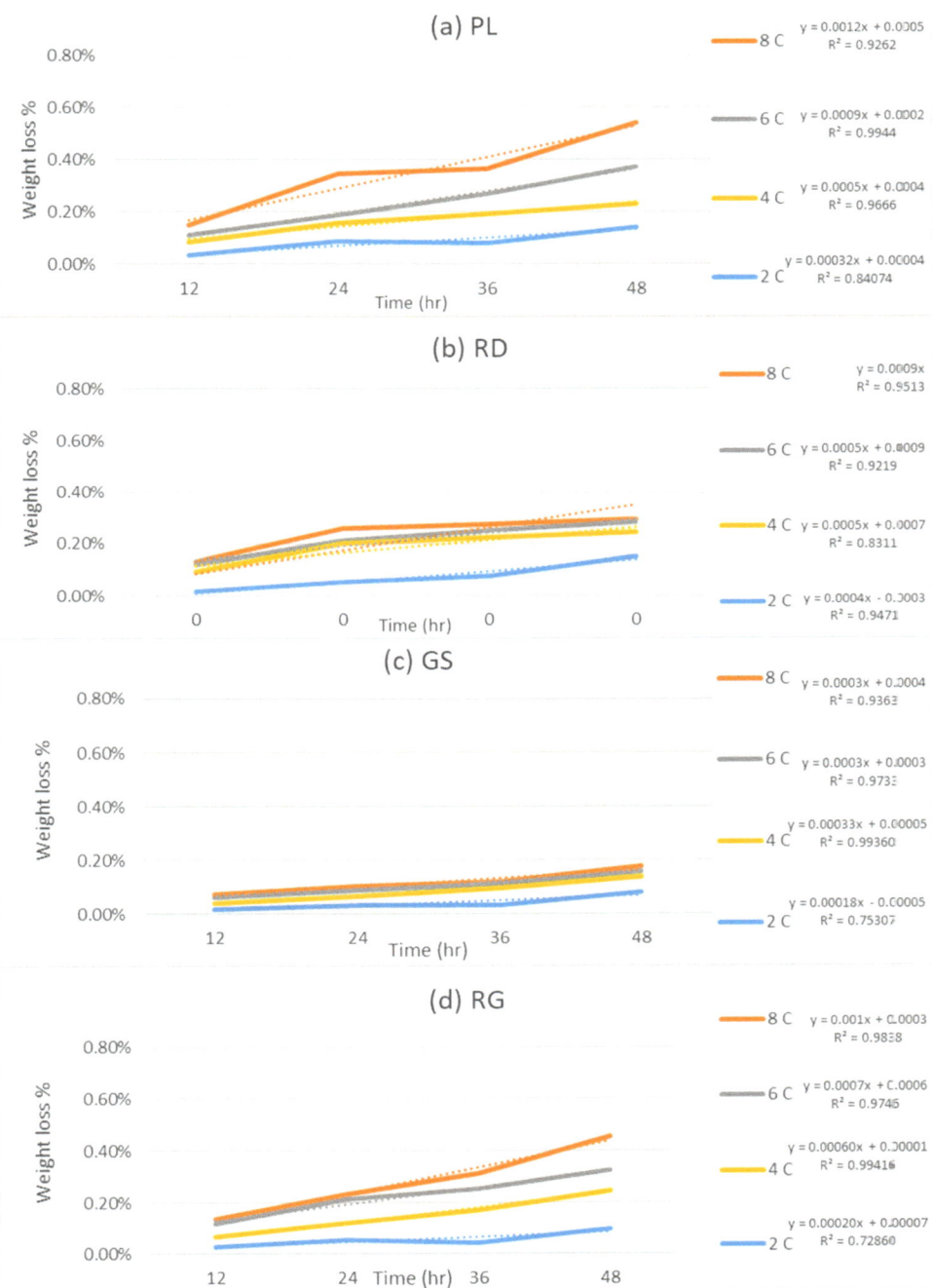

Figure 4. Change in weight loss with time at different temperatures (**a**): Pink Lady, (**b**): Red Delicious, (**c**): Granny Smith, (**d**): Royal Gala.

These findings hold significant implications for the apple industry, enabling stakeholders such as growers, distributors, and retailers to make informed decisions regarding postharvest handling practices. By recognizing the critical role of storage and transportation temperature in maintaining firmness and minimizing weight loss, industry can significantly benefit. However, cost of maintaining the lower storage temperature also need to be taken into consideration.

To further explore the relationship between temperature and the rate of firmness and weight loss, the slopes of the trends depicted in Figures 3 and 4 were calculated for each cultivar across the storage temperatures of 2 °C, 4 °C, 6 °C, and 8 °C. These slopes provide valuable insights into the rate at which firmness and weight are affected by temperature variations. The results of these calculations are plotted in Figures 5 and 6, showcasing the rate of firmness and weight loss with respect to temperature for each cultivar. The coefficient of determination (R^2) for each relationship is indicated in both figures, providing an assessment of the strength of the correlations.

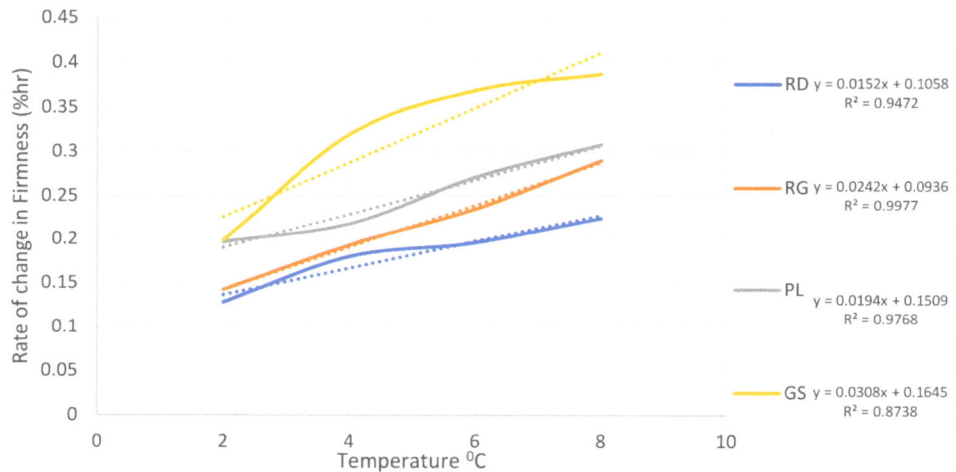

Figure 5. Rate of firmness loss as a function of temperature.

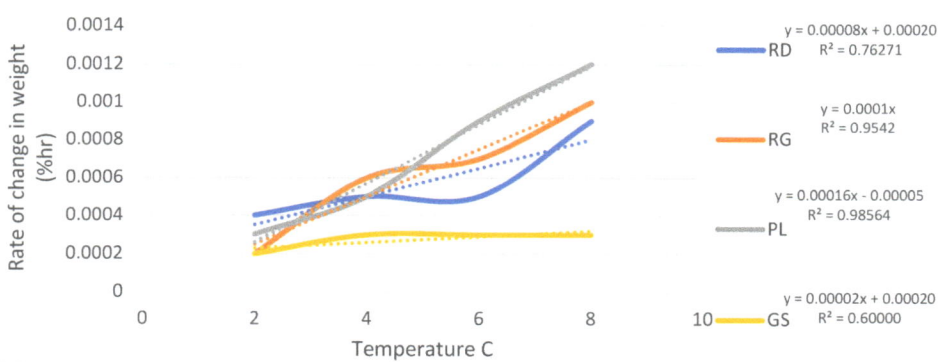

Figure 6. Rate of weight loss as a function of temperature.

Figure 5 demonstrates the rate of firmness change as a function of temperature, illustrating the varying slopes for each cultivar. The R^2 values associated with these relationships further confirm the significance of temperature in influencing firmness degradation. Meanwhile, Figure 6 presents the rate of weight loss with temperature for the different cultivars. As observed, the slopes vary among the cultivars, indicating different rates of weight loss in response to temperature fluctuations. The coefficient of determination (R^2) values in Figure 6 quantify the extent to which temperature affects weight loss for each cultivar.

Considering that firmness and weight loss are influenced by both storage time and temperature, it becomes imperative to develop an empirical prediction model that encompasses both factors. By utilizing the experimental datasets obtained for firmness and weight loss, a prediction model can be established to estimate the impact of time and temperature on these quality indicators.

The development of such a model holds considerable value, as it would enable industry professionals to make informed decisions regarding postharvest handling practices. By employing the empirical prediction model, stakeholders in the apple industry would gain the ability to optimize storage conditions, anticipate the rate of firmness and weight loss, and enhance overall product quality and shelf life. The experimental data sets serve as the foundation for this predictive model, allowing for accurate estimations and insights into the interplay between storage time, temperature, and the resulting changes in firmness and weight loss.

The general formula for the multiple regression model is:

$$Y = b_0 + b_1X_1 + b_2X_2 + \ldots\ldots b_nX_n \tag{6}$$

For firmness function f(t, T):

$$F = a + \alpha t + \beta T \tag{7}$$

where F is the firmness, a is constant, α is the estimated regression coefficient of time t, and β is the estimated regression coefficient of temperature T.

After fitting the data into regression model using SPSS software, version 28.0.1.0 (142), the firmness function of time and temperature is as follows:

$$F(PL) = 8.98 - 0.023t - 0.061T \tag{8}$$

$$F(GS) = 8.952 - 0.028t - 0.076T \tag{9}$$

$$F(RG) = 8.44 - 0.019t - 0.069T \tag{10}$$

$$F(RD) = 8.11 - 0.016t - 0.053T \tag{11}$$

Similar to the firmness fitting, the weight loss is also predicted using the linear regression model. The function of weight loss (WL) based on both temperature and time are as the following equations:

$$W(GS) = -0.087 + 0.003t + 0.021T \tag{12}$$

$$WL(PL) = -0.262 + 0.006t + 0.058T \tag{13}$$

$$WL(RG) = -0.274 + 0.006t + 0.056T \tag{14}$$

$$WL(RD) = -0.128 + 0.005t + 0.037T \tag{15}$$

Validation of the Models

The proposed models were validated using a new set of experimental data using apples stored at a different temperature (10 °C) with the same time durations. The results from the experiments were plotted and compared with the results from the predicted models from Equations (8)–(15) (Figures 7 and 8).

Figure 7 is dedicated to illustrating the observed changes in firmness for each apple cultivar during a 48-h storage period at a temperature of 10 °C. Similarly, Figure 8 presents the plots showcasing the changes in weight loss for each cultivar under the same storage conditions. The data points represent the actual measurements obtained from the experiment, while the solid lines on the graphs represent the predictions derived from the multi-regression model based on Equations (8)–(15). The legend accompanying the figures provides information about the experimental conditions corresponding to each dataset. The inclusion of the prediction lines allows for a visual comparison between the observed and predicted values, enabling a comprehensive assessment of the model's accuracy.

In order to validate the accuracy of the multi-regression model, a comparison was made between the predicted values and the experimental values for both firmness and weight loss as the validation process is crucial in assessing the reliability and effectiveness of the model. The results of this comparison were plotted in Figures 9 and 10, respectively, for all the apple cultivars at the temperature of 10 °C.

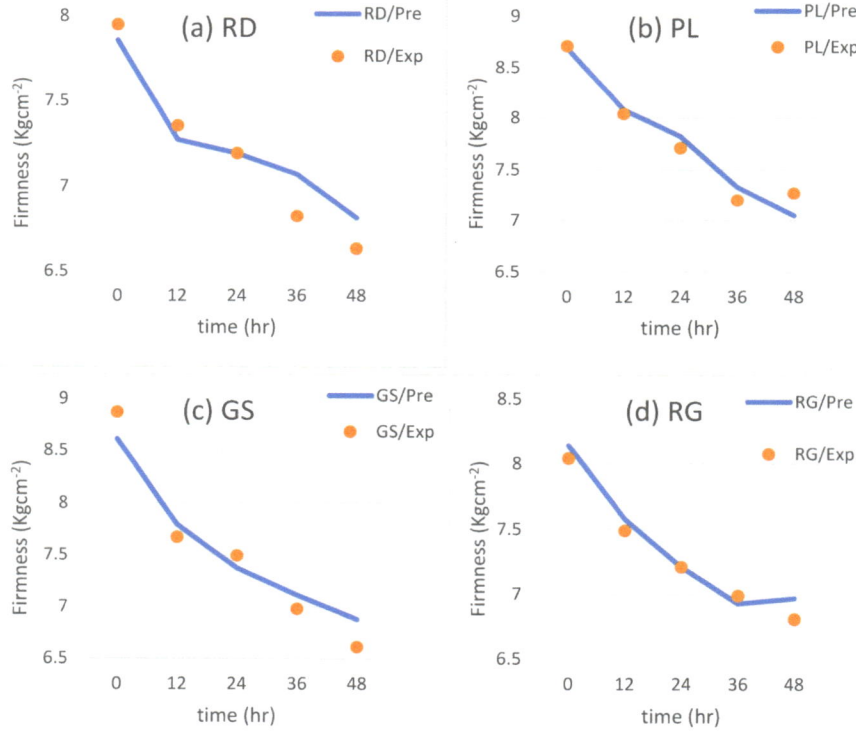

Figure 7. Validation of the firmness model: (**a**) RD, (**b**) PL, (**c**) GS, (**d**) RG.

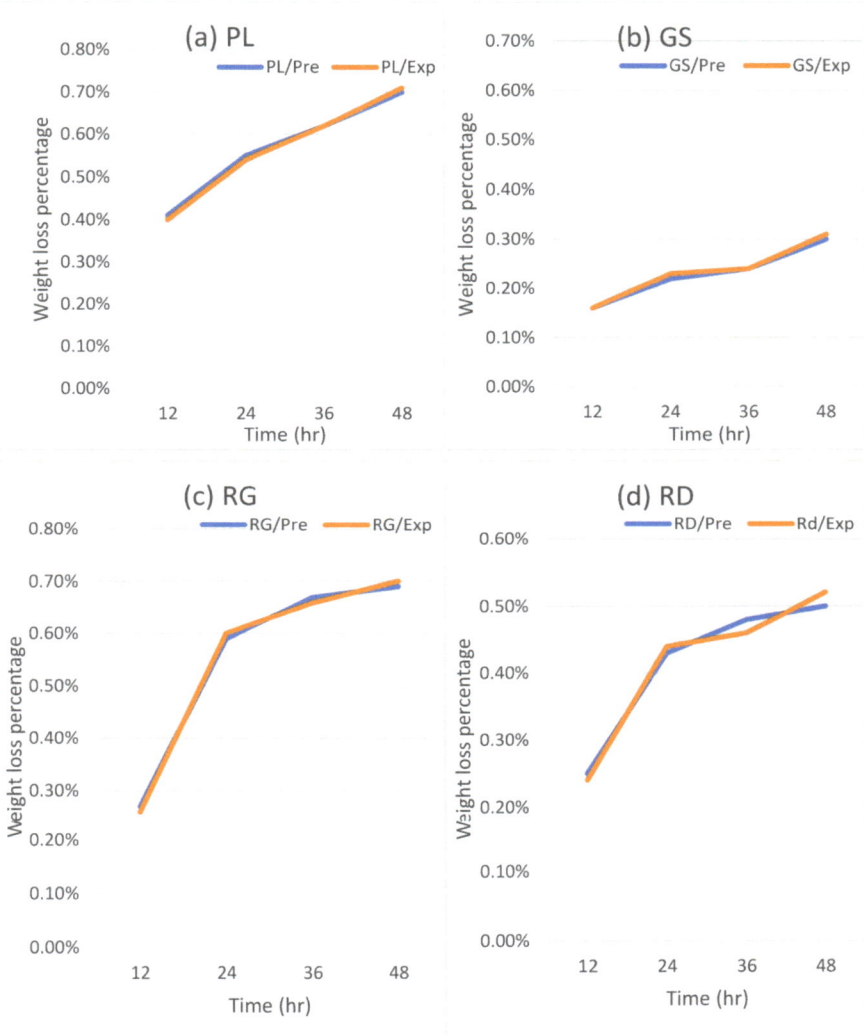

Figure 8. Validation of the weight loss model: (**a**) PL, (**b**) GS, (**c**) RG, (**d**) RD.

Upon inspection of the plotted data, a clear and close fit between the predicted and experimental values can be observed in both Figures 9 and 10. The data points align closely along the diagonal lines with a slope of 1.0, indicating a high degree of agreement between the predicted and experimental values. The linear regression equation derived from the firmness values yielded y = 1.0936x − 0.7377, with a slope of 1.0939 that is very close to 1.0. This close proximity to 1.0 indicates a highly accurate prediction of firmness. The coefficient of determination (R^2) was calculated to be 0.9544, highlighting the model's ability to accurately predict firmness under the tested conditions.

Similarly, the regression equation for weight loss, derived from the combined data of all apple cultivars, was found to be y = 1.007x − 0.00003. The slope of the equation was also very close to 1.0, reinforcing the model's accuracy in predicting weight loss values. The coefficient of determination (R^2) obtained for the weight loss predictions was 0.9964, signifying an exceptional level of accuracy in the model's predictions.

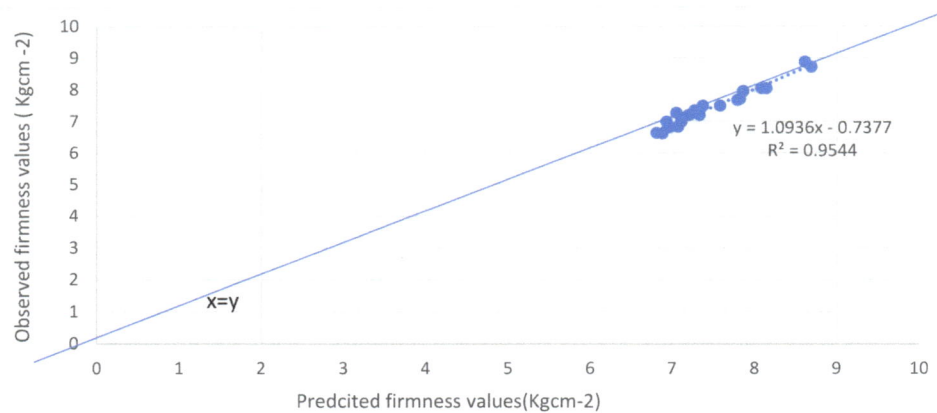

Figure 9. Predicted versus observed firmness values for all cultivars.

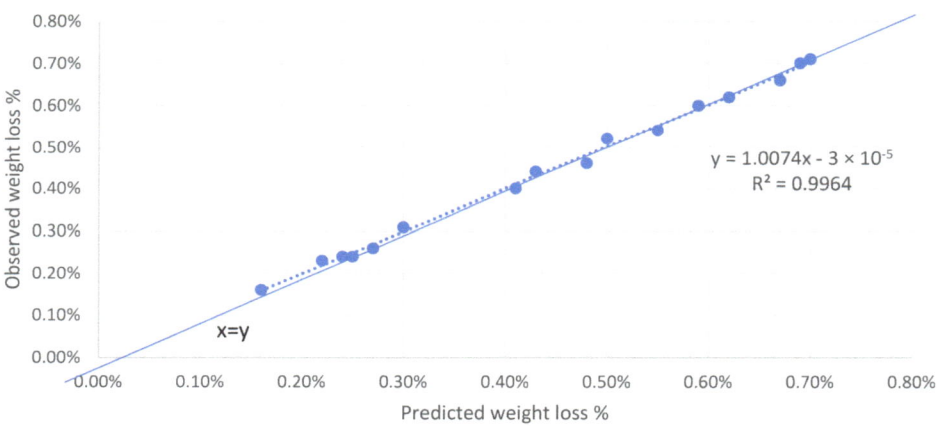

Figure 10. Predicted versus observed weight loss values for all cultivars.

From the figures it can noticed that the weight loss data shown better correlation than firmness data. Firmness is affected more by the transportation condition, as well as the maturity level at harvest is affecting firmness, and could be the reason behind the fitting of data [18]. Overall, the validation process demonstrates the robustness and reliability of the multi-regression model in predicting both firmness and weight loss for the tested apple cultivars at a temperature of 10 °C. The close fit between the predicted and experimental values, as evidenced by the alignment along the diagonal lines with a slope of 1.0, underscores the model's high level of accuracy and its ability to provide precise predictions for quality indicators under the specified storage conditions.

4. Conclusions

The objective of this study is to examine how storage conditions and duration affect some physical quality attributes of apple cultivars, and to develop empirical prediction models for weight loss and firmness under controlled temperature during transportation and short-term storage. The findings suggest that the firmness and weight loss is significantly impacted by storage temperature. Lowering the temperature can minimize changes in firmness and weight loss, thereby preserving the quality of the product during transportation and storage.

The empirical models developed in this study demonstrate a robust relationship between predicted and experimental results at 10 °C. These models can be utilized by producers and retailers to forecast product physical quality under different circumstances, enabling them to adjust storage and transportation conditions to maintain product quality over an extended period. Future research should focus on evaluating non-invasive microimaging techniques for assessing apple firmness to achieve more accurate outcomes. Furthermore, the models could be enhanced by incorporating relative humidity levels and their impact on the quality parameters examined. Although higher quality can be achieved by lowering the storage temperature, it also increases the storage cost. Future studies can be conducted to determine the optimum storage conditions taking quality target and cost into consideration. More studies are also required to investigate other quality attributes, including biochemical properties.

Author Contributions: Conceptualization, R.A.S.A. and A.K.; Methodology, R.A.S.A. and A.K.; Software, R.A.S.A.; Formal analysis, R.A.S.A.; Investigation, R.A.S.A.; Resources, P.Y.; Writing—original draft, R.A.S.A.; Writing—review & editing, P.Y. and A.K.; Supervision, P.Y. and A.K. All authors have read and agreed to the published version of the manuscript.

Funding: This study was funded by the Department of Education and Training (AUS) through a Research Training Program (Stipend) Domestic (RTP) scholarship and the ARC Grants LP200100493 and DP220103668.

Data Availability Statement: Data is contained within the article.

Acknowledgments: We would like to express our sincere appreciation to Vindya Thathsaranee Weligamafor her invaluable contribution in the statistical analysis in this paper.

Conflicts of Interest: The authors declare no conflict of interest.

Appendix A

Table A1. Change in firmness with time at different temperatures for Granny smith, Pink Lady, Red Delicious, and Royal Gala varieties with storage time (h).

Temperature (°C)	Time (h)	Firmness * ($kg \cdot cm^{-2}$) Values for Different Apple Varieties			
		Granny Smith	Pink Lady	Red Delicious	Royal Gala
2	0	8.61 ± 0.25	7.38 ± 0.26	7.86 ± 0.24	8.14 ± 0.28
	12	8.41 ± 0.36	8.71 ± 0.51	7.81 ± 0.49	8.13 ± 0.10
	24	8.07 ± 0.47	8.37 ± 0.33	7.75 ± 0.33	7.90 ± 0.15
	36	7.98 ± 0.72	8.60 ± 0.19	7.62 ± 0.50	7.72 ± 0.46
	48	7.83 ± 0.54	8.69 ± 0.16	7.32 ± 0.18	7.63 ± 0.51
4	0	8.61 ± 0.25	8.69 ± 0.45	7.86 ± 0.24	8.14 ± 0.28
	12	8.25 ± 0.23	8.38 ± 0.46	7.55 ± 0.22	8.02 ± 0.57
	24	7.89 ± 0.33	8.12 ± 0.14	7.54 ± 0.33	7.75 ± 0.33
	36	7.63 ± 0.28	7.99 ± 0.25	7.37 ± 0.47	7.61 ± 0.26
	48	7.32 ± 0.27	7.80 ± 0.33	7.05 ± 0.43	7.38 ± 0.80
6	0	8.61 ± 0.25	8.69 ± 0.45	7.86 ± 0.24	8.14 ± 0.28
	12	8.08 ± 0.26	8.29 ± 0.42	7.48 ± 0.32	7.90 ± 0.30
	24	7.63 ± 0.32	8.05 ± 0.55	7.43 ± 0.38	7.57 ± 0.52
	36	7.30 ± 0.34	7.81 ± 0.33	7.25 ± 0.33	7.37 ± 0.73
	48	7.16 ± 0.05	7.58 ± 0.48	7.00 ± 0.10	7.23 ± 0.79

Table A1. *Cont.*

Temperature (°C)	Time (h)	Firmness * (kg·cm^{-2}) Values for Different Apple Varieties			
		Granny Smith	Pink Lady	Red Delicious	Royal Gala
8	0	8.61 ± 0.25	8.69 ± 0.45	7.86 ± 0.24	8.14 ± 0.28
	12	7.98 ± 0.16	8.16 ± 0.20	7.32 ± 0.25	7.83 ± 0.31
	24	7.53 ± 0.43	7.90 ± 0.19	7.24 ± 0.58	7.31 ± 0.39
	36	7.22 ± 0.24	7.73 ± 0.08	7.07 ± 0.40	7.07 ± 0.40
	48	7.05 ± 0.24	7.37 ± 0.60	6.85 ± 0.28	7.07 ± 0.49
10	0	8.61 ± 0.25	8.69 ± 0.45	7.86 ± 0.24	8.14 ± 0.28
	12	7.79 ± 0.34	8.08 ± 0.40	7.32 ± 0.20	7.58 ± 0.51
	24	7.37 ± 0.05	7.82 ± 0.23	7.24 ± 0.24	7.21 ± 0.26
	36	7.11 ± 0.12	7.33 ± 0.50	7.07 ± 0.18	6.93 ± 0.28
	48	6.88 ± 0.31	7.05 ± 0.60	6.85 ± 0.26	6.97 ± 0.53

* The average values of firmness data obtained from four replicates are presented into the last two decimal points.

Table A2. Change in weight loss with time at different temperatures Granny Smith, Pink Lady, Red Delicious, and Royal Gala at different temperatures during the storage.

Temperature (°C)	Time (h)	Weight Loss * (%) Values for Different Apple Varieties			
		Granny Smith	Pink Lady	Red Delicious	Royal Gala
2	12	0.02 ± 0.01	0.02 ± 0.01	0.02 ± 0.01	0.03 ± 0.01
	24	0.03 ± 0.01	0.07 ± 0.04	0.05 ± 0.02	0.03 ± 0.04
	36	0.03 ± 0.03	0.07 ± 0.03	0.08 ± 0.02	0.06 ± 0.03
	48	0.08 ± 0.02	0.01 ± 0.04	0.15 ± 0.01	0.10 ± 0.02
4	12	0.04 ± 0.01	0.08 ± 0.01	0.09 ± 0.03	0.06 ± 0.01
	24	0.07 ± 0.01	0.16 ± 0.02	0.20 ± 0.08	0.12 ± 0.02
	36	0.10 ± 0.01	0.19 ± 0.01	0.22 ± 0.06	0.17 ± 0.01
	48	0.11 ± 0.07	0.23 ± 0.03	0.24 ± 0.07	0.25 ± 0.03
6	12	0.06 ± 0.01	0.11 ± 0.02	0.12 ± 0.05	0.12 ± 0.02
	24	0.09 ± 0.01	0.19 ± 0.03	0.21 ± 0.05	0.21 ± 0.03
	36	0.11 ± 0.01	0.27 ± 0.08	0.25 ± 0.06	0.25 ± 0.01
	48	0.12 ± 0.08	0.37 ± 0.04	0.28 ± 0.07	0.32 ± 0.05
8	12	0.08 ± 0.01	0.15 ± 0.06	0.13 ± 0.08	0.14 ± 0.02
	24	0.10 ± 0.02	0.34 ± 0.10	0.26 ± 0.12	0.23 ± 0.02
	36	0.12 ± 0.03	0.36 ± 0.10	0.27 ± 0.09	0.31 ± 0.03
	48	0.18 ± 0.01	0.54 ± 0.28	0.29 ± 0.04	0.45 ± 0.09
10	12	0.16 ± 0.02	0.41 ± 0.03	0.25 ± 0.05	0.27 ± 0.05
	24	0.22 ± 0.06	0.55 ± 0.11	0.43 ± 0.07	0.59 ± 0.11
	36	0.22 ± 0.01	0.62 ± 0.13	0.48 ± 0.10	0.67 ± 0.12
	48	0.30 ± 0.06	0.70 ± 0.08	0.48 ± 0.25	0.50 ± 0.36

* The average weight loss values have been calculated using four replicates during the experimentation.

References

1. Karim, M.A.; Hawlader, M.N.A. Mathematical modelling and experimental investigation of tropical fruits drying. *Int. J. Heat Mass Transf.* **2005**, *48*, 4914–4925. [CrossRef]
2. Welsh, Z.G.; Khan, M.I.H.; Karim, M.A. Multiscale modeling for food drying: A homogenized diffusion approach. *J. Food Eng.* **2021**, *292*, 110252. [CrossRef]
3. Kumar, C.; Karim, M.; Joardder, M.U. Intermittent drying of food products: A critical review. *J. Food Eng.* **2014**, *121*, 48–57. [CrossRef]
4. Brasiello, A.; Adiletta, G.; Russo, P.; Crescitelli, S.; Albanese, D.; Di Matteo, M. Mathematical modeling of eggplant drying: Shrinkage effect. *J. Food Eng.* **2013**, *114*, 99–105. [CrossRef]
5. McMinn, W.A.M.; Magee, T.R.A. Physical characteristics of dehydrated potatoes—Part II. *J. Food Eng.* **1997**, *33*, 49–55. [CrossRef]
6. Ali, S.; Anjum, M.A.; Ejaz, S.; Hussain, S.; Ercisli, S.; Saleem, M.S.; Sardar, H. Carboxymethyl cellulose coating delays chilling injury development and maintains eating quality of 'Kinnow' mandarin fruits during low temperature storage. *Int. J. Biol. Macromol.* **2021**, *168*, 77–85. [CrossRef]
7. Abanoz, Y.Y.; Okcu, Z. Biochemical content of cherry laurel (*Prunus laurocerasus* L.) fruits with edible coatings based on caseinat, Semperfresh and lecithin. *Turk. J. Agric. For.* **2022**, *46*, 908–918. [CrossRef]
8. Saleem, M.S.; Ejaz, S.; Anjum, M.A.; Ali, S.; Hussain, S.; Ercisli, S.; Ilhan, G.; Marc, R.A.; Skrovankova, S.; Mlcek, J. Improvement of Postharvest Quality and Bioactive Compounds Content of Persimmon Fruits after Hydrocolloid-Based Edible Coating Application. *Horticulturae* **2022**, *8*, 1045. [CrossRef]
9. Saran, E.Y.; Cavusoglu, S.; Alpaslan, D.; Eren, E.; Yilmaz, N.; Uzun, Y. Effect of egg white protein and agar-agar on quality of button mushrooms (*Agaricus bisporus*) during cold storage. *Turk. J. Agric. For.* **2022**, *46*, 173–181. [CrossRef]
10. Davey, S.S.; Richards, C. Supermarkets and private standards: Unintended consequences of the audit ritual. *Agric. Hum. Values* **2013**, *30*, 271–281. [CrossRef]
11. Grabska, J.; Beć, K.B.; Ueno, N.; Huck, C.W. Analyzing the Quality Parameters of Apples by Spectroscopy from Vis/NIR to NIR Region: A Comprehensive Review. *Foods* **2023**, *12*, 1946. [CrossRef]
12. Wang, Y.; Lu, Q.; Liu, Q.; Nicolescu, C.M.; Sun, Y. Quality Evaluation on Agri-Fresh Food Emergency Logistics Service. In *Advances in Intelligent Systems, Computer Science and Digital Economics IV*; Springer Nature: Cham, Switzerland, 2023; pp. 313–329.
13. Kyriacou, M.C.; Rouphael, Y. Towards a new definition of quality for fresh fruits and vegetables. *Sci. Hortic.* **2018**, *234*, 463–469.
14. Johnston, J.W.; Hewett, E.W.; Banks, N.H.; Harker, F.R.; Hertog, M.L.A.T.M. Physical change in apple texture with fruit temperature: Effects of cultivar and time in storage. *Postharvest Biol. Technol.* **2001**, *23*, 13–21. [CrossRef]
15. Mditshwa, A.; Fawole, O.A.; Vries, F.; van der Merwe, K.; Crouch, E.; Opara, U.L. Repeated application of dynamic controlled atmospheres reduced superficial scald incidence in 'Granny Smith' apples. *Sci. Hortic.* **2017**, *220*, 168–175. [CrossRef]
16. FAO; FAOSTAT. Crops and Livestock Products. 2019. Available online: http://www.fao.org/faostat/en/ (accessed on 19 May 2023).
17. Spencer, S. *Price Determination in the Australian Food Industry: A Report/Prepared by Whitehall Associates*; Canberra, A.C.T., Ed.; Australian Government Department of Agriculture, Fisheries and Forestry: Canberra, Australia, 2004.
18. Watkins, C.B.; Reid, M.S.; Harman, J.E.; Padfield, C.A.S. Starch iodine pattern as a maturity index for Granny Smith apples. *N. Zeal. J. Agric. Res.* **1982**, *25*, 587–592. [CrossRef]
19. Way, R.D.; Aldwinckle, H.S.; Lamb, R.C.; Rejman, A.; Sansavini, S.; Shen, T.; Watkins, R.; Westwood, M.N.; Yoshida, Y. Apples (*Malus*). *Genet. Resour. Temp. Fruit Nut Crops* **1991**, *290*, 3–46. [CrossRef]
20. DeEll, J.; Khanizadeh, S.; Saad, F.; Ferree, D. Factors affecting apple fruit firmness: A review. *J. Am. Pomol. Soc.* **2001**, *55*, 8–27.
21. Lana, M.M.; Tijskens, L.M.M.; van Kooten, O. Effects of storage temperature and fruit ripening on firmness of fresh cut tomatoes. *Postharvest Biol. Technol.* **2005**, *35*, 87–95. [CrossRef]
22. Rao, M.; Lund, D. Kinetics of thermal softening of foods—A review. *J. Food Process. Preserv.* **1986**, *10*, 311–329. [CrossRef]
23. Bourne, M.C. Effect of temperature on firmness of raw fruits and vegetables. *J. Food Sci.* **1982**, *47*, 440–444. [CrossRef]
24. Toledo, R.T.; Singh, R.K.; Kong, F. *Fundamentals of Food Process Engineering*; Springer: Berlin/Heidelberg, Germany, 2007; Volume 297.
25. Labuza, T.P. *Shelf-Life Dating of Foods*; Food & Nutrition Press, Inc.: Trumbull, CT, USA, 1982.
26. Thai, C.; Shewfelt, R.; Gamer, J. Tomato color changes under constant and variable storage temperatures: Empirical models. *Trans. ASAE* **1990**, *33*, 607–0614. [CrossRef]
27. Lu, L.; Zuo, W.; Wang, C.; Li, C.; Feng, T.; Li, X.; Wang, C.; Yao, Y.; Zhang, Z.; Chen, X. Analysis of the postharvest storage characteristics of the new red-fleshed apple cultivar 'meihong'. *Food Chem.* **2021**, *354*, 129470. [CrossRef] [PubMed]

Disclaimer/Publisher's Note: The statements, opinions and data contained in all publications are solely those of the individual author(s) and contributor(s) and not of MDPI and/or the editor(s). MDPI and/or the editor(s) disclaim responsibility for any injury to people or property resulting from any ideas, methods, instructions or products referred to in the content.

Article

Supercritical Fluid Extraction of Oils from Cactus *Opuntia ficus-indica* L. and *Opuntia dillenii* Seeds

Ghanya Al-Naqeb [1,2,*], Cinzia Cafarella [3], Eugenio Aprea [1], Giovanna Ferrentino [4,*], Alessandra Gasparini [4], Chiara Buzzanca [3], Giuseppe Micalizzi [3], Paola Dugo [3,5], Luigi Mondello [3,5,6] and Francesca Rigano [3]

1. Center Agriculture Food Environment (C3A), University of Trento, 38098 Trento, Italy
2. Department of Food Sciences and Nutrition, Faculty of Agriculture Food and Environment, University of Sana'a, Sana'a P.O. Box 1247, Yemen
3. Department of Chemical, Biological, Pharmaceutical and Environmental Sciences, University of Messina, 98168 Messina, Italy
4. Faculty of Science and Technology, Free University of Bozen-Bolzano, Piazza Università 5, 39100 Bolzano, Italy
5. Chromaleont s.r.l., c/o, Department of Chemical, Biological, Pharmaceutical and Environmental Sciences, University of Messina, 98168 Messina, Italy
6. Unit of Food Science and Nutrition, Department of Medicine, University Campus Bio-Medico of Rome, 00128 Rome, Italy
* Correspondence: ghanya.alnaqeb@unitn.it (G.A.-N.); giovanna.ferrentino@unibz.it (G.F.)

Abstract: This study aimed to assess the capability of supercritical fluid extraction (SFE) as an alternative and green technique compared to Soxhlet extraction for the production of oils from *Opuntia ficus-indica* (OFI) seeds originating from Yemen and Italy and *Opuntia dillenii* (OD) seeds from Yemen. The following parameters were used for SFE extraction: a pressure of 300 bar, a CO_2 flow rate of 1 L/h, and temperatures of 40 and 60 °C. The chemical composition, including the fatty acids and tocopherols (vitamin E) of the oils, was determined using chromatographic methods. The highest yield was achieved with Soxhlet extraction. The oils obtained with the different extraction procedures were all characterized by a high level of unsaturated fatty acids. Linoleic acid (≤62% in all samples) was the most abundant one, followed by oleic and vaccenic acid. Thirty triacylglycerols (TAGs) were identified in both OFI and OD seed oils, with trilinolein being the most abundant (29–35%). Vanillin, 4-hydroxybenzaldehyde, vanillic acid, and hydroxytyrosol were phenols detected in both OFI and OD oils. The highest γ-tocopherol content (177 ± 0.23 mg/100 g) was obtained through the SFE of OFI seeds from Yemen. Overall, the results highlighted the potential of SFE as green technology to obtain oils suitable for functional food and nutraceutical products.

Keywords: *Opuntia ficus-indica* L.; *Opuntia dillenii*; supercritical fluid extraction; antioxidant activity; chemical characterization

Citation: Al-Naqeb, G.; Cafarella, C.; Aprea, E.; Ferrentino, G.; Gasparini, A.; Buzzanca, C.; Micalizzi, G.; Dugo, P.; Mondello, L.; Rigano, F. Supercritical Fluid Extraction of Oils from Cactus *Opuntia ficus-indica* L. and *Opuntia dillenii* Seeds. *Foods* 2023, *12*, 618. https://doi.org/10.3390/foods12030618

Academic Editors: Gianluca Nardone, Rosaria Viscecchia and Francesco Bimbo

Received: 4 January 2023
Revised: 28 January 2023
Accepted: 29 January 2023
Published: 1 February 2023

Copyright: © 2023 by the authors. Licensee MDPI, Basel, Switzerland. This article is an open access article distributed under the terms and conditions of the Creative Commons Attribution (CC BY) license (https:// creativecommons.org/licenses/by/ 4.0/).

1. Introduction

Opuntia is a genus of plants belonging to the Cactaceae family, the latter counting around 1800 species. Among them, approximately 1400 species belong to the genus *Opuntia* and are distributed in Mediterranean countries, Mexico, Europe, and other areas [1]. *Opuntia ficus-indica* L. (OFI) and *Opuntia dillenii* (OD) are the most frequently used species for human consumption. In the traditional medicine of many countries, they are used to treat some diseases and disorders, such as whooping cough and eye inflammation, or as anti-ulcerogenic or antidiarrheal agents [2]. OFI, commonly called prickly pear, is cultivated in many areas of Yemen, where it is used as food, feed, and also as prepared products such as juice, jam, and cosmetics, thus making this plant extremely advantageous from an economic point of view [3,4]. The fruits of OFI contain about 9–10% of seeds [5], which are reported to contain 5–16% of oil [3,6]. Various researchers have reported the fatty acids composition of OFI oil, which is rich in unsaturated fatty acids (80–88%) such as linoleic acid (49.3–78.8%), oleic acid (12.8–25.3%), vaccenic acid (4.3–6.3%), and linolenic acid (0.23–1.1%).

The main saturated fatty acids are palmitic (9.3–14.3%) and stearic (2.2–4.3%) acids [7–10]. Seven different phenolic compounds have been identified in the OFI oil. They belong to three families: hydroxyl cinnamic acid derivates (p-coumaric acid, p-coumaric acid ethyl ester, ferulic acid), hydroxyl cinnamaldehyde derivates (furaldehyde), and hydroxybenzaldehyde derivates (4-hydroxy benzaldehyde, vanillin, syringaldehyde) with vanillin, syringaldehyde, and ferulaldehyde detected at the highest level [11]. The total amount of tocopherols in OFI oil was reported in the range between 3.9 and 50.0 mg/100 g [12], where γ-tocopherol showed 94.12% of the total vitamin E content [13]. The triacylglycerols (TAGs) profile of OFI oil was reported, with trilinolein and oleyldilinolein being the main ones, with an average percentage of 25.6% and 21.5%, respectively [13].

Concerning OD, it is a wild plant that can be found in different places in Yemen, especially in Taiz and Hodeida. The OD fruits have a sour taste and big seeds (Figure 1). Some people in Yemen make fresh and concentrated juice from the OD fruits, thus discarding a large number of seeds. OD seeds have been reported to contain 6.65–13.12% of oil [7–9], which are rich in bioactive molecules, such as mono- and polyunsaturated fatty acids (MUFAs and PUFAs) and vitamin E. At the moment, only a limited number of studies deal with a comprehensive chemical characterization of the oil. To the best of the authors' knowledge, the only work reporting a quite detailed investigation of the chemical composition of OD oil was carried out on a Moroccan species after a conventional solvent extraction procedure [8].

On the other hand, an impressive number of studies have recently shown that the seed oils of both OFI and OD possess good nutritional value and multiple health benefits, including antioxidant activity both in vivo [10] and in vitro [8,11], antimicrobial [12], antidiabetic [3,13,14], lipid-lowering effect [15], in vitro anticancer effect [16], anti-inflammatory, and antiulcer effects [10,11].

Figure 1. Appearance of *Opuntia ficus-indica* L. and *Opuntia dillenii* fruits and seeds.

Due to the multiple health benefits of both OFI and OD seed oils, an efficient extraction process is of prime importance to preserve the quality of these oils. Different extraction processes have been applied for OFI and OD seeds, including maceration [10], cold press [14,17,18], ultrasound-assisted extraction [19], and Soxhlet [20,21].

However, such methods are time-consuming, have environmental hazards, and have negative implications for operator safety. Moreover, they might have a threat to consumer health if the organic solvents are not completely removed. Recently, supercritical fluid extraction (SFE) has appeared to be a valid alternative for seed oil extraction and has received considerable attention. The SFE system has many advantages when carbon dioxide (CO_2) is used as a solvent, which is not toxic, environmentally friendly, non-explosive, and of food grade [17,22,23]. Furthermore, the oil can be extracted under low temperatures and oxygen-free conditions [20]. Studies have shown that SFE-extracted oils are enriched with antioxidant compounds [21,22].

There is quite a poor report of the literature data on the SFE extraction of both OFI and OD seeds with detailed chemical characterization, especially triacylglycerol profiles (TAGs), vitamin E, and phenolic compounds, have been found. Rather, only the total fatty acid (FA) composition was elucidated, and the antioxidant activity was assessed [9,23].

In the present study, oils from OFI seeds originating from Yemen (YV1) and Italy (I/S) and OD (YV2) originating from Yemen were extracted using the SFE system and compared to the one obtained from a Soxhlet extraction. The recovered oils were comprehensively characterized in terms of their chemical composition, including fatty acids, tocopherols (vitamin E), and polyphenols.

2. Materials and Methods

2.1. Chemicals

Analytical standards of trinonanoin, triundecanoin, tritridecanoin, tripentadecanoin, triheptadecanoin, trinonadecanoin, and the C4-C24 even carbon saturated FAMEs standard mixture (1000 µg/mL each in n-hexane), n-heptane, methanol, acetic acid (reagent grade), potassium hydroxide (KOH), acetonitrile, methanol, 2-propanol (LC-MS grade), n-hexane (HPLC grade) were purchased from Merck Life Science (Darmstadt, Germany).

2.2. Samples Preparation

Three varieties of prickly pear fruits were used in this study: OFI from Yemen (YV1), OD (YV2) from Yemen, and OFI from Italy/Sicily (I/S). Seeds were separated from the fruits, naturally dried at room temperature, packed in plastic bags, and stored at 4 °C until used. Dried seeds were milled using a hammer miller (Mill- LM3100, Perten Instruments, Sweden). The resulting seeds powder showed a particle size distribution of a diameter between 250 and 100 µm, determined by an orbital sieve shaker (Retsch GmbH, Verder Scientific, Haan, Germany).

2.3. Supercritical Fluid Extraction

SFE was performed in a semi-batch pilot system (Superfluidi s.r.l., Padova, Italy) as previously described [22] with some modifications. For each extraction run, the basket was filled with about 100 g of milled seeds. The following processing parameters were used: a pressure of 300 bar, a CO_2 flow rate of 1 L/h, and temperatures of 40 °C and 60 °C. The selection of the operative parameters was based on previous experience with SFE performed on other seeds [23,24]. The pressure was set at the highest level of the system. Two temperatures were tested to assess their effect on the yield. During each extraction, the oil was collected every 10 min from the separator while the CO_2 was released. The extraction time was equal to 2.30 h.

2.4. Soxhlet Extraction

For the Soxhlet extraction, n-hexane was used in order to recover oils from the powdered seeds. Batches of 20 g either of milled OFI or OD seeds were placed in each extraction thimble, and 150 mL of n-hexane was used. The extraction was carried out for 4 h. Then, the solvent was evaporated from the oil using a rotary evaporator (LABOROTA 4000, Heidolph, Schwabach, Germany). The oil yield was calculated on a dry weight basis as a ratio between the amount of oil and the number of ground seeds used for the extraction:

$$\text{Yield (\%)} = 100 \times \frac{\text{Weight of extracted oil}}{\text{Weight of initial used seeds}} \quad (1)$$

2.5. Chemical Characterization of OFI and OD Seeds Oil

2.5.1. Determination of the Total Fatty Acid Composition by GC Analyses

Fatty acid methyl esters (FAMEs) were obtained by a cold transesterification reaction as previously reported by Ciriminna et al. [24]. Briefly, 50 mg of oil was dissolved in 1 mL of n-heptane and added to 0.1 mL of a 2N solution of potassium hydroxide (KOH) in methanol. The mixture was stirred for 5 min at room temperature. The heptane upper layer was transferred into a 1.5 mL autosampler vial prior to GC-MS and GC-FID (FID, flame ionization detector) analyses for FAME identification and quantification, respectively.

Specifically, GC-MS analyses were performed using a GC-2010 (Shimadzu, Duisburg, Germany) gas chromatograph coupled to a single quadrupole mass spectrometer (QP2020, Shimadzu). An AOC-20i autosampler and split/splitless injector were installed on the GC system. The separation of the analytes was carried out by using an SLB-IL60 (ionic liquid, IL) 30 m × 0.25 mm ID × 0.20 μm df (Merck Life Science) capillary column, operated under the following temperature program: from 50 °C to 280 °C at 3 °C min^{-1}. Helium was utilized as a carrier gas at a constant linear velocity of 30 cm s^{-1} and initial inlet pressure of 26.6 kPa. The injection volume, injector temperature, and split ratio were 0.2 μL, 280 °C, and 1:100, respectively. A single quadrupole MS detector was operated in the scan mode (mass range of 40–550 m/z); the interface and ion source temperatures were 250 °C and 220 °C, respectively. The GC-MS solution software (version 4.50 Shimadzu) was used for data collection and handling. A dedicated mass spectra database, namely LIPIDS ver. 1.0 (Shimadzu) was used for the spectral match. Moreover, a lab-constructed linear retention index (LRI) database [25] was employed for LRI matching according to an automatic data processing method, by applying a dual-filter identification (minimum MS spectral similarity of 85% and LRI tolerance of ±10 units). In such respect, the C4-C24 FAMEs standard solution was injected prior to the analysis under the same experimental conditions for the calculation of LRIs for the detected peaks.

GC-FID analyses were carried out using a GC-2010 instrument (Shimadzu) equipped with a flame ionization detector (FID). The GC column, temperature program, carrier gas, linear velocity, and injection conditions were the same as described for the GC-MS system. The initial inlet pressure was 99.4 kPa. The FID temperature was set at 280 °C. The following FID gas flows were set: 40 mL min^{-1} for H_2, 30 mL min^{-1} for the make-up gas (N_2), and 400 mL min^{-1} for air. Data were collected and processed using the LabSolution software (version 5.92, Shimadzu, Duisburg, Germany). All samples were analyzed in triplicate.

2.5.2. Determination of the Triacylglycerol Composition by LC/MS Analysis

An amount of 10 mg of triundecanoin (C11C11C11) was added as an internal standard (IS) to 20 mg of oil, and the mixture was dissolved in 2-propanol, up to a final volume of 1 mL, prior to the LC-MS analysis. The instrumental setup consisted of a Nexera UHPLC system coupled to an LCMS-2020 spectrometer through an atmospheric pressure chemical ionization (APCI) source (Shimadzu, Duisburg, Germany). Separations were performed on an Ascentis Express C18 10 cm × 2.1 mm, 2.7 μm d.p. columns (Merck Life Science, Darmstadt, Germany). The employed mobile phases were acetonitrile (A) and 2-propanol (B), and the linear gradient program was as follows: 0–52 min, 0–70% B, held for 3 min. The flow rate was 0.5 mL min^{-1}, the oven temperature was 35 °C, and the injection volume was 5 μL.

The following MS parameters, through the APCI source in a positive (+) ionization mode, were employed: the interface, desolvation line, and heat block temperatures were set at 450 °C, 250 °C, and 300 °C, respectively; nebulizing gas and drying gas flow (N_2) were set at 1.5 L min^{-1} and 5 L min^{-1}, respectively; and the acquisition MS range was 250–1200 m/z with an event time of 1 s. Data were acquired by using LabSolution ver. 5.95 software (Shimadzu, Duisburg, Germany) and processed through the recently implemented ChromLinear software.

For qualitative purposes, a reference standard mixture of odd chain carbon number triacylglycerols (TAGs) from trinonanoin (C9C9C9) to trinonadecanoin (C19C19C19) was injected at the beginning and at the end of the analytical batch in order to automatically calculate LRI for all the peaks of the sample. The ChromLinear software was able to match the calculated LRI and acquired MS spectra for single TAGs with the previously built LRI database and MS spectral library according to an automatic and rapid dual-filter identification strategy, similar to GC-MS analyses. For quantitative purposes, the normalized peak areas (ratio between the analyte and IS peak areas) were used for relative quantification, considering a quite similar MS response for all the identified TAGs.

2.5.3. LC-FLD Analysis of Vitamin E

Vitamin E determination was carried out as previously described by Dugo et al. [26] with a slight modification. Briefly, 1 mL of oil was dissolved in n-hexane up to a final volume of 10 mL prior to LC-FLD analysis. The employed analytical system consisted of a Nexera-X2 system (Shimadzu, Milan, Italy) equipped with an RF-20AXS fluorescence detector, as described in previous work. Data acquisition and processing were performed by the LCMS solution Ver. 5.85 software (Shimadzu, Duisburg, Germany). The chromatographic separation was carried out on an Ascentis Si column, 250 × 4.6 mm L × I.D., with a particle size of 5 μm (Merck KGaA, Darmstadt, Germany), kept at 25 °C and operated in isocratic elution mode (n-hexane/isopropanol 99:1, v/v) at a flow rate of 1.7 mL min^{-1}. The injection volume was 5 μL. The peaks were recorded using 290 nm as the excitation wavelength and 330 nm as the emission wavelength. A previously validated method by Dugo et al. [26], based on external calibration, was used for the quali-quantitative determination of the detected vitamers.

2.5.4. LC Analyses of Phenolic Compounds

Phenolic compounds were recovered from the oil by liquid–liquid extraction, using methanol: water 80:20 (v:v) mixture, according to a 1:1 ratio sample: solvent (w:v), after the dissolution of the oil in n-hexane. Briefly, 0.5 g of oil was dissolved in 0.5 mL of n-hexane and added to 0.5 mL of methanol: water 80:20 (v:v). The mixture was sonicated for 5 min and centrifuged for 10 min at 3000 rpm. The remaining n-hexane phase was extracted two more times under the same conditions, and the three methanolic aqueous phases were pooled, washed with 0.5 mL of n-hexane, and concentrated to dryness under a nitrogen stream. The residue product was reconstituted in 200 μL of methanol prior to injection into the LC instrument, which was equipped with a photodiode array detector (PDA) SPD-M20A directly connected to the LC column outlet and serially coupled to an LCMS-2020 spectrometer via an electrospray (ESI) source for mass spectrometry (MS). The chromatographic separation was achieved on an Ascentis Express RP C18 (150 mm × 4.6 mm ID × 2.7 μm) column, operated at a temperature of 35 °C and at a flow rate of 1.0 mL min^{-1}. The mobile phase consisted of (A) water with 0.15% of acetic acid and (B) acetonitrile, according to the following gradient program: 0–30 min, 2–30% B, 30–40 min, 30–65%, held for 10 min; the injection volume was 5 μL. PDA detection was applied in the range of 200–700 nm with a sampling frequency of 4.1667 Hz and a time constant of 0.480 s. Chromatograms were extracted at 280 nm. ESI-MS acquisition was performed in the mass range 100–800 m/z with an event time of 1 s; nebulizing and drying gas (N$_2$) was set at 1.5 and 5.0 L/min, respectively. DL and heat block temperatures were set at 300 and 350 °C, respectively.

2.6. Antioxidant Activity by DPPH Assay

The antioxidant activity of the OFI and OD seed oils extracted by SFE and Soxhlet was evaluated using the 1,1-diphenyl-2-picrylhydrazil assay (DPPH) as described by Brand-Williams et al. [27]. The extracted oils (250 mg) from YV1, YV2, and I/S seeds were mixed with 1.5 mL of methanol and sonicated for 5 min at 25 °C three times. The methanolic extract was collected and used to perform the analysis. The DPPH reagent was prepared by dissolving 10 mg in 250 mL of methanol. The measurements were performed by transferring about 1.9 mL of DPPH solution into the cuvettes and adding 100 µL of the methanolic extracts. The samples were stored in the dark for 1 h at room temperature. The absorbance was measured at 515 nm with a spectrophotometer (Cary 100 Series UV–Vis Spectrophotometer, Agilent Technologies, Milano, Italy). The antioxidant activity of the oils was determined using a standard calibration curve based on Trolox (6-hidroxy-2,5,7,8-tetramethylchroman-2-carboxylic acid, from Sigma–Aldrich Co, St. Louis, EUA) as a reference antioxidant. The analysis was performed in triplicate, and the results were expressed as the Trolox equivalent value per gram of oil (mg Trolox/g oil).

2.7. Statistical Analysis

All parameters obtained for fatty acids, TAGs, and phenolic compounds were statistically analyzed to detect significant differences ($\alpha = 0.001$) among the samples by a one-way ANOVA performed through the XLSTAT software. The post-hoc HSD Tukey test, when appropriate, was applied to find out which specific groups' means (compared with each other) were different. Data obtained from oil yield and antioxidant activity were analyzed using one-way ANOVA GraphPad Prism 7. The number of independent experiments, details on statistical comparisons, and levels of significance were indicated in the captions of the respective figures and tables.

3. Results and Discussion

3.1. Oil Yields

The yield of the seed oils from the three different samples of OFI and OD seeds was highly dependent on the extraction method, the temperature applied during SFE, and the varieties. The results showed that the highest yields were obtained through Soxhlet extraction with 12 ± 1.35% for YV1, 10 ± 0.89% for YV2, and 11.8 ± 1.34% for I/S compared to 6.2 ± 1.27, 8.3 ± 1.34% for YV1, 5.7 ± 0.96, 7.4 ± 1.21% for YV2 and 6.3 ± 0.76, and 7.7 ± 1.31% for I/S by SFE extraction at 40 °C and 60 °C, respectively.

For the SFE method at a fixed pressure and different temperatures, the extraction yields increased from 6.2 ± 1.27% to 8.3 ± 1.34% in YV1, changing the temperature from 40 °C to 60 °C. Basically, the YV1 sample originated from Yemen was characterized by the highest yield when extracted by Soxhlet and the SFE method at 60 °C. Moreover, for all the samples, the extraction yields significantly increased by increasing the SFE temperature from 40 °C to 60 °C.

The selection of SFE processing parameters was based on the previous experience of extraction with other seeds [22,28]. These findings were in agreement with other previous studies using the SFE system at fixed pressure and different temperature from other seeds, such as *Nigells satvia* seeds [28,29] and *Swietenia mahagoni* seeds [30].

Data on the SFE extraction of OFI seed oil were limited in the literature [16,23]. The SFE extraction of Tunisian OFI oil at pressures of 180 bar, temperatures of 40 °C and extraction time of 135 min with a CO_2 flow rate of 15 mL·s^{-1} was reported to provide a yield of 3.4% and 1.94% for spiny and thornless varieties, respectively [23]. Both results were lower than those of the present study at the same temperature (40 °C). One reason could be the difference in the pressure (180 vs. 300 bar) or the sample variability. The SFE extraction of oil from OD seeds has been reported only by Liu et al. [9]. They found that both the temperature and pressure of the SFE process affected the extraction yield. The yield first increased, reached a maximum value of around 45 °C, and then decreased at higher temperatures, while it increased more linearly by increasing the pressure. They obtained a maximum yield of 6.65%, despite the use of higher pressure with respect to the present study.

Several factors were reported to affect the oil yield, including extraction methods, geographical origin and the harvest period of the samples, fruit ripeness, and type of solvents [6,23]. In this study, the oil yield obtained by SFE was lower than the one obtained by Soxhlet extraction due to the polarity of CO_2, a solvent suitable for the recovery of non-polar lipids. On the other side, n-hexane showed higher extractability extracting some other polar lipids and compounds in addition to the non-polar lipids. The results of the present study were in agreement with previous findings. In this context, SFE was applied to OFI from two Tunisian prickly pear seed types—the spiny (wild) and thornless (cultivated). The yielded result was significantly higher (10.32% (wild) and 8.91%) for Soxhlet compared to SFE (3.4% (spiny) and 1.94% (thornless)) [26]. In another study, the extraction of OFI using SFE was reported with a yield of 6.5% [30–33].

3.2. Chemical Characterization of Seed Oils

3.2.1. Fatty Acid Composition

The fatty acid (FA) composition of OFI (YV1 from Yemen and I/S from Sicily) and OD (YV2) from Yemen) seed oils extracted using SFE at 40 °C, 60 °C, and Soxhlet are reported in Table 1, while the typical GC chromatographic profile is depicted in Figure S1 (Supplementary material), in which only one chromatogram is shown since no qualitative differences were observed between the different samples. The results indicated that OFI and OD seed oils were characterized by a high level of unsaturated FAs (around 83%). They were mainly composed of linoleic acid (C18:2w6), which accounted for a percentage higher than 62% in all the samples. In the present study, oleic acid (C18:1w9) content ranged from a minimum of 9.88% in YV2 extracted by SFE at 60 °C to a maximum of 13.97% in YV1 extracted by Soxhlet, immediately followed by YV1 seed oil extracted by SFE at 60 °C and the I/S sample extracted by SFE at both tested temperatures. Such results clearly pinpointed the FA composition of both *Opuntia* spp. OFI and OD seed oils were not particularly affected by the extraction method, while it was most affected by the botanical species of the investigated seeds.

The main saturated fatty acids were palmitic and stearic acids in the three samples. Palmitic acid (C16:0) ranged from a minimum of 11.8% in YV2 and YV1 obtained by Soxhlet and SFE at lower temperatures, respectively, to a maximum of 12.5–12.6% in I/S and YV1 (both from OFI seeds) obtained by SFE at 40 °C and 60 °C, respectively. As for minor FAs, the omega-3 α-linolenic acid (C18:3w3) was quantified in a higher amount in the SFE extracts of YV2 seeds, immediately followed by the SFE extracts of YV1 and I/S seeds at 60 °C and 40 °C, respectively, highlighting a good capability of supercritical CO_2 to extract more polar FAs, even if present at low levels.

Generally, the FA quali-quantitative profile of both OFI and OD seed oils determined in the present study was in accordance with the previous literature [3,31,32], which reported linoleic acid to be the most abundant compound in the range 57.54–66.57%, followed by oleic acid in the range 15.2–24.3%, thus representing almost 80% of the total FAs and contributing to a favorable ratio UFA/SFA > than four.

3.2.2. HPLC/MS Analysis of Intact Lipids

Conversely, from the total FA composition, TAGs, which represent the native lipid composition of the seeds, were rarely investigated in OFI seed oils, while no studies are reported in the literature about the TAGs composition of OD samples. Indeed, the intact lipid composition could also be affected by the extraction method since the way in which single FAs are combined in the more complex TAG structure significantly affects the polarity of the molecule compared to their affinity to supercritical CO_2.

Table 2 reports the list of the identified compounds, while the typical LC chromatographic profile is depicted in Figure S2 (Supplementary material), in which only one chromatogram is shown since no qualitative differences were observed between different samples. The profile of the sample was reported in comparison with one of the reference standard mixtures used for LRI calculation, according to the equation reported in the insert of the figure. It is quite clear that TAGs are eluted according to their increasing hydrophobicity, expressed as partition number (PN=CN-2DB, where CN is the carbon number and DB is the number of double bonds of the FAs bound to the glycerol backbone). The use of the LRI identification strategy, in combination with the search into the homemade MS spectral library, allowed the reliable identification of 30 TAGs, many of them identified for the first time in *Opuntia* spp. seed oils.

Table 1. Fatty acid (FA) composition of YV1, YV2, and I/S seed oils extracted using SFE and Soxhlet methods.

FA	YV1 (Area%)			YV2 (Area%)			I/S (Area%)		
	SFE 40 °C	SFE 60 °C	Soxhlet	SFE 40 °C	SFE 60 °C	Soxhlet	SFE 40 °C	SFE 60 °C	Soxhlet
C14:0	0.08 ± 0.00 [b]	0.09 ± 0.00 [ab]	0.09 ± 0.01 [ab]	0.09 ± 0.01 [a]	0.09 ± 0.01 [a]	0.09 ± 0.01 [ab]	0.09 ± 0.01 [a]	0.09 ± 0.01 [a]	0.09 ± 0.01 [ab]
C16:0	11.81 ± 0.01 [d]	12.62 ± 0.10 [a]	12.35 ± 0.01 [b]	11.97 ± 0.01 [c]	12.06 ± 0.01 [c]	11.76 ± 0.01 [d]	12.52 ± 0.02 [b]	12.39 ± 0.02 [b]	11.86 ± 0.01 [d]
C16:1w7	0.67 ± 0.01 [e]	0.74 ± 0.00 [a]	0.74 ± 0.01 [a]	0.7 ± 0.01 [bcd]	0.72 ± 0.01 [ab]	0.69 ± 0.01 [cde]	0.71 ± 0.01 [abc]	0.68 ± 0.01 [de]	0.7 ± 0.01 [bcd]
C16:1w5	0.05 ± 0.01 [c]	0.07 ± 0.01 [b]	0.08 ± 0.01 [ab]	0.04 ± 0.01 [cd]	0.03 ± 0.01 [d]	0.03 ± 0.01 [d]	0.08 ± 0.01 [ab]	0.09 ± 0.01 [a]	0.04 ± 0.01 [cde]
C16:2w4	0.08 ± 0.01 [c]	0.07 ± 0.01 [e]	0.07 ± 0.01 [cde]	0.11 ± 0.01 [b]	0.13 ± 0.01 [a]	0.13 ± 0.01 [a]	0.08 ± 0.01 [cd]	0.07 ± 0.00 [de]	0.11 ± 0.01 [b]
C17:0	0.04 ± 0.01 [bc]	0.04 ± 0.01 [c]	0.04 ± 0.01 [bc]	0.05 ± 0.01 [abc]	0.05 ± 0.01 [a]	0.05 ± 0.01 [a]	0.05 ± 0.01 [ab]	0.04 ± 0.01 [c]	0.05 ± 0.02 [ab]
C18:0	3.1 ± 0.01 [e]	2.69 ± 0.01 [h]	2.7 ± 0.01 [h]	3.43 ± 0.01 [d]	3.88 ± 0.01 [b]	3.91 ± 0.01 [a]	2.75 ± 0.01 [g]	2.92 ± 0.01 [f]	3.69 ± 0.01 [c]
C18:1w9	11.7 ± 0.01 [d]	13.8 ± 0.01 [b]	13.97 ± 0.01 [a]	11.22 ± 0.01 [e]	9.88 ± 0.01 [h]	10.02 ± 0.01 [g]	13.65 ± 0.01 [c]	13.82 ± 0.01 [b]	10.76 ± 0.01 [f]
C18:1w7	4.82 ± 0.01 [f]	5.41 ± 0.01 [b]	5.49 ± 0.01 [a]	4.82 ± 0.02 [f]	4.65 ± 0.01 [g]	4.83 ± 0.01 [f]	5.16 ± 0.01 [d]	5.2 ± 0.01 [d]	4.97 ± 0.01 [e]
C18:1w3	0.11 ± 0.01 [b]	0.16 ± 0.01 [a]	0.19 ± 0.01 [a]	0.1 ± 0.01 [bc]	0.06 ± 0.01 [c]	0.08 ± 0.01 [bc]	0.2 ± 0.01 [a]	0.19 ± 0.02 [a]	0.1 ± 0.01 [bc]
C18:2w6	66.09 ± 0.01 [b]	62.78 ± 0.04 [e]	62.7 ± 0.03 [e]	65.79 ± 0.02 [c]	66.59 ± 0.01 [a]	66.52 ± 0.02 [a]	63.12 ± 0.03 [d]	62.77 ± 0.03 [e]	65.83 ± 0.02 [c]
C18:3w3	0.23 ± 0.01 [e]	0.37 ± 0.01 [b]	0.28 ± 0.01 [d]	0.42 ± 0.01 [a]	0.42 ± 0.01 [a]	0.34 ± 0.00 [c]	0.39 ± 0.01 [b]	0.34 ± 0.01 [c]	0.32 ± 0.01 [c]
C20:0	0.33 ± 0.01 [d]	0.27 ± 0.01 [f]	0.3 ± 0.01 [e]	0.34 ± 0.01 [c]	0.4 ± 0.01 [ab]	0.41 ± 0.01 [a]	0.29 ± 0.01 [e]	0.34 ± 0.02 [cd]	0.38 ± 0.02 [b]
C20:1w9	0.16 ± 0.01 [b]	0.15 ± 0.01 [b]	0.16 ± 0.01 [b]	0.16 ± 0.01 [b]	0.19 ± 0.01 [a]	0.21 ± 0.01 [a]	0.14 ± 0.01 [b]	0.14 ± 0.01 [b]	0.2 ± 0.01 [a]
C20:1w7	0.13 ± 0.01 [d]	0.17 ± 0.01 [bc]	0.18 ± 0.01 [ab]	0.14 ± 0.01 [cd]	0.12 ± 0.01 [d]	0.13 ± 0.01 [d]	0.17 ± 0.01 [ab]	0.19 ± 0.01 [a]	0.14 ± 0.01 [d]
C22:0	0.22 ± 0.01 [cd]	0.23 ± 0.01 [c]	0.27 ± 0.01 [c]	0.2 ± 0.01 [de]	0.27 ± 0.01 [bc]	0.29 ± 0.01 [a]	0.18 ± 0.01 [f]	0.22 ± 0.01 [c]	0.27 ± 0.01 [b]
C22:1w7	0.25 ± 0.01 [de]	0.16 ± 0.01 [g]	0.19 ± 0.01 [ef]	0.25 ± 0.01 [d]	0.27 ± 0.01 [bc]	0.3 ± 0.01 [a]	0.25 ± 0.01 [d]	0.29 ± 0.01 [ab]	0.29 ± 0.02 [a]
C24:1w9	0.06 ± 0.01 [bc]	0.07 ± 0.01 [ab]	0.06 ± 0.01 [ab]	0.05 ± 0.01 [c]	0.06 ± 0.01 [bc]	0.06 ± 0.01 [ab]	0.06 ± 0.01 [bc]	0.07 ± 0.02 [a]	0.06 ± 0.01 [ab]
SFA	15.67 ± 0.38 [f]	15.97 ± 0.29 [d]	15.8 ± 0.22 [e]	16.15 ± 0.42 [c]	16.87 ± 0.11 [a]	16.66 ± 0.21 [b]	15.99 ± 0.11 [d]	16.15 ± 0.22 [c]	12.79 ± 0.11 [g]
MUFA	17.93 ± 0.97 [d]	20.74 ± 0.95 [b]	21.14 ± 0.83 [a]	17.49 ± 0.64 [e]	16.00 ± 0.62 [h]	16.32 ± 0.71 [g]	20.42 ± 0.82 [c]	20.67 ± 0.72 [b]	17.26 ± 0.82 [f]
PUFA	66.4 ± 0.91 [c]	63.22 ± 0.94 [f]	63.06 ± 0.93 [g]	66.32 ± 0.93 [cd]	67.13 ± 0.91 [a]	66.99 ± 0.81 [b]	63.59 ± 0.83 [e]	63.18 ± 0.73 [f]	66.26 ± 0.82 [d]
PUFA/SFA	4.24 ± 0.02 [b]	3.96 ± 0.02 [f]	3.99 ± 0.01 [e]	4.11 ± 0.02 [c]	3.98 ± 0.01 [ef]	4.02 ± 0.01 [d]	3.98 ± 0.01 [ef]	3.91 ± 0.01 [g]	5.18 ± 0.04 [a]

Values are presented as means ± SD ($n = 3$). Different letters within a row indicate significant differences, a–h at $p < 0.01$.

Table 2. TAG and DAG composition of composition of YV1, YV2, and I/S seeds oils extracted using SFE and Soxhlet.

Compound	YV1 (Area%)			YV2 (Area%)			I/S (Area%)		
	SFE 40 °C	SFE 60 °C	Soxhlet	SFE 40 °C	SFE 60 °C	Soxhlet	SFE 40 °C	SFE 60 °C	Soxhlet
LL	1.44 ± 0.20 [c]	1.30 ± 0.10 [c]	1.03 ± 0.47 [c]	2.99 ± 0.14 [b]	3.62 ± 0.25 [a]	2.70 ± 0.20 [b]	1.03 ± 0.12 [c]	0.95 ± 0.15 [c]	2.51 ± 0.12 [b]
OL+LP	1.18 ± 0.10 [c]	1.02 ± 0.10 [cd]	0.89 ± 0.40 [cd]	1.91 ± 0.16 [ab]	2.21 ± 0.12 [a]	1.66 ± 0.07 [b]	0.74 ± 0.12 [d]	0.88 ± 0.06 [cd]	1.60 ± 0.16 [b]
LnLnLn	0.10 ± 0.02 [b]	0.14 ± 0.02 [b]	0.10 ± 0.04 [b]	0.56 ± 0.24 [b]	0.27 ± 0.04 [b]	0.25 ± 0.03 [b]	0.51 ± 0.08 [b]	1.23 ± 0.19 [a]	0.45 ± 0.07 [b]
LLnLn	0.49 ± 0.07 [e]	0.51 ± 0.12 [de]	0.25 ± 0.07 [e]	1.60 ± 0.32 [ab]	1.02 ± 0.11 [cd]	0.60 ± 0.06 [de]	1.33 ± 0.16 [bc]	2.07 ± 0.19 [a]	0.41 ± 0.03 [e]
LnLnP	0.18 ± 0.03 [bcd]	0.15 ± 0.02 [cd]	0.14 ± 0.05 [d]	0.40 ± 0.34 [ab]	0.36 ± 0.07 [abc]	0.44 ± 0.09 [a]	0.30 ± 0.06 [abcd]	0.42 ± 0.04 [a]	0.34 ± 0.08 [abc]
OLnLn	0.57 ± 0.10 [c]	0.61 ± 0.04 [c]	0.62 ± 0.02 [c]	1.02 ± 0.09 [b]	0.69 ± 0.04 [c]	0.60 ± 0.02 [c]	1.04 ± 0.09 [b]	1.33 ± 0.04 [a]	0.56 ± 0.07 [c]
LLL	34.08 ± 0.41a [bcd]	31.99 ± 0.68 [bcd]	31.63 ± 0.98 [cd]	32.72 ± 1.10 [abcd]	34.34 ± 0.70 [abc]	35.24 ± 0.56 [a]	31.08 ± 0.61 [de]	29.44 ± 0.92 [e]	35.00 ± 1.56 [abc]
LnPPo+OLLn	0.38 ± 0.07 [c]	0.28 ± 0.10 [c]	0.46 ± 0.14 [bc]	0.31 ± 0.01 [c]	0.49 ± 0.03 [cb]	0.78 ± 0.07 [ab]	0.46 ± 0.18 [bc]	0.88 ± 0.06 [a]	0.52 ± 0.08 [bc]
OLL	19.36 ± 0.54 [bcd]	21.28 ± 0.10 [ab]	22.10 ± 0.88 [a]	18.20 ± 0.19 [cde]	16.57 ± 0.99 [e]	17.49 ± 0.38 [de]	21.17 ± 0.15 [ab]	19.82 ± 0.26 [bc]	18.47 ± 0.34 [cde]
LLP	18.80 ± 0.57	18.70 ± 0.23	17.80 ± 1.08	18.25 ± 0.46	18.77 ± 0.08	18.45 ± 1.02	19.01 ± 1.31	18.22 ± 1.11	17.62 ± 0.57
OOL+GLL	5.29 ± 0.20 [bc]	6.19 ± 0.10 [a]	6.62 ± 0.32 [a]	4.63 ± 0.12 [cd]	3.94 ± 0.17 [d]	4.19 ± 0.06 [d]	6.05 ± 0.17 [ab]	6.05 ± 0.30 [ab]	4.51 ± 0.29 [cd]
SLL+OLP	10.18 ± 0.12	9.44 ± 0.20	9.84 ± 0.35	9.46 ± 0.42	9.59 ± 0.60	10.10 ± 0.34	9.35 ± 0.26	9.85 ± 0.74	9.70 ± 0.25
LPP	1.49 ± 0.11	1.67 ± 0.09	1.67 ± 0.19	1.46 ± 0.12	1.37 ± 0.10	1.30 ± 0.12	1.59 ± 0.31	1.75 ± 0.23	1.17 ± 0.14
GOL+C22:1LL	0.57 ± 0.07	0.57 ± 0.01	0.67 ± 0.089	0.68 ± 0.13	0.49 ± 0.05	0.66 ± 0.11	0.58 ± 0.08	0.71 ± 0.07	0.71 ± 0.17
SOL+OOP	3.45 ± 0.23 [ab]	3.71 ± 0.11 [a]	3.61 ± 0.23 [a]	2.89 ± 0.15 [b]	2.89 ± 0.23 [b]	3.06 ± 0.22 [ab]	3.38 ± 0.27 [ab]	3.52 ± 0.18 [ab]	2.91 ± 0.04 [b]
SLP	1.04 ± 0.08	1.15 ± 0.14	1.14 ± 0.21	1.25 ± 0.15	1.24 ± 0.19	1.11 ± 0.16	1.00 ± 0.07	1.377 ± 0.16	0.97 ± 0.11
C22:1OL+C24:1LL	0.29 ± 0.02	0.25 ± 0.06	0.21 ± 0.04	0.33 ± 0.03	0.17 ± 0.01	0.25 ± 0.09	0.29 ± 0.13	0.31 ± 0.01	0.17 ± 0.02
C22:1LP	0.13 ± 0.40 [ab]	0.12 ± 0.01 [ab]	0.11 ± 0.10 [ab]	0.06 ± 0.01 [b]	0.11 ± 0.02 [ab]	0.17 ± 0.01 [a]	0.14 ± 0.04 [ab]	0.15 ± 0.05 [ab]	0.21 ± 0.02
AOL+BLL	0.60 ± 0.30	0.55 ± 0.16	0.53 ± 0.05	0.63 ± 0.10	0.79 ± 0.23	0.69 ± 0.11	0.60 ± 0.15	0.69 ± 0.19	0.63 ± 0.07
SOO+C20:1OP+SSI	0.40 ± 0.30	0.37 ± 0.05	0.32 ± 0.045	0.44 ± 0.14	0.40 ± 0.16	0.43 ± 0.06	0.37 ± 0.02	0.39 ± 0.03	0.45 ± 0.13

Mean values with different letters are significantly different, a–e at $p < 0.01$. Fatty acid legend: L = linoleic acid C18:2n6; O = oleic acid C18:1n9; P = palmitic acid C16:0; Ln = α-linolenic acid C18:3n3; Po = palmitoleic acid C16:1n7; S = stearic acid C18:0; A = arachidic acid C20:0; B = behenic acid C22.

According to the results of the ANOVA test reported in Table 2, major quantitative differences between the samples were obtained for diacylglycerols (DAGs) and Ln-containing TAGs ($p < 0.01$), apart from significant differences observed for some Linoleic (L)-containing TAGs, such as trilinolein (LLL), oleyl-dilinolein (OLL) and dioleyl-linolein (OOL), the latter partially coeluted with gondoyl-dilinolein (GLL).

Particularly, DAGs were detected at a significantly higher percentage in the YV2 sample extracted by SFE at 60 °C, while Ln-containing TAGs were quantified at higher levels in the Sicilian sample (I/S) obtained through SFE at 60 °C. Such results appeared in contrast with GC results of the FA composition, which reported the highest content of Ln in YV2 extracted by SFE. However, the GC analysis provided results regarding the total FA composition, including free FAs and monoacylglycerols, not detected through the employed LC method. Then, it could be supposed that both YV2 and I/S seed oils had a satisfactory content of Ln, extracted in major amounts through SFE compared to Soxhlet, and independently from the native form in which it was present in the seeds (TAGs, free FA or MAG).

Generally, the quali-quantitative TAGs profile of the seed oils here investigated was similar to the composition previously reported for OFI seed oil originating from Tunisia and extracted by Soxhlet using hexane, with LLL and OLL as the main components with an average percentage of 25.6% and 21.5, respectively [33]. The analyses of the present study showed a higher content of LLL, near or higher than 30% in all the samples, while OLL was contained at a lower percentage, even less than 20%, in some of the investigated samples. Such difference in the TAG composition reflects the lower content of oleic acid found from the GC analyses. On the other hand, a higher content of dilinoleyl-palmitin (LLP) was obtained, comparable to OLL (17.62–19.01%). Additionally, many Ln-containing TAGs were identified at percentages of around 0.5%, as well as other minor TAGs, not present in the work by Mannoubi et al. [33], which reported a remarkably higher content of another oleic acid (O)- containing TAG, namely OOL (11.40 vs. 3.94–6.62% of the present study).

3.2.3. LC-FLD Analysis of Vitamin E

The LC-FLD analyses of vitamin E revealed the presence of γ-tocopherol as the only vitamer contained in all the OFI and OD seed oils. Such a result is in accordance with a previous work in the literature on Moroccan OFI and OD seed oils extracted using hexane, in which only γ-tocopherol was detected [34]. Other authors reported the presence of α-tocopherol and δ-tocopherol with a content lower than 2 and 10 mg/100, respectively [5,33,35]. However, γ-tocopherol was still the main vitamer, with a content ranging from 67.5 to 91.7 mg/100 g.

As shown in Table 3, γ-tocopherol ranged between 73 and 178 mg/100 g. More in detail, the Yemen OD seed oils (sample YV2) showed only a moderate variation in the γ-tocopherol content based on the extraction procedure. Conversely, the Yemen OFI seed oil (sample YV1) and the Sicilian I/S sample showed a significantly higher content ($p < 0.01$) of γ-tocopherol in the SFE extract at 60 °C. Generally, for both YV1 and I/S seed oils, the tocopherol content was affected by the extraction method, as was reported previously by Regalado-Rentería et al. [36], who compared cold pressing and maceration extraction for different *Opuntia* species. They found that the γ-tocopherol content was significantly higher in cold-pressed oils than those extracted with solvents. Additionally, the vitamin E content was highly dependent on the botanical species since YV2 seed oils contained a significantly lower content ($p < 0.01$) of γ-tocopherol than YV1. This was in agreement with previous findings [34] in which OFI seeds were richer in vitamin E than OD seeds.

Table 3. Vitamin E (γ-tocopherol) content in YV1, YV2, and I/S seeds oils extracted using SFE and Soxhlet.

	YV1			YV2			I/S		
	SFE 40 °C	SFE 60 °C	Soxhlet	SFE 40 °C	SFE 60 °C	Soxhlet	SFE 40 °C	SFE 60 °C	Soxhlet
Vitamin E (mg/100 g)	108.1 ± 0.82 [c]	177.5 ± 0.23 [a]	135.40 ± 0.55 [b]	81.6 ± 0.16 [g]	89.4 ± 0.47 [f]	88.9 ± 1.32 [e]	73.0 ± 0.27 [h]	105.2 ± 0.51 [d]	89.7 ± 0.07 [i]

Mean values with different letters are significantly different, a–i at $p < 0.01$.

3.2.4. Phenolic Compounds

The extracts from the oils were analyzed by LC-PDA/ESI-MS in order to obtain a chromatographic profile to correlate with the results of the antioxidant activity. Moreover, the identification of single phenolic compounds resulted in an added value to the present work as it allowed us to identify the molecules responsible for specific biological activities. Within this context, most works in the literature reported mainly the total phenolic and/or flavonoid content, while few data were published about the detailed phenolic composition of these samples. Among them, the results of the present study were not far from those previously reported [37]. In a published study, the phenolic composition of OFI cold-pressed oils from six different locations in Morocco was investigated. Seven phenolic compounds were identified, belonging to three families, the hydroxyl cinnamic acid derivatives (p-coumaric acid, p-coumaric acid ethyl ester, ferulic acid), the hydroxyl cinnamaldehyde derivatives (feruladehyde) and the hydroxybenzaldehyde derivatives (4-hydroxybenzaldehyde, vanillin, syringaldehyde) with vanillin, syringaldehyde, and feruladehyde being the most abundant ones.

In the present study, a total of 14 phenols were identified with quali-quantitative differences among different samples and/or extraction methods. They belong to the same chemical classes reported in the literature [37] with the addition of two simple phenols, namely tyrosol and hydroxytyrosol, which are characterized by a well-known high antioxidant power [38], three benzoic acid derivatives, namely syringic acid, vanillic and homovanillic acids, and one coumarin derivative that is the 4-hydroxycoumarin. As for the benzoic acids, they could derive from vanillin oxidation (with vanillin confirmed as the most representative phenol in both prickly pear seed oils) or could be natural flavoring agents with antimicrobial properties [39]. Among cinnamic acid derivatives, p-coumaric acid and its ethyl ester were not detected, while a moderate content of cinnamic acid was observed in most of the samples under investigation. Its presence was quite conceivable since it is a central intermediate in the phenylpropanoid biosynthetic pathway, leading to the synthesis of other hydroxycinnamic derivatives, flavonoids, isoflavones, lignin, and coumarins (e.g., 4-hydroxycoumarin) [40]. Finally, among cinnamaldehyde derivatives, synapaldehyde was also identified for the first time in both OFI and OD seed oils. It is also involved in the biosynthesis of lignin [41,42], then it could be reasonably contained in the seed oil. Indeed, published studies [43] detected glucoside derivatives of both synapoyl and feruloyl acids in the ethanolic extracts of the seeds, thus supporting the tentative identification of the present study. All these findings are summarized in Table S1 (Supplementary material), which lists the tentatively identified phenols, their chemical family, and their percentage in the analyzed oils.

To the best of the authors' knowledge, there were no previous studies on the detailed profiling of phenolic compounds in OD seed oil. Actually, the YV2 sample showed the richest phenolic profile, especially when Soxhlet extraction was used, followed by SFE at 60 °C. The HPLC-PDA chromatograms of YV2 seed oils extracted through the three methods are shown in Figure 2D–F, which point out the different signal intensities of the samples obtained through Soxhlet (intensity scale multiplied by five compared to the SFE). Soxhlet extraction provided the most intense signal of phenolic compounds in the case of the Sicilian sample (I/S) (Figure 2G–I highlighted an intensity scale multiplied by 4 compared to the corresponding SFE). Strangely, similar profiles were obtained for the YV1 samples (Figure 2A–C), independently from the extraction procedure. As for the quali-quantitative composition, the SFE extracts were dominated by the vanillin content, which decreased in the Soxhlet extract due to the simultaneous extraction of a major number of compounds (e.g., vanillin isomer, syringic, ferulic, and cinnamic acids) with a higher overall yield (details on the quali-quantitative composition of phenolic compounds in OFI and OD seed oil was shown in Supplementary material, Table S1).

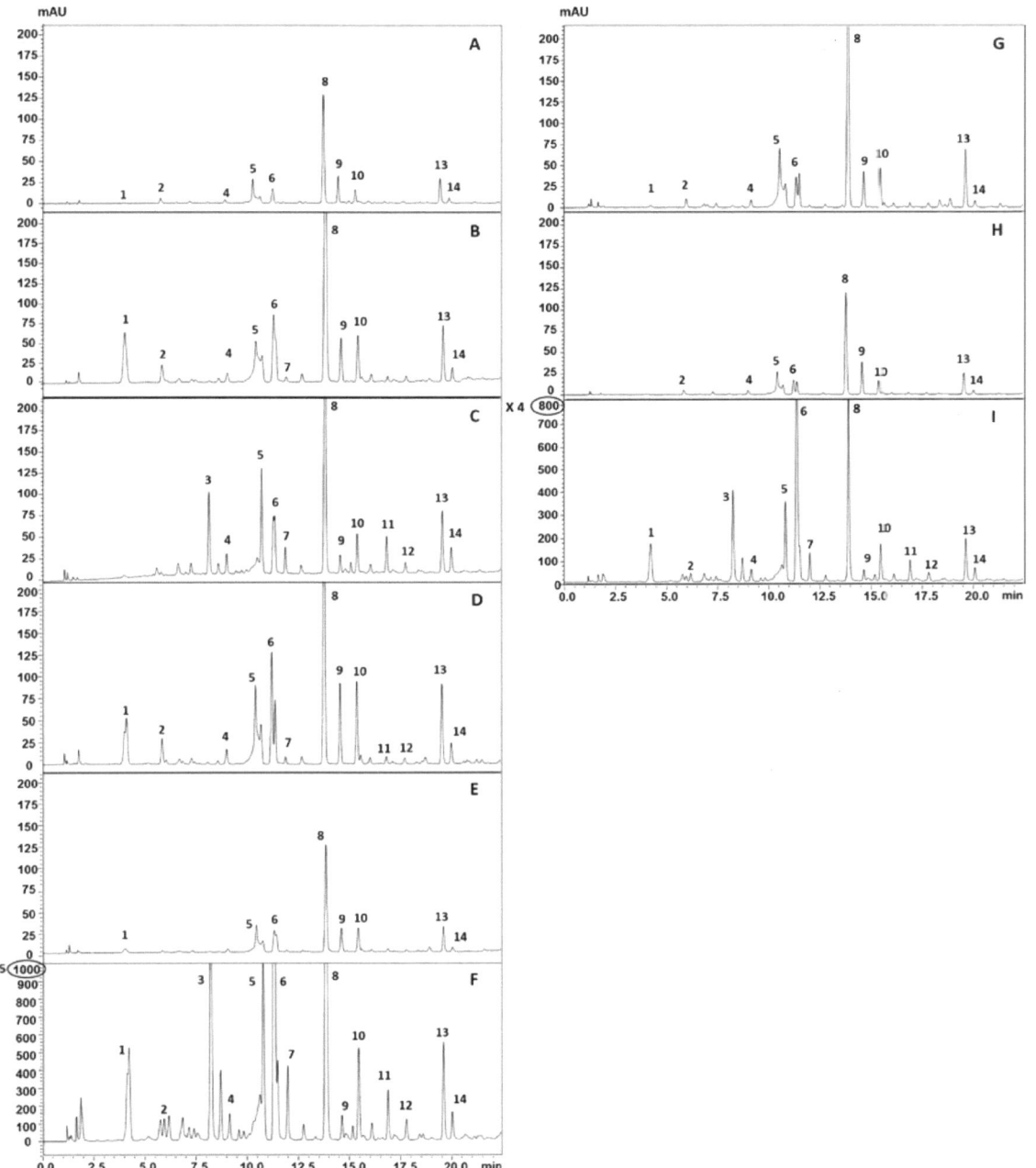

Figure 2. HPLC-PDA chromatograms of phenolic compounds in YV1 (**A–C**), YV2 (**D–F**), and I/S (**G–I**) seed oils extracted using SFE at 60 °C (**A,D,G**), 40 °C (**B,E,H**), and Soxhlet extraction (**C,F,I**).

3.3. Antioxidant Activity of OFI and OD Seeds Oils

The total antioxidant activity of YV1, YV2, and I/S extracted oils using SFE at different temperatures, and Soxhlet was assessed spectrophotometrically using the DPPH assay.

The antioxidant activity was equal to 47.78 ± 0.41, 34.96 ± 5.323, and 42.16 ± 1.452 mg Trolox eq./g oil in YV1 oil samples extracted by Soxhlet, SFE at 40 °C and 60 °C, respectively (Figure 3). Conversely, YV2 showed significant differences in the antioxidant activity depending on the extraction with 52.86 ± 1.526, 30.21 ± 2.78, and 42.67 ± 0.1718 mg Trolox eq./g oil for the Soxhlet, SFE at 40 and 60 °C samples, respectively.

The Soxhlet extraction of the I/S sample provided the highest antioxidant activity (61.54 ± 1.8) compared to all other samples, despite no significant differences ($p > 0.01$) observed with respect to YV2 oil extracted by Soxhlet. The differences in antioxidant activity among different samples extracted by Soxhlet might be ascribed to the geographical location where the seeds were harvested. Liu et al. [44] also reported that the geographical location significantly affected the phenolic composition and antioxidant activities of the extracts.

For the antioxidant activity, differences between the three extraction methods were observed, with Soxhlet appearing more efficient than SFE at 60 °C, which was more efficient than SFE at 40 °C for both YV2 and I/S samples, while similar profiles were obtained for YV1 with all the extraction procedures.

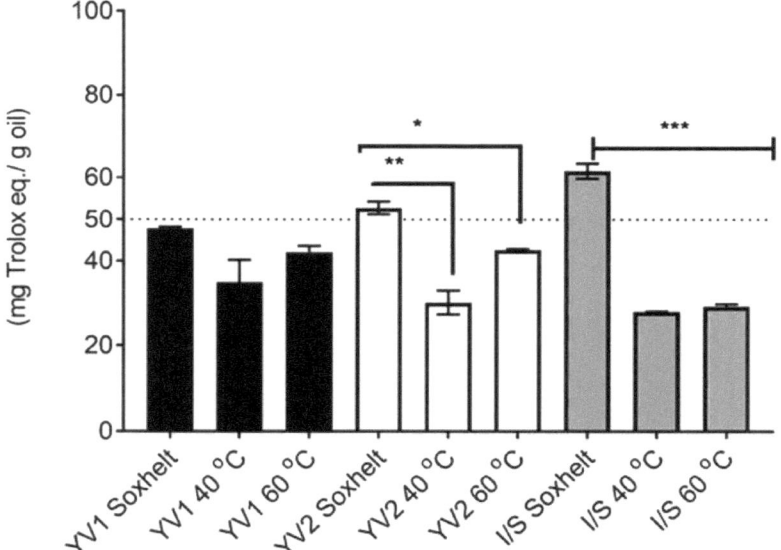

Figure 3. Antioxidant activity (mg Trolox eq./g oil) of OFI and OD oils extracted by SFE at 40 °C, 60 °C, and Soxhlet. Among each variety, different symbols (* $p < 0.05$, ** $p < 0.01$, *** $p < 0.001$) indicate significant differences compared to Soxhlet extraction.

The results of the present study on the antioxidant activity of OFI seeds were comparable to previous findings, in which another extraction method, i.e., cold pressing, was used for the production of Algerian seed oil [45]. It was reported that Soxhlet extraction provided seed oils with higher antioxidant activity, as also shown in other studies if the extraction solvent was carefully selected [5]. For instance, the results of the free radical scavenging activity of OFI originating from Yemen and extracted by different solvents, namely n-hexane, petroleum ether, and chloroform-methanol (2:1, v/v) showed that the oil extracted using chloroform-methanol (2:1, v/v), exhibited the highest antioxidant activity (87%) towards the DPPH radical, especially compared to petroleum ether (76%) [5]. Another work on the determination of phenolic compounds and antioxidant activity of

Algerian OFI seed oil indicated that the best results of the antioxidant capacity of seeds were obtained when 75% of acetone (rather than ethanol and methanol) were mixed with water and used as an extraction solvent, reaching an antioxidant capacity ranging from 50% to 95% by increasing the extraction temperature [46]. As for the oil extracted from OD seeds, a previous work highlighted a notable antioxidant activity of the seed oil obtained using SFE at 45 °C, 46.96 MPa, and 2.79 h of extraction time. The radical scavenging activity was nearly comparable to a reference ascorbic acid solution [9].

4. Conclusions

SFE and Soxhlet extraction was applied for the recovery of oils from OFI and OD seeds. Higher oil yields were obtained by Soxhlet extraction compared to SFE. Fatty acid profiles were investigated. Linoleic acid was the major compound, followed by oleic acid and palmitic acid for both SFE and Soxhlet extraction methods. SFE allowed for better extraction of linolenic acid compared to Soxhlet extraction. A total of 30 TAGs were identified in both OFI and OD seed oils for the first time. The main TAGs components were the same for the SFE and Soxhlet extraction. LLL was the major TAG, followed by OLL and SLL+OLP. The phenolic profiles showed 14 different compounds, with vanillin, 4-hydroxy benzaldehyde, vanillic acid, and hydroxytyrosol appearing as the most abundant phenols. The best results for OFI and OD SFE extracted oils were obtained at 60 °C, which provided a higher oil yield, antioxidant activity, linolenic acid, and γ-tocopherol contents compared to the oils extracted at 40 °C. In contrast, Soxhlet allowed for better extraction of total phenolic compounds with higher antioxidant activity. The results clearly highlight that an in-deep study on the optimization of SFE processing conditions is still needed to obtain oils with a higher yield, phenolic profile, and high antioxidant activity. This would allow the application of SFE as a green extraction technology to obtain oils that are useful for the formulation of functional food or nutraceutical products.

Supplementary Materials: The following supporting information can be downloaded at: https://www.mdpi.com/article/10.3390/foods12030618/s1, Figure S1. GC-FID chromatogram of FA profile of Yemen OFI seeds oil (YV1) extracted using SFE at 40 °C; Figure S2. HPLC-MS chromatogram (A) and reference standard mixture (B) of YV1 sample; Table S1. List of identified phenols in YV1, YV2 and I/S samples extracted by SFE and Soxhlet.

Author Contributions: G.A.-N.: conceptualization, data curation, investigation, visualization, Writing—Original draft, C.C.: investigation. E.A.: Writing—Review and editing, supervision, resources, G.F.: conceptualization, data curation, investigation, visualization, Writing—Review and editing. A.G., investigation. C.B.: investigation. G.M.: data curation, validation. P.D.: Supervision, resources, visualization. L.M.: resources. F.R.: data curation, Writing—Original draft, methodology, software, formal analysis. All authors have read and agreed to the published version of the manuscript.

Funding: This work was supported by the Open Access Publishing Fund of the Free University of Bozen-Bolzano. Part of this article was supported by the Scholar at Risk (SAR), University of Trento fund number Fondo ricerca AR SAR-finanz. CARITRO.

Data Availability Statement: Data are contained within the article or supplementary material.

Acknowledgments: The first author is grateful to the University of Trento, Italy, and Scholar at Risk (SAR), Italy Section at the University of Trento for their support. The authors thank Shimadzu Corporation and Merck Life Science for their continuous support.

Conflicts of Interest: The authors declare that they have no known competing financial interests or personal relationships that could have appeared to influence the work reported in this paper.

References

1. Guerrero, P.C.; Majure, L.C.; Cornejo-Romero, A.; Hernández-Hernández, T. Phylogenetic Relationships and Evolutionary Trends in the Cactus Family. *J. Hered.* **2019**, *110*, 4–21. [CrossRef] [PubMed]
2. Sharma, C.; Rani, S.; Kumar, B.; Kumar, A.; Raj, V.V.; Vihar, V. Plant *Opuntia dillenii*: A Review on Its Traditional Uses; Phytochemical and Pharmacological Properties. *Pharm. Sci.* **2015**, *1*, 29–43.
3. Al-Naqeb, G.; Fiori, L.; Ciolli, M.; Aprea, E. Prickly Pear Seed Oil Extraction; Chemical Characterization and Potential Health Benefits. *Molecules* **2021**, *26*, 5018. [CrossRef]
4. Al-Naqeb, G. Effect of prickly pear cactus seeds oil on the blood glucose level of streptozotocin-induced diabetic rats and its molecular mechanisms. *Int. J. Herb. Med.* **2015**, *3*, 29–34.
5. Taoufik, F.; Zine, S.; El Hadek, M.; Idrissi Hassani, L.M.; Gharby, S.; Harhar, H.; Matthäus, B. Oil content and main constituents of cactus seed oils *opuntia ficus indica* of different origin in Morocco. *Med. J. Nutr. Metab.* **2015**, *8*, 85–92. [CrossRef]
6. Karabagias, V.K.; Karabagias, I.K.; Gatzias, I.; Badeka, A.V. Prickly Pear Seed Oil by Shelf-Grown Cactus Fruits: Waste or Maste? *Processes* **2020**, *8*, 132. [CrossRef]
7. Loukili, E.A.; Abrigach, F.; Bouhrim, M.; Bnouham, M.; Fauconnier, M.L.; Ramdani, M. Chemical Composition and Physicochemical Analysis of *Opuntia dillenii* Extracts Grown in Morocco. *J. Chem.* **2021**, 8858929. [CrossRef]
8. Ghazi, Z.; Ramdani, M.; Fauconnier, M.L.; El Mahi, B.; Cheikh, R. Fatty Acids Sterols and Vitamin E Composition of Seed Oil of *Opuntia Ficus Indica* and *Opuntia Dillenii* from Morocco. *J. Mater. Environ. Sci.* **2013**, *4*, 967–972.
9. Liu, W.; Fu, Y.; Zu, Y.; Tong, M.; Wu, N.; Liu, X.; Zhang, S. Supercritical carbon dioxide extraction of seed oil from *Opuntia dillenii* Haw. and its antioxidant activity. *Food Chem.* **2009**, *114*, 334–339. [CrossRef]
10. Bardaa, S.; Turki, M.; Ben Khedir, S.; Mzid, M.; Rebai, T.; Ayadi, F.; Sahnoun, Z. The Effect of Prickly Pear, Pumpkin, and Linseed Oils on Biological Mediators of Acute Inflammation and Oxidative Stress Markers. *Biomed. Res. Int.* **2020**, *2020*, 5643465. [CrossRef]
11. Khémiri, I.; Essghaier Hédi, B.; Sadfi Zouaoui, N.; Ben Gdara, N.; Bitri, L. The Antimicrobial and Wound Healing Potential of *Opuntia ficus indica* L. inermis Extracted Oil from Tunisia. *Evid.-Based Complement. Altern. Med.* **2019**, *2019*, 9148782. [CrossRef] [PubMed]
12. Ramírez-Moreno, E.; Cariño-Cortés, R.; Cruz-Cansino, N.D.; Delgado-Olivares, L.; Ariza-Ortega, J.A.; Montañez-Izquierdo, V.Y.; Hernández-Herrero, M.M.; Filardo-Kerstupp, T. Antioxidant and antimicrobial properties of cactus pear (*Opuntia*) seed oils. *J. Food Qual.* **2017**, *2017*, 3075907. [CrossRef]
13. Benattia, F.K.; Arrar, Z.; Khabba, Y. Evaluation the Hypo and Antihyperglycemic Activity of Cactus Seeds Extracts (*Opuntia ficus-indica* L.). *PhytoChem. BioSub. J.* **2017**, 11. [CrossRef]
14. Bouhrim, M.; Ouassou, H.; Loukili, E.; Ramdani, M.; Mekhfi, H.; Ziyyat, A.; Legssyer, A.; Aziz, M.; Bnouham, M. Antidiabetic Effect of *Opuntia Dillenii* Seed Oil on Streptozotocin-Induced Diabetic Rats. *Asian Pac. J. Trop. Biomed.* **2019**, *9*, 381–388.
15. Ennouri, M.; Fetoui, H.; Bourret, E.; Zeghal, N.; Guermazi, F.; Attia, H. Evaluation of some biological parameters of *Opuntia ficus indica*. 2. Influence of seed supplemented diet on rats. *Bioresour. Technol.* **2006**, *97*, 2136–2140. [CrossRef]
16. Becer, E.; Kabaday, H.; Meriçli, F.; Meriçli, A.H.; Kıvançl, B.; Vatansever, S. Apoptotic Effects of *Opuntia ficus indica* L. Seed Oils on Colon Adenocarcinoma Cell Lines. *Proceedings* **2018**, *2*, 1566.
17. Gaaffar, I.F.; Zainuddin, N.A.M.; Zainal, S. Comparison of Identified Compounds from Extracted Pelargonium Radula Leaves by Supercritical Fluid Extraction and Commercial Geranium Essential Oil IOP Conference Series: Materials Science and Engineering. In Proceedings of the International Conference on Chemical and Material Engineering (ICCME 2020), Semarang, Indonesia, 6–7 October 2020; Volume 1053.
18. Sodeifian, G.; Ghorbandoost, S.; Sajadian, S.A.; Saadati Ardestani, N. Extraction of oil from Pistacia khinjuk using supercritical carbon dioxide: Experimental and modeling. *J. Supercrit. Fluids.* **2016**, *110*, 265–274. [CrossRef]
19. Idris, S.A.; Markom, M.; Abd Rahman, N.; Ali, J.M. Prediction of overall yield of Gynura procumbens from ethanol-water + supercritical CO_2 extraction using artificial neural network model. *Case Stud. Therm. Eng.* **2022**, *5*, 100175. [CrossRef]
20. Bada, J.C.; León-Camacho, M.; Copovi, P.; Alonso, L. Characterization of apple seed oil with denomination of origin from Asturias, Spain. *G Rasas Aceites* **2014**, *65*, e027.
21. Ferrentino, G.; Ndayishimiye, J.; Haman, N.; Scampicchio, M. Functional activity of oils from brewer's spent grain extracted by supercritical carbon dioxide. *Food Bioproc Tech.* **2019**, *12*, 789–798. [CrossRef]
22. Ferrentino, G.; Giampiccolo, S.; Morozova, K.; Haman, N.; Spilimbergo, S.; Scampicchio, M. Supercritical fluid extraction of oils from apple seeds: Process optimization; chemical characterization and comparison with a conventional solvent extraction. *IFSET* **2020**, *64*, 102428. [CrossRef]
23. Yeddes, N.; Chérif, J.K.; Jrad, A.; Barth, D.; Trabelsi-Ayadi, M. Supercritical SC-CO(2) and Soxhlet n-Hexane Extract of Tunisian *Opuntia ficus indica* Seeds and Fatty Acids Analysis. *J. Lipids.* **2012**, *2012*, 914693. [CrossRef] [PubMed]
24. Ciriminna, R.; Bongiorno, D.; Scurria, A.; Danzì, C.; Timpanaro, G.; Delisi, R.; Avellone, G.; Pagliaro, M. Sicilian *Opuntia ficus-indica* Seed Oil: Fatty Acid Composition and Bio-Economical Aspects. *Eur. J. Lipid Sci. Technol.* **2017**, *120*, 1870029. [CrossRef]
25. Micalizzi, G.; Ragosta, E.; Farnetti, S.; Dugo, P.; Tranchida, P.Q.; Mondello, L.; Rigano, F. Rapid and miniaturized qualitative and quantitative gas chromatography profiling of human blood total fatty acids. *Anal. Bioanal. Chem.* **2020**, *412*, 2327. [CrossRef]

26. Dugo, L.; Russo, M.; Cacciola, M.; Mandolfino, F.; Salafia, F.; Vilmercati, A.; Fanali, C.; Casale, M.; De Gara, L.; Dugo, P.; et al. Determination of the Phenol and Tocopherol Content in Italian High-Quality Extra-Virgin Olive Oils by Using LC-MS and Multivariate Data Analysis. *Food Anal. Methods* **2020**, *13*, 1027. [CrossRef]
27. Brand-Williams, W.; Cuvelier, M.; Berset, C. Use of a free radical method to evaluate antioxidant activity. *LWT-Food Sci. Technol.* **1995**, *28*, 25–30. [CrossRef]
28. Al-Naqeep, G.; Ismail, M.; Allaudin, Z. Regulation of low-density lipoprotein receptor and 3-hydroxy-3-methylglutaryl coenzyme A reductase gene expression by thymoquinone-rich fraction and thymoquinone in HepG2 cells. *Lifestyle Genom.* **2009**, *2*, 163–172. [CrossRef]
29. Solati, Z.; Baharin, B.S.; Hossein Bagheri, H. Supercritical carbon dioxide ($SC-CO_2$) extraction of *Nigella sativa* L. oil using full factorial design. *Ind. Crops. Prod.* **2012**, *36*, 519–523. [CrossRef]
30. Hartati, H.; Md Salleh, L.; Aziz, A.; Che Yunus, M.A. The Effect of Supercritical Fluid Extraction Parameters on the Swietenia Mahagoni Seed Oil Extraction and its Cytotoxic Properties. *J. Teknol.* **2014**, *69*, 51–53. [CrossRef]
31. Touil, A.; Chemkhi, S.; Zagrouba, F. Physico-Chemical Characterization of *Opuntia dillenii* Fruit. *Int. J. of Food Eng.* **2010**, *6*, 5. [CrossRef]
32. Kolniak-Ostek, J.; Kita, A.; Miedzianka, J.; Andreu-Coll, L.; Legua, P.; Hernandez, F. Characterization of Bioactive Compounds of *Opuntia ficus-indica* (L.) Mill. Seeds from Spanish Cultivars. *Molecules* **2020**, *25*, 5734. [CrossRef] [PubMed]
33. El Mannoubi, I.; Barrek, S.; Skanji Casabianca, H.; Zarrouk, H. Characterization of *Opuntia ficus indica* seed oil from Tunisia. *Chem. Nat. Compd.* **2009**, *45*, 616. [CrossRef]
34. Albergamo, A.; Potortí, A.G.; Di Bella, G.; Amor, N.B.; Lo Vecchio, G.; Nava, V.; Rando, R.; Ben Mansour, H.; Lo Turco, V. Chemical Characterization of Different Products from the Tunisian Opuntia ficus-indica (L.) Mill. *Foods* **2022**, *11*, 155. [CrossRef] [PubMed]
35. Gharby, S.; Ravi, H.K.; Guillaume, D.; Vian, M.A.; Chemat, F.; Charrouf, Z. 2-methyloxolane as alternative solvent for lipid extraction and its effect on the cactus (*Opuntia ficus-indica* L.) OCL Oilseeds and fats crops and lipids. *EDP* **2020**, *27*, hal-03139032.
36. Regalado-Rentería, E.; Aguirre-Rivera, J.; González-Chávez, M.R.; Sánchez-Sánchez, R.M.; Martínez-Gutiérrez, F.; Juárez-Flores, B.I. Assessment of Extraction Methods and Biological Value of Seed Oil from Eight Variants of Prickly Pear Fruit (*Opuntia* spp.). *Waste Biomass Valori* **2020**, *11*, 1181–1189. [CrossRef]
37. Chbani, M.; Matthäus, B.; Charrouf, Z.; El Monfalouti, H.; Kartah, B.; Gharby, S.; Willenberg, I. Characterization of Phenolic Compounds Extracted from Cold Pressed Cactus (*Opuntia ficus-indica* L.) Seed Oil and the Effect of Roasting on Their Composition. *Foods* **2020**, *9*, 1098. [CrossRef]
38. Martínez, L.; Ros, G.; Nieto, G. Hydroxytyrosol: Health Benefits and Use as Functional Ingredient in Meat. *Medicines* **2018**, *5*, 13. [CrossRef]
39. Patra, A.K. An overview of antimicrobial properties of different classes of phytochemicals. In *Dietary Phytochemicals and Microbes*; Patra, A., Ed.; Springer: Dordrecht, The Netherlands, 2012.
40. Chouhan, S.; Sharma, K.; Zha, J.; Guleria, S.; Koffas, M.A.G. Recent Advances in the Recombinant Biosynthesis of Polyphenols. *Front. Microbiol.* **2017**, *8*, 2259–2274. [CrossRef]
41. Li, L.; Popko, J.L.; Umezawa, T.; Chiang, V.L. 5-hydroxyconiferyl aldehyde modulates enzymatic methylation for syringyl monolignol formation; a new view of monolignol biosynthesis in angiosperms. *Biol. Chem.* **2000**, *275*, 6537–6545. [CrossRef]
42. Richet, N.; Tozo, K.; Afif, D.; Banvoy, J.; Legay, S.; Dizengremel, P.; Cabané, M. The response to daylight or continuous ozone of phenylpropanoid and lignin biosynthesis pathways in poplar differs between leaves and wood. *Planta* **2012**, *236*, 727–737. [CrossRef]
43. Chougui, N.; Tamendjari, A.; Hamidj, W.; Hallal, S.; Barras, A.; Richard, T.; Larbat, R. Oil composition and characterisation of phenolic compounds of *Opuntia ficus-indica* seeds. *Food Chem.* **2013**, *139*, 796–803. [CrossRef] [PubMed]
44. Liu, Y.; Chen, P.; Zhou, M.; Wang, T.; Fang, S.; Shang, X.; Fu, X. Geographic variation in the chemical composition and antioxidant properties of phenolic compounds from *Cyclocarya paliurus* (Batal) Iljinskaja leaves. *Molecules* **2018**, *23*, 2440. [CrossRef] [PubMed]
45. Brahmi, F.; Haddad, S.; Bouamara, K.; Yalaoui-Guellal, D.; Prost-Camus, E.; Barros, J.P.; Prost, M.; Atanasov, A.; Madani, K.; Boulekbache-Makhlouf, L. Comparison of chemical composition and biological activities of Algerian seed oils of *Pistacia lentiscus* L.; *Opuntia ficus indica* (L.) mill. and *Argania spinosa* L. Skeels. *Ind. Crops Prod.* **2020**, *151*, 112456. [CrossRef]
46. Chaalal, M.; Touati, N.; Louaileche, H. Extraction of phenolic compounds and in vitro antioxidant capacity of prickly pear seeds. *Acta Bot. Gall.* **2012**, *159*, 467–475. [CrossRef]

Disclaimer/Publisher's Note: The statements, opinions and data contained in all publications are solely those of the individual author(s) and contributor(s) and not of MDPI and/or the editor(s). MDPI and/or the editor(s) disclaim responsibility for any injury to people or property resulting from any ideas, methods, instructions or products referred to in the content.

Regenerative Food Innovation: The Role of Agro-Food Chain By-Products and Plant Origin Food to Obtain High-Value-Added Foods

Charles Stephen Brennan

School of Science, RMIT University, Melbourne, VIC 3000, Australia; charles.brennan@rmit.edu.au

Abstract: Food losses in the agri-food sector have been estimated as representing between 30 and 80% of overall yield. The agro-food sector has a responsibility to work towards achieving FAO sustainable goals and global initiatives on responding to many issues, including climate pressures from changes we are experiencing globally. Regenerative agriculture has been discussed for many years in terms of improving our land and water. What we now need is a focus on the ability to transform innovation within the food production and process systems to address the needs of society in the fundamental arenas of food, health and wellbeing in a sustainable world. Thus, regenerative food innovation presents an opportunity to evaluate by-products from the agriculture and food industries to utilise these waste streams to minimise the global effects of food waste. The mini-review article aims to illustrate advancements in the valorisation of foods from some of the most recent publications published by peer-reviewed journals during the last 4–5 years. The focus will be applied to plant-based valorised food products and how these can be utilised to improve food nutritional components, texture, sensory and consumer perception to develop the foods for the future.

Keywords: regeneration; food innovation; sustainability; biomolecules; nutrition; wellbeing; valorisation

1. Food Waste and Sustainability Impact

Food losses due to the processing and production operations of the agri-food sector have been estimated to represent between 30 and 80% of overall yield [1], and this can be up to 1.3 billion metric tons of food material wasted each year, as illustrated in Figure 1. The majority of the losses have been calculated as originating from waste issues in the fruit and vegetable industry (possibly due to the high perishability of such products). However, the aqua food industry and meat industries closely follow the overall amount of production and processing loss related to supply chain issues [2].

Figure 1. Pictorial representation of total amount of estimated food waste and loss per year based on the primary sectors.

1.1. Distribution of Origins of Food Waste and Loss across the Globe

Many researchers have illustrated that this issue of waste material originates from processing, supply chain and consumer use stages of the lifecycle of food rather than production losses specifically derived from farming practices [2]. This is also dependent on culture and governmental influence. For instance, in Australasia, the amount of food loss from grocery supply chains accounts for 5–6% of all food loss, whereas in Europe, the figure is close to 16%, whilst in Asia, this can be between 20 and 30% [3]. The research from Martindale et al. [2] illustrates the need for an integrated approach to achieve the goals of sustainability, which we have been addressing for a while, and that this needs to reflect strategies to reduce the impact of climate change on the supply and security of the food and beverage industry. Table 1 illustrates a recent estimate of global food waste produced on behalf of each household in major countries throughout the world. What it illustrates is that there is a significant variation in the estimated food waste per capita each year depending on regionality and supply chain procedures [3]. What is of interest is the per capita differentiation between countries, possibly due to cultural influence.

Table 1. Estimated scale of food waste per household on an annual basis in selected countries data obtained from [3].

Country	Total Annual Food Waste (Tonnes)	Food Waste per Capita (kg per Person)
China	91,646,213	64
India	68,760,163	50
United States	19,359,951	59
Japan	8,159,891	64
Germany	6,263,775	75
Australia	2,562,110	102

1.2. Potential of Digital Technology in Guiding Us through Process Optimisation

The development of the Internet of Things (IoT) will aid the digitalisation of the food processing and production sectors, and understanding the data that is within such studies would help improve our insight into the potential methodologies we can employ to reduce waste as well as optimise industry processing. For instance, Caldeira et al. [4] approached the issue from a mass balance exercise across multiple stages of the lifecycle analysis within the food industries in Europe and broke the potential of waste generation into specific areas across the lifecycle of food. Such a high level of waste has been subjected to evaluation in relation to the FAO sustainable goals and global initiatives in harnessing food ingredients from secondary side streams of the agri-food industry [5]. The focus of the analysis by Caldeira et al. [4] was on the opportunity to develop strategies that could address Sustainable Development Goal 12, which emphasises ensuring sustainable consumption and production patterns across society and which has led to governmental policy documents outlining a proposal to reduce levels of waste by up to a half by 2030.

1.3. Sustainability, Food Waste and Environmental Concerns on Food Valorisation

Figure 2 is a simplistic diagrammatic representation of the complexity of the factors involved in food waste utilisation in the food industry. Researchers have identified that food waste quantities vary across sectors and regions, with differences in practices, infrastructure and cultural norms impacting the amount of waste generated.

Figure 2. A simplified pictorial of the independent components that interact when considering food sustainability and security. In particular, there is a dependency on supply chain, processing and production systems in reducing waste generation in order to protect the environment for the future.

The environmental impact of food waste is considered a high priority from a governmental policy framework as it contributes to greenhouse gas emissions, as decomposing food releases methane in landfills. The FAO has reported that the carbon footprint of food waste could be as much as 3.3 billion tons of carbon a year, having a significant effect on the levers of climate change [5]. The impact of high levels of food waste on sustainability and resource use cannot be underestimated, especially in the food production and processing industries, which consume resources including water, land and energy. As such, this waste production exerts economic implications as loss of resources invested in production, transportation and distribution. It affects profitability for businesses and can lead to increased costs for consumers as well as food scarcity [3,4].

So, while the wasting of foods across the supply chain raises ethical concerns in a world where millions of the population face hunger, it remains a possibility that redirecting food waste to those in need can address some aspects of food insecurity issues. This emphasis on employing measures to reduce waste production in a short period of time has also created intense interest in the opportunity to recover and reuse waste from production and process operations [6–9]. Much of the focus has been on improving the recovery rates from plant- [10] and meat-based material [11] to reduce the impact on the environment and create added-value products that can be used as ingredients in the circular economy within food ingredient utilisation. These initiatives have a beneficial effect in terms of recovering potentially powerful bioactive compounds, which may be effective in the improvement of human health, with research illustrating the potential to utilise waste streams from plant-based products as enhancers to the nutritional quality of processed foods [9,11–14].

It is undeniable that as the growth in the world's population steadily progresses toward the figure of 9 billion individuals, the stress and strain applied to the food production and processing systems upon a delicate global environmental fabric will need to be addressed rapidly [2,4,5]. Reflecting on this from a purely academic viewpoint, we, as thought leaders, have a responsibility to act as stewards for the future and develop a sense of governance of our land and waters, together with food systems regionally, nationally and globally. A word that springs to mind when considering this situation is that of curation. We have the duty to be mature curators of our future whereby, being respectful of the past, we can establish a new vision for the future with the principles of stewardship and governance securely at the centre of our decision-making processes.

The way that we move forward in the integration of technology, industry and systems will create the impact that is essential for us to fulfil the ambitious sustainable development goals that are highlighted by so many academics and policy makers. Part of this move will be towards understanding geopolitical and human-centred activities, as many of these are so important in directing our attention to how we reinvent the future. Part of this is also how we can apply meaningful technological innovation alongside advances in the comprehension of the role of big data and Industry 4.0 and 5.0 to respond to the opportunity of regenerative food innovation systems [15].

1.4. Can Valorisation Be the Answer to Sustainability and Regeneration

This could include the role of waste recovery and reutilisation to address issues of sustainability and food insecurity, with researchers indicating that up to 2% of global food consumers are facing real-life food insecurity issues, which means that up to 2 billion people are witnessing the effects of malnutrition [16]. The signposts are clear and evident for all of us reading academic literature as well as social media content, namely that things have to change. The term regenerative food innovation attempts to create a vision whereby researchers, policy makers, industry stakeholders and the global consumer body can move not only to return to what was considered appropriate in sustainability and security systems of the last decade or two but to reinvent these systems to create newness and a more sustainable and secure future for food production, consumption and innovation. Tittonell et al. [17] recently reviewed the importance of regenerative agriculture in providing agroecological solutions to sustainability across diverse cultures and governments. The authors highlight that the proponents of regenerative agriculture aim for outcomes beyond what sustainability can provide by establishing a balance between agricultural use and practices of the land, which is deeply connected to the concept of governance of land and areas. This could be a mixture of cultural–historical practices, socioeconomic pressures and sovereignty of lands and waters in a complex interconnection of themes (Figure 3).

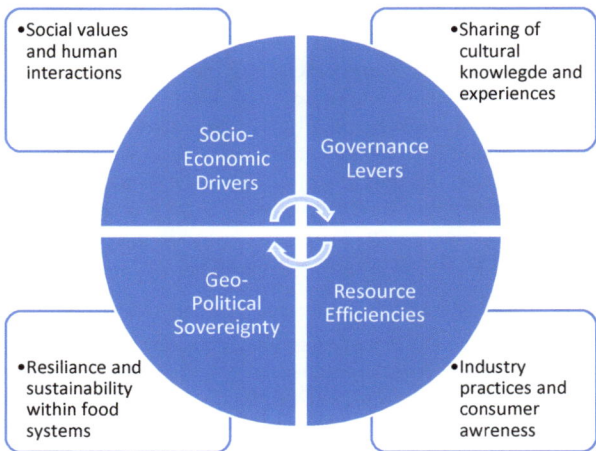

Figure 3. Simplistic representation of factors interplaying with the socioeconomic and geopolitical cross-cultural drivers for regenerative food systems.

There is undoubtable a link here between understanding production practices and process operations in order to achieve some of the future benefits we are seeking. For instance, Das et al. [18] have evaluated the impact of food production and storage processes in order to enhance both food security and sustainability in emerging nations. A good example of this is the recent study that examined the effects of climate change and maritime security in the Indo-pacific regions, as this has direct effects on smaller nations where land-based agricultural systems are only part of the story for sustainability and security issues, as they rely heavily on maritime resources for trade, economic development and basic provision of foods [19]. At the same time, Bhatkar et al. [8,9] illustrated that concerns related to production, processing and post farm gate systems are essential in creating efficiencies in the overall supply chain of food production. Careful consideration of each step in the food production and processing cycle will illustrate potential savings to be obtained at the pre-farm gate stages, harvesting and preparation, ingredient generation and food production operations, as well as cold chain systems and consumer utilisation. Small gains from each of these stages, when combined, could create large wholescale benefits in terms of sustainability savings. Indeed, supply chain dynamics and improving the resilience of blockchains are major subjects in addressing the sustainable development goals placed on food systems [20,21].

2. Consumers and Their Focus

At the same time, we need to focus squarely on the consumer without wholescale take up from consumers. Any of the advances we achieve through enhancements in production and processing units of operations and in cleaner processes that conserve energy and address the reclamation of waste products will not create an impact in terms of global sustainability and security. It is evident that consumers are increasingly prioritizing sustainability and environmental impact when considering their choice of food products and that foods made from waste streams align with these values by reducing waste and utilizing resources efficiently.

2.1. Consumer Perception and Food Valorisation

Consumer perception of food materials recovered from waste streams often revolves around the safety and nutritional quality of foods made from waste. There are concerns about hygiene and the potential presence of contaminants in valorised food materials, which need addressing. Similarly, there is hesitation among consumers to use valorised food ingredients from a taste and textural perception of these foods. The recent research evaluating the perspective of repurposing food waste streams in the Netherlands throws light on the behavioural practices of the actors involved in promoting and applying regenerative food systems to recapture redundant food materials post-food loss and waste [22]. What was interesting about this article was the interview with key participants in the process, highlighting that repurposing food waste attracts great appeal if carried out on a local basis and that this diminished the distance between the source of valorised food products and reutilization products. There is a lot of research that is required to determine this exact relationship in terms of the eyes of the consumer and the perception of improving the local vs. the global waste problems.

2.2. Levers of Adaptation Related to Consumer Taste and Texture

To understand the levers that contribute to consumer perception of repurposing food materials into food products, researchers have endeavoured to tackle the issues around the consumption of alternative food products, whether in unusual tastes, appearances or the actual labelling of the products. For instance, Crawford, Low and Newman [23] examined the specific issue of understanding the barriers that consumers, in this case children, put in place to avoid eating unfamiliar foods. From a psychological viewpoint, this research is fascinating as it explores conceptual aspects of triggers to sensory aversion, which is central to consumer preference and acceptance of foods. The psychological barrier to consumer engagement is the reluctance to try new foods that either look or smell unappealing at first glance. It is not only the possibility of new foods from reclaimed sources of ingredients that may cause consumer hesitancy, but the use of nontraditional plant-based sources of proteins can prove hard for the consumers to engage with, as the study of Munialo [24] illustrated when evaluating different forms of plant protein sources. The impact of animal husbandry on climate change and the sustainability of food systems is called into question a lot; however, there are drawbacks to an over-reliance on arable cropping systems. For instance, whilst plant-based foods are often rich in proteins [9,24], one of the challenges for food researchers and the industry as a whole is how to understand the potential of these ingredients in modern food factory applications where the consistency of ingredients is imperative in order to maintain efficiency and reduce production loss. The study of Munialo [24] explored these issues in relation to the extraction of proteins and the requirement to characterise the functionality of these ingredients in order for them to be used in commonly consumed food products from both a consumer and a cultural identity. This is something that requires further research as we progress with the concepts of utilising novel food materials.

2.3. Examples of the Application of Valorised Material into Food Produce

A good example of using recovered by-products in food systems is the consumer view of basic food items such as bread and what is culturally perceived as the norm, especially in relation to the health concerns raised from the excessive consumption of starchy foods such as breads, biscuits and extruded snack products. For instance, from a nutrition and wellbeing viewpoint, the use of whole grains in bread (and other cereal products) has a tremendous advantage in terms of their nutritional diversity and the impact that the range of bioactive ingredients, from whole-grain material (phenolic compounds and enhanced fibre content, to name just two main components) can exert on overall health outcomes of an individual compared to excessively refined flours and the products made with these [25]. Whilst the high composition of dietary fibre and phenolic compounds are of great significance in enhancing human nutrition, the taste and texture of fibre-rich foods are not universally accepted, as too are the flavours that are generated from phenolic-rich ingredients (tending to be bitter due to their high phenolic content). This relates to consumer perceptions of novel or different foods, as previously discussed in the article from Crawford, Low and Newman [23], and whilst that may have had a focus on perceptions of children to different food products, consumer perceptions of foods tend to be formed relatively early in our development. Ross [26] cleverly illustrated the significance of food structure in consumption behaviour when examining the impact of variations in food texture on the perceived ability of individuals to masticate and, hence, enjoy foods. In this work, Ross [26] delved into the importance of adjusting food textures to suit the physiological requirements of consumers using swallowing-compromised cohorts and relating this to the ability to consume. The work, although based on select cohorts, has ramifications across all consumer appreciation of foods (with regard to texture, mouthfeel and enjoyment) and should be considered when evaluating barriers to the consumption of valorised food by-products in the food industry.

One of the ways to overcome consumer hesitation to engage in the consumption of food waste materials is to evaluate their potential as ingredients of the future. Providing clear and accurate information about the source of ingredients, production processes and safety measures is crucial in influencing consumer perception of the benefits of such ingredients, not only in terms of global sustainability but also in advancing human nutrition. A plethora of research has focused on how some waste streams can be adapted in order to capture the bioactive compounds within by-products and upcycle these as ingredients to supply added nutritional benefits [27–29]. Extraction processes have come under intense research for a number of reasons, which include the ease of extraction of the beneficial bioactive ingredients, the thermal and enzymic stability of the molecular structure of the recovered compounds, and also the drive for a clean label ingredient and hence the requirement for what we can collectively call green technology for their retrieval [30–33]. From a traditional perspective, the most common ways of recovering bioactive fractions from agricultural waste products have tended to be based on the use of organic solvents such as methanol, ethanol or acetone. The mining of waste streams for potentially valuable ingredients that could be used in the food industry has been central to the idea of value-added commodities from valorisation. Melini et al. [34] illustrated that the recovery of phenolic compounds from agricultural by-products could be valuable in creating nutrient ingredients to be incorporated into foods such as bread and enhance the antioxidant levels of the product, thus creating a potential benefit for the consumer in terms of nutritional profiles. In a similar vein, Nunes et al. [35] illustrated the potential of recovering by-products from the olive oil industry as therapeutic ingredients. We understand that olive oil has numerous benefits as a food ingredient and as part of the Mediterranean diet, associated with exceptional lifestyle advantages and longevity of individuals based on its links in reducing cholesterol levels and blood pressure, attenuating glycaemic response and aiding the establishment of a gut microbiota diversity [36,37]. However, the olive oil industry produces significant levels of by-products, which, whilst high in residual antioxidants, fibre and other associated bioactive co-passengers, are regarded as low value. Hence, the attention to recovering these materials and their molecular compounds has intensified from a pharmaceutical avenue as well as applying them to food systems such as bread and cereal-based foods [38]. This research into the waste minimisation of waste products crosses numerous plant products, such as the work conducted by Romano et al. [34] on the recovery of bioactive compounds from the citrus industry, another high waste-producing manufacturing operation.

3. Refining Extraction Techniques to Enhance Safety and Biological Activities

3.1. Food Safety and Environmental Considerations for Waste Recovery

Previous research earlier in this mini-review illustrated the substantial amount of waste that is derived from the by-products and trimmings of fruits, vegetables, grains and even the meat industry. One of the aspects that is commonly observed in food processing is that it often involves the removal of the outer layers of seeds and fruits as well as the core products. This leads to substantial waste production, especially in industries that involve pulping, juicing or refining fruits and vegetables. It needs to be recognised that some of the reasons behind the removal of these components from the edible food streams, as well as the antinutritional/toxic elements, are important to consider when endeavouring to recover as much material as possible from waste streams [8,9]. Bartkiene et al. [39] reported on the potential safety implications when the recovery of food by-products is employed to manufacture food-grade ingredients. What they illustrated was that the choice of extraction technique was crucial in terms of maintaining the safety and functionality of the compounds recovered [40]. Most notably, maintaining the bioactivity and, hence, functionality of these ingredients is a major challenge when considering the efficacy of including these value-added ingredients in food materials. There is also concern in relation to the environmental pressure associated with some extraction processes using harsh chemical treatment in order to recover functional components. For this reason, many researchers have focused on extraction technology, which is often referred to as 'green' extraction or processing.

3.2. Extraction Processes Which May Be Regarded as Novel

Enzyme-assisted extraction of waste streams has been used as an enzyme that can be used to enhance the release of phenolic compounds from plant matrices, improving extraction efficiency and maintaining the biological structure of the extracted compounds [30,32]. In addition, microbial fermentation of plant waste streams can aid in breaking down complex structures that are prominent in the cellular components of waste streams. Such a disruption to the structure of the cellular components helps in releasing phenolics contained therein. One of the challenges we face is determining the most efficient and cost-effective extraction method for specific plant waste streams, which is essential while ensuring the purity and stability of extracted phenolic compounds for various applications, including food, pharmaceuticals and cosmetics [32]. The establishment of economically feasible processes for phenolic recovery from plant waste streams needs to consider the financial costs associated with extraction and purification, as well as the environmental impact [30,33]. Implementing sustainable practices to minimise the environmental footprint of extraction processes, such as reducing solvent use or utilising ecofriendly methods, is crucial to benefit the environment rather than create more impacts that need mitigating [33,41].

For this reason, there has been increased attention on what could be regarded as novel extraction techniques that rely on cellular disruption of waste materials to aid the overall separation and recovery of functional food components. Notably, Bartkiene et al. [39] found that the combination of ultrasonication and fermentation enhanced the rate of recovery of bioactive ingredients. Speed and ease of recovery of biologically active components useful for the functional food business form an exciting opportunity to speed up the application of food waste streams into food products. Bartkiene et al. [39] also concentrated on ingredient safety, noting that attention still needs to be specific in relation to biogenic amines, mycotoxins and micro- and macroelements. Indeed, contamination of these recovered ingredients can present a number of challenges with heavy metals, residual chemicals and microbial contaminants [42,43].

Despite these potential safety issues, the valorisation of the active compounds from agricultural and food processing by-products is important for the future sustainability of the food industry and the upcycling of potentially valuable nutritional compounds [29,34,44]. Lima et al. [45] illustrated that the bioactive compounds extracted from Amazonian fruit and vegetable by-products could be used to significantly change the nutritional profile of foods and thus potentially be used to lower risk factors associated with cancer and metabolic diseases. Phenolic compounds from South American material have been of great interest, as per the research from Beltrán-Medina et al. [46], who have investigated the potential of coffee by-product utilisation in foods, this time including them in extruded snack products and therefore upcycling waste material into a ready-to-eat food product acceptable for consumers.

Similarly, Park et al. [47] evaluated the potential of spent coffee material as a functional food ingredient and the most economical extraction technique to be used in the recovery of the material for functional food purposes. The researchers illustrated the benefits of sequential extraction techniques and pretreatment methods in enhancing the recovery of high-value products from spent coffee grounds using both supercritical carbon dioxide and ultrasonic treatment. Most notably, ultrasound pretreatment improved the antioxidant activities (DPPH and ABTS assays) of the valorised spent grain compounds into high-value products, illustrating increased efficiency possible through combinations of extraction methods.

To that end, Tien et al. [48] investigated the efficiency in the recovery of waste material using deep eutectic solvent and sequential microwave-ultrasound-assisted approaches and found that this could be a solution going forward as a method to ensure the breakdown of cellular material and hence the release of bioactive functional compounds from waste products. However, Velusamy, Rajan and Radhakrishnan [49] investigated the use of pulse electric field technology (PEF) as a relatively nonthermal mechanism to break the cellular components of plant walls and enhance the potential of leaching bioactive ingredients during extraction processes. The key themes behind these recent publications are to optimise the recovery rates of beneficial ingredients from waste streams whilst also minimising the impact of the processes employed in the recovery methods. The use of environmental pollutants or high energy-requiring processes flies against the basic tenets of regeneration and sustainability. As mentioned before, this is where computer optimisation and machine learning can help with the green recovery of compounds through the use of technology advancements being created in our development of Industry 4.0 and 5.0 platforms [2,9,14].

3.3. Valorised Food Materials and Their Impact on Consumer Health and Wellbeing

One of the common compounds recovered from waste streams in plant-based systems is dietary fibres (and their co-passengers), which are, in turn, beneficial for colonic functionality and gut health [50]. Indeed, the recovery of dietary fibre compounds from by-products has been a constant feature of research interest during the last 5 years [31,51–56]. Some fibres act as prebiotics, nourishing beneficial gut bacteria, such as inulin found in chicory root and oligosaccharides in legumes. One premise is that dietary fibres resist digestion in the stomach and small intestine due to resistance to human enzymes. As they reach the large intestine, they become available for fermentation by gut microbiota, which ferment these undigested fibres, breaking them down into short-chain fatty acids (SCFAs) such as acetate, propionate and butyrate. SCFAs support a healthy gut environment by nourishing colon cells, enhancing gut barrier function and regulating pH levels.

Extending this research into a physiological aspect, these SCFAs influence glucose and lipid metabolism, aiding in blood sugar regulation and reducing cholesterol levels, with further research indicating a potential influence on gene regulation from a cellular basis. Thus, the focus on the recovery of bioactive ingredients that enhance the nutritional quality of foods through enhancing microbial populations and functional gut responses holds the potential for the future [50,53,54].

3.4. Practical Applications of Isolated Food Components on Consumer Nutrition

There is an issue in terms of how we can ensure that these recovered ingredients can be incorporated into foods, as the enhancements of foods (such as bread and biscuits [53–56], which are common cereal food that is used in these studies) can affect the texture, structure, mouthfeel and, hence, consumer acceptability of foods using the recovered ingredients, even if the fibre components recovered could be of great help in manipulating gut functionality and chronic diseases, as illustrated by the work of Pansai et al. [57] when recovering fibre compounds from pitya fruit (dragon fruit) and evaluating their mode of improving gut functionality. The work of Hsu, Chang and Shiau [58] is another good example of this, where they endeavoured to recover components from pitya material and incorporate it into bread, finding that the modification of the texture was a challenge to maintain, and whilst it was possible to achieve a product which resembled a control (standard) product, artefacts such as colour and hence taste, also need to be evaluated when incorporating antioxidant-rich, phenolic-containing, substances. This speaks to the requirement for careful attention to consumer preference and how consumer acceptance of novel foods (using these novel ingredients) needs to be considered.

While Figure 1 illustrates that a major sector for the production of food waste and loss was the fruit and vegetable industries, the meat, fish and dairy sectors are also important to consider. Recently, Rana et al. [59] reviewed the specific potential of fish waste reutilisation in food innovation and suggested a number of products that could be derived from aquatic sources. The FAO [60] has estimated that fish production globally was 214 million tonnes in 2020 and that potentially up to 60% of the weight of the commercial catch is wasted in the form of potential by-products derived from carcass, head, skin and bones [61–64]. These materials are rich in digestible proteins as well as structural proteins, in addition to minerals and valuable oils. Enzymatic and fermentation processes have been used to recover proteins from fish materials, reducing the use of solvent extraction requirements [62,63]. Such recovered components from fish sources have been used in a wide range of edible food products to the benefit of consumer nutrition and sensory perception, not only in the sense of providing added bioactive components in innovative food products [64–66] but also by providing texturizing agents to enhance structure and composition of foods [67–70].

What the research illustrates is the potential of bioactive compounds—abundant in waste streams of various plant-based products as well as animal sources—that can have significant influences on protein and carbohydrate digestion of novel food products as well as the textural composition for consumer preference. These compounds can lead to the creation of value-added food products from the recovered waste lines [53]. There are many other considerations to be made in relation to regenerative food innovation, such as the utilisation of these resources in clean label ingredients for food processors and exploration of not only the use of the food waste products but material that may be carried over in terms of food processing technologies [71]. Future reviews should focus on these aspects in much more detail and provide a deeper meta-analysis of the key research originating across the world so as to ensure innovation without boundaries occurring in foods.

4. Conclusions

So, where does this leave us in searching for a new way to employ sustainable protocols to create a future for regenerative food innovation?

Certainly, there exists tremendous potential in using both the phenolic compounds of by-products from food production and processing as well as recovering proven nutritional enhancing components within the bioactive complex that presents itself as dietary fibre and its associated molecules. Implementing efficient production practices, optimising processing techniques to minimise waste generation, and adopting innovative technologies for resource recovery are also of paramount concern. The essential ethos of regenerating is the opportunity to create a new outcome from historical practices. This is where it all gets quite exciting, especially from an academic viewpoint—who are intrinsically connected with applying knowledge for the benefit of society—as we have the potential to use our experience of the past and innovations of the present to develop a future that will relish in the thought development of how to use our vision of technology to create a better future. Thus, redirecting surplus food to food banks, repurposing waste as animal feed or compost, and creating value-added products from by-products has the potential to reduce the overall global impact of waste generation during the production and processing of foods. In order to do this, we need to collaborate across boundaries and disciplines, push our vision to the limits and, if we need to, shake the laws of physics (or science, at least). So, let us enjoy the journey of exploitation and actually utilise the knowledge we have gained from the biological characterisation of the chemical composition of waste materials and the potential new ways of enhancing recovery yields. In order to ensure a true impact is delivered, we need to work in collaboration with industries that are consumer-focused and develop platform products that are acceptable for large-scale consumption. This way, we will move from a pure recovery and utilisation of waste materials to the ability to enhance food innovation systems of the future—regenerative food innovation in practice.

Funding: This research received no direct funding from outside sources.

Institutional Review Board Statement: Not applicable.

Informed Consent Statement: Not applicable.

Data Availability Statement: The original contributions presented in the study are included in the article, further inquiries can be directed to the corresponding author.

Acknowledgments: The author expresses their gratitude to the experts for their help reviewing this mini-review.

Conflicts of Interest: The author declares no conflicts of interest.

References

1. Read, Q.D.; Muth, M.K. Cost-effectiveness of four food waste interventions: Is food waste reduction a "win–win"? *Resour. Conserv. Recycl.* **2021**, *168*, 105448. [CrossRef]
2. Martindale, W.; Hollands, T.Æ.; Jagtap, S.; Hebishy, E.; Duong, L. Turn-key research in food processing and manufacturing for reducing the impact of climate change. *Int. J. Food Sci. Technol.* **2023**, *58*, 5568–5577. [CrossRef]
3. UNEP. *Food Waste Index Report 2021*; UN Environment Programme: Nairobi, Kenya, 2021; p. 100, ISBN 978-92-807-3868-1.
4. Caldeira, C.; De Laurentiis, V.; Corrado, S.; van Holsteijn, F.; Sala, S. Quantification of food waste per product group along the food supply chain in the European Union: A mass flow analysis. *Resour. Conserv. Recycl.* **2019**, *149*, 479–488. [CrossRef]
5. FAO. Sustainable Development Goals. 2020. Available online: https://www.fao.org/sustainable-development-goals-data-portal/data/ (accessed on 16 January 2024).
6. Costello, C.; Birischi, E.; McGarvey, R.G. Food waste in campus dining operations: Inventory of pre- and post-consumer mass by food category, and estimation of embodied greenhouse gas emissions. *Renew. Agric. Food Syst.* **2015**, *31*, 191–201. [CrossRef]
7. Ben-Othman, S.; Jõudu, I.; Bhat, R. Bioactives from agri-food wastes: Present insights and future challenges *Molecules* **2020**, *25*, 510. [CrossRef] [PubMed]
8. Bhatkar, N.S.; Shirkole, S.S.; Brennan, C.; Thorat, B.N. Pre-processed fruits as raw materials: Part I—Different forms, process conditions and applications. *Int. J. Food Sci. Technol.* **2022**, *57*, 4945–4962. [CrossRef]
9. Bhatkar, N.S.; Shirkole, S.S.; Brennan, C.; Thorat, B.N. Pre-processed fruits as raw materials: Part II—Process conditions, demand and safety aspects. *Int. J. Food Sci. Technol.* **2022**, *57*, 4918–4935. [CrossRef]

10. Pop, C.; Suharoschi, R.; Pop, O.L. Dietary Fiber and Prebiotic Compounds in Fruits and Vegetables Food Waste. *Sustainability* **2021**, *13*, 7219. [CrossRef]
11. Moslemy, N.; Sharifi, E.; Asadi-Eydivand, M.; Abolfathi, N. Review in edible materials for sustainable cultured meat: Scaffolds and microcarriers production. *Int. J. Food Sci. Technol.* **2023**, *58*, 6182–6191. [CrossRef]
12. Sadhukhan, J.; Dugmore, T.I.J.; Matharu, A.; Martinez-Hernandez, E.; Aburto, J.; Rahman, P.K.S.M.; Lynch, J. Perspectives on "Game Changer" Global Challenges for Sustainable 21st Century: Plant-Based Diet, Unavoidable Food Waste Biorefining, and Circular Economy. *Sustainability* **2020**, *12*, 1976. [CrossRef]
13. Ali, A.; Wei, S.; Liu, Z.; Fan, X.; Sun, Q.; Xia, Q.; Liu, S.; Hao, J.; Deng, C. Non-thermal processing technologies for the recovery of bioactive compounds from marine by-products. *LWT* **2021**, *147*, 111549. [CrossRef]
14. Conrad, Z.; Blackstone, N.T. Identifying the links between consumer food waste, nutrition, and environmental sustainability: A narrative review. *Nutr. Rev.* **2021**, *79*, 301–314. [CrossRef]
15. Hassoun, A.; Prieto, M.A.; Carpena, M.; Bouzembrak, Y.; Marvin, H.J.P.; Pallares, N.; Barba, F.J.; Punia, S. Exploring the role of green and Industry 4.0 technologies om achievingsustainable development goals in food sectors. *Food Res. Int.* **2022**, *162*, 112068. [CrossRef]
16. Lai, M.; Rangan, A.; Grech, A. Enablers and barriers of harnessing food waste to address food insecurity: A scoping review. *Nutr. Rev.* **2020**, *80*, 1836–1855. [CrossRef]
17. Tittonell, P.; El Mujtar, V.; Georges, F.; Kebede, Y.; Laborda, L.; Luján, S.R.; de Vente, J. Regenerative agriculture—Agroecology without politics? *Front. Sustain. Food Syst.* **2022**, *6*, 844261. [CrossRef]
18. Das, S.; Barve, A.; Sahu, N.C.; Muduli, K.; Kumar, A.; Luthra, S. Analysing the challenges to sustainable food grain storage management: A path to food security in emerging nations. *Int. J. Food Sci. Technol.* **2023**, *58*, 5501–5509. [CrossRef]
19. Brennan, J.; Germond, B. A methodology for analysing the impacts of climate change on maritime security. *Clim. Chang.* **2024**, *177*, 15. [CrossRef]
20. Galanakis, C. (Ed.) Food waste valorization opportunities for different food industries. In *The Interaction of Food Industry and Environment*; Academic Press: Cambridge, MA, USA, 2020; pp. 341–422, ISBN 9780128164495. [CrossRef]
21. Zaraska, M. Fighting food waste. *New Sci.* **2021**, *245*, 42–45. [CrossRef]
22. Rao, M.; Bast, A.; de Boer, A. Understanding the phenomenon of food waste valorisation from the perspective of supply chain actors engaged in it. *Agric. Econ.* **2023**, *11*, 40. [CrossRef]
23. Crawford, B.; Low, J.Y.Q.; Newman, L. Understanding barriers of eating unfamiliar fruits and vegetables in children using 'Sensory Play': A narrative review. *Int. J. Food Sci. Technol.* **2023**, *58*, 4075–4087. [CrossRef]
24. Munialo, C.D. A review of alternative plant protein sources, their extraction, functional characterisation, application, nutritional value and pinch points to being the solution to sustainable food production. *Int. J. Food Sci. Technol.* **2023**, *59*, 462–472. [CrossRef]
25. Allai, F.M.; Azad, Z.; Gul, K.; Dar, B.N. Wholegrains: A review on the amino acid profile, mineral content, physicochemical, bioactive composition and health benefits. *Int. J. Food Sci. Technol.* **2022**, *57*, 1849–1865. [CrossRef]
26. Ross, C.F. The texture factor: Food product development for discrete populations including orally compromised elderly consumers and children with Down syndrome. *Int. J. Food Sci. Technol.* **2023**, *58*, 6151–6157. [CrossRef]
27. Kumar, V.; Sharma, N.; Umesh, M.; Selvaraj, M.; Al-Shehri, B.M.; Chakraborty, P.; Duhan, L.; Sharma, S.; Pasrija, R.; Awasthi, M.K.; et al. Emerging challenges for the agro-industrial food waste utilization: A review on food waste biorefinery. *Bioresour. Technol.* **2022**, *362*, 127790. [CrossRef] [PubMed]
28. Oluwole, O.; Fernando, W.B.; Lumanlan, J.; Ademuyiwa, O.; Jayasena, V. Role of phenolic acid, tannins, stilbenes, lignans and flavonoids in human health—A review. *Int. J. Food Sci. Technol.* **2022**, *57*, 6326–6335. [CrossRef]
29. Bianchi, F.; Tolve, R.; Rainero, G.; Bordiga, M.; Brennan, C.S.; Simonato, B. Technological, nutritional and sensory properties of pasta fortified with agro-industrial by-products: A review. *Int. J. Food Sci. Technol.* **2021**, *56*, 4356–4366. [CrossRef]
30. Panzella, L.; Moccia, F.; Nasti, R.; Marzorati, S.; Verotta, L.; Napoltano, A. Bioactive Phenolic Compounds from Agri-Food Wastes: An Update on Green and Sustainable Extraction Methodologies. *Front. Nutr. Food Sci. Technol.* **2020**, *7*, 60. [CrossRef]
31. Villacís-Chiriboga, J.; Elst, K.; Van Camp, J.; Vera, E.; Ruales, J. Valorization of byproducts from tropical fruits: Extraction methodologies, applications, environmental and economic assessment—A Review (Part 1: General overview of the byproducts, traditional biorefinery practices and possible applications). *Compr. Rev. Food Sci. Food Saf.* **2020**, *19*, 405–447. [CrossRef]
32. Morejón Caraballo, S.; Rohm, H.; Struck, S. Green solvents for deoiling pumpkin and sunflower press cakes: Impact on composition and technofunctional properties. *Int. J. Food Sci. Technol.* **2023**, *58*, 1931–1939. [CrossRef]
33. Banerjee, J.; Singh, R.; Vijayaraghavan, R.; MacFarlane, D.; Patti, A.F.; Arora, A. Bioactives from fruit processing wastes: Green approaches to valuable chemicals. *Food Chem.* **2017**, *225*, 10–22. [CrossRef] [PubMed]
34. Melini, V.; Melini, F.; Luziatelli, F.; Ruzzi, M. Functional Ingredients from Agri-Food Waste: Effect of Inclusion Thereof on Phenolic Compound Content and Bioaccessibility in Bakery Products. *Antioxidants* **2020**, *9*, 1216. [CrossRef]
35. Nunes, A.; Marto, J.; Gonçalves, L.; Martins, A.M.; Fraga, C.; Ribeiro, H.M. Potential therapeutic of olive oil industry by-products in skin health: A review. *Int. J. Food Sci. Technol.* **2022**, *57*, 173–187. [CrossRef]
36. Farràs, M.; Martinez-Gili, L.; Portune, K.; Arranz, S.; Frost, G.; Tondo, M.; Blanco-Vaca, F. Modulation of the Gut Microbiota by Olive Oil Phenolic Compounds: Implications for Lipid Metabolism, Immune System, and Obesity. *Nutrients* **2020**, *12*, 2200. [CrossRef] [PubMed]

37. Jimenez-Lopez, C.; Carpena, M.; Lourenço-Lopes, C.; Gallardo-Gomez, M.; Lorenzo, J.M.; Barba, F.J.; Prieto, M.A.; Simal-Gandara, J. Bioactive Compounds and Quality of Extra Virgin Olive Oil. *Foods* **2020**, *9*, 1014. [CrossRef] [PubMed]
38. Cedola, A.; Cardinali, A.; D'Antuono, I.; Conte, A.; Del Nobile, M.A. Cereal foods fortified with by-products from the olive oil industry. *Food Biosci.* **2020**, *33*, 100490. [CrossRef]
39. Bartkiene, E.; Bartkevics, V.; Pugajeva, I.; Borisova, A.; Zokaityte, E.; Lele, V.; Sakiene, V.; Zavistanaviciute, P.; Klupsaite, D.; Zadeike, D.; et al. Challenges Associated with Byproducts Valorization—Comparison Study of Safety Parameters of Ultrasonicated and Fermented Plant-Based Byproducts. *Foods* **2020**, *9*, 614. [CrossRef] [PubMed]
40. Rao, M.; Bast, A.; de Boer, A. Valorized Food Processing By-Products in the EU: Finding the Balance between Safety, Nutrition, and Sustainability. *Sustainability* **2021**, *13*, 4428. [CrossRef]
41. Romano, R.; De Luca, L.; Aiello, A.; Rossi, D.; Pizzolongo, F.; Masi, P. Bioactive compounds extracted by liquid and supercritical carbon dioxide from citrus peels. *Int. J. Food Sci. Technol.* **2022**, *57*, 3826–3837. [CrossRef]
42. Tylewicz, U.; Tappi, S.; Nowacka, M.; Wiktor, A. Safety, Quality, and Processing of Fruits and Vegetables. *Foods* **2019**, *8*, 569. [CrossRef]
43. Azinheiro, S.; Carvalho, J.; Prado, M.; Garrido-Maestu, A. Application of Recombinase Polymerase Amplification with Lateral Flow for a Naked-Eye Detection of *Listeria monocytogenes* on Food Processing Surfaces. *Foods* **2020**, *9*, 1249. [CrossRef]
44. Chai, T.T.; Tan, Y.N.; Ee, K.Y.; Xiao, J.; Wong, F.C. Seeds, fermented foods, and agricultural by-products as sources of plant-derived antibacterial peptides. *Crit. Rev. Food Sci. Nutr.* **2019**, *59* (Suppl. S1), S162–S177. [CrossRef]
45. Lima, R.S.; de Carvalho, A.P.A.; Conte-Junior, C.A. Health from Brazilian Amazon food wastes: Bioactive compounds, antioxidants, antimicrobials, and potentials against cancer and oral diseases. *Crit. Rev. Food Sci. Nutr.* **2020**, *63*, 12453–12475. [CrossRef]
46. Beltrán-Medina, E.A.; Guatemala-Morales, G.M.; Padilla-Camberos, E.; Corona-González, R.I.; Mondragón-Cortez, P.M.; Arriola-Guevara, E. Evaluation of the Use of a Coffee Industry By-Product in a Cereal-Based Extruded Food Product *Foods* **2020**, *9*, 1008. [CrossRef]
47. Park, J.-S.; Nkurunziza, D.; Roy, V.C.; Ho, T.C.; Kim, S.-Y.; Lee, S.-C.; Chun, B.-S. Pretreatment processes assisted subcritical water hydrolysis for valorisation of spent coffee grounds. *Int. J. Food Sci. Technol.* **2022**, *57*, 5090–5101. [CrossRef]
48. Tien, N.N.T.; Le, N.L.; Khoi, T.T.; Richel, A. Characterisation of dragon fruit peel pectin extracted with natural deep eutectic solvent and sequential microwave-ultrasound-assisted approach. *Int. J. Food Sci. Technol.* **2022**, *57*, 3735–3749. [CrossRef]
49. Velusamy, M.; Rajan, A.; Radhakrishnan, M. Valorisation of food processing wastes using PEF and its economic advances—Recent update. *Int. J. Food Sci. Technol.* **2023**, *58*, 2021–2041. [CrossRef]
50. Ratanpaul, V.; Stanley, R.; Brennan, C.; Eri, R. Manipulating the kinetics and site of colonic fermentation with different fibre combinations—A review. *Int. J. Food Sci. Technol.* **2023**, *58*, 2216–2227. [CrossRef]
51. Hossain, A.K.M.M.; Brennan, M.A.; Mason, S.L.; Guo, X.; Zeng, X.A.; Brennan, C.S. The Effect of Astaxanthin-Rich Microalgae "Haematococcus pluvialis" and Wholemeal Flours Incorporation in Improving the Physical and Functional Properties of Cookies. *Foods* **2017**, *6*, 57. [CrossRef] [PubMed]
52. Benito-González, I.; Martínez-Sanz, M.; Fabra, M.J.; López-Rubio, A. Health effect of dietary fibers. In *Dietary Fiber: Properties, Recovery, and Applications*; Elsevier: Amsterdam, The Netherlands, 2019; pp. 125–163.
53. Bordiga, M.; Travaglia, F.; Locatelli, M. Valorisation of grape pomace: An approach that is increasingly reaching its maturity—A review. *Int. J. Food Sci. Technol.* **2019**, *54*, 933–942. [CrossRef]
54. Cui, J.; Lian, Y.; Zhao, C.; Du, H.; Han, Y.; Gao, W.; Xiao, H.; Zheng, J. Dietary Fibers from Fruits and Vegetables and Their Health Benefits via Modulation of Gut Microbiota. *Compr. Rev. Food Sci. Food Saf.* **2019**, *18*, 1514–1532. [CrossRef] [PubMed]
55. Grasso, S.; Omoarukhe, E.; Wen, X.; Papoutsis, K.; Methven, L. The Use of Upcycled Defatted Sunflower Seed Flour as a Functional Ingredient in Biscuits. *Foods* **2019**, *8*, 305. [CrossRef]
56. Hsu, C.T.; Chang, Y.H.; Shiau, S.Y. Color, antioxidation, and texture of dough and Chinese steamed bread enriched with pitaya peel powder. *Cereal Chem.* **2019**, *96*, 76–85. [CrossRef]
57. Pansai, N.; Detarun, P.; Chinnaworn, A.; Sangsupawanich, P.; Wichienchot, S. Effects of dragon fruit oligosaccharides on immunity, gut micrbiome, and their metabolites in healthy adults—A randomized double-blind placebo controlled study. *Food Res. Int.* **2023**, *167*, 112657. [CrossRef]
58. Schmid, V.; Trabert, A.; Schäfer, J.; Bunzel, M.; Karbstein, H.P.; Emin, M.A. Modification of Apple Pomace by Extrusion Processing: Studies on the Composition, Polymer Structures, and Functional Properties. *Foods* **2020**, *9*, 1385. [CrossRef]
59. Rana, S.; Singh, A.; Surasani, V.K.R.; Kapoor, S.; Desai, A.; Kumar, S. Fish processing waste: A novel source of non-conventional functional proteins. *Int. J. Food Sci. Technol.* **2023**, *58*, 2637–2644. [CrossRef]
60. FAO. The State of World Fisheries and Aquaculture 2022. 2022. Available online: https://www.fao.org/publications/sofia/2022 (accessed on 16 January 2024).
61. Ananey-Obiri, D.; Tahergorabi, R. Development and characterization of fish-based superfoods. In *Current Topics on Superfoods*; Shiomi, N., Ed.; IntechOpen: London, UK, 2018; Volume 395, pp. 116–124.
62. Bhuimbar, M.V.; Bhagwat, P.K.; Dandge, P.B. Extraction and characterization of acid soluble collagen from fish waste: Development of collagen-chitosan blend as food packaging film. *J. Environ. Chem. Eng.* **2019**, *7*, 102983. [CrossRef]
63. Blanco, M.; Vázquez, J.A.; Pérez-Martín, R.I.; Sotelo, C.G. Hydrolysates of fish skin collagen: An opportunity for valorizing fish industry byproducts. *Mar. Drugs* **2017**, *15*, 131. [CrossRef]

64. Desai, A.S.; Brennan, M.; Gangan, S.S.; Brennan, C. Utilization of fish waste as a value-added ingredient: Sources and bioactive properties of fish protein hydrolysate. In *Sustainable Fish Production and Processing 2022*; Galanakis, C.M., Ed.; Academic Press: London, UK, 2022; pp. 203–225.
65. Tahergorabi, R.; Matak, K.E.; Jaczynski, J. Fish protein isolate: Development of functional foods with nutraceutical ingredients. *J. Funct. Foods* **2015**, *18*, 746–756. [CrossRef]
66. Villamil, O.; Váquiro, H.; Solanilla, J.F. Fish viscera protein hydrolysates: Production, potential applications and functional and bioactive properties. *Food Chem.* **2017**, *224*, 160–171. [CrossRef] [PubMed]
67. Wasswa, J.; Tang, J.; Gu, X. Utilization of fish processing by-products in the gelatin industry. *Food Rev. Int.* **2007**, *23*, 159–174. [CrossRef]
68. Lin, L.; Regenstein, J.M.; Lv, S.; Lu, J.; Jiang, S. An overview of gelatin derived from aquatic animals: Properties and modification. *Trends Food Sci. Technol.* **2017**, *68*, 102–112. [CrossRef]
69. Mirzapour-Kouhdasht, A.; Moosavi-Nasab, M. Shelf-life extension of whole shrimp using an active coating containing fish skin gelatin hydrolysates produced by a natural protease. *Food Sci. Nutr.* **2019**, *8*, 214–223. [CrossRef] [PubMed]
70. Ali, A.M.M.; Kishimura, H.; Benjakul, S. Physicochemical and molecular properties of gelatin from skin of golden carp (*Probarbus jullieni*) as influenced by acid pretreatment and prior-ultrasonication. *Food Hydrocoll.* **2018**, *82*, 164–172. [CrossRef]
71. Brennan, C.S. Regenerative food innovation delivering foods for the future: A viewpoint on how science and technology can aid food sustainability and nutritional well-being in the food industry. *Int. J. Food Sci. Technol.* **2024**, *59*, 1–5. [CrossRef]

Disclaimer/Publisher's Note: The statements, opinions and data contained in all publications are solely those of the individual author(s) and contributor(s) and not of MDPI and/or the editor(s). MDPI and/or the editor(s) disclaim responsibility for any injury to people or property resulting from any ideas, methods, instructions or products referred to in the content.

Review

A Conceptual Model Relationship between Industry 4.0—Food-Agriculture Nexus and Agroecosystem: A Literature Review and Knowledge Gaps

Chee Kong Yap [1,*] and Khalid Awadh Al-Mutairi [2]

[1] Department of Biology, Faculty of Science, Universiti Putra Malaysia, Serdang 43400 UFM, Selangor, Malaysia
[2] Department of Biology, Faculty of Science, University of Tabuk, Tabuk P.O. Box 741, Saudi Arabia; kmutairi@ut.edu.sa
* Correspondence: yapchee@upm.edu.my

Abstract: With the expected colonization of human daily life by artificial intelligence, including in industry productivity, the deployment of Industry 4.0 (I4) in the food agriculture industry (FAI) is expected to revolutionize and galvanize food production to increase the efficiency of the industry's production and to match, in tandem, a country's gross domestic productivity. Based on a literature review, there have been almost no direct relationships between the I4—Food-Agriculture (I4FA) Nexus and the agroecosystem. This study aimed to evaluate the state-of-the-art relationships between the I4FA Nexus and the agroecosystem and to discuss the challenges in the sustainable FAI that can be assisted by the I4 technologies. This objective was fulfilled by (a) reviewing all the relevant publications and (b) drawing a conceptual relationship between the I4FA Nexus and the agroecosystem, in which the I4FA Nexus is categorized into socio-economic and environmental (SEE) perspectives. Four points are highlighted in the present review. First, I4 technology is projected to grow in the agricultural and food sectors today and in the future. Second, food agriculture output may benefit from I4 by considering the SEE benefits. Third, implementing I4 is a challenging journey for the sustainable FAI, especially for the small to medium enterprises (SMEs). Fourth, environmental, social, and governance (ESG) principles can help to manage I4's implementation in agriculture and food. The advantages of I4 deployment include (a) social benefits like increased occupational safety, workers' health, and food quality, security, and safety; (b) economic benefits, like using sensors to reduce agricultural food production costs, and the food supply chain; and (c) environmental benefits like reducing chemical leaching and fertilizer use. However, more studies are needed to address social adaptability, trust, privacy, and economic income uncertainty, especially in SMEs or in businesses or nations with lower resources; this will require time for adaptation to make the transition away from human ecology. For agriculture to be ESG-sustainable, the deployment of I4FA could be an answer with the support of an open-minded dialogue platform with ESG-minded leaders to complement sustainable agroecosystems on a global scale.

Keywords: Industry 4.0; agriculture; food industry; social; economy; environment

Citation: Yap, C.K.; Al-Mutairi, K.A. A Conceptual Model Relationship between Industry 4.0—Food-Agriculture Nexus and Agroecosystem: A Literature Review and Knowledge Gaps. *Foods* **2024**, *13*, 150. https://doi.org/10.3390/foods13010150

Academic Editors: Rosaria Viscecchia, Francesco Bimbo and Gianluca Nardone

Received: 20 November 2023
Revised: 18 December 2023
Accepted: 19 December 2023
Published: 1 January 2024

Copyright: © 2024 by the authors. Licensee MDPI, Basel, Switzerland. This article is an open access article distributed under the terms and conditions of the Creative Commons Attribution (CC BY) license (https://creativecommons.org/licenses/by/4.0/).

1. Introduction

The topics of Agriculture-Industry 4.0 (I4) [1–21] and Foods-I4 [22–89] can be found in the literature. Additionally, precision agriculture (PA), which incorporates elements of I4, has also been widely reported [90–164]. However, there has been a lack of discussion on the direct relationships between the I4—Food-Agriculture (I4FA) Nexus and the agroecosystem.

Before this review paper discusses the above topic of concern in Section 5, the basic understanding of the agroecosystem [165–168], I4 for PA [90–164,169–172], and I4 for current and future sustainable food agriculture [20,173–175] is introduced in the following opening sections.

1.1. The Agroecosystem

An ecosystem's existence is and should be supported by nutrient cycling, both spatially and temporally. In an ecosystem, heterotrophs, autotrophs, and decomposers (microbes) recycle nutrients from nutrient pools (Figure 1). Thus, their interactions make up a basic ecosystem's functioning. Overall, the nutrient–producer–consumer–decomposer nexus can be considered the mother of sustainability. The agroecosystem is no exception to the addition of human activities.

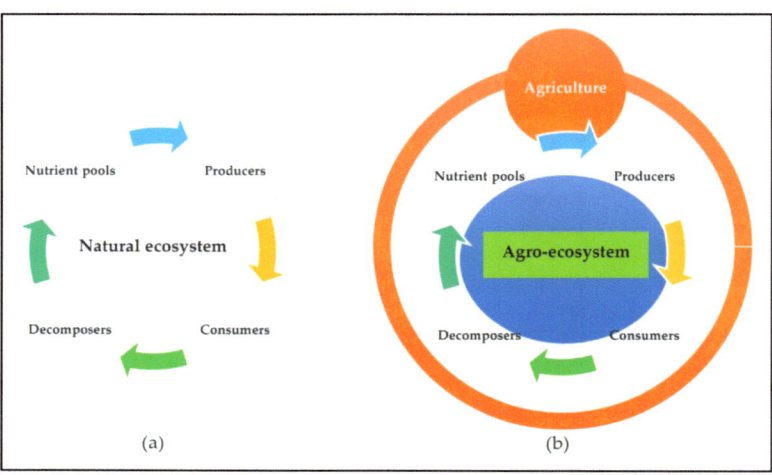

Figure 1. Four basic components between (**a**) a natural ecosystem (where there are renewable natural resources from the ecological ecosystem) and (**b**) an agroecosystem (where all nutrients and planting conditions are regulated and controlled by man) [165,166].

The factors of human control and net productivity are higher in agroecosystems when compared to those in the natural ecosystem. However, species, genetic diversity, and stability are higher in the natural ecosystem when compared to those in the agroecosystem. Furthermore, the trophic interactions and habitat heterogeneity are simple (or linear) in agroecosystems, whereas in the natural ecosystem they are complicated (or complex). The nutrient cycles are a closed system in the natural ecosystem, but the agroecosystem is an open one [165,166]. Agroecosystems are coexisting human–natural production systems that supply the rising human population's need for food, fuel, and fiber [165–168]. Agroecosystems are also frequently linked to higher nutrient input, most of which escapes from the farming area and thus causes eutrophication in nearby ecosystems that are not directly involved in agriculture. The common question is, "How can we sustain the agroecosystem to cater to increasing food demand and security?" The significant challenges now are the changes brought about by the natural and human-induced processes that impact how they can operate sustainably in a nutrient-cycling model between a natural ecosystem and a man-impacted agroecosystem (Figure 1).

1.2. Industry 4.0 for Precision Agriculture

Industrial revolutions are technological developments that impact society, development, and the environment. The steam engine and broad energy availability started the first industrial revolution; the assembly line and mass manufacturing started the second; and robots to perform work started the third [8]. The fourth industrial revolution is being discussed (Figure S1). The production systems should speak informally and make decisions based on system facts. I4 involves digitization, food supply chain (FSC) management analytics tools for monitoring, tracking, and analysis, and operational competence and efficiency. I4 accelerates assembly digitalization by employing sensors and other electronics across all assembly segments and products [15].

Sustainable business practices in energy-efficient building and smart manufacturing with low-carbon emission industrialization are supported by I4 technology [169]. Since 2020, an increasing number of articles have related I4 to sustainability [169–172]. The rising research linking I4 and sustainability shows that smart factories are built on sustainability [172]. In the future, I4 technologies will be widely used in socio-economic and environmental (SEE) sectors. This entails developing and improving innovative digital tools and instruments for the massive data collecting driven by unforeseen industry developments [1–3]. Multiple I4 sustainability functions have complicated the preceding linkages, according to Ghobakhloo [170]. I4's immediate outcomes prepare the way for its socio-environmental sustainability functions, such as increased social welfare, sustainable energy, and harmful emissions reduction. Ghobakhloo [170] also believed the digital revolution would promote sustainability. Thus, they worked together to guarantee that I4.0 effectively, reasonably, and equally fulfilled global sustainability plans.

Mobile technology significantly influences sustainability in all industries, whereas nanotechnology significantly impacts cars and electronics [171]. Technical, social, and structural development and networked and cooperative digitalization are expected in I4 [10]. High agricultural production is needed to meet the growing food demand. Cyber–physical systems (CPSs), the Internet of Services (IoS), the Internet of Things (IoT), cloud computing (CC), and big data are I4 technologies that might digitize agricultural FSC [17]. I4 is the most notable technological innovation that might assist businesses and entrepreneurs in meeting these challenges [8].

1.3. Industry 4.0 for Current and Future Sustainable Food Agriculture

The literature on sustainable agriculture can be found [90–164]. However, not all publications mention the use of I4 directly. However, the elements of I4 are proposed indirectly and are already implemented in PA's sustainability [90–164].

Farmers should expect more profits from PA technology. PA should improve society's sustainability [93]. PA is growing more popular worldwide as a dynamic manufacturing method. In assessing its environmental and economic sustainability, this approach's ability to reduce pesticide use by controlling land parcel-level pesticide application and boosting profitability and incomes was considered. PA has been linked to social collective action, but little is known regarding the actor and education roles [96].

The agriculture sustainability issues include nitrogen management. PA approaches instead of regular tillage may boost nitrogen cycle efficiency, benefiting the environment, crops, and soils [109]. Nanomaterials in agriculture are used in crop production, soil and water management, diagnostic measures, controlled chemical usage, and plant protection due to their properties, tiny size, and surface-to-volume ratio [142]. PA's usage of nanotechnology advanced with nano-based insecticides, herbicides, fertilizers, and early disease diagnoses [142]. The major method for ensuring the sustenance and economic growth of a nation is agriculture. PA's rapid advancement has helped agriculture and related industries to adapt to big data and machine intelligence. Machine learning offers useful analytical and computational approaches for integrating datasets from several sources [149].

The two fundamental agricultural concerns consist of the growing of nutritious food while lowering crop production's negative consequences on the land, water, and climate [115]. Controlling plant infections can help solve these problems since plant diseases reduce crop productivity and profitability, which feeds a large portion of the globe. New methods and technology are needed to sustain agricultural production systems and manage plant diseases [115]. PA advances greener agriculture. Many farmers have the equipment for on-site operation but rarely use it, limiting I4 utilization [119]. Sahoo et al. [152] stated that sustainable agriculture is essential to all life on Earth since the world still needs food. Sustainable agriculture involves holistic livestock, crop, and fisheries management to make farming more self-sustaining over time [152].

Sustainable agriculture using I4 has been reported since 2019 [20,173–175]. Trivelli et al. [20] suggested I4 for PA in the agri-food business. I4 technologies may help accomplish the UN SDGs and assist the agricultural FSC [74]. Unmanned vehicles detect insect migration, identify species, estimate damage, and apply pesticides on the spot for precision control in digital agriculture [175]. Smallholder farmers may benefit from the creation of a digital platform that addresses their issues throughout the farming cycle and brings all the relevant parties together at the national level to promote sustainable agriculture and cutting-edge digital technology [173]. Santiteerakul et al. [174] stated that a plant factory using intelligence technology might increase product quality, productivity, crop yield by year, food safety, resource efficiency, and staff quality of life. If the food processing business understands I4, the digital–physical framework will spur global food sector advances. This may inspire all organizations to provide innovative food and develop greater competition around the food agricultural industry (FAI) expansion [74].

According to the Literature, I4 technologies will boost agricultural output today and in the future. Hence, they should be linked. Two reasons explain the link.

First, food comes from agriculture. Studies and talks on deploying I4 in agriculture to increase food security for the growing population have been well reported [1–21]. That was a smart move. The rising demand for agricultural commodities, notably processed meals, meat, dairy, and seafood, will strain food production and delivery networks. This study examines whether the technologies that underpin these two PA paradigms are similar [20]. Digital technologies have a similar function. Agriculture and allied activities must support all human pursuits for future food security. However, population increase and resource competition continue to threaten agricultural supply networks, threatening sustainable agriculture. To address agricultural sustainability, PA and FSC coordination must improve [9]. These issues are becoming increasingly sophisticated in agricultural supply networks and production systems. I4 for agriculture is likely to be the answer.

Second, I4 and related technologies might make food agriculture firms more competitive in the digital age [4]. Agriculture 4.0 is typified by the growing use of digital technology in food [5]. Agriculture and livestock are vital to social and economic stability. FSC management benefits from increased visibility, provenance, digitalization, disintermediation, and smart contracts [62]. Prasad et al. [3] reported that the IoT links many items, technologies, and devices in a network to speed up processes, eliminate information loss, and enable device–cloud/device–device communication. The fundamental question is how IoT will assist food and agriculture. I4 smart agriculture uses IoT in urban and rural areas [4].

As there is a lack of study on the direct relationships between the I4FA Nexus and the agroecosystem, the objectives of this study are (a) to evaluate the state-of-the-art relationships between the I4FA Nexus and the agroecosystem and (b) to discuss the challenges and knowledge gaps in the sustainable FAI that the I4 technologies can assist. The purpose is fulfilled by (a) reviewing all the relevant publications from the Scopus database and (b) drawing a conceptual relationship between the I4FA Nexus and the agroecosystem, in which the I4FA Nexus is categorized into economic, societal, and environmental perspectives.

2. Methodology
Literature Collection

Instead of a wide standard literature review, a systematic literature review (SLR) is more suitable. Thus, in the current review study, the SLR technique of Preferred Reporting Items for Systematic Reviews and Meta-Analyses (PRISMA) by Moher et al. [176] was employed to add to the body of information already available in "Industry 4.0" and "Food". The evidence-based reporting standard of PRISMA is helpful for critical evaluation. Overall, Figure 2 depicts the systematic process stages that were modified for this review paper. As Elsevier's Scopus is the world's largest abstract and citation database of peer-reviewed scientific literature journals, books, and conference proceedings and covers research topics across all scientific, technical, and medical disciplines [177], it was chosen for the literature analysis in the present study.

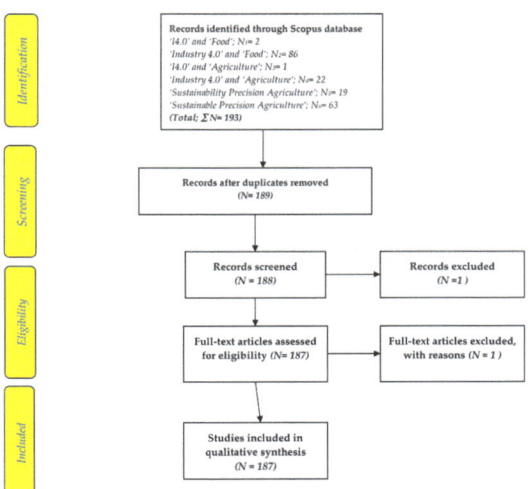

Figure 2. Flowchart of Preferred Reporting Items for Systematic Reviews and Meta-Analyses (PRISMA) used in the present study, adapted from Moher et al. [176].

This study assessed the scholarly distributions on "I4.0" and "Food" found in the Scopus bibliographic database, which was chosen for its size and the variety of its distributions. On 10 December 2023, by using the keywords 'I4.0' or 'Industry 4.0' and 'Food', which must be found in the title of the papers under the Scopus database, a total of 88 papers arrived. After removing 4 duplicated papers and 1 irrelevant paper, a total of 84 papers from the Scopus database were found. With the keywords 'Industry 4.0', or 'I4.0' and 'Agriculture', a total of 23 papers were found (Figure 2).

In addition, the topics on 'sustainability (or sustainable) precision agriculture', which had to appear in the article title, were found in 82 papers, of which 19 papers had the keywords 'sustainability precision agriculture', and 63 papers had 'sustainable precision agriculture' (one was discarded due to its being a 'correction' article). Therefore, a total of 187 papers are included in the present review study (Figure 2).

Bibliometric analyses are an established method to evaluate research literature, particularly in the scientific fields that benefit from computational data treatment and that have witnessed increased scholarly output [178]. VOSviewer (version 1.6.20) is software that generates a clear graphical representation of bibliometric maps, especially for extensive datasets [179]. To highlight the trends of the studies conducted on the topic of 'Industry 4.0' and 'Food' from 2016 to 2024 (on 111 papers from the Scopus database), we performed a bibliometric analysis using the VOSviewer software (VOS stands for visualization of similarities—see www.vosviewer.com; accessed on 5 December 2023). Separately, other visualizations were performed based on 'sustainability (or sustainable) precision agriculture' from 1995 to 2023 (on 82 papers from the Scopus database). Scopus comprises many significant research papers and offers integrated analysis tools for creating informative visual representations [177]. VOSviewer was employed to analyze each keyword, calculating links, total link strengths, and co-occurrences with other keywords.

3. Results

The studies and discussions on the use of I4 in the food sectors in particular have been reported in the literature [22–89] (with a total of 67 papers) (Figure 2) and represent significantly more in terms of the number of publications than 'I4' plus 'agriculture' per se [1–21] (with a total of 21 papers). This is because food items are part and parcel of the human needs and life requirements that are necessary for the continued survival of humankind. At the same time, agriculture is the center of activities where human food is provided. There are 14 countries/regions (Table S1) selected based on the relevancy of adopting I4 into the FAI, based on a literature search on the keywords 'I4' or 'Industry 4.0' and 'Food' found in different regions or countries.

After carefully examining each paper, the reviewed articles can be specifically categorized based on the focus of the studies/review. The order of the decreasing number of the categories is socio-economy > SEE > social > sustainability > economy > socio-environment > economic environment > environment [1]. This indicates that the reviewed papers with the I4FA Nexus are mainly concerned with SEE and sustainability. The following discussion is therefore weighted on the social, economic, and environmental categories under the I4FA Nexus.

Using VOSviewer software, visualizations of the paper network-based data confirmed the main themes of research based on the documents and sources using clustering patterns, which are presented in Figures 3–6.

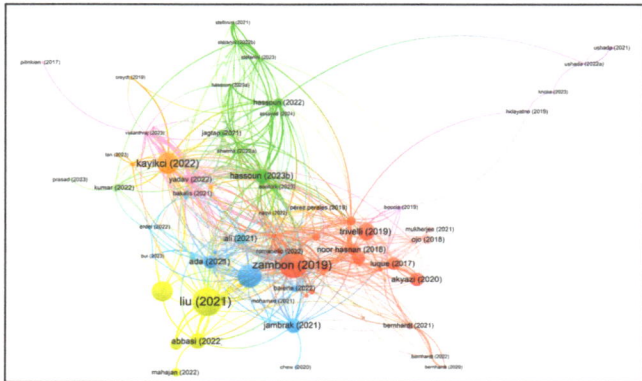

Figure 3. A bibliometric analysis of research that created a visualization of the paper network confirmed the main themes of research based on 84 papers (documents) (out of 111 papers) with 11 clusters. The literature is based on the Scopus database; the keywords 'I4.0' or 'Industry 4.0' and 'Food', which had to appear in the article title, were found in 111 papers. The papers ranged from 2016 to 2024 and 2019 to 2023.

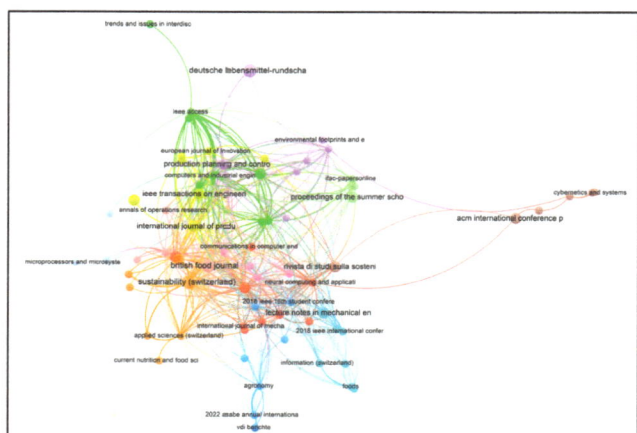

Figure 4. A bibliometric analysis of research that created a visualization of the paper network confirming the main themes of research, based on 72 journals (sources) with 12 clusters. The literature is based on the Scopus database; the keywords 'I4.0' or 'Industry 4.0' and 'Food', which had to appear in the article title, were found in 111 papers. The papers ranged from 2016 to 2024 and 2019 to 2023.

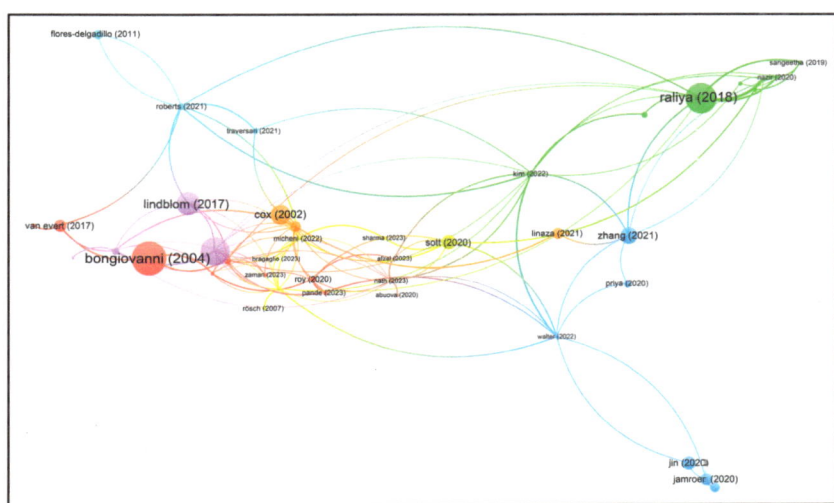

Figure 5. A bibliometric analysis of research created a visualization of the paper network confirming the main research themes based on 43 papers (documents) with 8 clusters. The literature is based on the Scopus database on the topics of 'sustainability (or sustainable) precision agriculture', which had to appear in the article title; 82 papers were found. The papers ranged from 1995 to 2023 and 2002 to 2023.

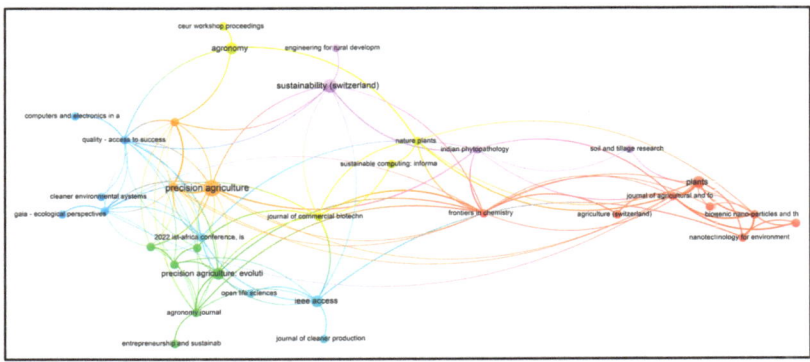

Figure 6. A bibliometric analysis of research created a visualization of the paper network confirming the main research themes based on 35 sources (journals) with 7 clusters. The literature is based on the Scopus database on the topics of 'sustainability (or sustainable) precision agriculture', which had to appear in the article title; 82 papers were found. The papers ranged from 1995 to 2023 and 2002 to 2023.

Based on the keywords 'I4' and 'Food', the authors have mainly published their papers since 2018, according to the visualization (Figure 3). This indicates increasing numbers of papers, sometimes with similar authors or co-authors, specializing in similar topics to satisfy the current and future knowledge needs regarding I4 and food.

From Figure 4, the visualization shows that at least 72 different journals have been published on the topics of 'I4' and 'Food' since 2016. The journals include *Sustainability* (Basel), *British Food Journal*, *Applied Sciences*, *Information*, *ACM International Conference Proceeding Series*, *Advances in Intelligent Systems and Computing*, *Deutsche Lebensmittel-Rundschau*, *E3S Web of Conferences*, *Engineering Proceedings*, and others, as shown in Table S2.

Based on the keywords 'sustainability (or sustainable) precision agriculture', the authors have mainly published their papers since 2002, according to the visualization (Figure 5). This indicates increasing numbers of papers, sometimes with similar authors or co-authors, specializing in similar topics to satisfy the current and future knowledge needs regarding sustainable agriculture.

In Figure 6, the visualization shows that at least 35 different journals have been published on the topics of 'sustainability (or sustainable) precision agriculture' since 2002. The journals include *Agriculture* (Switzerland), *Agronomy Journal*, *American Journal of Alternative Agriculture*, *Biochemical Systematics and Ecology*, *Biomaterials Advances*, *Biosystems Engineering*, and others, as shown in Table S3.

4. Discussion

The following discussions will focus on the four major observations based on the literature reviewed in the present study. They are: Section 4.1. The adoption of I4 in the agriculture and food sectors has been constantly growing since 2011 and is expected to increase in the future; Section 4.2. Good prospects for the I4 implementation into food agricultural production; Section 4.3. The challenges of the sustainable agricultural food industry in adopting Industry 4.0; and Section 4.4. The knowledge gaps for future studies.

4.1. The Adoption of I4 in the Agriculture and Food Sectors Has Been Constantly Growing since 2011 and Is Expected to Increase in the Future

This is well supported by the literature review from three critical points, namely: (a) the expected higher number of publications on the topics of I4 food agricultural production in the future; (b) the social perspective on the growth of the human population; and (c) the number of countries that have started using I4 in their food agricultural industries.

4.1.1. Expectedly Higher Number of Publications on the Topics of I4 Food Agricultural Production in the Future

It is expected that a higher number of publications started earlier on the topic of I4 (since 2012) (Figure 7a) than on the topic of I4FA (since 2016) (Figure 7b). It is expected that the number of publications on I4 will increase to over 30,000 papers by 2065 (Figure 7a), while that on I4FA will reach over 800 papers by 2065 (Figure 7b). This is logically acceptable, considering that the I4 topic covers all study disciplines, ranging from sea to land to outer space. It is interesting to see that a higher number of publications started earlier on the topic of PA (since 1982) (Figure 7c) than on the topic of sustainable (or sustainability) PA (since 1995) (Figure 7d). It is expected that the number of publications on PA will increase to over 1500 papers by 2065 (Figure 7c), while that on sustainable (or sustainability) PA will reach over 100 papers by 2065 (Figure 7d). Notably, when the keyword 'Sustainable' (or 'Sustainability') is included in the topic of a scientific paper, the specialization of a niche discipline of a reach study is triggered, with ample potential for research topics and opportunities. Moreover, the UNSDG timeframe is only until 2030, but the sustainability effort in SEE is continuous. The UNSGDs' 2030 deadline needs an extension to an unlimited time frame when climate change is taken into consideration.

These positive increasing trends are in line with those of (a) 'Population size' (Pop), (b) 'carbon dioxide emission' (CO_2), and (c) 'Energy per capita' from the same periods, based on data cited from the OurWorldInData.org, as shown Figure 8. This expected higher number of publications on the topics of I4 food agricultural production in the future contributes to the massive paradigm shift of I4 implementation in the future of food agricultural production, which is discussed in terms of the social aspect in Section 4.1.2 below.

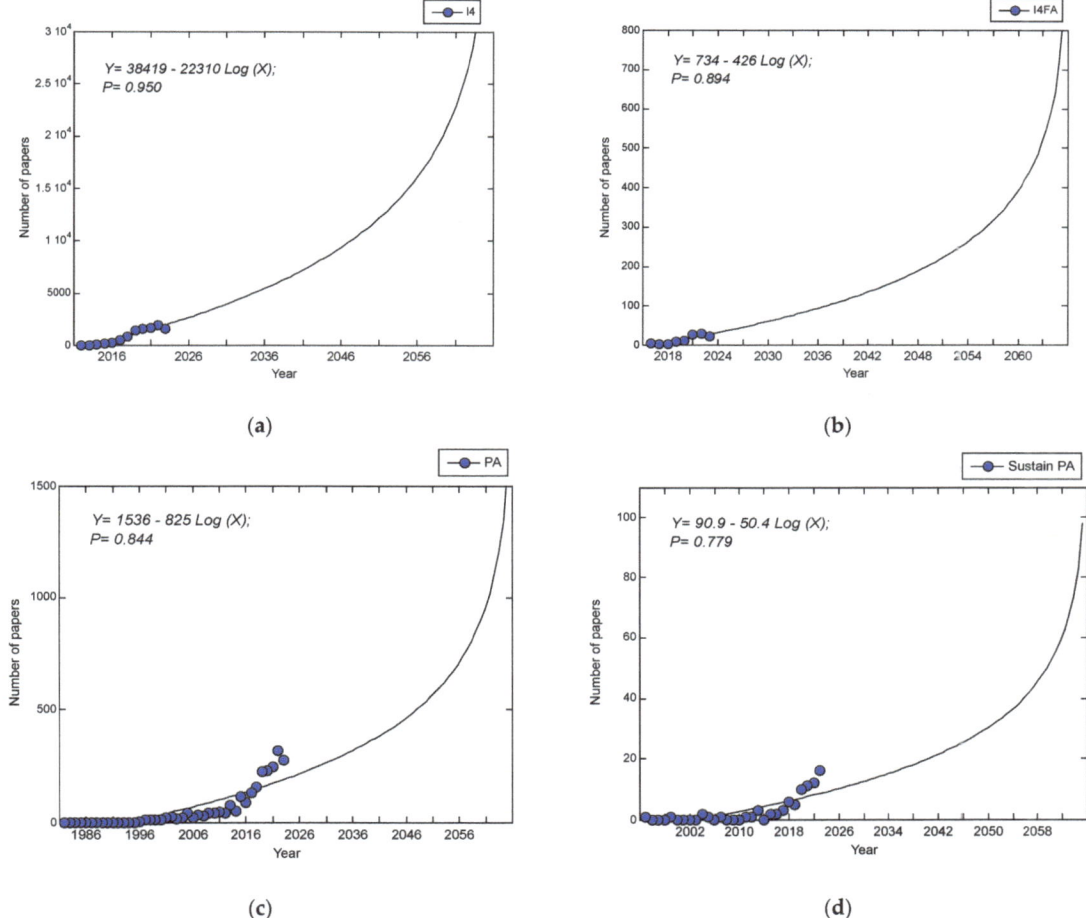

Figure 7. Numbers of papers with keywords of (**a**) 'Industry 4.0' (I4; from 2012 to 2023); (**b**) 'Industry 4.0 Food Agriculture' (I4FA; from 2016 to 2023); (**c**) 'Precision Agriculture' (PA; from 1982 to 2023); (**d**) 'Sustainable (or sustainability) Precision Agriculture (Sustain PA; from 1995 to 2023), based on Scopus database. In addition, the number of papers is extrapolated to 2065 using logarithmic equations for the four graphs. Note: 1 10^4 indicates 1×10^4; similarly applying to others.

4.1.2. Social Perspective on the Increment of the Human Population

The positive increasing trend could be explained from a social perspective. There have been increasing numbers of countries employing I4 in their food agricultural activities to cost-effectively supply the increasing demand for food among their increasing population sizes (Figure 8a). The following papers [22–85] indicate the connectivity between food and I4. However, the following discussion is focused on the close connection between food agriculture and I4 from a socio-economic perspective.

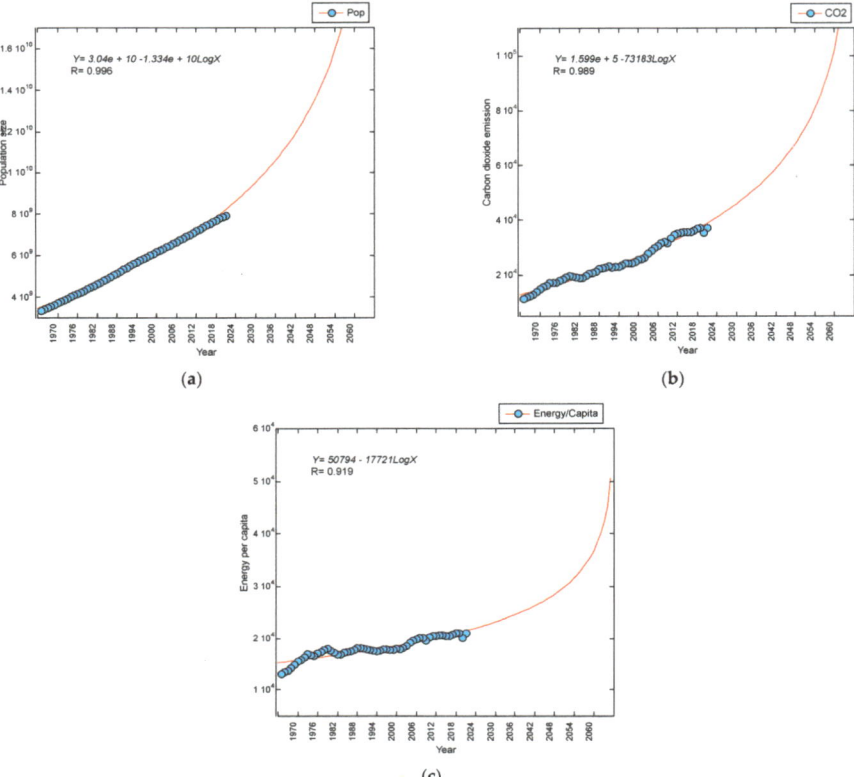

Figure 8. Increasing trends from 1965 to October 2023 for (**a**) 'Population size' (Pop), (**b**) 'carbon dioxide emission' (CO_2), and (**c**) 'Energy per capita' from 1965 to October 2023, based on data cited from the OurWorldInData.org. In addition, their increasing trends are extrapolated to 2065 using logarithmic equations for the three graphs. Note: Annual total production-based carbon dioxide (CO_2) emissions, excluding land-use change, measured in million tonnes. Primary energy consumption per capita (energy per capita), measured in kilowatt-hours per person per year. $1\ 10^4$ indicates 1×10^4; similarly applying to others.

Many such studies have been reported in the literature [35,71]. Hidayatno et al. [71] found that financial benefit increases I4 adoption in Indonesia's food and beverage sector because an FSC utilizing an I4 innovation will determine the economy's management capabilities. Kumar et al. [38] found 12 CE-related barriers to I4 implementation in SFSC. Cause–effect analysis and obstacle prominence evaluation were performed using Rough-DEMATEL. Managers, practitioners, and planners can benefit from knowing and overcoming the study's findings.

Akyazi et al. [63] offered an industry-driven proactive plan for the food sector's digital transformation. To achieve this purpose, they established the essential competencies and abilities needed for each food business's professional profile. To achieve this, they established an automated database of current and prospective careers, skills, and talents. This database might guide the industry through I4's revisions [63]. Academics and politicians believe that the I4FA Nexus supports the ecosystem's societal growth. Kayikci et al. [41] established a blockchain-enabled FSC architecture, covering prospects and current barriers, based on a thorough literature analysis and semi-structured case interviews from emerging economies. They examined whether blockchain technology can solve FAI challenges, including traceability, trust, and accountability. Their work paved the way for future academics to address technological and human difficulties in the I4 age to lessen food business challenges. They gave instances of blockchain technology in I4, prompting more research and warning of the potential risks. The I4FA Nexus may cover and advance the ecosystem's social (trust) progress.

Enarevba et al. [51] investigated combining Lean Six Sigma with I4 technologies in sub-Saharan Africa to decrease pre- and post-harvest food waste. The UN predicts a 33% worldwide population increase by 2050 and a 99% increase in sub-Saharan Africa. These expected trends will raise food security concerns, with sub-Saharan Africa facing the greatest demand growth. This I4FA Nexus covers the ecosystem's socio-economic development. This I4 strategy is ideal for the Barranquilla food business since it meets logistics needs like FSC transparency and integrity management.

4.1.3. Many Countries Have Started I4 in Their Food Agricultural Industries

Those countries that have started implementing I4 elements into their food agricultural production (in the phases from planting to marketing) included India [13,16,17,31], Russia [12,64], the UK [67,87], Italy [74], Australia [28,57], Indonesia [71], the European Union (EU) [79], Malaysia [88], Poland [42], Spain [44], Poland and Israel [19], Moldova [89], China [27], and the United Arab Emirates (UAE) [85] (Figures 9 and 10).

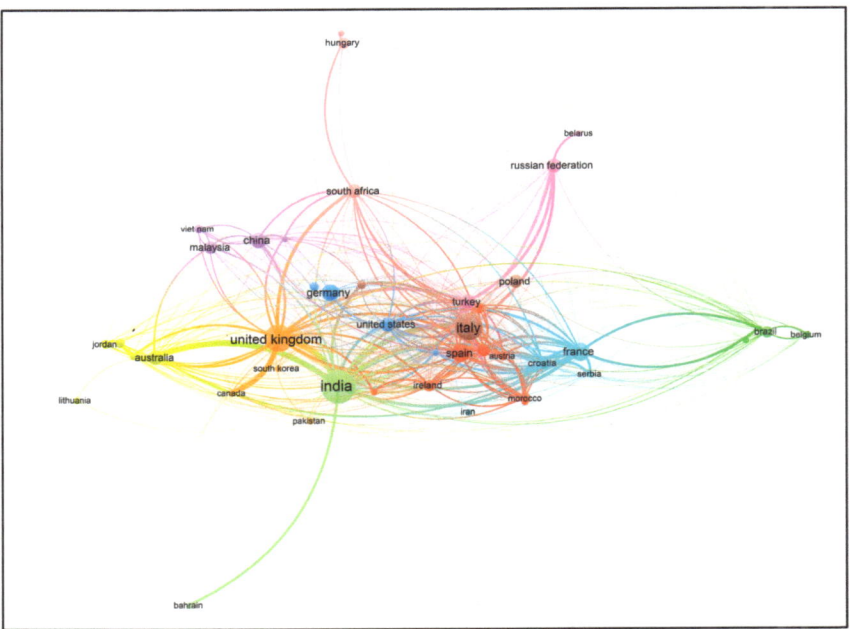

Figure 9. A bibliometric analysis of research created a visualization of the paper network, confirming the main themes of research, based on 49 countries with 11 clusters. The literature is based on the Scopus database; the keywords were 'I4.0' or 'Industry 4.0' and 'Food' and had to appear in the article title; 111 papers were found. The papers ranged from 2016 to 2024 and 2019 to 2023.

Using VOSviewer software (version 1.6.20), the visualizations of the paper network-based data by clustering patterns confirmed the main themes of research based on countries; these visualizations are presented in Figures 9 and 10.

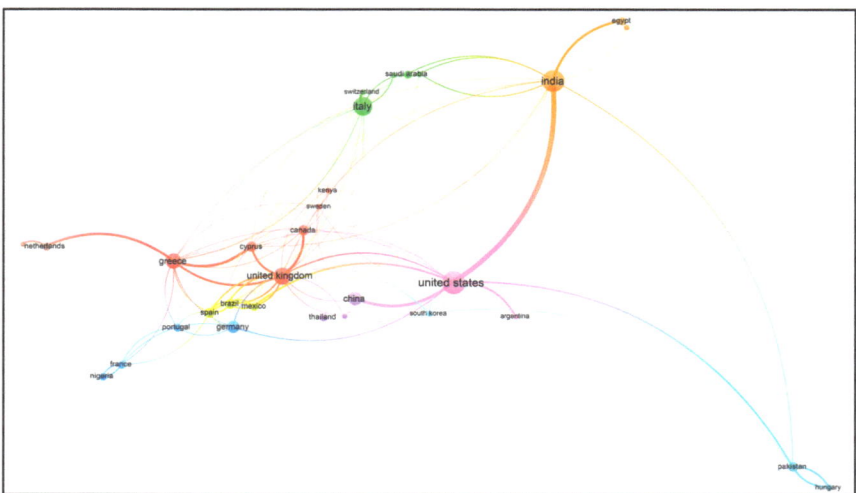

Figure 10. A bibliometric analysis of research created a visualization of the paper network, confirming the main themes of research, based on 36 countries with 9 clusters. The literature is based on the Scopus database on the topics of 'sustainability (or sustainable) precision agriculture', which had to appear in the article title; 82 papers were found. The papers ranged from 1995 to 2023 and 2002 to 2023.

Italy

Boccia et al. [74] recognized the potential of new technologies for food firms and their sustainability and management. The concept of the I4FA Nexus is thought to cover and support the ecosystem's social development. In this Italian case, the function of I4 was to advance assistance towards the flexible chaining of the boards in worldwide frameworks and to perceive the possibilities of new advances for food organizations, their manageability, and their executives [74]. Romanello and Veglio [35] investigated the causes and effects of adopting I4 technologies in the context of an Italian food processing business. Their study emphasized the factors that influence and provide obstacles to the adoption of various I4 technologies. This I4FA Nexus is thought to encompass and contribute to the ecosystem's social advancement. The effects of PA on nitrogen management were examined by Marinello et al. [109], who considered a 52 ha experimental site at a private farm in a typical Po Valley field in northeastern Italy. Using sustainable PA, they identified the crucial plantation area of corn (*Zea mays*), which can aid in defining the corrective measures that should be taken to lessen and minimize the effects of agriculture on the environment.

India

The I4 and circular economy (CE) adoption hurdles in the Indian agriculture FSC have also been highlighted by Kumar et al. [13]. They stated substantial barriers to implementing the I4-CE model, including a lack of government backing and incentives, regulations, and procedures. They reported that the stakeholders in the FSC will benefit from the research findings as they plan the strategic deployment of I4-CE. Arora [16] assessed the areas of I4 applicability to the Indian agricultural industry and how considerable advantages may be given to the farmers. Using the Delphi approach, the list of these digital use cases was then improved and prioritized while considering the use cases' economic importance to Indian farmers and the simplicity of their implementation. A framework for cyber–physical agricultural systems (CPASs), which intelligently integrates CPS, IoT, CC, and big data with agricultural systems, was suggested by Sharma et al. [17]. CPAS may be used to increase productivity and leverage agricultural supply networks. The IoT-based global agricultural production management and control system, according to the I4 idea, was introduced by Szewczyk et al. [19].

Chatterjee et al. [31] studied how food and beverage firms will be affected by digital transformation in the employment of I4 technology in India's post-COVID-19 era. As part of their path towards digital transformation in the post-COVID-19 situation, they reported that there was a substantial market for the food and beverage industries employing I4 technology.

The UK

The US and Europe have certain cultural commonalities and some cultural distinctions. For European farmers, one issue is the safeguarding of data as they are sent across platforms. Farmers in the US are less critical of this. The direct impacts of I4 technological capabilities (I4TC) and FSC integration on the efficiency of the sustainable agriculture FSC were studied by Sharma et al. [9]. Through the research of UK-based food and beverage producers, Kobnick et al. [67] have shown how deploying I4 activities is mostly tactical and hence divorced from the enterprises' business models. They found that businesses must constantly develop their business models to apply I4. This I4FA Nexus is thought to cover and support the ecosystem's social (manufacturing) advancement.

The digitalization of assembly through I4 operations, according to Koebnick and McFarlane [87], will impact all businesses, including the food and beverage industry. The analysis of UK-based food and beverage companies used in this article demonstrates that the usage of I4 exercises is typically strategic and, as a result, detached from the organizations' action plans. This results from a lack of critical thinking about how I4 will affect their entire organization and about the prevalence of proficiency-arranged corporate societies.

Russia

The investigation of I4 principles was the focus of a study by Filatov et al. [64] to boost the competitiveness of the Russian Federation's food and processing sector. The new revolution's major challenge is less related to the technology than it is to the knowledge and training required to employ it. The extra benefits of multidisciplinary research and development are overlooked since the development of the new industrial revolution's separate components is unpredictable. The social environment will alter significantly when production enters a new phase [64]. According to this I4FA Nexus, the ecosystem's socio-economic development is covered. Aleksandrov et al. [12] analyzed digital transformations in agro-industrial complexes and identified the potential and risks for long-term socio-economic growth. They considered business cases for the effective digitization of agriculture by evaluating the economic impacts of digital technologies.

Australia

Ali and Aboelmaged [28] examined the perceived motivations and challenges associated with adopting FSC I4 in the food and beverage sector through interviews with top managers from the Australian beverage and FSC. They found that the key drivers for implementing FSC I4 are decreasing consumer needs, supply–demand misalignment, cost optimization, and the threat of legal penalties. They presented a fresh approach to qualitative data analysis that advances the field of FSC management's methodology. This I4FA Nexus is thought to cover and contribute to the ecosystem's socio-economic development (more job possibilities). Ali et al. [57] used a sample of 302 replies from senior managers in the Australian food processing sector as the basis for empirical testing. They discovered that the leading causes of FSC disruptions are supply–demand mismatches, process risks, and transportation risks. They alerted management to the adverse effects of FSC interruptions and the need for I4 technology to overcome the challenges. According to this I4FA Nexus, the ecosystem's socio-economic development is covered.

Other Countries

Hidayatno et al. [71] conceptualized a systemic connection structure that could highlight the interactions between the policies and important factors influencing the adoption of technology 4.0 in the Indonesian food and beverage sector. Ushada et al. [33] simulated the trust-based decision-making process used by Indonesian small and medium-sized enterprise (SME) groups while adopting I4, namely ergonomic machines and e-commerce technologies. They used the Java, Sumatera, and Nusa Tenggara groups to develop the best trust and decision-making approaches. Ichsan et al. [72] showed the situation of the food and beverage manufacturing industry in Indonesia and the framework for implementing digital transformation in the direction of I4. According to this I4FA Nexus, the ecosystem's socio-economic development is covered.

Oltra-Mestre et al. [44] studied the impact of I4 as a set of enabling technologies related to the core process innovation practice and product innovation of the FAI. They offered case studies of two Spanish businesses that processed fresh foods and competed in the meat, fruit, and vegetable industries, which are two significant industrial subsectors. This offered a framework for understanding how I4 technologies help researchers and management achieve competitive results by facilitating key innovation processes. According to this I4FA Nexus, the ecosystem's socio-economic development is covered.

Pilinkien et al. [79] created a case study of the EU food sector by simulating several logistic network scenarios. They designed a competitiveness strategy based on the I4 idea and the lean philosophy. They demonstrated that a sustainable FSC, with minimal management costs and the visibility of the entire food chain, can be accomplished by deploying a logistic cluster in the EU and employing the devised competitiveness strategy. This I4FA Nexus is thought to cover and support the ecosystem's social development.

In Poland, Kafel et al. [42] examined the official information on I4 and the digitalization elements offered by the Polish food organizations in response to I4 operations. Microsoft Excel forms were used to create charts utilizing the retrieved data. The data were then examined using both quantitative and qualitative content analysis. They found that activities carried out by Polish food organizations listed on the stock market increasingly showed signs of I4 and digitalization. Because of this, the top management boards are more confident and more interested in modernizing their I4-based operations. The ecosystem's sociological (food organizations) and economic (stock exchange) advancement is thought to be covered and contributed to by this I4FA Nexus.

To propel its socio-economic progress and to achieve high-income nation status, Malaysia quickly grasped the reception of I4, according to Bujang and Abu Bakar [88]. Food and agribusiness production were identified as key factors in achieving this. The company as a whole, as well as the Andalusian food sector in particular, must implement the method suggested by I4. According to Luque et al. [80], it should be seen as an unusual advancement opportunity for the area. It is expected that, along with other industries, the food and beverage sector will embrace flexible and individualized manufacturing techniques [80].

Using an IoT-based approach, the broad rural creation of the board and control, as being necessary to the I4 notion, was proposed by Szewczyk et al. [19]. The four levels of the proposed framework—choice assistance, information handling, information collecting and transmission, and sensors—will be tested in Poland and Israel. An effective and efficient information procurement layer is essential for activities to succeed in the nation's territory. Perciun et al. [89] analyzed the idea of I4 in agriculture through the current national and international expertise in digital technologies in Moldova. To assess the technical, human, and financial feasibility of using digital technologies to simplify agricultural firms' management and to assure the sustainable growth of the national economy, they examined the situation of the Moldovan agro-industrial sector. They used digital technology to identify viable areas for agricultural growth and to assess the potential impact of their adoption on the production cycle and on raising the quality and competitiveness of domestic farm goods.

In China, the research of Sun et al. [27] found that the IoT significantly improved the activities of the CE. The practices of the CE included circular design, green manufacturing, re-manufacturing, and recycling. These environmentally friendly business practices complemented the company's efforts to improve its environmental performance while boosting its economic performance. In the UAE, Kurdi et al. [85] empirically evaluated the effects of FSC I4 and FSC risk on organizational performance in the food manufacturing sector. They concluded that for food manufacturing enterprises to be competitive, efficient, and productive, they should start and develop their transition to FSC I4.

Therefore, all of the above literature reviewed points related to the fact that deploying I4 in the FAI in many countries is ever-expanding, now and in the future.

4.2. Good Prospects for the I4 Implementation in Food Agricultural Production

There are good prospects for the I4 technologies in PA for food agricultural production; these may be considered by looking at the social, economic, and environmental benefits.

4.2.1. Social Benefits

Based on the literature review, the significant social benefits were (a) increased occupational safety and workers' health and (b) increased food quality, security, and safety.

Increased Occupational Safety and Workers' Health

Many studies indicated that the use of I4 in PA has benefited the food and agricultural sectors in terms of increased occupational safety and workers' health [1,2,25,27,29,44,46,56,152,180–182].

When employed as seed priming agents, nanoparticles increase the seed germination rate, which benefits the plant's overall development. Using insecticides and fertilizers with nanocapsules has revolutionized agricultural and animal health without harming the environment. The application of nanotechnology can effectively integrate various agricultural practices with sustainable production. Despite the various potential advantages of nanotechnology, it is crucial to consider the environmental safety risks carefully. Nanotechnology enhances their performance and sufficiency by boosting viability and security and reducing social insurance costs [152].

The potential of I4 in agriculture was previously covered by Knoke et al. [2]. Agritech Business 4.0 was updated using I4 technology by Sivakumar et al. [1] in 2021. I4 technologies are aligned with Agritech Business 4.0's core components, including crop management, soil management, pest and disease management, water conservation, protection of farmers' health, increased productivity, food safety, and FSC and the bolstering of the ties between urban and rural areas.

The main reason why human food is connected to I4 is because I4 is expected to assist and complement the FSC, food security, and food sustainability. Due to the fast-paced corporate climate, technology improvements, client preferences, growing competitive pressure, globalization of FSC, and environmental disturbances, the globe is witnessing technological disruptions. Digitalization initiatives have been increasing in the agri-food sector [29]. They must adopt new technology to ensure efficient and effective administration of their responsibilities. Although I4 technologies can provide chances for process innovation, how they affect innovation practices in the FAI needs more research output [44], which has been heavily challenged by climate change and population expansion [56].

To create smart factories with a strong focus on sustainability, Jambrak et al. [46] highlighted the need to consider the implementation of smart sensors, artificial intelligence (AI), big data, and additive technologies with nonthermal technologies. SWOT analysis revealed the potential for energy savings during food processing, optimized overall environmental performance, reduced manufacturing costs, the production of eco-friendly goods, improved working conditions, and a greater degree of health and safety during food processing. Advanced thermal and nonthermal technologies can be sustainable methods that comply with the United Nations Sustainable Development Goals (UN-SDGs). According to this I4FA

Nexus, the SEE development of the ecosystem is covered. According to Senturk et al. [182], they employed a variety of digital devices in our daily lives, and these changes have been rather drastic. The usage of these technologies in diverse applications has recently been investigated in the agricultural and food industries. They suggested using these technologies, particularly IoT-based systems, to address the industry's longstanding issues with food safety, mycotoxin contamination, pesticide residues, and growing waste. Sun et al. [27] stated that the IoT significantly enhanced CE activities, practices, and policies. They also significantly enhanced green manufacturing, circular design, remanufacturing, and recycling practices. An improvement in environmental performance can significantly impact a company's success. By integrating IoT-based I4 technology into CE practices, their [27] research offered the framework for contributing nations/companies to achieve economic and long-term sustainability goals simultaneously. This I4FA Nexus is thought to cover and contribute to the ecosystem's advancement on the economic and environmental fronts.

To identify the existence of abnormalities in the operation of industrial systems, Tancredi et al. [25] presented a structured approach that combines digital twin models, machine learning algorithms, and I4's IoT. The suggested remedy has been created to be implementable in manufacturing facilities and is not explicitly intended for I4 applications. They found that two of the three machine learning algorithms were shown to be sufficiently successful in forecasting abnormalities [25] and recommended their deployment for the boosting of worker safety at industrial facilities. This I4FA Nexus is thought to cover and contribute to societal advancement of the environment (and the employees' safety).

Increased Food Quality, Security and Safety

Many studies indicated that using I4 in PA has benefitted the food and agricultural sectors by increasing food quality, security, and safety [22,98,116,154,161].

Yadav et al. [22] examined these important agriculture FSC technologies by considering five research axes: information system management, traceability and food safety, food waste, decision making and agribusiness control and monitoring, and other ad hoc applications. They proposed that integrating the technologies they had evaluated might be more beneficial for offering affordable solutions and enhancing sustainability in agriculture FSC. Additionally, blockchain has the potential to revolutionize how food security and safety are achieved. This I4FA Nexus is thought to aid in the ecosystem's socio-economic development. Regarding product quality, environmental concerns, and the welfare of humans and cattle, Cox [98] examined the technological advancements that boost agricultural and livestock output worldwide. They examined the methods for obtaining, using, and disseminating the necessary information. These stages are associated with the PA idea, which generally applies to crop and livestock production.

Preserving and responsibly using arable land resources are essential to ensuring global food security. Soil resources are under tremendous strain due to competition for land use from urbanization and commercial land use [116]. Land erosion and desertification are already causing the world's arable land to decline, and our attempts to guarantee commercial land availability are worsening the situation. In addition to ensuring that the land is used as efficiently as possible, PA can improve the possibility of the global agriculture sectors being restored. One way to think of integrated nutrient and pest management is as future-proof land and water conservation, along with zero tillage, organic farming, and vertical planting [116].

According to Zhang et al. [154], global food security is being threatened by climate change, population growth, conflicting needs for land to develop biofuels, and deteriorating soil quality. There are excellent prospects for sustainable food production because of the convergence of PA, where farmers use artificial intelligence and nanotechnology to react in real time to changes in crop development [154]. To optimize targeting, uptake, delivery, nutrient capture, and the long-term impacts on soil microbial communities, it is possible to design nanoscale agrochemicals that combine optimal and functionality profiles by combining current nutrient cycling and crop productivity models with nano informatics approaches [154].

Food markets have been more globalized in recent decades as a result of trade agreements that have reduced protectionist laws, according to Saeys and De Baerdemaeker [161]. While this has made a more fantastic range of food goods more accessible to customers at lower costs, governments, merchants, and consumers worldwide increasingly worry about the safety and quality of their food products. PA technology can assist producers in meeting good agricultural practice standards and can relieve them of the administrative burden associated with demonstrating compliance, in addition to giving governments, merchants, and consumers the information they need to ensure food quality and safety [161].

4.2.2. Economic Benefits

Based on all the literature reviews, the major economic benefits were (a) the use of sensors (IoT) to reduce the costs of agricultural production; (b) the reduction in costs via the FSC; and (c) the reduction in the costs of food production using the green technology of I4.

Use of Sensors (IoT) to Reduce Costs of Agricultural Production

Many studies considered the use of sensors (IoT) to reduce the costs of agricultural production [14,15,97,99,128,135,144,146,149,150,155,162,164].

By approaching sustainable intensification in agriculture to strike a balance between environmental stewardship and agricultural yield, PA has grown in popularity. Improving agricultural output while reducing adverse environmental effects is the goal of sustainable intensification. Using cutting-edge technology like IoT, GPS, GIS, sensors, drones, and machine learning has made it possible to complement this. This technology allows farmers to cultivate their land more precisely and efficiently [99].

To enhance the information layer and communication processes in the I4 architecture, Manogaran et al. [14] developed an information scheduling and optimization framework. Through the use of this framework, process delay and stagnancy are reduced through the best possible scheduling and classification of agricultural information A smart farm's control flexibility is calculated using the latency and stagnancy towards the end of yields. The classification component sorts data based on processing and completion times using offloading to remove backlogs. Mukherjee et al. [15] addressed the impact of I4 on the agricultural FSC. I4 examined how the agriculture FSC may benefit, and it completed a thorough examination of the literature. The agriculture FSC industry is one of them. It also shows how I4 in FSC management for agriculture may be applied to boost productivity, customer satisfaction, and efficiency. Their study may help forecast the future interactions between I4 and agriculture FSC management, bringing I4 and Agriculture 4.0 together.

PA makes a significant contribution to sustainable agriculture [144]. Technological advancements were the foundation for the multidisciplinary conversation and the creation of these novel approaches. PA became conceivable with the global positioning system and the new sensor systems made available by information technology. Farm automation, site-specific farming, fleet management, and field robots are all made feasible by the applications of these technologies. This can be carried out by optimizing farm, plant, machine, and job management [144]. Spatial planning for agriculture growth can be aided by implementing web-based information systems, an essential component of IoT technology [146].

According to Patel et al. [162], the Indian agricultural sector has significant challenges in achieving food and environmental security in the new millennium, as indicated by the country's rapidly growing population and diminishing production. In terms of improving the land's carrying capacity sustainably, PA technologies may be the best choice [162]. The arrival of new ICT technologies within the broader IoT framework has made PA adoption possible [128]. The design of networks for PA can be supported by formal software engineering models and procedures, according to Bodei et al. [128].

The PA technique maximizes the production of high-quality crops by monitoring the environment and field conditions while reducing environmental pollution with little input (e.g., fertilizer, herbicides, and pesticides) [135]. However, a fundamental barrier to the widespread adoption of PA is still the absence of data; these data are a crucial aspect of the achievement of PA. Additionally, Kim and Lee thoroughly examined and described electrochemical sensors—such as those that track soil, plant development, and environmental factors [135].

To understand machine learning's use in agriculture, Priya et al. [149] offered the fundamental idea of the technology as well as the systematic procedures. Sangeetha et al. [150] stated that nanoparticles are a potential medium for drug administration because of their ability to pass through this barrier with ease and without the need for outside assistance. So, by using genetic engineering, nanoparticles may be able to transfer biomolecules to plants. When fertilizers and pesticides are used carelessly, the environment is contaminated, and biodiversity is threatened [150].

In ornamental nursery production, over-fertilization is a widespread practice [155]. Fertilizer treatment estimations are often inaccurate since visual inspection is frequently utilized to assess plant nutrition levels. Two non-destructive sensors, Soil Plant Analysis Development (SPAD-502) and GreenSeekerTM, were investigated by Freidenreich et al. [155] to determine whether they were suitable for detecting the absorption of nutrients into plant tissue. As an efficient and non-destructive instrument for sustainable fertilizer management practices in the ornamental plant business, their technique might be used as a reference for nursery producers and landscaping staff.

According to Fountas et al. [97], who studied the methodologies and implementation of PA throughout the previous 25 years, the acceptance of technology and its impacts on crop management, the environment, and the sustainability of agricultural systems are all related. For each field at the site-specific level, the farm manager may obtain data on soil, yield spatial distribution, weather, crop scouting, remote sensing, and yield collecting methods. Enhancing productivity and profitability while mitigating environmental impacts will be possible with new sensors that identify anomalous responses in the soil or crops [97]. A state-of-the-art chemical sensor system was created to analyze Thai sustainable PA chemically at a reasonable cost for use in rural Thailand and other locations [164].

Reduction in Costs via the Food Supply Chain

Many studies indicated that using I4 in PA has benefited the food and agricultural sectors in reducing costs via FSCs [22,24,26,28,36,48,50,52,55,56,66,74,75,122].

These FSCs and agricultural productions may be improved with I4 solutions, resulting in higher product quality, greater food output, and optimized operations, among other advantages. Using the production of and market for chicken as an example, the interconnections between FSC resilience, I4, and sustainability are examined [56]. The UN estimated global food losses and waste in 2011 at 1.3 billion tonnes annually [66].

The objective of Perez Perales et al. [75] was to classify these technologies according to the two following standards: the primary subjects to be addressed in each objective and the FSC participant where it is performed specify the environmental or social goal to be achieved [75]. They focused on technologies that address environmental and social sustainability because economic sustainability will rely on the particular characteristics of the company (an FSC using a certain I4 may be successful while others are not). The social evolution of the ecosystem is assumed to be covered by and supported by this I4FA Nexus.

Lopes et al. [55] offered a technique for using CE business models to solve losses and waste in the FSC. Initially, a comprehensive literature review was conducted to determine how CE is used at the cutting edge of the food waste industry. In terms of management contributions, they suggested deploying CE business models more widely to solve food losses and waste while accounting for the retail tier's participation. This I4FA Nexus covers the socio-economic evolution of the ecosystem.

Mohajeri et al. [52] offered a model of the advantages of operations for the food reverse FSC by putting the I4 concept into practice. A device that recycles household waste was introduced as an example of the I4. Electric cars have also been considered for delivery and pickup by I4. Recyclable stations have defined the rate of progress. Many methods for recycling food waste using different technologies have been selected and assessed based on the I4 indicators. Food waste is sent to recycling stations, which are establishments maintained, operated, or used to purchase, sell, or store it before recycling it by using the appropriate machinery. The model's several goals minimize the negative effects of environmental degradation and transportation costs while maximizing the benefits of recycling and consumer response. In this work, the whale optimization approach is applied. They provided a comprehensive reverse supply chain management method for food waste based on the I4. According to this I4FA Nexus, the ecosystem's SEE evolution is covered.

I4 represents a group of CPSs. It supports the idea of "smart factories", wherein machinery is given online access, linked to a production process monitoring system, and empowered to make decisions independently. This framework showcases the emerging innovations that are happening all around the world, especially in Europe. It might lead all enterprises to more competitive outcomes and to intriguing results in the years to come. In this context, there seems to be substantial room for growth, even for the most innovative component of FSC management—the service sector [74]. Nevertheless, this potential needs to be adequately exploited and backed by targeted investments. The FAI is undergoing significant change due to I4's increasing digitalization. Smart technology is altering the dynamics of the FAI, requiring increased automation. Thanks to the new automation phase, the sector may now enjoy streamlined, dependable, and efficient processes, services, and products, but it also requires new professional capabilities from its personnel. It is critical to identify the near-term skill requirements for the sector to close the skill gaps between the labor force and the industry demands [48].

I4 is required across the FSC to handle the rising global demand for food products and the concerns about food security and safety [22]. Green technologies have drawn significant interest in many food applications, even though I4 technologies are changing various production and consumption sectors, including the food and agriculture industries [24]. Poor food quality and safety lead to foodborne illnesses and costly food crises, eroding consumer confidence, and reducing the effectiveness of cold food chains. I4-related modifications to food traceability systems, using automatic identification and sensor technologies instead of manual paper-based record-keeping, can improve data transfer and self-monitoring to reduce issues related to food quality. Before selecting a technology to meet a certain need, it is important to assess its performance with regard to many considerations [36]. The two main issues that industrial firms grapple with are adopting I4 technologies, which automate and boost plant productivity, and evaluating more environmentally friendly items and processes [26]. FSC 4.0-related research on the factors influencing investment in these technologies is still in its early stages despite a notable increase in knowledge related to I4 technologies [28]. Companies that produce food and packaging will need to transition from a linear to a CE by implementing policies that can increase the sustainability of their operations and products from an environmental, social, and economic perspective. Thus, food companies must reinvent themselves to stay competitive in the market by using innovative methods and tools to boost the efficiency and output of their establishments [50].

Regarding employment, turnover, and added value, the food and beverage industry was once again recognized as the largest manufacturing sector in the European Union in 2017, according to Bucci et al. [122]. Nonetheless, most businesses are SMEs, which exhibit a sluggish pace of innovation and PA technology adoption. With the arrival of the digital era, agri-food SMEs are finding new ways to apply technology advancements throughout the FSC—from farm to fork—to boost their competitiveness. Their [122] study, which addresses the state of the art, affirms that technology applications in food production are essential for guaranteeing sustainable farming systems.

Reduction in Costs of Food Production Using Green Technology of Industry 4.0

Many studies have indicated that the deployment of I4 can reduce the costs of food production [26,79,141].

Concerns about food security, climate change, and population expansion are causing agriculture to undergo a digital revolution. Information technology affects agriculture in ways that lower costs while increasing productivity and sustainability. To help with the field identification of pests, plant diseases, and inadequate plant nutrition, PA uses IoT, deep learning, predictive analytics, and AI-based technologies. The following are the goals of the study: (1) To examine the function of smart technologies and how they affect the sustainability of PA; (2) to consider the usual use of deep learning and IoT data analytics in PA; and (3) to look at the obstacles to the adoption of sustainable PA. For an in-depth study, IoT devices gather data and send them to data analytics and deep learning [141]. According to Micheni et al. [141], the data help farmers to manage crop diversity, phenotypes and selection, crop performance, soil quality, pH level, irrigation, and the amount of fertilizer applied. Their analysis focused on important PA success elements and common application domains. Technology adoption is influenced by cost, privacy, safety, and legal and technological concerns. The research will be useful to government agencies, academic institutions, individual farmers, and agricultural authorities.

In the European Union, SME ranchers provide 49.6% of the food consumed in the region. Although the administration is committed to paying for the excess, the Common Agricultural Policy was designed to guarantee a market for SME ranchers. Food is deemed trash when it has passed its expiration date or is supplied to other company sectors at substantially reduced pricing. This problem prompted the authors to construct a contextual study of the food business in the European Union by showcasing various well-calculated organizational scenarios and implementing an intensity system based on the I4 concept and lean methodology [79].

Technological, environmental, economic, and social considerations were considered in Stefanini and Vignali's [26] assessment of automated guided vehicles (AGV) as an I4 application. The systems' environmental and economic impacts were compared using life cycle costing and the life cycle assessment approach, which was performed using the SimaPro 9.1 application. Social concerns concerning the workers' working circumstances were considered in the 4.0 scenario. The evaluation's conclusions can benefit companies considering using AGVs for material handling and can contribute to the corpus of scientific knowledge. The question of whether adopting AGVs will lead to more sustainable end logistics processes in the food company was addressed with their foundation. According to this I4FA Nexus, the SEE development of the ecosystem is considered.

4.2.3. Environmental Benefits

PA can help manage food agricultural production inputs in an environmentally friendly way. Based on all the literature reviews [90–164], the major environmental benefits were (a) the reduction in chemical leaching, avoiding excessive fertilizer application; (b) the increase in energy efficiency; and (c) the reduction in food wastes (recycling) using green technology.

Reduction in Chemical Leaching Avoiding Excessive Fertilizer Application

Based on the present literature review, many studies have focused on PA using a more environmentally friendly fertilization application [90–92,94,99,102,109–111,117,121,139].

PA enhances field-level management for sustainable food production. Sustainable farm production includes the alignment of agricultural practices to soil fertility, crop demands, and environmental circumstances [121]. PA aims to increase farm profits by (1) efficient resource management through the variable-rate application of nutrients, agrochemicals, and water; (2) reducing crop yield losses during harvesting; (3) minimizing environmental risks (e.g., greenhouse gas emissions and nutrient leaching); and (4) optimizing farming input footprints. Site-specific agricultural inputs are needed to maximize farm earnings and safeguard the environment with PA technology [121]. PA is involved in food security, environmental preservation, sustainable resource utilization, and economic advantages. Yield monitoring, remote sensing, and efficient fertilizer, water, and pesticide delivery to crops are covered. Thus, food production and resource efficiency may be maximized without waste or environmental damage from excessive fertilizer or pesticide use [94].

PA aids the environment by targeting inputs to decrease losses from excess applications, nutrient imbalances, weed escapes, insect damage, etc. Reduced pesticide resistance is another benefit. Few publications have analyzed the measured environmental variables directly, such as by leaching with soil sensors. Most of the calculated indirect environmental benefits are derived by assessing chemical loading reduction [90]. PA technologies for food security and sustainability are vital resources that review PA research across disciplines. It also addresses innovative tools and approaches to improve system implementation. Engineering and computer science are used in PA research to enhance crop health, irrigation, and fertilizer use [102].

Farm management today must satisfy ecological, economic, and social needs. Due to various legislation, farmers must achieve sustainability and environmental protection standards. More commonly, they must record, archive, and validate data [117]. Comprehensive planning modules allow graphical planning and execution of PA activities, including cultivation, cropping, fertilization, pest management, and harvesting. Fertilizer application, including PA, illustrates planning, execution, and graphical and tabular (database) assessment [117]. Van Evert et al. [93] examined how conventional PA practices boost profitability and sustainability. They calculated each scenario's output, input, and environmental values. This allowed us to compute profit and social profit, which is revenues minus conventional expenses minus external production costs. Sustainability may be measured by social profit. PA boosts olive sustainability and potato profitability and sustainability. Nath [99] envisioned sustainable intensification and examined PA's role. PA practices, such as precision irrigation, fertilizers, pest and disease control, and animal farming, are highlighted in this review. Thus, technology innovation, sustainable farming, data analytics, and legislative interventions will shape sustainable PA.

Peerlinck and Sheppard [102] optimized winter wheat crop yield output to boost farmers' production. Optimization might lead to poor sustainability if too much fertilizer is applied or the farming equipment is overworked. Therefore, they included sustainability targets that directly address these issues. A novel multi-objective factored evolutionary algorithm solves multi-objective optimization using overlapping subpopulations. Their results showed that multi-objective optimization with overlapping subpopulations improves objective space exploration. PA is used in olive orchards (*Olea europaea* L.) to manage agronomic variability and give plants the correct input quantity without loss [111]. Roma et al. [111] developed a GIS platform employing GEOBIA algorithms to create prescription maps for variable rate (VRT) nitrogen fertilizer treatment in olive orchards.

Dubos et al. [139] compared the optimal N and K rates advised by each approach in adult oil palm using long-term fertilization experiments. Leaf analyses (LA) yielded modest rates relative to nutritional balance. LA showed each block's prospective yield clearly. They concluded that this perfectible technology was more environmentally friendly and did not reduce yields or soil mineral reserves.

Increase in Energy Efficiency

Based on the present literature review, many studies have focused on PA by increasing energy efficiency during food agricultural production [113,123,126,145,147].

Agricultural irrigation has attracted attention to the boosting of agricultural production and the conservation of water. Traditional irrigation uses water and electricity to schedule irrigation [123]. A fuzzy-based intelligent irrigation scheduling system employing a low-cost wireless sensor network was proposed by Jamroen et al. [123]. A cost study verified the irrigation scheduling system's economic viability. Energy-intensive cereal-based farming techniques in South Asia's Indo-Gangetic Plains distort agricultural income and the environment [126].

Achour et al. [147] reviewed recent greenhouse technology used in hardware design, environmental monitoring, dynamics modelling, microclimate control, energy optimization, green energy integration, and storage system implementation. Renewable energies like solar and geothermal have become extensively adopted as ecologically benign alternatives, making greenhouse energy self-sufficient and able to exchange electricity with the grid. The Agri.q for PA can map, monitor, manipulate, and collect small soil and crop samples in unstructured agricultural environments due to its modular articulated mechanical structure and specific sensors and tools, according to Botta and Cavallone [145]. Sustainable 5G PA is hindered by sensor node (SN) battery capacity [113]. Chien et al. [113] proposed a system for charging SNs and gathering sensory data using unmanned aerial vehicles to overcome this challenge.

Reduction in Food Waste (Recycling) Using Green Technology

Based on the present literature review, many studies have reported a reduction in food waste (recycling) using green technology during food agricultural production [67,80,183–185]. This is also well indicated in Figure S1, where the industrial revolutions complemented the agricultural revolutions, from Industry 1.0 to I4 [11].

The digitization of manufacturing through I4 initiatives will influence several industries, including the food and beverage industry [67]. The integration of PA with smart grid technology has been proposed by Odara et al. [157] as a potential strategy for augmenting the capacity of sustainable energy supply. Agriculture is a significant burden due to its many chores, including irrigation, crop harvesting, and processing. Integrating such practices holds promise for the enhancement of agricultural systems by mitigating input expenses, especially those associated with waste management. Moreover, the use of carbon-neutral fuels might potentially have good environmental outcomes.

The predicted adoption of I4 is expected to enhance production capacity, increase output value, and assist the government in attaining its economic objectives [71]. The food sector holds a prominent role in the economy of Andalusia, owing to its significance, advantages, and potential. Consequently, it poses a substantial challenge within the region's economic framework. Implementing the framework proposed by I4 is of utmost importance for the whole sector and specifically for the FAI in Andalusia. It should be regarded as a significant opportunity for businesses to progress. It is expected that, as with other sectors, the food and beverage industry will emerge as a frontrunner in using adaptable and personalized manufacturing techniques [80].

Logically, there are always reasons for the connectivity between the I4FA Nexus and the agroecosystem (Figure 11). The primary reason behind this is the increasing pattern in electricity production worldwide [183] (Figure S2). The per capita (kilocalorie) supply from all foods per day [184] (Figure S3) in the past few decades has shown a very positive increment, and this pattern is expected to continue in the foreseeable future. Overall, the per capita calorie supply has steadily risen worldwide. However, these patterns differ in different parts of the world. In recent decades, the caloric supply in Asia and Africa has increased significantly. For the previous several decades, there has been convergence in the worldwide trends in caloric supply due to the greater growth in the world's poorer regions. However, there was an inverse pattern for the renewable freshwater resources (RFRs) per capita [185] (Figure S4). The overall amount of renewable flows and the population density determine per capita RFRs. Per capita, renewable withdrawals will decrease if RFRs diminish, which can often happen in nations with significant yearly rainfall variation, such as during monsoon seasons. Similarly, if total RFRs stay the same, the per capita levels may decrease if a nation's population increases. Population growth is causing many countries' per capita RFRs to decrease.

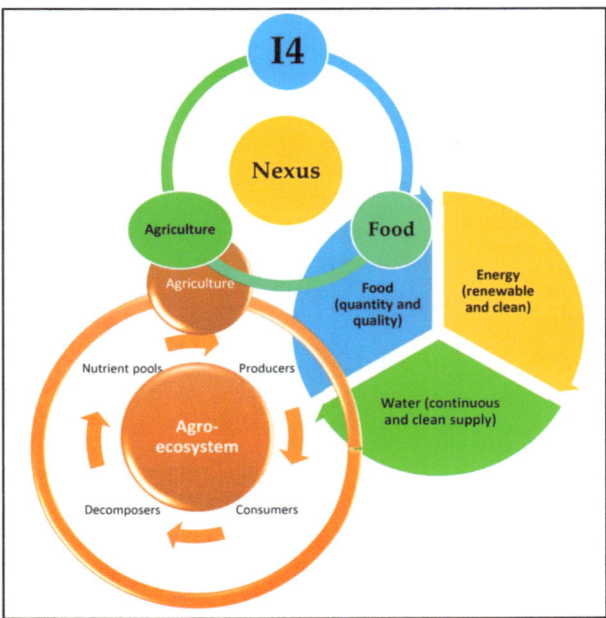

Figure 11. An idealistic conceptual relationship between Industry 4.0 (I4)—Food-Agriculture Nexus and agroecosystem (with three related to United Nations Sustainable Development Goals), with foods (in abundance in terms of quantity and quality, as indicated by deep green color) in the society (present and future) when environmental stress (climate change) factor is taken into account.

4.3. The Challenges of Sustainable Agricultural Food Industry in Adopting Industry 4.0

On the other hand, the I4 technology-driven nature combined with the relatively early phases of the I4 technologies life cycle implies and raises several concerns. Based on all the literature reviews, the major issues of the sustainable FAI in adopting I4 were (a) the question of preparedness, (b) lack of trust, (c) privacy issues, and (d) economic revenue uncertainties.

Feeding a rising population, ensuring farmers' livelihoods, and conserving the environment are the three difficulties facing the world's food systems, according to Brooks et al. [186]. The three challenges in the incorporation of a socio-economic environment must be faced simultaneously if lasting progress in any of them is to be achieved. Thus, this is crucial. Given the scope and complexity of these issues, policymakers may need to try out new techniques to create a set of answers that appeal to all stakeholders.

Food agriculture as a sustainable and non-substitutable resource has been well supported and should not become a debatable issue now and in the future. Several facts and figures in the past have justified this agricultural sustainability from the point of view of the ecological agroecosystem and, lastly, from a socio-economic perspective. A population predicted to increase depends on the global food system to feed its members with safe and nourishing food. In addition to additional mouths to feed, the demand for meat, fish, and dairy will increase as wages rise in emerging and developing nations [186]. The advancement of existing technology and fresh suggestions concerning switching from outdated methods to more effective ones for manufacturing nutrient-dense foods are presented [187]. To solve all of the above problems, the use of I4 in PA is recommended.

I4 may make managing agroecosystems easier in order to produce secure food and nutrition. Theoretically or fundamentally, agroecosystem management is essential to preserve ecological stability, social equality, economic viability, and cultural vitality [92]. Also, it is in tandem with the UN-SDGs, especially with regard to Zero Hunger (Goal

No. 2), Responsible Consumption and Production (Goal No. 12), Life Below Water (Goal No. 14), and Life on Land (Goal No. 15).

The impacts of an increased amount of atmospheric CO_2 emissions that could potentially adversely affect our future agricultural output and food quality were considered. The agroecosystem now places a fresh emphasis on the hot subject of climate change. These risks can alter the environment abruptly or gradually, hurting biotic processes and deteriorating abiotic circumstances. Although the I4FA Nexus complementing the agroecosystem is an almost perfect approach, determining how many hurdles and obstacles first need to be overcome by the less privileged countries or industries is again a never-ending discussion. Below are some of the potential challenges of incorporating I4 in PA.

4.3.1. Social Predicaments

These involve changes in human behavior and mindsets during the paradigm shift. The social issues included in the following discussion are (a) the question of the preparedness of small industries (social adaptations), (b) lack of trust, and (c) privacy issues [188–190].

The Question of Preparedness

This mainly involves industrial modifications. I4 manageability with economic practices has received a significant financial commitment. Implementation will have SEE effects. A significant financial investment, time for adaptation, especially in less-privileged nations or businesses, and a shift from the human ecological paradigm are needed.

SME businesses often employ century-old machinery. Konur et al. [59] presented a unique case study of switching a traditional food producer to I4. The article describes their development and transition challenges. They showed smart production control CPSs. The system's novel data collection, information extraction, and intelligent monitoring services had increased productivity and consistency while lowering operational expenses. Similar food production and SME industries can benefit from the approach and learning. To avoid mass technological unemployment, a social ecosystem for seamless technology adoption with social design is needed [64]. This I4FA Nexus covers ecosystem socio-economic development.

Farm production is moving towards IoT-based smart systems with smart items as the world becomes digital. This trend is expected to accelerate as AI-powered devices and smart technologies grow more widespread. Smart objects detect conditions and respond intelligently without human intervention. Real-time agricultural field monitoring saves money, manages resources, and informs choices. The IoT, a key I4 enabler, has enabled innovative agriculture technology for cost reductions and output increases and improved big data analytics for future choices. However, limited-resource agriculture struggles to modify production to suit present needs [7]. I4 might modernize smart farming by improving productivity and reducing human intervention. This smart paradigm automates planting and output yield using innovative methods. Farm systems are improved by adding the I4 paradigm to intelligent computer and communication technology [14].

Pérez Perales et al. [16] focused on natural and socially sustaining developments. Manageability and economic practices are crucial in most organizations' internationally flexible supply networks [68,191]. Most companies utilize this to manage production, services, and corporate social responsibility. Most companies employ manageability to meet client requirements in the present supportable social consciousness, which includes food production. Ojo et al. [68,77] linked I4 to food-producing FSC standards. Thus, economic practices for I4 deployment in the agricultural food business may make managing it difficult.

These business models have created new FAI labor skill needs [63]. Precision pest control is being introduced in the developed world using artificial neural network-based machine learning (pheromone-based visual traps for insect identification) and electric nose technology-based automatic machines or sensing devices for hotspot (infestation area) identification. These technologies are expensive and sophisticated. Therefore, resource-poor farmers are reluctant to use them [175]. Thus, the efficacy of using I4 smart technology for insect pest control and precise pest management is still debated [175]. According to Furstenau et al. [172], the scientific community focuses on economic and environmental conditions while overlooking social issues. Thus, many discussions and debates have always concerned I4 and sustainability research challenges, perspectives, and concepts. Facchini et al. [23] examined the competitiveness risks and opportunities by determining the "readiness degree" of agri-food enterprises to employ smart technology. They used smart technologies to measure the company's economic, social, and sustainable competitiveness. This I4FA Nexus may help the ecosystem's socio-economic growth.

De Carolis et al. [188] accurately anticipated that digital technology will drive manufacturing transformation in I4. In practice, such technologies allow firms to find ways to turn increased complexity into long-term competitiveness and profitable growth. However, the practice still affects industrial implementation. According to Cotrino et al. [189], I4 technologies like the IoT, virtual reality, and CC are changing company structures in manufacturing and small SMEs. A literature analysis found that most large companies have investment strategies, some of which are reviewed in this paper. The major projections show that the major enterprises' I4 investments exceed the SMEs' yearly revenues, making it difficult for SMEs to obtain these technologies. The study found two gaps: the newest literature study does not explore I4's practical use in SMEs, and there are no I4 implementation roadmaps for SMEs. SME finance cannot pick the finest technology, design the best strategy, and pay for extensive consultancy help. They showed SMEs how to access I4 technology with inexpensive investments.

Hizam-Hanafiah et al. [28] discovered 30 I4-ready models with 158 dimensions. The prevalence of technology among these six most prominent qualities suggests more research on I4 preparation. Mechanized farming displaced indigenous farming during the first two industrial revolutions, and PA is new. Industrial farming increases productivity, but some challenges have become increasingly important. I4 is expected to accelerate the fourth agricultural revolution [11]. Climate change, resource limitations, changing customer demands, and rigorous regulations are continually on stakeholders' minds in the FAI, which utilizes many resources. The food business has implemented I4. Improving transparency through AR experiences is a key focus. Although I4 technologies are used more in the FAI, AR is still underutilized [53]. I4, the current industrial revolution, has transformed the dynamics of the industry as a whole, causing the food business to evolve quickly.

This digital revolution is real, but which digital technology will benefit each business field is uncertain [5]. Baierle et al. [5] analyzed the adoption of digital technologies in several industrial sectors to see which of them may boost agricultural system performance. They analyzed industrial sectors to create a digital transformation framework to boost FAI competitiveness in Agriculture 4.0. The food sector frequently uses only one digital technology. Therefore, they showed the need for concurrent and joint investments in the other technologies addressed in this research. Public policy must stimulate the FAI's digital technology development [5]. Arora et al. [6] analyzed the use of these technologies in agriculture and created a priority ranking based on how effectively they overcome these difficulties. Two steps were taken in their research. First, I.4.0 technologies and agricultural FSC bottlenecks were identified. A discussion follows on the proposed framework, which blends data envelopment analysis with analytical hierarchy. They found that agricultural technology can improve FSC management. Their research prioritized options based on final weights. This ranking system can help farmers and the government choose the best technologies to automate the agricultural FSC.

Naqvi et al. [7] converted conventional agriculture into IoT-enabled smart systems to address quality issues. According to Bernhardt et al. [8], some agricultural regions need improvement. Bernhardt et al. [10] investigated whether there were techniques and whether these structures were suitable for agriculture. I4's approaches help agriculture, they reported. Agriculture has different structures; thus, they must be changed. Liu et al. [11] investigated industrial agriculture's contemporary situation and the lessons learned from industrialized agricultural production patterns, processes, and the agri-FSC. They focused on the critical scientific issues and agricultural applications of these technologies.

Lack of Trust

Societies may only encourage more sustainable farming systems by developing policies that incorporate SEE concerns [180–192]. When advising SME businesses in the food and beverage industry to implement I4, policymakers consider trust a key factor. According to their knowledge, familiarity, agreement, and preferences, the SMEs' degree of trust in executing I4 is described as their belief in using the right technology for I4. Several Kansei terms, or factors relating to human thinking, are included in the complicated concept of trust [58]. I4 is the most prominent example of a technical breakthrough that may help businesses and entrepreneurs address these difficulties in such a scenario [20]. Digital technologies also play a similar role in the PA sector. Therefore, although I4 is a future paradigm shift in the FAI, its deployment faces many socio-economic consequences that need time for adaptations and mindset adjustments.

Ushada et al. [34] aimed to deploy I4 in food and beverage industry SMEs by modelling group preference decision making. The travelling salesman problem-based decision-making process was modelled using an ant colony optimization approach. They showed that equipment and tools were the most popular choices for I4 implementation. When choosing the first characteristic, adaptability was the top choice. They anticipated that the high confidence level in group choices would support I4's sustainability. The method adds to several already-existing theoretical frameworks for decision making based on group preferences and can help the management of SME's to implement I4.

Ushada et al. [58] utilized artificial neural network modelling to simulate SMEs' confidence in implementing I4. They found that the result was a categorization of trust as "overtrust", "trust", or "distrust". They showed that education, knowledge, familiarity, benefits, preference ranking, and linguistic components all impacted SMEs' levels of trust.

Privacy Issues

Prasad et al. [3] analyzed the significance of numerous applications, including smart agriculture, smart cities, smart healthcare, and smart medicine, as well as their features, security problems, and privacy concerns. Along with future study topics and breadth, frameworks for reducing the effect of security and privacy problems are also highlighted. An AI-based smart farming protocol was presented by Mahajan et al. [4] since AI techniques are crucial for enhancing the performance of I4 standards. Using clustering and routing methods, they created the lightweight clustering protocol for I4-enabled PA.

A broad framework is developed in Bigliardi's [49] thorough literature evaluation of the use of the I4 paradigm in the food business. A basic review of green and I4 technologies from a food viewpoint was presented by Hassoun et al. [24]. The UN-SDGs and I4 enablers (such as artificial intelligence, big data, smart sensors, robots, blockchain, and the IoT) will be connected to green food technologies (such as green preservation, processing, extraction, and analysis). These technologies promise to promote ecological and digital changes in food systems that will benefit society, the economy, and the environment. While the use of digital technologies and other I4 technology advancements in the FAI is still in its infancy, various green technologies have already offered creative solutions for significant changes in the food system.

4.3.2. Economic Revenue Uncertainties

This is especially due to the cost-intensive nature and difficulties involved in estimating full financial benefits and economic effectiveness, as indicated by some published studies [13,20,21,59,190].

The relationship between the domains of I4 and PA was considered by Trivelli et al. [20]; they examined the most prevalent technologies employed in each area to identify similar trends and technological overlaps. A method combining manual and automated analysis was created to do this. They discovered a lexicon of 324 words related to PA technologies, a graph outlining the relationships between the technologies, and a depiction of the major technology clusters observed. To provide thoughts and concerns for the future, Zambon et al. [21] analyzed retraces of the stages of the industrial and agricultural revolutions that have occurred up to the present. To enable the effective application of I4 principles, they examined the unique difficulties faced by agriculture throughout the FSC.

Agribusiness organizations have started implementing technology to create an FSC that is more sophisticated, customer-focused, and sustainable. Even if the adoption of linked new technologies and the CE concept poses many difficulties, they have already shown their utility in the industrial sector in achieving a sustainability goal [13]. The interaction between people, machines, and electronics in today's industries is considered to be a smart ecosystem that is necessary for the efficient production of goods. I4 is a group of technologies that serve as enablers for such intelligent ecosystems and enable the transformation of industrial processes. However, the need for modernization and automation at conventional factories makes it necessary to overcome several practical obstacles when adopting and implementing I4 [59].

4.3.3. Environmental Impacts of Industry 4.0 in Food Agriculture

The implementation of Industry 4.0 in food agriculture can have negative impacts on the environment. One negative impact is the increased use of resources, such as energy and water, due to the integration of advanced technologies. This increased resource consumption can contribute to the environmental degradation and stress already caused by limited resources. Additionally, digitizing and automating agricultural processes can lead to biodiversity loss.

By replacing manual labor with machines and robots, there is a potential decrease in the diversity of plant and animal species that traditionally coexist in agricultural ecosystems. Industry 4.0 and the digitization of agricultural processes can lead to a further intensification of production methods, which can negatively impact soil quality and cause water pollution and the excessive use of chemicals. Moreover, the reliance on digital technologies and connectivity in Industry 4.0 can also increase the vulnerability of food agriculture systems to cyber-attacks and data breaches. Furthermore, the increased reliance on technology and automation in Industry 4.0 can lead to the loss of traditional farming practices and local knowledge, potentially disrupting agricultural communities' cultural and social fabric. While Industry 4.0 offers numerous benefits for the food agriculture industry, such as increased efficiency and productivity, it is crucial to carefully consider and address the potential adverse environmental impacts to ensure sustainable and responsible implementation. Implementing Industry 4.0 in food agriculture can have positive and negative environmental impacts. The digitization and automation of agricultural processes in Industry 4.0 can decrease biodiversity by replacing manual labor with machines and robots. This can disrupt the balance of plant and animal species in agricultural ecosystems. Additionally, the intensification of production methods in Industry 4.0 can negatively impact soil quality and cause water pollution and the excessive use of chemicals. Furthermore, the increased reliance on digital technologies and connectivity in Industry 4.0 can increase the vulnerability of food agriculture systems to cyber-attacks and data breaches.

4.4. The Knowledge Gaps for Future Studies

This gives a holistic overview of the past research based on the keywords' co-occurrences with 'Industry 4.0' and 'Food' (Figure 12). The analysis reveals a discernible prominence reflecting three principal domains of investigation, namely three significant clusters that can be identified based on visualization in Figure 12 (top panel): (a) sustainable development, (b) food industries and the food supply chain, (c) the circular economy and the supply chain. Finally, recent studies focused on smart cities, emerging markets, agri-food competition, and machine learning (Figure 12 bottom panel).

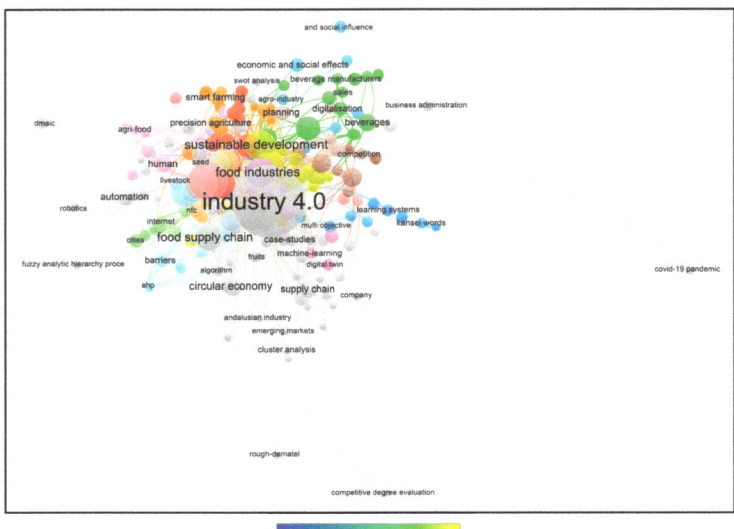

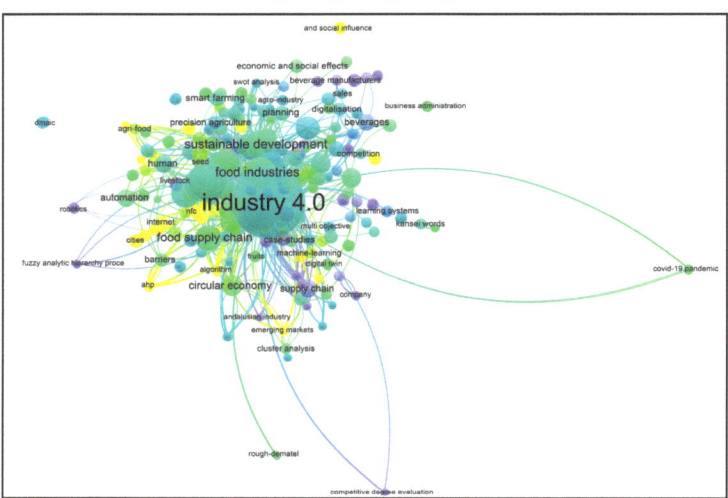

Figure 12. A bibliometric analysis of research themes on 'Industry 4.0' and 'Food'. Top panel: visualization of the paper network confirming the main themes of research. Bottom panel: evolution of research trends between 2019 and 2023. The colors in the top panel indicate the themes of research that are being discussed in the papers while in the bottom panel the colors indicate the year of publication.

Similarly, a holistic overview of the past research based on the keywords' co-occurrences with 'sustainability (or sustainable) precision agriculture' is presented in Figure 13. The analysis reveals a discernible prominence reflecting three principal domains of investigation, namely three major clusters that can be identified based on visualization in Figure 13 (top panel): (i) agriculture, crops, agricultural robots, and sustainable agriculture; (ii) environmental impact, alternative agriculture, and the Internet of Things; and (iii) information technology, precision agriculture technology, cultivation, crop yield, and agriculture production. Recent studies focus on food production, climate change, crop yield, blue economy, agricultural chemicals, and animals (Figure 13 bottom panel).

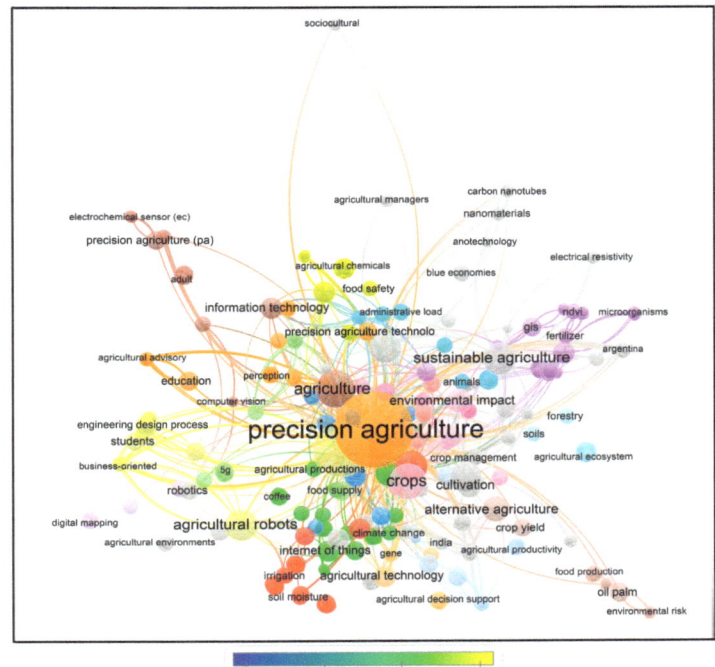

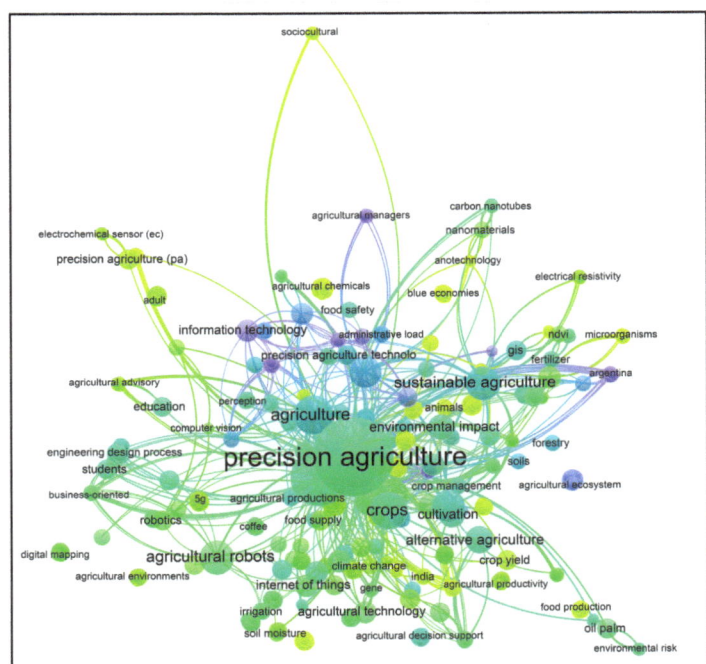

Figure 13. A bibliometric analysis of research themes on the 'sustainability (or sustainable) precision agriculture'. Top panel: visualization of the paper network confirming the main themes of research. Bottom panel: evolution of research trends between 2019 and 2023. The colors in the top panel indicate the themes of research that are being discussed in the papers while in the bottom panel the colors indicate the year of publication.

Interestingly, in both visualizations in Figures 12 and 13, there have been no studies on implementing ESG to effectively manage I4's deployment in the agricultural food sectors. Therefore, the ESG is highly under-studied. This is a visible knowledge gap between EGS and I4 implementation in the FAI. It is now receiving more and more of the attention from the agricultural food industries in their annual governance reports. Future studies should focus on ESG implementation as the solution for effective manageability of I4's deployment in the agricultural food sectors.

4.4.1. Importance of ESG in Precision Agriculture

Environmental, social, and governance (ESG) principles have been applied to the agricultural food industries' businesses [192–199]. The data on ESG are important for several reasons as they provide valuable insights into a company's sustainability, responsible practices, and long-term performance [194–199].

ESG considerations are crucial in agriculture. They strengthen local communities, encourage moral labor, reduce environmental damage, and improve governance [192]. ESG principles may help agriculture preserve the environment, progress society, and sustain the economy. ESG considerations may help agribusinesses attract ethical investors, fulfil customer demand for sustainable and ethical products, and reduce resource shortage and climate change concerns [193,194]. ESG integration improves risk management, sustainability, stakeholder confidence, and customer demand for ethical and environmentally friendly products [195]. Sustainable and ethical practices improve social inclusion, economic stability, FSC resilience, and environmental protection [196–199].

This study is relevant because agricultural enterprises require a new management culture that considers global environmental threats to humanity. At a time when Russia's green (responsible) finance sector is just starting to grow, agriculture is one of the most attractive areas for capital investment to sustain development, preserve biocapacity, and lead the globe. The issues are defined. Due to its conservative management and state regulatory monopoly, the agriculture business is unattractive for venture capital and green finance from banks, which hinders innovation and sustainable growth [192]. Agriculture and forestry are key businesses. As ESG grows, stakeholders are more interested in its impact on agricultural and forestry company performance [193].

Dorashka et al. [192] systematically studied the global green financing of agribusiness enterprises and the Russian ESG financing market to develop specific proposals for the involvement of agribusiness enterprises and financial institutions in financing sustainable development projects as an objective necessity for life on Earth. Zeng and Jiang [193] used two-stage least squares to examine the theoretical and empirical implications of ESG for corporate performance in 156 listed agricultural and forestry enterprises. They stated that (1) ESG and corporate performances are strongly correlated and that higher ESG ratings improve corporate performance; (2) social and governance performances are better at encouraging business performance growth than environmental performance; and (3) there are no discernible differences between listed firms in forestry and agriculture with regard to how ESG affects corporate performance. They also advised listed companies to promote green growth. Their findings helped listed agriculture and forestry firms to boost ESG performance and corporate success.

Buallay [194] examined how sustainability reporting affects agricultural operations and financial and market performances. According to their statistics, ESG has no substantial relationship with operational, financial, or market performance. Governance transparency positively affects market performance when each ESG element is separately regressed against performance, which is surprising. Hrebicek et al. [195] examined an organization's environmental, social, economic, and governance (ESG) variables in examining corporate sustainability reporting trends in the agricultural and food processing industries. The relationship between environmental and sustainability metrics and corporate sustainability reporting needs to be revised [195].

Business, environmental, economic, and social data are recorded, standardized, registered, and collated into key performance indicators [196]. The organization can acquire and incorporate these data in the corporate sustainability or environmental report if such requests arise [197,198]. The combined achievement of ESG performance metrics would measure business success in various economic activities. Sustainability performance is sometimes characterized as environmental, social, and economic/financial performance, ignoring governance [199]. The ESG and the indicators do not focus on the agriculture sector, which affects many food processing sustainability issues and all linkages in the FSC. The ESG in the Food Processing Sector Supplement includes food sector efforts to promote the environmental, social, and economic sustainability of food production chains, including agriculture [195]. The present literature on I4FA [200–226] supports this.

4.4.2. Environmental Factors in Agribusiness ESG

This is because environmental stresses are drivers of food supply deficiency. Figure 14 shows the conceptual relationships between the agroecosystem and three related UN-SDGs, where food items are deficient in quantity and quality in today's and future societies under the presence and impacts of environmental stresses (climate change) factors. Therefore, the connectivity between agriculture and I4 stems from the need for food sustainability and sustainable FSC [200–217].

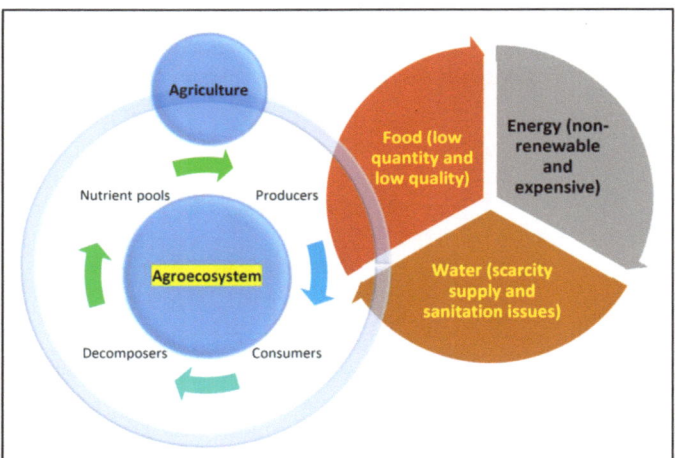

Figure 14. A conceptual relationship between agroecosystems, where food items are deficient in quantity and quality (as indicated by the deep blue color) in today's and future societies under the presence and impacts of environmental stresses (climate change) factor.

Pollution, climate change, unsustainable land use [165–167], unsustainable farming practices, and overexploitation of resources [168] are well-known factors that stress our global agroecosystem in the efforts to balance production demand with population growth. Agroecosystem stress reduction strategies are critically required. Switching to green food production is smart. Agroecosystem ecotoxicologists are studying climate change aspects to decrease human influence and preserve natural ecosystems. Agroecosystem management helps to meet the UN-SDGs, including Zero Hunger, Life on Land, Responsible Consumption and Production, and Life Below Water. More specifically, the UN-SDGs' success depends on agroecosystem sustainability.

I4 technology is growing in popularity, but how it may be conceptually integrated to supplement the agroecosystems' renewable resources remains a knowledge gap and a global conversation. This review paper seeks to connect and discuss how idealistic conceptual relationships between the I4FA Nexus and the agroecosystem can be logically connected and married. This food (in abundance and quality) in society (present and future) is under environmental climate change stress.

Population increase, climate change, food waste, and pandemics have hampered global food security [39]. Understanding how to preserve the agroecosystem to keep an ever-supplied food quantity under the iconic CE of I4 applications will be crucial to developing a sustainable FSC [57]. This complete approach to sustainable production and consumption with limited and contaminated natural resources has made the CE idea popular globally [43].

A healthy PA agroecosystem with I4 will provide high-quality, sustainable food (Figure S1). Due to ESG implementation, sustainable land and water management, natural resource preservation, and biodiversity preservation were achieved. Agriculture needs biodiversity preservation to survive. It involves safeguarding natural ecosystems, native species, and biodiversity-friendly farming practices and avoiding pesticide and chemical fertilizer applications [192–199].

ESG implementation provided climate change mitigation and adaptation plans, as described above. Agribusinesses must reduce greenhouse gas emissions and adapt. To minimize fossil fuel use, agro-forestry, PA, soil carbon absorption, and renewable energy can be used [192–199].

Thus, ESG improves risk management and sustainability in agriculture. Agricultural risk management could be improved using ESG to identify and mitigate resource shortages, climate change, and regulatory changes. Morality and adaptation to changing conditions promote long-term viability. Overall, one of the proposed ways of understanding the knowledge gaps in the food agroecosystem is ESG implementation (Figure 15).

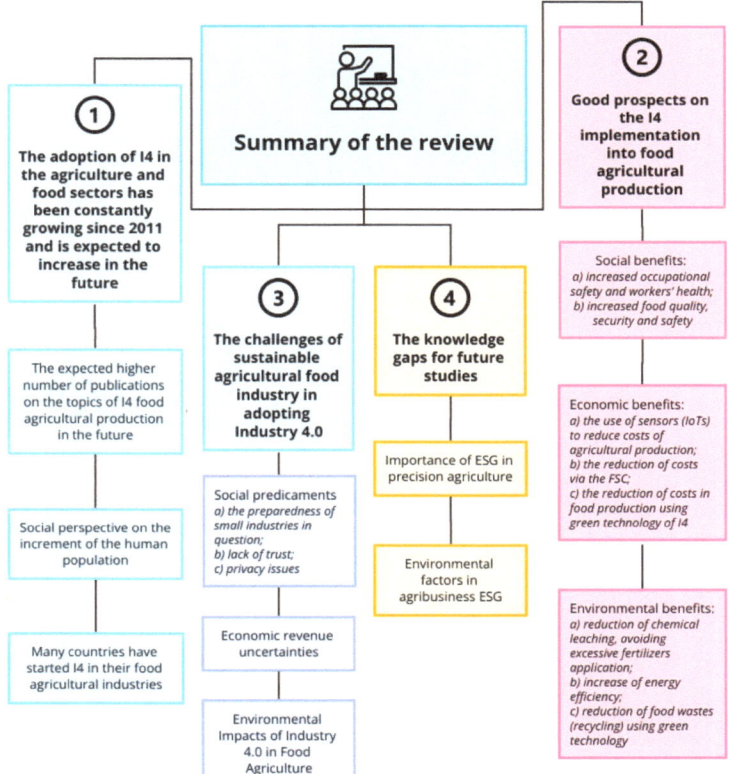

Figure 15. Overall review outcomes and the knowledge gaps from the present review study.

5. Concluding Remarks

The above analysis throws light on four fascinating issues to consider. To begin, the use of I4 technology in the agriculture and food industries is expected to continue to increase both now and in the future. Second, there are good prospects for the I4 implementation in food agricultural production. This paper discussed the social benefits, including increased occupational safety, workers' health, and increased food quality, security, and safety. The economic benefits were the use of sensors (IoT) to reduce the costs of agricultural production, the reduction in costs via the FSC, and the reduction in the costs of food production using the green technology of I4. The environmental benefits included the reduction in chemical leaching, the avoidance of excessive fertilizer application, the increase in energy efficiency, and the reduction in food wastes (recycling) using green technology.

Third, there are always challenges facing the sustainable FAI in adopting I4. This paper discussed the challenges related to the preparedness of small industries (social adaptations), lack of trust, privacy issues, economic revenue uncertainties, and some environmental impacts. Even though I4 is anticipated to be a paradigm shift in the future of food agriculture, its implementation will have several SEE impacts. These impacts will require time for adaptation, particularly in industries or countries with fewer resources, as well as a significant financial commitment and a shift in thinking away from a human ecology frame of mind.

Fourth, the knowledge gaps for future studies were identified as the ESG to be proposed as the solution for effectively managing I4's deployment in the agricultural food sectors. For agriculture to be considered sustainable with regard to ESG, it is essential that all aspects of sustainability, including social, economic, and environmental sustainability, cooperate. In addition, there is always the need to provide reasons for the relationship between the I4FA Nexus and the agroecosystem.

From this review, the concept of I4FA might be brought into the real world with an open-minded conversation platform with ESG-minded leaders that could help complement sustainable agroecosystems worldwide.

Supplementary Materials: The following supporting information can be downloaded at: https://www.mdpi.com/article/10.3390/foods13010150/s1, Table S1: A literature search on keywords 'I4.0' and 'Food' found in different regions or countries, Table S2: The names of journals and their numbers (found 111 papers) published based on Scopus database with keywords 'Industry 4.0' and 'Food', ranging from 2016 to 2024, Table S3: The names of journals and their numbers (found 82 papers) published based on Scopus database with keywords "sustainability (or sustainable) precision agriculture", ranging from 2015 to 2023, Figure S1: Diagram of development roadmap of industrial revolutions and agricultural revolutions. Diagram cited from Liu et al. [11], Figure S2: Electricity production (TWh) worldwide from 1990 to 2022 [183], Figure S3: Per capita kilocalorie supply from all foods per day from 1961 to 2020 from different regions and worldwide [184], Figure S4: Renewable freshwater resources per capita of different regions in the world from 1961 to 2019 [185].

Author Contributions: Conceptualization, C.K.Y. and K.A.A.-M.; methodology and validation, C.K.Y. and K.A.A.-M.; formal analysis, C.K.Y.; investigation, C.K.Y.; resources, K.A.A.-M.; data curation, C.K.Y.; writing—original draft preparation, C.K.Y.; writing—review and editing, C.K.Y. and K.A.A.-M. All authors have read and agreed to the published version of the manuscript.

Funding: This research received no external funding.

Acknowledgments: The authors thank the reviewers for improving the draft.

Conflicts of Interest: The authors declare no conflicts of interest.

References

1. Sivakumar, E.; Ganesan, G.; Ragavi. Harnessing I4.0 Technologies for Climate Smart Agriculture and Food Security. In Proceedings of the 5th International Conference on Future Networks & Distributed Systems, Dubai, United Arab Emirates, 15–16 December 2021; pp. 504–510. [CrossRef]
2. Knoke, T.; Gandorfer, M.; Henkel, A. Industry 4.0–potential and challenges in forestry and agriculture. *Schweiz. Z. Fur Forstwes.* **2023**, *174*, 19–23. [CrossRef]

3. Prasad, S.; Samimalai, A.; Rani, S.R.; Kumar, B.P.P.; Hegde, N.; Banu, S. Information Security and Privacy in Smart Cities, Smart Agriculture, Industry 4.0, Smart Medicine, and Smart Healthcare. *Lect. Notes Netw. Syst.* **2023**, *528*, 621–635.
4. Mahajan, H.B.; Junnarkar, A.A.; Tiwari, M.; Tiwari, T.; Upadhyaya, M. LCIPA: Lightweight clustering protocol for industry 4.0 enabled precision agriculture. *Microprocess. Microsyst.* **2022**, *94*, 104633. [CrossRef]
5. Baierle, I.C.; da Silva, F.T.; de Faria Correa, R.G.; Schaefer, J.L.; Da Costa, M.B.; Benitez, G.B.; Benitez Nara, E.O. Competitiveness of Food Industry in the Era of Digital Transformation towards Agriculture 4.0. *Sustainability* **2022**, *14*, 11779. [CrossRef]
6. Arora, C.; Kamat, A.; Shanker, S.; Barve, A. Integrating agriculture and industry 4.0 under "agri-food 4.0" to analyze suitable technologies to overcome agronomical barriers. *Br. Food J.* **2022**, *124*, 2061–2095. [CrossRef]
7. Naqvi, S.A.H.; Kazmi, R.; Iftikhar, E. Quality Assessment Framework for IoT Based Systems for Agriculture Industry 4.0. *Comm. Comp. Infor. Sci.* **2022**, *1615*, 134–142.
8. Bernhardt, H.; Treiber, M.; Flores, P.; Sun, X.; Schumacher, L. Opportunities for Agriculture through Industrial Internet of Things/Industry 4.0—A comparison between US and Europe. In Proceedings of the ASABE Annual International Meeting, Houston, TX, USA, 17–20 July 2022.
9. Sharma, R.; Kamble, S.; Mani, V.; Belhadi, A. An Empirical Investigation of the Influence of Industry 4.0 Technology Capabilities on Agriculture Supply Chain Integration and Sustainable Performance. *IEEE Transac. Eng. Manag.* **2022**, 1–21. [CrossRef]
10. Bernhardt, H.; Bozkurt, M.; Brunsch, R.; Colangelo, E.; Herrmann, A.; Horstmann, J.; Kraft, M.; Marquering, J.; Steckel, T.; Tapken, H.; et al. Challenges for agriculture through industry 4.0. *Agronomy* **2021**, *11*, 1935. [CrossRef]
11. Liu, Y.; Ma, X.; Shu, L.; Hancke, G.P.; Abu-Mahfouz, A.M. From Industry 4.0 to Agriculture 4.0: Current Status, Enabling Technologies, and Research Challenges. *IEEE Transac. Ind. Inform.* **2021**, *17*, 4322–4334. [CrossRef]
12. Aleksandrov, I.; Daroshka, V.; Isakov, A.; Chekhovskikh, I.; Ol, E.; Borisova, E. Agriculture sphere in the era of Industry 4.0: The world experience and Russian practice of the digital business model building in the agroindustry. *E3S Web Conf.* **2021**, *258*, 06058. [CrossRef]
13. Kumar, S.; Raut, R.D.; Nayal, K.; Kraus, S.; Yadav, V.S.; Narkhede, B.E. To identify industry 4.0 and circular economy adoption barriers in the agriculture supply chain by using ISM-ANP. *J. Clean. Prod.* **2021**, *293*, 126023. [CrossRef]
14. Manogaran, G.; Hsu, C.-H.; Rawal, B.S.; Muthu, B.; Mavromoustakis, C.X.; Mastorakis, G. ISOF: Information Scheduling and Optimization Framework for Improving the Performance of Agriculture Systems Aided by Industry 4.0. *IEEE Int. Things J.* **2021**, *8*, 3120–3129. [CrossRef]
15. Mukherjee, S.; Baral, M.M.; Chittipaka, V.; Srivastava, S.C.; Pal, S.K. Discussing the Impact of Industry 4.0 in Agriculture Supply Chain. In *Recent Advances in Smart Manufacturing and Materials*; Lecture Notes in Mechanical Engineering; Springer: Berlin/Heidelberg, Germany, 2021; pp. 301–307.
16. Arora, D. Demand prognosis of industry 4.0 to agriculture sector in India. *Int. J. Knowl.-Based Intell. Eng. Syst.* **2021**, *25*, 129–138. [CrossRef]
17. Sharma, R.; Parhi, S.; Shishodia, A. Industry 4.0 Applications in Agriculture: Cyber-Physical Agricultural Systems (CPASs). In *Advances in Mechanical Engineering*; Lecture Notes in Mechanical Engineering; Springer: Berlin/Heidelberg, Germany, 2021; pp. 807–813.
18. Bernhardt, H.; Bozkurt, M.; Colangelo, E.; Horstmann, J.; Kraft, M.; Marquering, J.; Steckel, T.; Tapken, H.; Westerkamp, C.; Weltzien, C. Industry 4.0 and agriculture 4.0—the same or different? *VDI Berichte* **2020**, *2374*, 167–173.
19. Szewczyk, R.; Petruk, O.; Kamiński, M.; Kłoda, R.; Piwiński, J.; Winiarski, W.; Stańczyk, A.; Szałatkiewicz, J. Universal Data Acquisition Module PIAP-UDAM for INDUSTRY 4.0 Application in Agriculture. *Adv. Intell. Syst. Comp.* **2020**, *920*, 278–285.
20. Trivelli, L.; Apicella, A.; Chiarello, F.; Rana, R.; Fantoni, G.; Tarabella, A. From precision agriculture to Industry 4.0: Unveiling technological connections in the agrifood sector. *Br. Food J.* **2019**, *121*, 1730–1743. [CrossRef]
21. Zambon, I.; Cecchini, M.; Egidi, G.; Saporito, M.G.; Colantoni, A. Revolution 4.0: Industry vs. agriculture in a future development for SMEs. *Processes* **2019**, *7*, 36. [CrossRef]
22. Yadav, V.S.; Singh, A.R.; Raut, R.D.; Mangla, S.K.; Luthra, S.; Kumar, A. Exploring the application of Industry 4.0 technologies in the agricultural food supply chain: A systematic literature review. *Comp. Ind. Eng.* **2022**, *169*, 108304. [CrossRef]
23. Facchini, F.; Digiesi, S.; Mossa, G.; Mummolo, G. Evaluating the I4.0 Transformation Readiness of Agri-Food Companies: From Factories to 'Smart' Factories. Proceedings Summer School "Francesco Turco". 2018; pp. 409–416. Available online: https://iris.poliba.it/handle/11589/160907 (accessed on 19 November 2023).
24. Hassoun, A.; Prieto, M.A.; Carpena, M.; Bouzembrak, Y.; Marvin, H.J.P.; Pallarés, N.; Barba, F.J.; Punia Bangar, S.; Chaudhary, V.; Ibrahim, S.; et al. Exploring the role of green and Industry 4.0 technologies in achieving sustainable development goals in food sectors. *Food Res. Int.* **2022**, *162*, 112068. [CrossRef]
25. Tancredi, G.P.; Vignali, G.; Bottani, E. Integration of Digital Twin, Machine-Learning and Industry 4.0 Tools for Anomaly Detection: An Application to a Food Plant. *Sensors* **2022**, *22*, 4143. [CrossRef]
26. Stefanini, R.; Vignali, G. Environmental and economic sustainability assessment of an industry 4.0 application: The AGV implementation in a food industry. *Int. J. Adv. Manuf. Technol.* **2022**, *120*, 2937–2959. [CrossRef]
27. Sun, X.; Wang, X. Modeling and Analyzing the Impact of the Internet of Things-Based Industry 4.0 on Circular Economy Practices for Sustainable Development: Evidence from the Food Processing Industry of China. *Front. Psychol.* **2022**, *13*, 866361. [CrossRef] [PubMed]

28. Ali, I.; Aboelmaged, M.G.S. Implementation of supply chain 4.0 in the food and beverage industry: Perceived drivers and barriers. *Int. J. Prod. Perform. Manag.* **2022**, *71*, 1426–1443. [CrossRef]
29. Sharma, R. Industry 4.0 technologies in agri-food supply chains: Key performance indicators. *Adv. Ser. Manag.* **2022**, *27*, 179–187.
30. Stefanini, R.; Vignali, G. The Environmental, Economic and Social Impact of Industry 4.0 in the Food Sector: A Descriptive Literature Review. *IFAC-PapersOnLine* **2022**, *55*, 1497–1502. [CrossRef]
31. Chatterjee, S.; Chaudhuri, R.; Vrontis, D.; Galati, A. Digital transformation using industry 4.0 technology by food and beverage companies in post COVID-19 period: From DCV and IDT perspective. *Eur. J. Innov. Manag.* **2022**. [CrossRef]
32. Mostaccio, A.; Bianco, G.M.; Amendola, S.; Marrocco, G.; Occhiuzzi, C. RFID for Food Industry 4.0—Current Trends and Monitoring of Fruit Ripening. In Proceedings of the 2022 IEEE 12th International Conference on RFID Technology and Applications, RFID-TA, Cagliari, Italy, 12–14 September 2022; pp. 109–112. [CrossRef]
33. Ushada, M.; Trapsilawati, F.; Amalia, R.; Putro, N.A.S. Modeling Trust Decision-Making of Indonesian Food and Beverage SME Groups in the Adoption of Industry 4.0. *Cybern. Syst.* **2022**. [CrossRef]
34. Ushada, M.; Amalia, R.; Trapsilawati, F.; Putro, N.A.S. Group preference decision-making for the implementation of Industry 4.0 in food and beverage SMEs. *Technol. Anal. Strateg. Manag.* **2022**. [CrossRef]
35. Romanello, R.; Veglio, V. Industry 4.0 in food processing: Drivers, challenges and outcomes. *Br. Food J.* **2022**, *124*, 375–390. [CrossRef]
36. Islam, S.; Manning, L.; Cullen, J.M. Selection criteria for planning cold food chain traceability technology enabling industry 4.0. *Procedia Comp. Sci.* **2022**, *200*, 1695–1704. [CrossRef]
37. Erdei, E.; Kossa, G.; Kovács, S.; Popp, J.; Oláh, J. Xamining The Correlations Between Industry 4.0 Assets, External and Internal Risk Factors and Business Performance among Hungarian Food Companies. *Amfiteatru Econ.* **2022**, *24*, 143–158.
38. Kumar, A.; Mangla, S.K.; Kumar, P. Barriers for adoption of Industry 4.0 in sustainable food supply chain: A circular economy perspective. *Int. J. Prod. Perform. Manag.* **2022**. [CrossRef]
39. Hassoun, A.; Aït-Kaddour, A.; Abu-Mahfouz, A.M.; Rathod, N.B.; Bader, F.; Barba, F.J.; Biancolillo, A.; Cropotova, J.; Galanakis, C.M.; Jambrak, A.R.; et al. The fourth industrial revolution in the food industry—Part I: Industry 4.0 technologies. *Crit. Rev. Food Sci. Nutr.* **2022**. [CrossRef]
40. Adamik, A.; Liczmańska-Kopcewicz, K.; Pypłacz, P.; Wiśniewska, A. Involvement in renewable energy in the organization of the ir 4.0 era based on the maturity of socially responsible strategic partnership with customers—An example of the food industry. *Energies* **2022**, *15*, 180. [CrossRef]
41. Kayikci, Y.; Subramanian, N.; Dora, M.; Bhatia, M.S. Food supply chain in the era of Industry 4.0: Blockchain technology implementation opportunities and impediments from the perspective of people, process, performance, and technology. *Prod. Plan. Control* **2022**, *33*, 301–321. [CrossRef]
42. Kafel, P.; Nowicki, P. Industry 4.0 Aspects in Official Statements of Selected Food Sector Organizations Operating on the Polish Stock Market. In Proceedings of the 2021 IEEE International Conference on Technology and Entrepreneurship, ICTE, Kaunas, Lithuania, 24–27 August 2021. [CrossRef]
43. Ada, N.; Kazancoglu, Y.; Sezer, M.D.; Ede-Senturk, C.; Ozer, I.; Ram, M. Analyzing barriers of circular food supply chains and proposing industry 4.0 solutions. *Sustainability* **2021**, *13*, 6812. [CrossRef]
44. Oltra-Mestre, M.J.; Hargaden, V.; Coughlan, P.; Segura-García del Río, B. Innovation in the Agri-Food sector: Exploiting opportunities for Industry 4.0. *Creat. Innov. Manag.* **2021**, *30*, 198–210. [CrossRef]
45. Borowski, P.F. Innovative processes in managing an enterprise from the energy and food sector in the era of industry 4.0. *Processes* **2021**, *9*, 381. [CrossRef]
46. Jambrak, A.R.; Nutrizio, M.; Djekić, I.; Pleslić, S.; Chemat, F. Internet of nonthermal food processing technologies (Iontp): Food industry 4.0 and sustainability. *Appl. Sci.* **2021**, *11*, 686. [CrossRef]
47. Bakalis, S.; Gerogiorgis, D.; Argyropoulos, D.; Emmanoulidis, C. Food Industry 4.0: Opportunities for a digital future. *Food Eng. Innov. Across Food Supply Chain* **2021**, 357–368. [CrossRef]
48. Goti, A.; Akyazi, T.; Alberdi, E.; Oyarbide, A.; Bayon, F. Future skills requirements of the food sector emerging with industry 4.0. In *Innovation Strategies in the Food Industry: Tools for Implementation*, 2nd ed.; Academic Press: Cambridge, MA, USA, 2021; pp. 253–285.
49. Bigliardi, B. Industry 4.0 Applied to Food. In *Sustainable Food Processing and Engineering Challenges*; Academic Press: Cambridge, MA, USA, 2021; pp. 1–23. [CrossRef]
50. Stefanini, R.; Vignali, G. Food engineering systems in the next future: A compromise between sustainability and Industry 4.0. In *Proceedings Summer School Francesco Turco*; Associazione Italiana Docenti Impianti Industriali: Rome, Italy, 2021; 7p.
51. Enarevba, D.R.; Okwu, M.O.; Tartibu, L.K. Addressing food production waste in Africa: Integration of lean six sigma and Industry 4.0 technologies. In Proceedings of the 30th International Conference of the International Association for Management of Technology, IAMOT 2021—MOT for the World of the Future, Cario, Egypt, 19–23 September 2021; pp. 1181–1190. [CrossRef]
52. Mohajeri, S.; Harsej, F.; Sadeghpour, M.; Nia, J.K. Integrated reverse supply chain model for food waste based on industry 4.0 revolutions: A case study of producing the household waste recycling machine. *Qual. Assur. Saf. Crop Foods* **2021**, *13*, 70–83. [CrossRef]
53. Jagtap, S.; Saxena, P.; Salonitis, K. Food 4.0: Implementation of the Augmented Reality Systems in the Food Industry. *Procedia CIRP* **2021**, *104*, 1137–1142. [CrossRef]

54. Rangel, C.; Otero, J.; Antequera, F.; Bonadiez, Y.; Riquett, M.; Regalao-Noriega, C.J. A Look at the Literature Review of the Impact of Industry 4.0 on the Logistics Processes of the Food Sector in Barranquilla. In *Lecture Notes of the Institute for Computer Sciences, Social-Informatics and Telecommunications Engineering (LNICST)*; Springer: Cham, Switzerland, 2021; Volume 393, pp. 252–258.
55. de Sousa Jabbour, A.B.L.; Frascareli, F.C.D.O.; Santibanez Gonzalez, E.D.R.; Chiappetta Jabbour, C.J. Are food supply chains taking advantage of the circular economy? A research agenda on tackling food waste based on Industry 4.0 technologies. *Prod. Plan. Control* 2021, *34*, 967–983. [CrossRef]
56. Mohamed, A.H.A.; Menezes, B.C.; AL-Ansari, T. Interplaying of food supply chain resilience, industry 4.0 and sustainability in the poultry market. *Comp. Aided Chem. Eng.* 2021, *50*, 1815–1820.
57. Ali, I.; Arslan, A.; Khan, Z.; Tarba, S.Y. The Role of Industry 4.0 Technologies in Mitigating Supply Chain Disruption: Empirical Evidence From the Australian Food Processing Industry. *IEEE Transac. Eng. Manag.* 2021. [CrossRef]
58. Ushada, M.; Wijayanto, T.; Trapsilawati, F.; Okayama, T. Modeling SMEs' trust in the implementation of industry 4.0 using kansei engineering and artificial neural network: Food and beverage SMEs context. *J. Eng. Technol. Sci.* 2021, *53*. [CrossRef]
59. Konur, S.; Lan, Y.; Thakker, D.; Morkyani, G.; Polovina, N.; Sharp, J. Towards design and implementation of Industry 4.0 for food manufacturing. *Neural Comp. Appl.* 2023, *35*, 23753–23765. [CrossRef]
60. Barrientos-Avendaño, E.; Areniz-Arevalo, Y.; Coronel-Rojas, L.A.; Cuesta-Quintero, F.; Rico-Bautista, D. Industry foray model 4.0 applied to the food company your gourmet bread sas: Strategy for rebirth in the COVID-19 (SARS-CoV-2) pandemic. *RISTI-Rev. Iber. Sist. Tecnol. Inf.* 2020, *E34*, 436–449.
61. Musti, K.S.S. Industry 4.0-based large-scale symbiotic systems for sustainable food security in Namibia. In *Impacts of Climate Change on Agriculture and Aquaculture*; IGI Global: Hershey, PA, USA, 2020; pp. 186–206. [CrossRef]
62. Khan, P.W.; Byun, Y.-C.; Park, N. IoT-blockchain enabled optimized provenance system for food industry 4.0 using advanced deep learning. *Sensors* 2020, *20*, 2990. [CrossRef] [PubMed]
63. Akyazi, T.; Goti, A.; Oyarbide, A.; Alberdi, E.; Bayon, F. A guide for the food industry to meet the future skills requirements emerging with industry 4.0. *Foods* 2020, *9*, 492. [CrossRef] [PubMed]
64. Filatov, V.; Mishakov, V.; Osipenko, S.; Artemyeva, S.; Kolontaevskaya, I. Industry 4.0 concept as an incentive to increase the competitiveness of the food and processing industries of the Russian Federation. *E3S Web Conf.* 2020, *208*, 03040. [CrossRef]
65. Chew, K.W.; Leong, H.Y.; Show, P.L. Advanced food process technologies: Bridging conventional practices to industry 4.0. *Curr. Nutr. Food Sci.* 2020, *16*, 1286.
66. Ruggieri, R.; Vinci, G.; Ruggeri, M.; Sardaryan, H. Food losses and food waste: The Industry 4.0 opportunity for the sustainability challenge. *Riv. Studi Sulla Sostenibilita* 2020, *2020*, 159–177. [CrossRef]
67. Kobnick, P.; Velu, C.; McFarlane, D. Preparing for industry 4.0: Digital business model innovation in the food and beverage industry. *Int. J. Mechatron. Manuf. Syst.* 2020, *13*, 59–89. [CrossRef]
68. Ojo, O.O.; Shah, S.; Coutroubis, A. Impacts of Industry 4.0 in sustainable food manufacturing and supply chain. *Int. J. Integr. Supply Manag.* 2020, *13*, 140–158. [CrossRef]
69. Polyakov, R.K.; Gordeeva, E.A. Industrial enterprises digital transformation in the context of "Industry 4.0" growth: Integration features of the vision systems for diagnostics of the food packaging sealing under the conditions of a production line. *Adv. Intell. Syst. Comp.* 2020, *908*, 590–608.
70. Creydt, M.; Fischer, M. Traceability 4.0: Digitalization in the food industry. *Dtsch. Lebensm.-Rundsch.* 2020, *115*.
71. Hidayatno, A.; Rahman, I.; Rahmadhani, A. Understanding the systemic relationship of industry 4.0 adoption in the Indonesian food and beverage industry. In Proceedings of the 5th International Conference on Industrial and Business Engineering, Hong Kong, China, 27–29 September 2019; pp. 344–348. [CrossRef]
72. Ichsan, M.; Dachyar, M.; Farizal. Readiness for Implementing Industry 4.0 in Food and Beverage Manufacturer in Indonesia. *IOP Conf. Ser. Mat. Sci. Eng.* 2019, *598*, 012129. [CrossRef]
73. Addy, R. Industry 4.0 Pilot to Benefit Five Food Firms. Food Manufacture. 2019. Available online: https://www.scopus.com/inward/record.uri?eid=2-s2.0-85070971020&partnerID=40&md5=acc44c031f112f9fb1ecf450ad25589e (accessed on 1 May 2023).
74. Boccia, F.; Covino, D.; Di Pietro, B. Industry 4.0: Food supply chain, sustainability and servitization. *Riv. Studi Sulla Sostenibilita* 2019, *1*, 77–92. [CrossRef]
75. Perez Perales, D.; Verdecho, M.-J.; Alarcón-Valero, F. Enhancing the Sustainability Performance of Agri-Food Supply Chains by Implementing Industry 4.0. *IFIP Adv. Inform. Comm. Technol.* 2019, *568*, 496–503.
76. Noor Hasnan, N.Z.; Yusoff, Y.M. Short review: Application Areas of Industry 4.0 Technologies in Food Processing Sector. In Proceedings of the 2018 IEEE Student Conference on Research and Development (SCOReD), Selangor, Malaysia, 26–28 November 2018; p. 8711184. [CrossRef]
77. Ojo, O.O.; Shah, S.; Coutroubis, A.; Jimenez, M.T.; Ocana, Y.M. Potential Impact of Industry 4.0 in Sustainable Food Supply Chain Environment. In Proceedings of the 2018 IEEE International Conference on Technology Management, Operations and Decisions (ICTMOD), Marrakech, Morocco, 21–23 November 2018; Volume 8691223, pp. 172–177.
78. Simon, J.; Trojanova, M.; Zbihlej, J.; Sarosi, J. Mass customization model in food industry using industry 4.0 standard with fuzzy-based multi-criteria decision making methodology. *Adv. Mech. Eng.* 2018, *10*. [CrossRef]
79. Pilinkienė, V.; Gružauskas, V.; Navickas, V. Lean thinking and industry 4.0 competitiveness strategy: Sustainable food supply chain in the European Union. In *Trends and Issues in Interdisciplinary Behavior and Social ScienceProceedings of the 5th International*

Congress on Interdisciplinary Behavior and Social Science (ICIBSoS 2016), Jogjakarta, Indonesia, 5–6 November 2016; Rouledge: Abingdon, UK, 2017; pp. 15–20.
80. Luque, A.; Peralta, M.E.; de las Heras, A.; Córdoba, A. State of the Industry 4.0 in the Andalusian food sector. *Procedia Manuf.* **2017**, *13*, 1199–1205. [CrossRef]
81. De Silva, P.C.P.; De Silva, P.C.A. Ipanera: An Industry 4.0 based architecture for distributed soil-less food production systems. In Proceedings of the 2016 Manufacturing & Industrial Engineering Symposium (MIES), Colombo, Sri Lanka, 22 October 2016; p. 7780266.
82. Smethurst, E. Food Manufacturers Advised to Plan for Industry 4.0. Food Manufacture. 2016. Available online: https://www.scopus.com/inward/record.uri?eid=2-s2.0-84975048578&partnerID=40&md5=2f840c5bfe0b80190aac41bc3a8c0b16 (accessed on 19 November 2023).
83. Bücking, M.; Hengse, A. Fo od industry 4.0: The food industry must be innovative. *Dtsch. Lebensm.-Rundsch.* **2016**, *112*, 256–260.
84. Creydt, M.; Fischer, M. Food industry 4.0: Opportunities and strategies for the food sector. *Dtsch. Lebensm.-Rundsch.* **2016**, *112*, 22–28.
85. Kurdi, B.A.; Alzoubi, H.M.; Alshurideh, M.T.; Alquqa, E.K.; Hamadneh, S. Impact of supply chain 4.0 and supply chain risk on organizational performance: An empirical evidence from the UAE food manufacturing industry. *Uncertain Suppl. Chain Manag.* **2023**, *11*, 111–118. [CrossRef]
86. Vlachopoulou, M.; Ziakis, C.; Vergidis, K.; Madas, M. Analyzing agrifood-tech e-business models. *Sustainability* **2021**, *13*, 5516. [CrossRef]
87. Corallo, A.; Latino, M.E.; Menegoli, M. From industry 4.0 to agriculture 4.0: A framework to manage product data in agri-food supply chain for voluntary traceability. *Int. J. Nutr. Food Eng.* **2018**, *12*, 146–150.
88. Bujang, A.; Abu Bakar, B. Agriculture 4.0: Data-Driven Approach to Galvanize Malaysia's Agro-Food Sector Development. In Proceedings of the FFTC-RDA International Symposium on "Developing Innovation Strategies in the Era of Data-driven Agriculture", Jeonju, Republic of Korea, 29 October 2019; Volume 29, p. 1631.
89. Perciun, R.; Amarfii-Railean, N.; Nataliia, S. Industry 4.0 versus agriculture: Development perspectives of agriculture in the Republic of Moldova by assimilating digital technologies. *Cogito* **2020**, *12*, 178–200.
90. Bongiovanni, R.; Lowenberg-Deboer, J. Precision agriculture and sustainability. *Precis. Agric.* **2004**, *5*, 359–387. [CrossRef]
91. Abd El-Kader, S.M.; El-Basioni, B.M.M. *Precision Agriculture Technologies for Food Security and Sustainability*; IGI Global: Hershey, PA, USA, 2020; pp. 1–437.
92. Bowen, B.; Kallmeyer, A.R.; Erickson, H.H. Research experiences for teachers in precision agriculture and sustainability. In Proceedings of the 2017 ASEE Annual Conference & Exposition, Columbus, OH, USA, 25–28 June 2017.
93. van Evert, F.K.; Gaitán-Crema Schi, D.; Fountas, S.; Kempenaar, C. Can precision agriculture increase the profitability and sustainability of the production of potatoes and olives? *Sustainability* **2017**, *9*, 1863. [CrossRef]
94. Oliver, M.A.; Bishop, T.F.A.; Marchant, B.P. *Precision Agriculture for Sustainability and Environmental Protection*; Routledge: Abingdon, UK, 2013; pp. 1–283.
95. Bragaglio, A.; Romano, E.; Brambilla, M.; Bisaglia, C.; Lazzari, A.; Giovinazzo, S.; Cutini, M. A comparison between two specialized dairy cattle farms in the upper Po Valley. Precision agriculture as a strategy to improve sustainability. *Clean. Environ. Syst.* **2023**, *11*, 100146. [CrossRef]
96. Kountios, G.; Ragkos, A.; Bournaris, T.; Papadavid, G.; Michailidis, A. Educational needs and perceptions of the sustainability of precision agriculture: Survey evidence from Greece. *Precis. Agric.* **2018**, *19*, 537–554. [CrossRef]
97. Fountas, S.; Aggelopoulou, K.; Gemtos, T.A. Precision agriculture: Crop management for improved productivity and reduced environmental impact or improved sustainability. In *Supply Chain Management for Sustainable Food Networks*; John Wiley & Sons: Hoboken, NJ, USA, 2015; pp. 41–65. [CrossRef]
98. Cox, S. Information technology: The global key to precision agriculture and sustainability. *Comp. Electr. Agric.* **2002**, *36*, 93–111. [CrossRef]
99. Nath, S. A Vision of Precision Agriculture: Balance between Agricultural Sustainability and Environmental Stewardship. *Agron. J.* **2023**, early view. [CrossRef]
100. Balakuntala, M.V.; Ayad, M.; Voyles, R.M.; White, R.; Nawrocki, R.; Sundaram, S.; Priya, S.; Chiu, G.; Donkin, S.; Min, B.-C.; et al. Global sustainability through closed-loop precision animal agriculture. *Mech. Eng.* **2018**, *140*, 19–23. [CrossRef]
101. Sanches, G.M.; Bordonal, R.D.O.; Magalhães, P.S.G.; Otto, R.; Chagas, M.F.; Cardoso, T.D.F.; Luciano, A.C.D S. Towards greater sustainability of sugarcane production by precision agriculture to meet ethanol demands in south-central Brazil based on a life cycle assessment. *Biosyst. Eng.* **2023**, *229*, 57–68. [CrossRef]
102. Peerlinck, A.; Sheppard, J. Addressing Sustainability in Precision Agriculture via Multi-Objective Factored Evolutionary Algorithms. *Lect. Notes Comp. Sci.* **2023**, *13838*, 391–405.
103. Clapp, J.; Ruder, S.-L. Precision technologies for agriculture: Digital farming, gene-edited crops, and the politics of sustainability. *Glob. Environ. Politics* **2020**, *20*, 49–69. [CrossRef]
104. Ooi, C.L.; Kamil, N.N.; Mohd Salleh, K.; Leslie Low, E.-T.; Ong-Abdullah, M.; Lakey, N.; Ordway, J.M.; Garner, P.A.; Nookiah, R.; Sambanthamurthi, R.; et al. Improving oil palm sustainability with molecular-precision agriculture: Yield impact of SHELL DNA testing in the Malaysian oil palm supply chain. *Sci. Hortic.* **2023**, *321*, 112305.

105. Walter, S.; Boden, B.; Gunter, K.; Paul, B.; Lukas, F.; Lea, H. Analyze the relationship among information technology, precision agriculture, and sustainability. *J. Commer. Biotechnol.* **2022**, *27*, 158–168.
106. Sott, M.K.; Furstenau, L.B.; Kipper, L.M.; Giraldo, F.D.; Lopez-Robles, J.R.; Cobo, M.J.; Zahid, A.; Abbasi, Q.H.; Imran, M.A. Precision Techniques and Agriculture 4.0 Technologies to Promote Sustainability in the Coffee Sector: State of the Art, Challenges and Future Trends. *IEEE Access* **2020**, *8*, 149854–149867. [CrossRef]
107. Kumar, A.; Jnanesha, A.C.; Lal, R.K.; Chanotiya, C.S.; Venugopal, S.; Swamy, Y.V.V.S. Precision agriculture innovation focuses on sustainability using GGE biplot and AMMI analysis to evaluate GE interaction for quality essential oil yield in Eucalyptus citriodora Hook. *Biochem. Syst. Ecol.* **2023**, *107*, 104603. [CrossRef]
108. Bowen, B.; Kallmeyer, A.R.; Erickson, H.H. The impact of a research experience for teachers program in precision agriculture and sustainability for rural stem educators. In Proceedings of the 2019 ASEE Annual Conference & Exposition, Tampa, FL, USA, 15 June 2019. Available online: https://www.scopus.com/inward/record.uri?eid=2-s2.0-85078723431&partnerID=40&md5=e7 36a27e015874881cb9b08bec2cdb1b (accessed on 19 November 2023).
109. Marinello, F.; Gatto, S.; Bono, A.; Pezzuolo, A. Determination of local nitrogen loss for exploitation of sustainable precision agriculture: Approach description. *Eng. Rural Dev.* **2017**, *16*, 713–718.
110. Roy, T.; George, J.G. Precision Farming: A Step Towards Sustainable, Climate-Smart Agriculture. In *Global Climate Change: Resilient and Smart Agriculture*; Springer: Berlin/Heidelberg, Germany, 2020; pp. 199–220.
111. Roma, E.; Laudicina, V.A.; Vallone, M.; Catania, P. Application of Precision Agriculture for the Sustainable Management of Fertilization in Olive Groves. *Agronomy* **2023**, *13*, 324. [CrossRef]
112. Rosch, C.; Dusseldorf, M. Precision agriculture: How innovative technology contributes to a more sustainable agriculture. *GAIA–Ecol. Persp. Sci. Soc.* **2007**, *16*, 272–279.
113. Chien, W.-C.; Hassan, M.M.; Alsanad, A.; Fortino, G. UAV-Assisted Joint Wireless Power Transfer and Data Collection Mechanism for Sustainable Precision Agriculture in 5G. *IEEE Micro* **2022**, *42*, 25–32. [CrossRef]
114. Blackmore, B.S.; Wheeler, P.N.; Morris, J.; Morris, R.M.; Jones, R.J.A. The role of precision farming in sustainable agriculture: A European perspective. In *Site-Specific Management for Agricultural Systems*; John Wiley & Sons: Hoboken, NJ, USA, 1995; pp. 777–793. [CrossRef]
115. Roberts, D.P.; Short, N.M., Jr.; Sill, J.; Lakshman, D.K.; Hu, X.; Buser, M. Precision agriculture and geospatial techniques for sustainable disease control. *Ind. Phytopathol.* **2021**, *74*, 287–305. [CrossRef]
116. Farooqi, Z.U.R.; Ayub, M.A.; Nadeem, M.; Shabaan, M.; Ahmad, Z.; Umar, W.; Iftikhar, I. Precision Agriculture to Ensure Sustainable Land Use for the Future: Precision Agriculture and Arable Land Use. In *Research Anthology on Strategies for Achieving Agricultural Sustainability*; IGI Global: Hershey, PA, USA, 2022; pp. 1295–1315.
117. Mohr, S.; Schrenk, L.; Littmann, W. Opportunities of precision farming and verification management for sustainable agriculture—Application of a digital agro management system for planning, realisation, recording and analysis of crop production. *VDI Berichte* **2005**, 457–463.
118. Dash, P.B.; Naik, B.; Nayak, J.; Vimal, S. Socio-economic factor analysis for sustainable and smart precision agriculture: An ensemble learning approach. *Comput. Commun.* **2022**, *182*, 72–87. [CrossRef]
119. Lindblom, J.; Lundström, C.; Ljung, M.; Jonsson, A. Promoting sustainable intensification in precision agriculture: Review of decision support systems development and strategies. *Precis. Agric.* **2017**, *18*, 309–331. [CrossRef]
120. Shoub, Y. Sustainable precision agriculture—SPA—The revolutionary approach to irrigation and fertigation. *Int. Wat. Irrig.* **2016**, *36*, 14–15.
121. Zaman, Q.U. Precision agriculture technology: A pathway toward sustainable agriculture. In *Precision Agriculture: Evolution, Insights and Emerging Trends*; Academic Press: Cambridge, MA, USA, 2023; pp. 1–17. [CrossRef]
122. Bucci, G.; Bentivoglio, D.; Finco, A. Precision agriculture as a driver for sustainable farming systems: State of art in litterature and research. *Qual.-Access Success* **2018**, *19*, 114–121.
123. Jamroen, C.; Komkum, P.; Fongkerd, C.; Krongpha, W. An intelligent irrigation scheduling system using low-cost wireless sensor network toward sustainable and precision agriculture. *IEEE Access* **2020**, *8*, 172756–172769. [CrossRef]
124. Aubert, B.A.; Schroeder, A.; Grimaudo, J. IT as enabler of sustainable farming: An empirical analysis of farmers' adoption decision of precision agriculture technology. *Decis. Support Syst.* **2012**, *54*, 510–520. [CrossRef]
125. Tyagi, P.K.; Arya, A.; Ramniwas, S.; Tyagi, S. Editorial: Recent trends in nanotechnology in precision and sustainable agriculture. *Front. Plant Sci.* **2023**, *14*, 1256319. [CrossRef]
126. Parihar, C.M.; Meena, B.R.; Nayak, H.S.; Patra, K.; Sena, D.R.; Singh, R.; Jat, S.L.; Sharma, D.K.; Mahala, D.M.; Patra, S.; et al. Co-implementation of precision nutrient management in long-term conservation agriculture-based systems: A step towards sustainable energy-water-food nexus. *Energy* **2022**, *254*, 124243. [CrossRef]
127. Van Schilfgaarde, J. Is precision agriculture sustainable? *Am. J. Altern. Agric.* **1999**, *14*, 43–46. [CrossRef]
128. Bodei, C.; Degano, P.; Ferrari, G.-L.; Galletta, L. Sustainable precision agriculture from a process algebraic perspective: A smart vineyard. *Atti Della Soc. Toscana Sci. Nat. Mem. Ser. B* **2018**, *125*, 39–43.
129. Essl, L.; Atzberger, C.; Sandén, T.; Spiegel, H.; Blasch, J.; Vuolo, F. Multidisciplinary studies on sustainable nitrogen fertilisation considering the potential of satellite-based precision agriculture. *Bodenkultur* **2021**, *72*, 45–56.

130. Pande, C.B.; Moharir, K.N. Application of Hyperspectral Remote Sensing Role in Precision Farming and Sustainable Agriculture Under Climate Change: A Review. In *Climate Change Impacts on Natural Resources, Ecosystems and Agricultural Systems*; Springer Climate; Springer: Berlin/Heidelberg, Germany, 2023; pp. 503–520.
131. Goh, K.C.; Sim, S.Y.; Goh, H.H.; Bilal, K.; Sam, T.H.; Teoh, T.Y.; Tey, J.S. Evolution of precision agriculture computing towards sustainable oil palm industry. *Ind. J. Electr. Eng. Comp. Sci.* **2018**, *11*, 725–732. [CrossRef]
132. Flores-Delgadillo, L.; Fedick, S.L.; Solleiro-Rebolledo, E.; Palacios-Mayorga, S.; Ortega-Larrocea, P.; Sedov, S.; Osuna-Ceja, E. A sustainable system of a traditional precision agriculture in a Maya homegarden: Soil quality aspects. *Soil Tillage Res.* **2011**, *113*, 112–120. [CrossRef]
133. Shu, L.; Hancke, G.P.; Abu-Mahfouz, A.M. Guest Editorial: Sustainable and intelligent precision agriculture. *IEEE Transac. Ind. Inform.* **2021**, *17*, 4318–4321. [CrossRef]
134. Afzal, A.; Bell, M. Precision agriculture: Making agriculture sustainable. In *Precision Agriculture: Evolution, Insights and Emerging Trends*; Academic Press: Cambridge, MA, USA, 2023; pp. 187–210. [CrossRef]
135. Kim, M.-Y.; Lee, K.H. Electrochemical Sensors for Sustainable Precision Agriculture—A Review. *Front. Chem.* **2022**, *10*, 848320. [CrossRef]
136. Bournaris, T.; Mattas, A.; Michailidis, A.; Andujar, D.; Correia, M.; de Pascale, V.; Díaz, M.; Diezma, B.; Guadagni, A.; Karamouzi, E.; et al. SPARKLE e-Learning platform for sustainable precision agriculture. *CEUR Workshop Proc.* **2020**, *2761*, 334–339.
137. Abuova, A.B.; Tulkubayeva, S.A.; Tulayev, Y.V.; Somova, S.V.; Kizatova, M.Z. Sustainable development of crop production with elements of precision agriculture in Northern Kazakhstan. *Entrepr. Sustain. Issues* **2020**, *7*, 3200–3214. [CrossRef]
138. Jin, X.-B.; Yu, X.-H.; Wang, X.-Y.; Bai, Y.-T.; Su, T.-L.; Kong, J.-L. Deep learning predictor for sustainable precision agriculture based on internet of things system. *Sustainability* **2020**, *12*, 1433. [CrossRef]
139. Dubos, B.; Baron, V.; Bonneau, X.; Dassou, O.; Flori, A.; Impens, R.; Ollivier, J.; Pardon, L. Precision agriculture in oil palm plantations: Diagnostic tools for sustainable N and K nutrient supply. *OCL-Oilseeds Fats Crops Lipids* **2019**, *26*, 26.
140. Kayode, O.T.; Aizebeokhai, A.P.; Odukoya, A.M. Geophysical and contamination assessment of soil spatial variability for sustainable precision agriculture in Omu-Aran farm, Northcentral Nigeria. *Heliyon* **2022**, *8*, e08976. [CrossRef] [PubMed]
141. Micheni, E.; MacHii, J.; Murumba, J. Internet of Things, Big Data Analytics, and Deep Learning for Sustainable Precision Agriculture. In Proceedings of the IST-Africa 2022 Conference, Virtual Event, 16–20 May 2022. Available online: https://www.scopus.com/inward/record.uri?eid=2-s2.0-85137529817&doi:10.23919/IST-Africa56635.2022.9845510&partnerID=40&md5=5266637ba15a0cfb4f54fcff4856dd30 (accessed on 5 May 2023).
142. Nazir, R.; Ayub, Y.; Tahir, L. Green-nanotechnology for precision and sustainable agriculture. In *Biogenic Nano-Particles and Their Use in Agro-Ecosystems*; Springer: Berlin/Heidelberg, Germany, 2020; pp. 317–357.
143. Grandi, L.; Oehl, M.; Lombardi, T.; de Michele, V.R.; Schmitt, N.; Verweire, D.; Balmer, D. Innovations towards sustainable olive crop management: A new dawn by precision agriculture including endo-therapy. *Front. Plant Sci.* **2023**, *14*, 1180632. [CrossRef]
144. Demmel, M. Precision farming—New technologies for sustainable agriculture and their requirements on sensor systems. *VDI Berichte* **2004**, *1829*, 27–35.
145. Botta, A.; Cavallone, P. Robotics Applied to Precision Agriculture: The Sustainable Agri.q Rover Case Study. *Mech. Mach. Sci.* **2022**, *108*, 41–50.
146. Amelia, V.; Sinaga, S.; Bhermana, A. Web-based spatial information system to support land use planning in achieving sustainable and precision agriculture at regional scale. *IOP Conf. Ser. Earth Environ. Sci.* **2022**, *1005*, 012032. [CrossRef]
147. Achour, Y.; Ouammi, A.; Zejli, D. Technological progresses in modern sustainable greenhouses cultivation as the path towards precision agriculture. *Renew. Sustain. Energy Rev.* **2021**, *147*, 111251. [CrossRef]
148. Traversari, S.; Cacini, S.; Galieni, A.; Nesi, B.; Nicastro, N.; Pane, C. Precision agriculture digital technologies for sustainable fungal disease management of ornamental plants. *Sustainability* **2021**, *13*, 3707. [CrossRef]
149. Priya, R.; Ramesh, D. ML-based sustainable precision agriculture: A future generation perspective. *Sustain. Comp. Inform. Sys.* **2020**, *28*, 100439. [CrossRef]
150. Sangeetha, J.; Sarim, K.M.; Thangadurai, D.; Amrita Gupta, R.; Mundaragi, A.; Sheth, B.P.; Wani, S.A.; Baqual, M.F.; Habib, H. Nanoparticle-mediated plant gene transfer for precision farming and sustainable agriculture. In *Nanotechnology for Agriculture*; Springer: Berlin/Heidelberg, Germany, 2019; pp. 263–284.
151. Raliya, R.; Saharan, V.; Dimkpa, C.; Biswas, P. Nanofertilizer for Precision and Sustainable Agriculture: Current State and Future Perspectives. *J. Agric. Food Chem.* **2018**, *66*, 6487–6503. [CrossRef]
152. Sahoo, A.; Sethi, J.; Satapathy, K.B.; Sahoo, S.K.; Panigrahi, G.K. Nanotechnology for precision and sustainable agriculture: Recent advances, challenges and future implications. *Nanotechnol. Environ. Eng.* **2013**, *8*, 775–787. [CrossRef]
153. Lin, N.; Wang, X.; Zhang, Y.; Hu, X.; Ruan, J. Fertigation management for sustainable precision agriculture based on Internet of Things. *J. Clean. Prod.* **2020**, *277*, 124197. [CrossRef]
154. Zhang, P.; Guo, Z.; Ullah, S.; Melagraki, G.; Afantitis, A.; Lynch, I. Nanotechnology and artificial intelligence to enable sustainable and precision agriculture. *Nat. Plants* **2021**, *7*, 864–876. [CrossRef]
155. Freidenreich, A.; Barraza, G.; Jayachandran, K.; Khoddamzadeh, A.A. Precision agriculture application for sustainable nitrogen management of *Justicia brandegeana* using optical sensor technology. *Agriculture* **2019**, *9*, 98. [CrossRef]

156. Sarri, D.; Lombardo, S.; Lisci, R.; De Pascale, V.; Vieri, M. AgroBot Smash a Robotic Platform for the Sustainable Precision Agriculture. *Lect. Notes Civ. Eng.* **2020**, *67*, 793–801.
157. Odara, S.; Khan, Z.; Ustun, T.S. Integration of Precision Agriculture and SmartGrid technologies for sustainable development. In Proceedings of the 2015 IEEE Technological Innovation in ICT for Agriculture and Rural Development (TIAR), Chennai, India, 10–12 July 2015; Volume 7358536, pp. 84–89.
158. Carter, P.G.; Young, S.L. Applications of remote sensing in precision agriculture for sustainable production. *Precis. Agric. Sustain. Environ. Prot.* **2013**, *9780203128329*, 82–98.
159. Linaza, M.T.; Posada, J.; Bund, J.; Eisert, P.; Quartulli, M.; Döllner, J.; Pagani, A.; Olaizola, I.G.; Barriguinha, A.; Moysiadis, T.; et al. Data-driven artificial intelligence applications for sustainable precision agriculture. *Agronomy* **2021**, *11*, 1227. [CrossRef]
160. Kountios, G. The role of agricultural consultants and precision agriculture in the adoption of good agricultural practices and sustainable water management. *Int. J. Sustain. Agric. Manag. Inform.* **2022**, *8*, 144–155.
161. Saeys, W.; De Baerdemaeker, J. Precision agriculture technology for sustainable good agricultural practice. In Proceedings of the 5th International Conference, TAE 2013: Trends in Agricultural Engineering, Prague, Czech Republic, 3–6 September 2013; pp. 19–24.
162. Patel, N.R.; Pander, L.M.; Roy, P.S. Precision farming technologies for sustainable Agriculture in India—Current status and prospects. *Int. J. Ecol. Environ. Sci.* **2004**, *30*, 299–308.
163. Kumar, A.; Rani, M.; Aishwarya Kumar, P. Drone Technology in Sustainable Agriculture: The Future of Farming Is Precision Agriculture and Mapping. *Agric. Livest. Prod. Aquac. Adv. Smallhold. Farming Syst.* **2022**, *2*, 3–12.
164. Apichai, S.; Saenjum, C.; Pattananandecha, T.; Phojuang, K.; Wattanakul, S.; Kiwfo, K.; Jintrawet, A.; Grudpan, K. Cost-effective modern chemical sensor system for soil macronutrient analysis applied to Thai sustainable and precision agriculture. *Plants* **2021**, *10*, 1524. [CrossRef] [PubMed]
165. Caldwell, C.D. Natural Ecosystems Versus Agroecosystems. In *Introduction to Agroecology*; Caldwell, C., Wang, S., Eds.; Springer: Singapore, 2020. [CrossRef]
166. Tivy, J. Nutrient cycling in agro-ecosystems. *Appl. Geogr.* **1987**, *7*, 93–113. [CrossRef]
167. Dubey, P.K.; Singh, A.; Merah, O.; Abhilash, P.C. Managing agroecosystems for food and nutrition security. *Curr. Res. Environ. Sustain.* **2022**, *4*, 100127. [CrossRef]
168. Dubey, P.K.; Singh, A.; Chaurasia, R.; Pandey, K.K.; Bundela, A.K.; Dubey, R.K.; Abhilash, P.C. 2021. Planet friendly agriculture, farming for people and the planet. *Curr. Res. Environ. Sustain.* **2021**, *3*, 100041. [CrossRef]
169. Jamwal, A.; Agrawal, R.; Sharma, M.; Giallanza, A. Industry 4.0 technologies for manufacturing sustainability: A systematic review and future research directions. *Appl. Sci.* **2021**, *11*, 5725. [CrossRef]
170. Ghobakhloo, M. Industry 4.0, digitization, and opportunities for sustainability. *J. Clean. Prod.* **2020**, *252*, 119869. [CrossRef]
171. Bai, C.; Dallasega, P.; Orzes, G.; Sarkis, J. Industry 4.0 technologies assessment: A sustainability perspective. *Int. J. Prod. Econ.* **2020**, *229*, 107776. [CrossRef]
172. Furstenau, L.B.; Sott, M.K.; Kipper, L.M.; Machado, E.L.; Lopez-Robles, J.R.; Dohan, M.S.; Cobo, M.J.; Zahid, A.; Abbasi, Q.H.; Imran, M.A. Link between sustainability and industry 4.0: Trends, challenges and new perspectives. *IEEE Access* **2020**, *8*, 140079–140096. [CrossRef]
173. Mushi, G.E.; Di Marzo Serugendo, G.; Burgi, P.Y. Digital technology and services for sustainable agriculture in Tanzania: A literature review. *Sustainability* **2022**, *14*, 2415. [CrossRef]
174. Santiteerakul, S.; Sopadang, A.; Yaibuathet Tippayawong, K.; Tamvimol, K. The role of smart technology in sustainable agriculture: A case study of wangree plant factory. *Sustainability* **2020**, *12*, 4640. [CrossRef]
175. Shah, F.M.; Razaq, M. From agriculture to sustainable agriculture: Prospects for improving pest management in industrial revolution 4.0. In *Handbook of Smart Materials, Technologies, and Devices: Applications of Industry 4.0*; Springer: Edinburgh, UK, 2020; pp. 1–18.
176. Moher, D.; Liberati, A.; Tetzla_, J.; Altman, D.G.; The PRISMA Group. Preferred Reporting Items for Systematic Reviews and Meta-Analyses: The PRISMA Statement. *PLoS Med.* **2009**, *6*, e10000972009. [CrossRef]
177. Guz, A.N.; Rushchitsky, J.J. Scopus: A system for the evaluation of scientific journals. *Int. Appl. Mech.* **2009**, *45*, 351–362. [CrossRef]
178. Ellegaard, O.; Wallin, J.A. The bibliometric analysis of scholarly production: How great is the impact? *Scientometrics* **2015**, *105*, 1809–1831. [CrossRef] [PubMed]
179. Van Eck, N.; Waltman, L. Software survey: VOSviewer, a computer program for bibliometric mapping. *Scientometrics* **2010**, *84*, 523–538. [CrossRef] [PubMed]
180. Ghildiyal, S.; Joshi, K.; Rawat, G.; Memoria, M.; Singh, A.; Gupta, A. Industry 4.0 Application in the Hospitality and Food Service Industries. In Proceedings of the 2022 7th International Conference on Computing, Communication and Security, ICCCS 2022, Seoul, Republic of Korea, 3–5 November 2022.
181. Brodt, S.; Six, J.; Feenstra, G.; Ingels, C.; Campbell, D. Sustainable Agriculture. *Nat. Educ. Knowl.* **2011**, *3*, 1.
182. Senturk, S.; Senturk, F.; Karaca, H. Industry 4.0 technologies in agri-food sector and their integration in the global value chain: A review. *J. Clean. Prod.* **2023**, *408*, 137096. [CrossRef]
183. Ritchie, H.; Roser, M.; Rosado, P. Energy. Published online at OurWorldInData.org. 2022. Available online: https://ourworldindata.org/energy (accessed on 2 October 2023).

184. Roser, M.; Ritchie, H.; Rosado, P. Food Supply. Published online at OurWorldInData.org. 2013. Available online: https://ourworldindata.org/food-supply (accessed on 2 October 2023).
185. Ritchie, H.; Roser, M. Water Use and Stress. Published online at OurWorldInData.org. 2017. Available online: https://ourworldindata.org/water-use-stress (accessed on 2 October 2023).
186. Brooks, J.; Deconinck, K.; Giner, C. Three Key Challenges Facing Agriculture and How to Start Solving Them. Organisation for Economic Co-operation and Development (OECD), 6 June 2019. 2019. Available online: https://www.oecd.org/agriculture/key-challenges-agriculture-how-solve/ (accessed on 5 January 2021).
187. Soosay, C.; Kannusamy, R. Scope for industry 4.0 in agri-food supply chain. In *The Road to a Digitalized Supply Chain Management: Smart and Digital Solutions for Supply Chain Management. Proceedings of the Hamburg International Conference of Logistics (HICL)*; GmbH: Berlin, Germany, 2018; Volume 25, pp. 37–56.
188. De Carolis, A.; Macchi, M.; Negri, E.; Terzi, S. *A Maturity Model for Assessing the Digital Readiness of Manufacturing Companies*; Springer: Berlin/Heidelberg, Germany, 2017; Volume 513, pp. 13–20.
189. Cotrino, A.; Sebastian, M.A.; González-Gaya, C. Industry 4.0 Roadmap: Implementation for Small and Medium-Sized Enterprises. *Appl. Sci.* **2020**, *10*, 8566. [CrossRef]
190. Ejsmont, K.; Gladysz, B.; Kluczek, A. Impact of Industry 4.0 on Sustainability—Bibliometric Literature Review. *Sustainability* **2020**, *12*, 5650. [CrossRef]
191. Schumacher, A.; Erol, S.; Shin, W. A maturity model for assessing Industry 4.0 readiness and maturity of manufacturing enterprises. *Procedia CIRP* **2016**, *52*, 161–166. [CrossRef]
192. Daroshka, V.; Aleksandrov, I.; Fedorova, M.; Chekhovskikh, I.; Ol, E.; Trushkin, V. Agriculture and ESG Transformation: Domestic and Foreign Experience of Green Agribusiness Finance. *Lect. Notes Netw. Syst.* **2023**, *575*, 2357–2368.
193. Zeng, L.; Jiang, X. ESG and Corporate Performance: Evidence from Agriculture and Forestry Listed Companies. *Sustainability* **2023**, *15*, 6723. [CrossRef]
194. Buallay, A. Sustainability reporting and agriculture industries' performance: Worldwide evidence. *J. Agribus. Devel. Emerg. Econ.* **2023**, *12*, 769–790. [CrossRef]
195. Hrebicek, J.; Popelka, O.; Štencl, M.; Trenz, O. Corporate performance indicators for agriculture and food processing sector. *Acta Univ. Agric. Silvic. Mendel. Brun.* **2021**, *60*, 121–132. [CrossRef]
196. Hrebicek, L.; Soukopova, J.; Stencl, M.; Trenz, O. Integration of Economic, Environmental, Social and Corporate Governance Performance and Reporting in Enterprises. *Acta Univ. Agric. Silvic. Mendel. Brun.* **2011**, *59*, 157–177. [CrossRef]
197. Ritschelova, I.; Sidorov, E.; Hajek, M.; Hrebicek, J. Corporate Environmental Reporting in the Czech Republic and its Relation to Environmental Accounting at Macro Level. In Proceedings of the 11th Annual EMAN Conference on Sustainability and Corporate Responsibility Accounting. Measuring and Managing Business Benefits, AULA, Budapest, Hungary, 6 October 2009; pp. 55–60.
198. Hodinka, M.; Stencl, M.; Hrebicek, J.; Trenz, O. Current trends of corporate performance reporting tools and methodology design of multifactor measurement of company overall performance. *Acta Univ. Agric. Silvic. Mendel. Brun.* **2012**, *2*, 85–90. [CrossRef]
199. Shaltegger, S.; Wagner, M. Integrative Management of Sustainability Performance, Measurement and Reporting. *Int. J. Account. Audit. Perform. Eval.* **2006**, *3*, 1–19. [CrossRef]
200. Lahane, S.; Paliwal, V.; Kant, R. Evaluation and ranking of solutions to overcome the barriers of Industry 4.0 enabled sustainable food supply chain adoption. *Clean. Log. Supply Chain* **2023**, *8*, 100116. [CrossRef]
201. Trollman, H.; Samoilyk, I. Preface of the International Conference on Industry 4.0 for Agri-food Supply Chains: Addressing Socio-economic and Environmental Challenges in Ukraine (IC4AFSC 2023). *Eng. Proceed.* **2023**, *40*, 24.
202. Dadhaneeya, H.; Nema, P.K.; Arora, V.K. Internet of Things in food processing and its potential in Industry 4.0 era: A review. *Trends Food Sci. Technol.* **2023**, *139*, 104109. [CrossRef]
203. Kopishynska, O.; Utkin, Y.; Sliusar, I.; Muravlov, V.; Makhmudov, K.; Chip, L. Application of Modern Enterprise Resource Planning Systems for Agri-Food Supply Chains as a Strategy for Reaching the Level of Industry 4.0 for Non-Manufacturing Organizations. *Eng. Proceed.* **2023**, *40*, 15.
204. Frederico, G.F. From supply chain 4.0 to supply chain 5.0: Findings from a systematic literature review and research directions. *Logistics* **2021**, *5*, 49. [CrossRef]
205. Krupitzer, C.; Stein, A. Unleashing the Potential of Digitalization in the Agri-Food Chain for Integrated Food Systems. *Ann. Rev. Food Sci. Technol.* **2023**, *15*. [CrossRef]
206. Kumari, S.; Venkatesh, V.G.; Shi, Y. The Sustainability roadmap for the food industry 4.0. In *Smart Food Industry: The Blockchain for Sustainable Engineering: Volume I—Fundamentals, Technologies, and Manage*; CRC Press: Boca Raton, FL, USA, 2023; pp. 42–49.
207. Telukdarie, A.; Munsamy, M.; Katsumbe, T.H.; Maphisa, X.; Philbin, S.P. Industry 4.0 Technological Advancement in the Food and Beverage Manufacturing Industry in South Africa—Bibliometric Analysis via Natural Language Processing. *Information* **2023**, *14*, 454. [CrossRef]
208. Tan, J.; Goyal, S.B.; Singh Rajawat, A.; Jan, T.; Azizi, N.; Prasad, M. Anti-Counterfeiting and Traceability Consensus Algorithm Based on Weightage to Contributors in a Food Supply Chain of Industry 4.0. *Sustainability* **2023**, *15*, 7855. [CrossRef]
209. Ranjith Kumar, S.; Ramachandran, N.; Sivasubramanian, R.; Dhiyaneswaran, J.; Reji, A.K. Compelling Forces and Challenges for the Food Processing Industry to Adopt Industry 4.0. In *Internet of Things: Technological Advances and New Applications*; Apple Academic Press: Palm Bay, FL, USA, 2023; pp. 1–15.

210. Vasanthraj; Kaur, A.; Potdar, V.; Agrawal, H. Industry 4.0 Adoption in Food Supply Chain to Improve Visibility and Operational Efficiency—A Content Analysis. *IEEE Access* **2023**, *11*, 73922–73958. [CrossRef]
211. Stefanini, R.; Vignali, G. The influence of Industry 4.0 enabling technologies on social, economic and environmental sustainability of the food sector. *Int. J. Prod. Res.* **2023**, 1–18. [CrossRef]
212. Despoudi, S.; Sivarajah, U.; Spanaki, K.; Charles, V.; Durai, V.K. Industry 4.0 and circular economy for emerging markets: Evidence from small and medium-sized enterprises (SMEs) in the Indian food sector. In *Annals of Operations Research*; Springer: Berlin/Heidelberg, Germany, 2023; pp. 1–39.
213. Bui, T.-D.; Tseng, J.-W.; Tran, T.P.T.; Ha, H.M.; Lim, M.K.; Tseng, M.-L. Circular supply chain strategy in Industry 4.0: The canned food industry in Vietnam. *Bus. Strat. Environ.* **2023**, *early view*. [CrossRef]
214. Varbanova, M.; de Barcellosa, M.D.; Kirova, M.; De Steur, H.; Gellynck, X. Industry 4.0 implementation factors for agri-food and manufacturing SMEs in Central and Eastern Europe. *Serbian J. Manag.* **2023**, *18*, 167–179. [CrossRef]
215. Ngan, N.T.; Khoi, B.H. Using Intention of Online Food Delivery Services in Industry 4.0: Evidence from Vietnam. *Lect. Notes Netw. Syst.* **2023**, *647*, 142–151.
216. Chataut, R.; Phoummalayvane, A.; Akl, R. Unleashing the Power of IoT: A Comprehensive Review of IoT Applications and Future Prospects in Healthcare, Agriculture, Smart Homes, Smart Cities, and Industry 4.0. *Sensors* **2023**, *23*, 7194. [CrossRef]
217. Zulfiqar, F.; Navarro, M.; Ashraf, M.; Akram, N.A.; Munné-Bosch, S. Nanofertilizer use for sustainable agriculture: Advantages and limitations. *Plant Sci.* **2019**, *289*, 11027. [CrossRef]
218. Assayed, A.; Haddad, J.; Kilani, H.; Abdallah, R.; Kumar, V. Industry 4.0 in Resource Efficient and Cleaner Production: A Case Study from the Food Sector in Jordan. *Environ. Footpr. Eco-Design Prod. Proc. PF1* **2024**, *487*, 61–71.
219. Hassoun, A.; Boukid, F.; Ozogul, F.; Aït-Kaddour, A.; Soriano, J.M.; Lorenzo, J.M.; Perestrelo, R.; Galanakis, C.M.; Bono, G.; Bouyahya, A.; et al. Creating new opportunities for sustainable food packaging through dimensions of industry 4.0: New insights into the food waste perspective. *Trends Food Sci. Technol.* **2023**, *142*, 104238. [CrossRef]
220. Demir, Y.; Dincer, F.I. The effects of Industry 4.0 on the food and beverage industry. *J. Tour.* **2020**, *6*, 133–145. [CrossRef]
221. Katsis, M.; Papadatos, P.; Rigou, M.; Sirmakessis, S.; Vossos, D. Harnessing Skills for Sustainable Development: A Skills Matchmaking System for Smart Cities, Green Energy, Blue Economy and Precision Agriculture. In Proceedings of the 3rd International Conference on Control, Artificial Intelligence, Robotics and Optimization, Crete, Greece, 11–13 April 2023; pp. 86–93.
222. Zhang, P.; Lynch, I.; White, J.C.; Handy, R.D. Nano-enabled Sustainable and Precision Agriculture. In *Nano-Enabled Sustainable and Precision Agriculture*; Elsevier: Amsterdam, The Netherlands, 2023; pp. 1–559.
223. Kasuga, T.; Mizui, A.; Koga, H.; Nogi, M. Wirelessly Powered Sensing Fertilizer for Precision and Sustainable Agriculture. *Adv. Sustain. Syst.* **2023**, *early view*. [CrossRef]
224. Sharma, A.; Sharma, A.; Tselykh, A.; Bozhenyuk, A.; Choudhury, T.; Alomar, M.A.; Sánchez-Chero, M. Artificial intelligence and internet of things oriented sustainable precision farming: Towards modern agriculture. *Open Life Sci.* **2023**, *18*, 20220713. [CrossRef] [PubMed]
225. Teixeira, S.C.; Gomes, N.O.; Calegaro, M.L.; Machado, S.A.S.; de Oliveira, T.V.; de Fátima Ferreira Soares, N.; Raymundo-Pereira, P.A. Sustainable plant-wearable sensors for on-site, rapid decentralized detection of pesticides toward precision agriculture and food safety. *Biomater. Adv.* **2023**, *155*, 213676. [CrossRef] [PubMed]
226. Goyal, V.; Rani, D.; Ritika, M.S.; Deng, C.; Wang, Y. Unlocking the Potential of Nano-Enabled Precision Agriculture for Efficient and Sustainable Farming. *Plants* **2023**, *12*, 3744. [CrossRef] [PubMed]

Disclaimer/Publisher's Note: The statements, opinions and data contained in all publications are solely those of the individual author(s) and contributor(s) and not of MDPI and/or the editor(s). MDPI and/or the editor(s) disclaim responsibility for any injury to people or property resulting from any ideas, methods, instructions or products referred to in the content.

MDPI AG
Grosspeteranlage 5
4052 Basel
Switzerland
Tel.: +41 61 683 77 34

Foods Editorial Office
E-mail: foods@mdpi.com
www.mdpi.com/journal/foods

Disclaimer/Publisher's Note: The title and front matter of this reprint are at the discretion of the Guest Editors. The publisher is not responsible for their content or any associated concerns. The statements, opinions and data contained in all individual articles are solely those of the individual Editors and contributors and not of MDPI. MDPI disclaims responsibility for any injury to people or property resulting from any ideas, methods, instructions or products referred to in the content.

www.ingramcontent.com/pod-product-compliance
Lightning Source LLC
LaVergne TN
LVHW072320090526
838202LV00019B/2318